Caution!
Risk of electric shock.

The circuits and subjects discussed in this book operate from or involve dangerously high voltages. It is the reader's responsibility to ensure that any equipment constructed or modified using the information provided in this book, is safe to use.

Designing
High-Fidelity
Valve Preamps

by
Merlin Blencowe, MSc

2016

Published by Merlin Blencowe.

ISBN 978-0-9561545-3-8

Contents

4. The Small-Signal Pentode   166

5. Noise, Hum and Microphony   188

8. Compound Amplifiers — 288

9. Controls — 315

10. The Phono Stage — 335

Preface

If you are reading this then perhaps you are one of the growing number of enthusiasts who embrace the anachronism that is valve audio technology. I certainly belong in that category, and despite the protestations of modernists, it is not too difficult to appreciate the appeal of valve amplifiers beyond mere nostalgia. Valve circuits tend to be simple, which makes them easy to understand, modify, and repair, and gives them an appealing longevity that is sadly missing from our modern age of disposable consumer electronics and software obsolescence. Valves also tend to be much more forgiving of mistakes and abuse than transistors; they are not nearly as fragile as their light-bulb semblance implies. In other words, valve circuits tend to be difficult to get wrong, and it seems to me that a bad-sounding valve amp is never quite as bad as a bad-sounding solid-state amp. Yes, solid-state amps will always have the advantages of low cost and high efficiency, but valve amps still compete in terms of humanistic appeal and mental design effort.

Admittedly, there is no shortage of critics who will point out that the very best valve amp cannot match the linearity of even a fairly basic solid-state amp, and plenty of enthusiasts see better linearity as the unending, impossible crusade of domestic hi-fi. Indeed, they are so dogmatically inflexible on this point that I wonder if they ever actually enjoy listening to music, or if they are just perpetually disappointed with the limitations of technology –as one wry internet commentator dryly put it: *"I'm just putting another record on the distortion analyser, dear."*

I do not share this attitude. Perfect linearity is not a requirement for home listening (as opposed the recording and mastering); surely what matters is *realism* and *enjoyment*, both of which are subjective rather than absolute metrics, and both of which valve amps seem to deliver quite easily despite being technically poorer than many solid-state offerings (although exactly *why* is a question that has never been satisfactorily answered). That said, you will find no popular subjectivism or audio religion in this book. No unobtanium capacitors with supra-physical properties; no quantum electron streamlining; no directional cables; only real engineering, but accepting of the fact that listening rooms aren't perfect, humans are *exceedingly* fallible, and valve amps just look way cooler than transistor amps. If you share this attitude, then this book it for you.

Yes, there are hundreds of electronics textbooks from the 1950 and 60s that can be bought for next to nothing nowadays, all of which cover formal valve circuit theory to a higher standard than I ever could. There is also a small but steadily growing number of modern textbooks dedicated to valve audio, but I have found they tend to fall into two categories. The first are project books containing completed designs for the reader to copy, but little in the way of an education. The second are books that attempt to explain valve circuit design in a purely theoretical way, much like the textbooks of old, but allowing for modern audio trends. Very few books occupy the middle ground between pure, armchair theory, and real-world practical results, so it is this gap I have attempted to fill.

The intention of this book, therefore, is to provide a deep understanding of how small-signal valve circuits work and how to design them from scratch, *and* how they behave in reality. In other words, I have tried to be pragmatic –to present useful theory *and* to test it. The instructive value of testing real circuits built with real components, with all their hidden imperfections, cannot be overstated. It leads to many interesting practical conclusions that are not obvious from theory alone, and should leave reader with a good grasp of which design choices really matter, which are niceties, which are theoretical speculation, and which are plain audio myths.

This book is not a beginner's guide to audio electronics. Nevertheless, I appreciate that many readers will be enthusiastic amateurs who may be self-taught and may have gaps in their general electronics knowledge. I have therefore done my best to introduce, at least semi-formally, many subjects which I think are essential and might otherwise be missed. I have not shied away from the mathematics, but neither do I want to beat the reader over the head with it. Most of it is algebra, not calculus, and is explained step by step. This should allow intelligent novices to catch up quickly. Other basic topics are covered incidentally as part of more general discussion without –I hope– patronising more experienced readers. Admittedly, by focusing on certain details I have had to sacrifice breadth of material. No doubt some readers will be disappointed by the absence of directly-heated triodes, coupling transformers, balanced topologies, power valves and so on. I can only apologise for this, and offer the humble defence that much of this book deals with 'first principles' which can be extrapolated into other areas of design. If it helps you even a little bit, I shall consider it a success!

Acknowledgements

The author would particularly like to thank Paul Fawcett for the Herculean effort he put into proof reading the manuscript, and Stephen Keller for his apparently supernatural ability to source reference material. In addition, this book would not have been possible without the help, encouragement and co-operation of AMS Neve Ltd., David Ivan James, Mallory Nicholls, Holly Smith, Lee Taylor, The University of Leeds Library, and The University of York Library.

Chapter 1: Essential Electronics

"...the ill and unfit choice of words wonderfully obstructs the understanding."
Francis Bacon, 1620.

Mathematics lies at the heart of electronics and engineering, and you cannot really achieve much in electronics without learning at least some of it. Indeed, electronics might be said to be the physical manifestation of maths, with its adders, differentiators, integrators and so on (music and architecture being artistic manifestations of maths). On the other hand, this is not a book of mathematics, and it is not even an electronics textbook in the formal academic sense. Really it is meant to occupy a pragmatic middle ground between pure theory and practical, intuitive, get-your-hands-dirty design.

There are many formal textbooks on circuit theory and analysis, and the older editions can often be bought for practically nothing on the second hand market. Those written before about 1970 will cover valves,[*] and all of them will go into greater detail and include more examples, proofs and explanations than I can hope to provide here. This chapter can only be a whistle-stop tour of some basic concepts in circuit theory, focusing on those tools that are most fundamental and useful for practical design work. This will be a case of accelerated learning, and some prior knowledge of electronics is assumed. Nevertheless, special attention will be paid to subjects that are often overlooked or misunderstood by beginners and self-taught hobbyists, or which more formal textbooks may take for granted.

Sometimes the best way to pick up knowledge is to see it used in practice, and no doubt some things which seem obscure even after explanation in this chapter will slowly begin to sink in after seeing them used again and again in context, in other parts of the book. On the other hand, some of the theory presented here will not actually be used in the rest of the book, at least not directly. Nevertheless, everything here is what the author considers essential, general electronics knowledge, of which the reader should at least be vaguely aware. If you are serious about designing your own circuits, then you're sure to use all of it sooner or later.

---

[*] *Foundations of Wireless* by M. G. Scroggie (3rd to 7th editions) is particularly good. *Essays in Electronics* and *Second Thoughts on Radio Theory*, by the same author, are also highly recommended.

1.1: Current

Some non-academic textbooks (and a few that should know better) state that current is defined as a flow of electrons. This is absolutely not true. Electric current is defined as the *rate of change of electric charge*:[*]

$$i = \frac{\text{Change in charge}}{\text{Change in time}} = \frac{dq}{dt} \quad \text{amps} \tag{1.1}$$

An amp is therefore a convenient shorthand way of saying a 'coulomb per second'. If the rate of change happens to be constant then this simplifies to:

$$I = Q / t \quad \text{amps} \tag{1.2}$$

Note that upper-case letters are used for steady values whereas lower-case letters represent instantaneous values.

The English language was never designed to cope with the bizarre world of quantum physics, and it is all too easy for beginners to be foiled by the misleading choice of words used by engineers. We often talk about current flowing from one place to another (I will ashamedly do it many times throughout this book) but, strictly speaking, electric current itself does not 'flow' anywhere –electrons do, and that is not the same thing. This confusion is sometimes compounded by saying that current is a bit like water in a pipe, and when it comes to water we are used to the word 'current' meaning a particular body of moving water, such as the Gulf Stream. But in electronics it does not mean quite the same thing: the water in the pipe actually represents the bulk electric charge, while the electric current is represented by the *rate* at which the water is being moved, e.g. in litres per second, and the poor electrons don't even come into it!

Furthermore, current is a **scalar** quantity, which means it is just a number, just like temperature or wind speed are merely numbers. Electric current is not 'made of' anything, not even charge, any more than wind speed is made of wind. It is true that electrons carry charge, and moving charge *results* in current, but other things carry charge too (protons, positrons, ions, etc.), so moving electrons around is not the only way to get electric current. Admittedly, electrons are the most common charge carrier in electronic circuits, so for the sake of brevity it is common to think of electrons and charge as being synonymous for the purposes of circuit theory –but do not fall into the trap of believing they are literally identical.

In school we are often lead to believe that an electron is something like a billiard ball and that the charge it carries is like the colour red, say, carried by the billiard ball. But in fact the charge carried by an electron (or anything else) is not something

---

[*] Mathematicians will object to my equating d/dt with something so trivial as 'a change', but we haven't time to dwell on the interpretation of derivatives here.

trapped within the boundaries of the particle itself.[*] Really it is a sphere of influence extending to infinity, growing ever weaker, like the light that spills out of a light bulb. If a light bulb is moved around in a room we could measure the *rate* at which the light intensity is changing at any point in the room –not just where the bulb is– and this figure would be analogous to electric current. This 'atmosphere' of charge outside the electron is called **electric flux**[†] and is the reason that current can exist between the plates of capacitor even though the electrons themselves cannot cross the gap. It is worth pointing out, however, that 'flux' here does not imply flow, it simply means 'stuff', like the stuff in multicore solder or the mysterious stuff stored in Doc Brown's flux capacitor.

Current can be represented on a circuit diagram as an arrow, often drawn alongside the wire, although I prefer to draw it on the wire itself. This arrow does not indicate any physical movement, it is just a symbol or reference, indicating the direction in which the current is 'notionally positive' for the purposes of calculation. An intuitive way to get used to this notation is to say that the arrow indicates the way we would insert an ammeter into the circuit, with the arrow entering the meter through the red lead and exiting through the black lead.

Whatever type of charge is carried in the same direction as the arrow, the electric current *reckoned in that direction* will have the

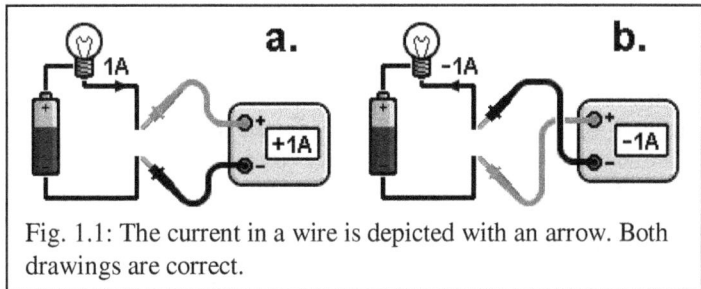

Fig. 1.1: The current in a wire is depicted with an arrow. Both drawings are correct.

same sign. In other words, if positive charge is carried in the same direction as the arrow, the current is positive. Conversely if positive charge is carried in the *opposite* direction to the arrow then the current is negative. For example, in fig. 1.1a electrons are pushed out of the negative terminal of the battery and carry negative charge anticlockwise around the circuit, in the opposite direction to the current arrow, so the current is mathematically positive. Equally, we could have drawn the arrow the other way around as in b., in which case the negative charge flows in the same direction as the arrow, so the current is negative. Thus there are two ways to depict exactly the same current, and the direction in which we draw the arrow is a completely free choice; *it is not meant to indicate the 'true' direction of anything!*

---

[*] Bear in mind that electrons are not really particles at all. Like all quantum phenomena they are something between a particle and a wave, and are rather more mysterious and 'fuzzy' than this book gives them credit for.
[†] Irritatingly, physicists use a different and less intuitive definition of electric flux from electronics engineers.

This sometimes causes consternation among students who dislike the fact that the current is positive when pointing in the opposite direction to electron flow, arguing that this so-called 'conventional current' must be backwards. The popularist explanation for this perceived problem is that the prolific scientist Benjamin Franklin, who was thinking in terms of an 'electric fluid', made a guess about its direction, and guessed wrong. Unfortunately, this explanation misses the point entirely. Electron flow and electric current are not two versions of the same thing, any more than traffic is the same thing as speed. Current is a scalar quantity so does not have *any* direction, only sign, which comes from a historical and entirely free choice of definition: dq/dt, or −dq/dt. Franklin happened to choose the former, with the rather harmonious result that a flow of *negative* charge means a *negative* current when reckoned in the same direction. Ultimately though, trying to argue that one is correct and the other is 'conventional' (a polite way of trying to say it is wrong) is like trying to argue which way is 'up' in outer space. Conventional current is not the embarrassing historical mistake that some make it out to be; it is in complete and perfect agreement with electron flow and the definitions of current, *if* all the definitions and symbols are interpreted correctly. Sadly, however, many authors fail to appreciate the relationships between electron flow, charge flow, and current (which does not flow).

In the previous discussion I made a very brief reference to electrons and how they were moving. This is *not* standard practice; it was only a simple circuit example for explaining the standard current-arrow notation. In almost any other situation we would talk only of current, voltage, charge, and signals, but *not* of subatomic particles. Trying to imagine what electrons are doing all the time is of little use and likely to lead to confusion. To drive the point home it is perhaps worth pointing out that in most cables electrical signals travel at about 2×10^8 m/s, or two thirds the speed of light, yet the electrons in the metal wires inside a cable move at a snail's pace called the **drift velocity.** They are like a toppling row of dominoes; each one does not have to move very far or very fast before it bumps into the next one, which shoves the next one, and the next, so the charge wavefront (i.e. the signal) propagates much faster down the line than the dominoes themselves can.

The drift velocity depends on the number of electrons in the material and the total amount of charge they transfer per second, that is, the current. In the sort of circuits we are interested in it will be a few centimetres per minute in metal, because there are trillions of free electrons packed into metal so they don't have to move very fast to transfer a lot of charge per second. But in the vacuum of a valve there are few electrons, so they have to move at thousands of kilometres per second to amount to the same level of current. In resistors the drift velocity will be something in between. And what if the signal happens to be AC? Then the electrons in a wire move back and forth, but they won't get very far before the current reverses, so really they vibrate on the spot. All this may be happening in one simple circuit, and if that is not enough to discourage people from trying to imagine the lives of electrons, then I don't know what is.

No doubt I have laboured the point. But the concept of electric current is quite abstract, and as children we are usually taught simplified concepts (i.e. lies) that become firmly embedded. These concepts may be good enough for children and laymen, but they lead to headaches for those who pursue the subject more deeply, so having such misconceptions broken is for your own good.

1.2: Voltage

It takes energy (work) to force electrons to move, and the potential energy needed to move an amount of charge from one point to the another is called the voltage 'between' or 'across' those two points.

$$v = \frac{\text{Change in work (energy)}}{\text{Change in charge}} = \frac{dw}{dq} \quad \text{volts} \tag{1.3}$$

A 'volt' is therefore a convenient shorthand way of saying a 'joule per coulomb'.

A voltage that *supplies* energy to the charge is given a special name: **electromotive force, or EMF.**[*] A battery is a prime example of a device that contains an EMF since it 'pumps' electrons from one terminal to the other, keeping current flowing when naturally it would not; the battery is the source of energy which keeps everything moving. As the electrons leave the battery and flow around the circuit, eventually making it back to the opposite terminal, they give up this energy in the various components they are forced through. The voltages measured across these other components correspond to energy being lost rather than supplied, and act in the opposite direction to the EMF voltage (see section 1.8). This sort of voltage is called **potential difference**. However, these distinctions are rarely needed for everyday electronics, so we simply call them all 'voltage'.

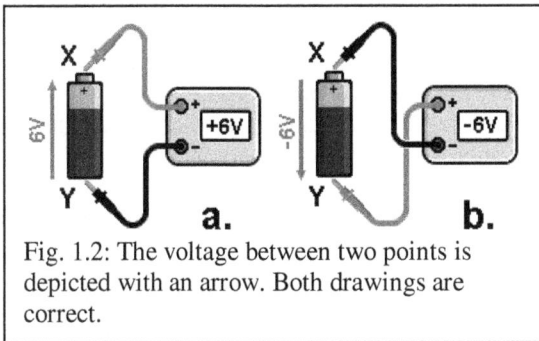

Fig. 1.2: The voltage between two points is depicted with an arrow. Both drawings are correct.

A voltage is represented on a diagram as an arrow. This arrow does not indicate any sort of physical movement, it is just a symbol or reference indicating which end is notionally positive (consider, a voltage can exist between two points even if there is no current, in which case there is definitely nothing moving, yet we still use an arrow).

The voltage is always measured from the head of the arrow, with respect to the tail. An intuitive way to get used to this notation is to say that the head of the arrow is

---

[*] But EMF it is not a literal force, it is a voltage, or force *per unit charge*, so it would be better to call it an electromotive *pressure*. This is just another example of the misleading use of words.

where we would connect the red lead of a voltmeter, and the tail is where we would connect the black lead. For example, in fig. 1.2a the potential at point X is 6V positive with respect to point Y, so we might say the voltage between them is +6V in the direction Y to X. Equally, we could have said that point Y is 6V *negative* with respect to point X, in which case the voltage between them is −6V in the direction X to Y, as shown in b. Thus there are two ways to depict exactly the same voltage, and the direction in which we draw the arrow is a completely free choice. After analysing the circuit (i.e. doing the maths) the sign of the voltage will become apparent.

1.3: Power

The rate at which energy is expended is called power. On planet Earth, lifting a 1kg weight by 1m requires 9.8 joules of energy, but if it is lifted very quickly then we need to supply that energy in a short space of time, so the power expended is greater than if it is lifted very slowly. Similarly, moving a lump of charge around a circuit requires energy, and moving it quickly (high current) requires more power than moving it slowly (low current). Power is given by:

$$p = \frac{\text{Change in work (energy)}}{\text{Change in time}} = \frac{dw}{dt} \text{ watts} \tag{1.4}$$

A watt is therefore a convenient shorthand way of saying a 'joule per second'.

Now, by using equations (1.1) and (1.3) we can write:

$$p = \frac{dw}{dq}\frac{dq}{dt} = iv \tag{1.5}$$

Again, lower-case letters indicate these are instantaneous values. The product of voltage and current is equal to the power being expended at a given moment. But we are usually more interested in the *average* (mean) power since this provides a more useful indication of how hot a component might get, or how much money the electric company will charge us. If we choose to plot a graph of the power as it varies over time then the average power is equal to the total area under the curve, divided by the time interval (T) we are looking at (see section 1.4). Students of calculus will know that to find an area we can use integration, so the average power is:

$$P = \frac{1}{T} \int_0^T p \, dt = \frac{1}{T} \int_0^T iv \, dt \tag{1.6}$$

But nobody likes integration, so here are some quick dodges. If the current and voltage are both constant (DC), then the average power is simply the product of the two:

$$P = IV \tag{1.7}$$

If the voltage is constant but the current is varying, then the average power is the product of the voltage and the *mean average* current.

$$P = VI_{av} \tag{1.8}$$

6

This rule will come in handy when finding the average power delivered by a battery or power supply, for example. The same principle applies if the current is constant but the voltage is varying, of course.

If both the current and voltage are both varying, *and they are in phase* (i.e. the circuit is resistive), then the average power is the product of the RMS values of each:

$$P = I_{rms} V_{rms}$$

(1.9)

1.4: Signals and Averages

Often in electronics we are interested in the long-term behaviour of signals and circuits, rather than what is happening at any particular instant. It is therefore natural to use average values in certain calculations, but it is important to be aware that there are different *kinds* of average, and they have different uses in circuit analysis. A few of these uses have already been mentioned in the previous section.

1.4.1: Mean Average

The mean average value of a waveform is equal to the DC 'content' of the signal, imagining it to be a pure AC signal 'riding on top' of a DC signal. In other words, it is proportional to the total charge that is transferred by the signal over a suitably long period of time. It is often referred to simply as the average value or mean value. Using a voltage signal as an example, the formal way to find the average is using integration:

$$V_{av} = \frac{1}{T} \int_0^T v \, dt$$

(1.10)

What this equation is saying is: allow all the positive areas under the waveform to cancel any negative areas, then divide the remaining area by the time interval you are interested in. Usually the period of time we are interested in (T) includes one whole cycle of the waveform, i.e. one time period, or a whole multiple thereof.

The process is simple enough that anyone can do it, whether or not you have heard of calculus, even if you have to draw the waveform on a bit of paper and start counting squares. With simple waveforms the average can often be found with just a little thought, or 'by inspection' as the mathematicians say. For example, it is

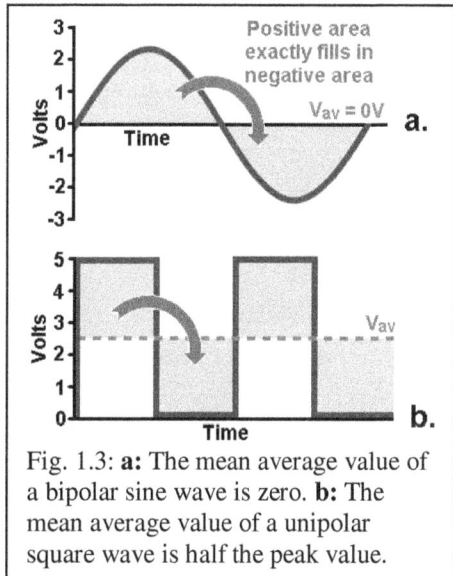

Fig. 1.3: **a:** The mean average value of a bipolar sine wave is zero. **b:** The mean average value of a unipolar square wave is half the peak value.

easy to see that the sine wave in fig. 1.3a has an average value of zero since the positive halves will exactly fill in the negative ones; it is pure AC. Each positive cycle transfers some charge which is exactly withdrawn again by each negative cycle, so the net movement of charge is zero. On the other hand, a square wave that switches between 0V and 5V must have an average value of 2.5V since half the positive areas will fill in half the empty space between each pulse, as shown in fig. 1.3b.* It can be thought of as a purely AC square wave riding on top of a 2.5V DC signal.

When talking about power dissipation it is nearly always the mean average power that is considered. This is because heat flows relatively slowly through a body, so the bulk temperature cannot follow rapid changes in dissipated power. It is therefore the long-term or mean average power that determines the running temperature of the component. Many components can withstand massive peaks in power dissipation provided they last for only a short time, since the long-term average works out to be much less.

1.4.2: Rectified Average

Another sort of average is the rectified average. This is found by rectifying the signal to make it completely positive and *then* finding the mean average (DC component) of the resulting waveform, in the same way as previously. Rectified average is often referred to simply as the average value, creating possible confusion with mean average. It is written mathematically:

$$V_{rav} = \frac{1}{T} \int_0^T |v| \, dt \qquad\qquad (1.11)$$

It can be seen that this is almost the same as equation (1.10) except for the two vertical lines sandwiching v. These lines indicate that it is the *magnitude* of v that must be used; the magnitude of a number ignores any minus sign and treats it as positive instead, which is equivalent to rectification. The rectified average value of

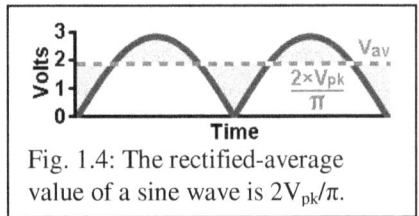

Fig. 1.4: The rectified-average value of a sine wave is $2V_{pk}/\pi$.

a voltage sine wave is $2V_{pk}/\pi$, as shown in fig. 1.4.

1.4.3: RMS Average

A voltage sine wave has a mean value of zero, but the mean *power* it delivers to a light bulb is certainly not zero. This is because both the positive and

---

* I like to imagine the waveform represents a cross section through some sand dunes. Finding the average is like allowing the sandy positive peaks to collapse to fill in the negative valleys, eventually forming a perfectly flat landscape at the level of the average.

negative parts of the sine wave deliver power equally well. This leads to another kind of average: the **root-mean-square** or **RMS**. This is a value that is equivalent to a DC signal which would have the same heating effect in a resistor as the varying signal does. In other words, it is proportional to the average power delivered by the varying signal into a *resistive* load. Using a voltage signal as an example, its RMS value is calculated using:

$$V_{rms} = \sqrt{\frac{1}{T} \int_0^T v^2 \, dt} \qquad (1.12)$$

The RMS value is found by first squaring the waveform (which makes it completely positive since any number multiplied by itself becomes positive), then finding the mean average of the resulting wave form, then taking the square root of the result. Note that squaring a signal is a multiplication, which is not the same thing as rectifying it. The RMS value of a voltage sine wave is $V_{pk}/\sqrt{2}$. RMS values are most useful when considering power circuits, which is why the mains supply is specified in terms of its RMS value. For example, a $230V_{rms}$ wall voltage is a sine wave with a peak value of $230 \times \sqrt{2} = 325V_{pk}$, or $650V_{pp}$.

It is also worth noting that most AC volt meters do not measure the RMS value of waveforms. Instead they measure the rectified average (which is easier to do) and then scale the result by a factor of 1.1 so that it reads the correct RMS value for a pure sine wave. But if the waveform is not a sine wave then this reading may be quite inaccurate. Meters that genuinely *do* measure RMS values are prominently labelled 'true RMS'.

1.5: Complex Numbers

At this point some readers may think that complex numbers are an advanced (and probably intimidating) mathematical topic, and that now is too soon to be talking about them. However, they are in fact extremely simple and very useful for dealing with sinusoidal waves, and many of the following topics will make a lot more sense after learning about complex numbers sooner rather than later. Nevertheless, we shall keep the treatment light.

Complex numbers are composed of two numbers called the **real** and **imaginary** parts. Now, it must be emphasised straight away that complex numbers are not complicated, and imaginary numbers are just as real as real numbers; mathematicians simply choose really awful names for things.[*]

One way of viewing a complex number is as the coordinates on a graph, known as an **Argand diagram** or **complex plane**, illustrated in fig. 1.5. Numbers along the horizontal axis are the real numbers (this is really just the number line we are taught at school) and numbers up the vertical axis are the imaginary numbers. Imaginary

---

[*] We have Carl Gauss and René Descartes to thank for these gems.

numbers are designated by the letter j.[†] Imaginary numbers are therefore no different from real numbers; they are simply the name used for the numbers on the other number line, just as the numbers on a map may be called longitude and latitude. A complex number is simply the coordinates of a point on the Argand diagram. The real part is the instruction to move however many places east or west from the origin, and the imaginary part is the instruction to move however many places north or south of the origin.

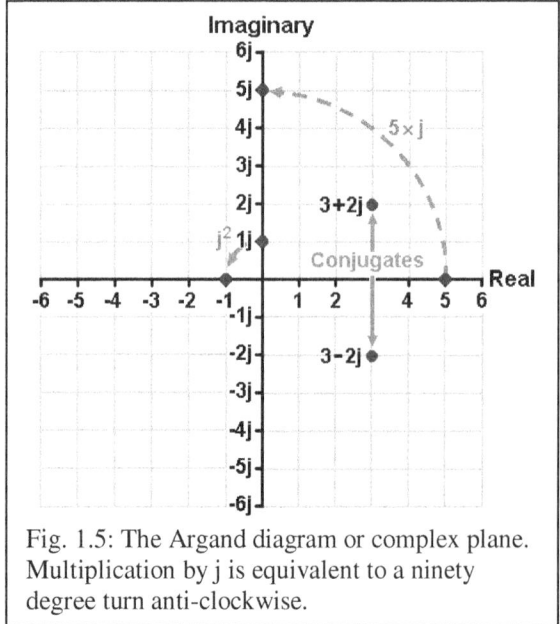

Fig. 1.5: The Argand diagram or complex plane. Multiplication by j is equivalent to a ninety degree turn anti-clockwise.

Now, suppose we start with a purely real number like 5, and multiply it by j, giving 5j (the j may be written before or after the number –it makes no difference). Looking at fig. 1.5 it is obvious that, relative to the origin, our initial point on the real axis has been turned anticlockwise by ninety degrees so that it now lies on the imaginary axis. Thus multiplying by j is equivalent to rotating (or phase shifting) something by ninety degrees. Multiplying by –j does the same thing but clockwise. This is why j appears in the formulae for capacitive and inductive reactance (discussed later) where current and voltage are ninety degrees out of phase with one another.

Alternatively, let us start at 1j and rotate by ninety degrees to arrive at –1. In other words, we have effectively multiplied 1j by another j, which is the same thing as j^2. Thus we discover that $j^2 = -1$, which is the formal definition of j! We might even take the square root of both sides and write $j = \sqrt{-1}$ which is another way of defining it, although mathematicians tend to get pedantic about this.

From this purely qualitative description we discover that j can be viewed either as a number –the square root of minus one– when doing calculations, or as a ninety degree shift when thinking about the physical meaning of those numbers. This makes it extremely useful and it appears throughout engineering mathematics as a result.

Before briefly going over the rules of complex arithmetic there is one extra thing we need to be aware of. Start with a complex number, say 3+2j, and reflect it about the real axis to create the number 3–2j. These two numbers are said to be **conjugates** of

---

[†] Mathematicians use the letter i, but engineers already use that letter for current.

one another. The conjugate of any complex number is always the same number but with an imaginary part of opposite sign. Such numbers have a special property that comes in handy when doing complex calculations: when they are multiplied together the result is always a purely real number. An example is shown shortly.

1.5.1: Complex Arithmetic

To add (or subtract) complex numbers we simply add (or subtract) the real and imaginary parts separately. For example, to add the two complex numbers $5 + 7j$ and $3 - 18j$:

First add the real parts: $5 + 3 = 8$

Then add the imaginary parts: $7 + (-18) = -11$

So the answer is: $(5 + 7j) + (3 - 18j) = 8 - 11j$

To multiply complex numbers we must expand them, turn any instances of j^2 into -1 and then group the terms into their real and imaginary parts again. For example, to multiply the same two complex numbers as previously, begin by expanding them:

$(5 + 7j)(3 - 18j) = 15 - 90j + 21j - 126j^2$

Now convert the j^2 into -1, which has the effect of turning -126 into $+126$, and collect the real and imaginary terms together:

$15 - 90j + 21j + 126 = 141 + 69j$

Also, let us try multiplying a complex conjugate pair, say 4+2j and 4–2j:

$(4 + 2j)(4 - 2j) = 16 - 8j + 8j - 4j^2$

$16 - 8j + 8j + 4 = 20 + 0j = 20$

The imaginary part has vanished, leaving only a real number, which is the special property of complex conjugates. In fact, the answer is the same as $4^2 + 2^2$, and by knowing this trick you can skip the expansion part and jump straight to the answer.

Dividing a complex number by a real number is easy; simply divide the real and imaginary parts separately. For example:

$$\frac{1 + 18j}{2} = \frac{1}{2} + \frac{18}{2}j = 0.5 + 9j$$

Dividing one complex number by another complex number is less straightforward. We first need to multiply top and bottom by the complex conjugate of the *bottom* number, which will turn the bottom into a real number. For example, to divide 4–7j by 3+5j, first multiply top and bottom by the conjugate 3–5j:

$$\frac{(4 - 7j)}{(3 + 5j)} \frac{(3 - 5j)}{(3 - 5j)} = \frac{(4 - 7j)(3 - 5j)}{3^2 + 5^2} = \frac{(4 - 7j)(3 - 5j)}{34}$$

Now carry out the multiplication on top in the usual way:

$$\frac{(4-7j)(3-5j)}{34} = \frac{12-20j-21j+35j^2}{34} = \frac{-23-41j}{34}$$

Now divide by the real number, which is easy.

$$\frac{-23-41j}{34} = -\frac{23}{34}-\frac{41}{34}j = -0.68-1.21j$$

1.6: Phasors

When dealing with sine waves we are often interested in two particular pieces of information: their magnitude (i.e. amplitude), and their phase. Drawing waveforms in full to compare them would be rather laborious, so there are various ways of conveying the important information of magnitude and phase in more compact notation. Of course, we could just write down the magnitude and phase as completely separate numbers, but that can make life difficult when doing calculations, so an alternative method is to imagine a rotating arrow or vector, called a **phasor**. The tail of the phasor is pinned to the origin of an Argand diagram while the other end traces out a circle like the hand of a clock, except this clock usually runs backwards.

What have circles to do with sine waves? One way to answer this question is by looking at a corkscrew or spring, as shown in fig. 1.6. If we look at it end-on we see a perfect circle, but when looking side-on we see a perfect sinusoidal wave. Now, suppose we mark a single point somewhere on the corkscrew with a spot of paint, and rotate it about its long axis. Looking end-on, the spot would revolve endlessly around a circular path. But looking side-on the spot

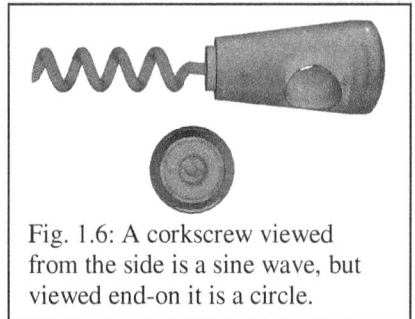

Fig. 1.6: A corkscrew viewed from the side is a sine wave, but viewed end-on it is a circle.

appears to bob up and down on top of a wave that travels past it. Thus a point that revolves around a circle can be made to trace out a sinusoidal wave over time, just by switching our point of view. The spot of paint represents the tip of a phasor.

By convention, the reference or 'starting position' is always with the phasor pointing to the right on the Argand diagram (the 3 o'clock position), and it rotates *anti-clockwise*. Now, if we look at the projection or 'shadow' that the tip makes on the real axis, it will again appear to bob back and forth as the phasor rotates. If you further imagine it to be attached to a pen then it will trace out a cosine wave if a sheet of paper is pulled downwards past it at a constant rate (like one of those old fashioned chart recorders used on polygraphs and seismometers). This process is illustrated in fig. 1.7. Alternatively, if we rotate the phasor by ninety degrees so that it is pointing upwards (12 o'clock) before we set it spinning then it will trace out a sine wave. Thus a sine wave is also a cosine wave that has been shifted forwards by ninety degrees, and *vice versa.*

The general equation for a voltage sine wave is:

$$v = V_{pk} \sin(2\pi ft) \tag{1.13}$$

Where v is the voltage at any particular instant in time designated by t. V_{pk} is the peak value or amplitude of the waveform and corresponds to the length of the phasor, while f is the frequency in hertz. It is also common to express the frequency in radians per second instead:

$$v = V_{pk} \sin(\omega t) \tag{1.14}$$

If we were interested in current rather than voltage then we could replace the v with i, and V_{pk} with I_{pk}. The same goes for any other quantity too, of course. Here we see that ω –which is called the **angular frequency**– is simply a shorthand way of writing 2πf. Angular frequency has the units of radians per second, which are like hertz (cycles per second) except for being 2π times larger. In other words, 1Hz is the same thing as 6.28 rad/sec.

Now, while we are tracing the wave on a bit of paper we could label its long axis according the angle through which the phasor has rotated at any given moment, relative to the starting position. Travelling all the way around the circle completes one cycle, so there are 360 degrees or 2π radians in a cycle. After completing one cycle we could continue to label the axis with ever increasing angle, or we could reset it to zero each time we cross the starting position; it is our choice since each cycle looks identical to the last.

But we could alternatively label the axis with the amount of *time* that has passed since the phasor started rotating. The rate at which the phasor rotates then determines the frequency of the wave, so one revolution per second produces a 1Hz signal, while one thousand revolutions per second produces a 1kHz signal, and so on. Thus an *angle* can correspond to an amount of *time*, and circuits that cause signals to be phase shifted therefore also cause time delays. Phase and time are not *always* equivalent, however, since phase may be said to reset back to zero after we pass 360 °, whereas time clearly

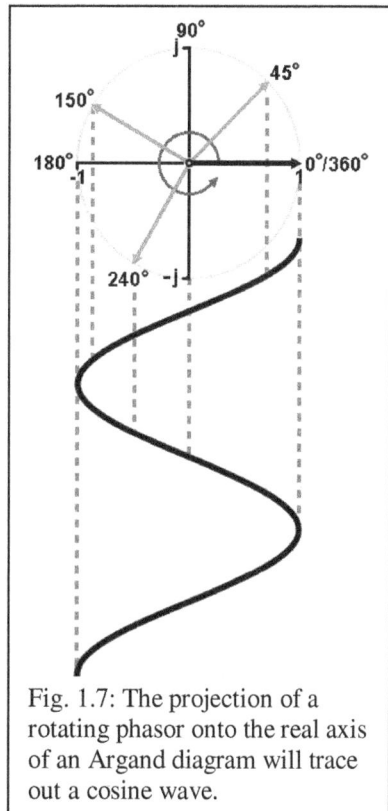

Fig. 1.7: The projection of a rotating phasor onto the real axis of an Argand diagram will trace out a cosine wave.

13

cannot be reset back to zero, much to the dismay of aged people everywhere. Nevertheless, provided we are careful about this, we can usually equate a phase lag with a time delay, and a phase lead with a time advance (but since time travel is impossible a 'time advance' in practice means only that one thing is less delayed relative to another thing).

Now, the position of the tip of the phasor can, at any time, be written down as a complex number, i.e. a pair of coordinates. Since these coordinates produce a right-angled triangle it is easy to find the length of the hypotenuse and its angle –i.e. the magnitude and phase– using trigonometry. Therefore, complex numbers can represent phasors frozen at a moment in time, and since phasors represent sinusoidal waves, so can complex numbers. This is why complex numbers feature so prominently in AC circuit theory.

Waveforms that are not sinusoidal can also be described by phasors but they require more than one, of various lengths, attached end-to-end, rotating at different rates. Each phasor then represents a different frequency that is present in the final waveform. This is better appreciated with an animated diagram (they can be found online), but it does lead us rather neatly to Fourier's theorem.

1.7: Fourier's Theorem

Fourier's theorem tells us that any periodic (i.e. repeating) waveform can be thought of as being built up from a series of pure sinusoidal frequencies, called the **Fourier series**. The lowest frequency present in the signal is called the **fundamental** or first **harmonic**, and this determines the rate at which the whole waveform pattern repeats, that is, its time period. All other component frequencies are exact multiples of the fundamental, so the second harmonic is twice the fundamental frequency; the third harmonic is three times the fundamental, and so on. Many signals also contain a 0Hz or DC component which could be thought of as the 0th harmonic, though this name is seldom used. The DC component is simply the mean average value of the waveform found by subtracting any negative areas from any positive areas, as described earlier in section 1.4.1.

Fig. 1.8 shows a classic example of the Fourier series of a square wave. The series contains only odd harmonics and decays rapidly. In fact, it is a universal result of Fourier's theorem that any waveform that is symmetrical about

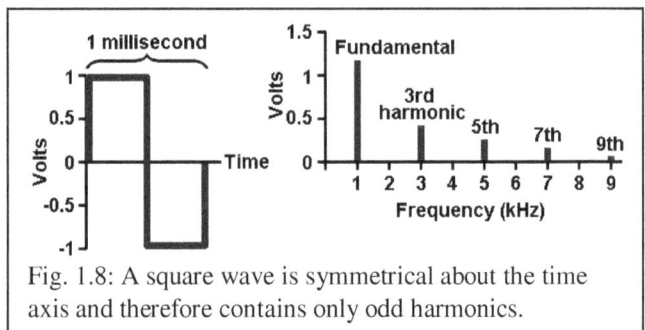

Fig. 1.8: A square wave is symmetrical about the time axis and therefore contains only odd harmonics.

the horizontal time axis contains *only* odd-order harmonics. Any non-symmetrical waveform will contain at least some even-order harmonics.

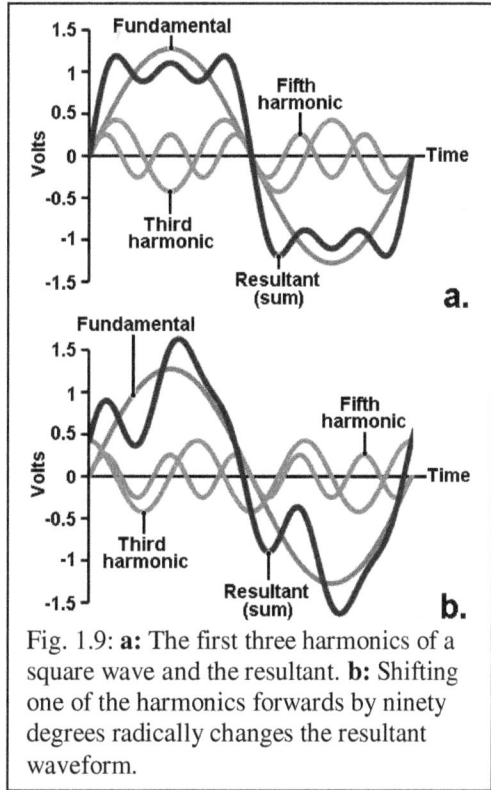

Strictly, it is both the amplitude *and* phase of each Fourier component that determines the final composite waveform. However, the human ear is not very sensitive to phase changes in mono sounds, so it is quite possible to get two signals that *sound* the same but *look* very different on an oscilloscope. This is because although they contain exactly the same amounts of each frequency, their phasings are different. For example, fig. 1.9a shows the first three harmonics of a square wave together with the resulting waveform (with only three harmonics the total is a rather pathetic excuse for a square wave). Fig. 1.9b shows the effect of phase shifting just one of the harmonics

Fig. 1.9: **a:** The first three harmonics of a square wave and the resultant. **b:** Shifting one of the harmonics forwards by ninety degrees radically changes the resultant waveform.

(in this case the third) by ninety degrees. The overall waveform now looks very different, yet to the ear both signals would sound the same.

The Fourier series applies only to periodic waveforms, but the idea can be extended to cover non-repeating waveforms too. It is then called the **Fourier transform** and is achieved by pretending that the signal does repeat, but that it takes an infinite amount of time to do so. The time period of a non-periodic waveform is therefore infinity, so its fundamental must be 0Hz and its harmonics can include any frequency at all, not just evenly-spaced discrete frequencies. The frequency content of a non-periodic waveform is called its **Fourier spectrum**, and the process of measuring it is called spectrum analysis.

Now, suppose we have a completely pure sine wave except that its amplitude decays with time, perhaps because someone is turning down a volume control. What is the frequency content of such a signal? Intuitively we might expect it to contain only the sine wave frequency; the fact that it is getting quieter over time seems irrelevant. But that is not so. The Fourier transform of a sine wave which decays exponentially is a peak with skirts (called a Lorentzian function), as shown in fig. 1.10. The centre of the peak corresponds to the sine frequency while the infinity of other frequencies occupying the skirts arise from the envelope of the signal.

15

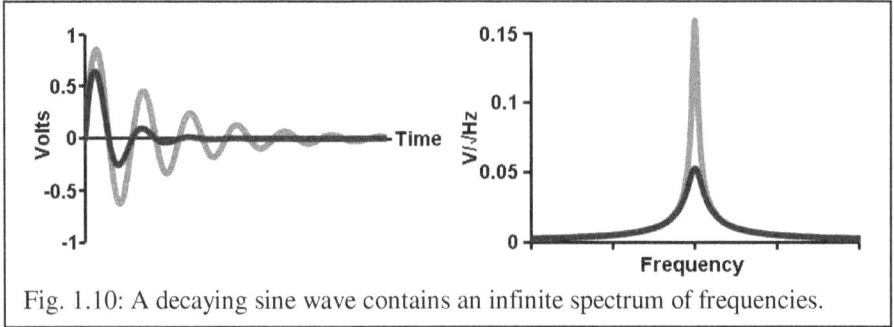

Fig. 1.10: A decaying sine wave contains an infinite spectrum of frequencies.

If the signal amplitude decays slowly –due to a volume control being turned down by hand– then these extra frequencies are present only in tiny amounts relative to the main sine frequency and are not noticed by the ear. In other words, the peak is narrow and more like the single spike that we would expect for a true, single frequency. But if the decay is rapid then the peak is reduced, that is, the skirts are larger *relative* to the amplitude of the sine frequency, and the effect may be audible, adding timbre to the otherwise pure tone. This is illustrated in fig. 1.10. Such signals are characteristic of many musical sounds including plucked strings and drums. This example is meant to emphasise the fact that the envelope of a signal contains frequency information just as much as the rest of it, so circuits which alter the envelope –such as audio compressors or amplifiers which sag under load– will have an inadvertent filtering effect too. This is of course undesirable in any high fidelity amplifier.

It is another universal rule of Fourier's theorem that the more sharp and peaky a signal is in the time domain, the shallower and broader is its response in the frequency domain, that is, the wider its bandwidth. Any waveform that exhibits high rates of change –i.e. steep edges and sharp corners– will contain a spectrum of high frequency components. Therefore, if an amplifier is overdriven so that it clips sharply then we can be sure the distorted waveform will contain a spray of high-order harmonics. By contrast, an amplifier that appears to 'compress' the waveform more gently when overdriven will introduce much less high frequency distortion. This is often put forward as a reason why gently clipping valve amplifiers may sound less unpleasant than most solid state amplifiers near overload.

One further corollary of the Fourier transform is the duality theorem which tells us that any waveform with a certain frequency response has a dual, and *vice versa*. For example, a DC signal in the time domain is a rather boring horizontal line, while in the frequency domain it is a single spike at 0Hz. According to the duality theorem we can apply this knowledge in reverse: a spike in the time domain must be a horizontal line in the frequency domain. In other words, by knowing the frequency response of a DC signal we now also know that an infinitely narrow voltage **impulse** must contain *all* frequencies with equal power. Such impulses are sometimes used for measuring the frequency response of systems for exactly this reason.

16

1.8: Resistance and Ohm's Law

Having ventured into a world of imaginary numbers, rotating phasors and Fourier transforms, let us return to some more down-to-earth electronics. If we apply an EMF to a circuit, electrons start moving. But if there is nothing pushing back then they will continue to accelerate forever and the current will grow to infinity. This does not happen in practice because the current encounters resistance that impedes its progress as it tries to surge forwards. You can think of this as an electrical version of Newton's third law: for every action there is an equal and opposite reaction. In other words, if we force a charge to flow in a material, a potential difference (voltage) is set up across the material to oppose us, as shown in fig. 1.11. This leads to the most famous of electrical laws –Ohm's law:[*]

$$v = iR \quad \text{volts} \tag{1.15}$$

Here, R is the resistance in ohms. The resulting voltage corresponds to the amount of energy we had to put in to move each unit of charge through the material. In the case of resistors this energy is converted to heat and lost, but in other components it may be stored. Additionally, if we rearrange Ohm's law into the form:

$$R = \frac{v}{i} \quad \text{ohms}$$

We see that an ohm is also a short hand way of saying a 'volt per amp'. This may seem like a redundant statement but it is a rock-solid principle to fall back on when we discover reactance, shortly.

Note in fig. 1.11 that the EMF arrow and current arrow point in the same direction at the battery, but at the resistor the voltage and current arrows are opposed. According to this standard sign convention, if a voltage and current pointing in the *same* direction are both positive or both negative, then we have a *source* of power, e.g. the battery. Any other combination is a sink of power, e.g. the resistor.

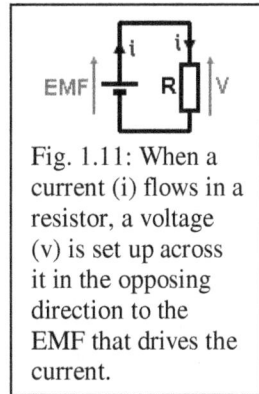

Fig. 1.11: When a current (i) flows in a resistor, a voltage (v) is set up across it in the opposing direction to the EMF that drives the current.

By combining Ohm's law with equation (1.5) we also obtain a couple of useful expressions for the power dissipated in a resistance:

$$p = i^2 R = \frac{v^2}{R} \quad \text{watts} \tag{1.16}$$

1.8.1: Resistors in Series and Parallel

When resistors are placed in series the current in each one is the same, since there is no other path for the current to take. And since the current must flow from

---

[*] Like so many things in the world of science the law credited to Georg Ohm was actually formulated even earlier by someone else: Henry Cavendish.

one resistor to the next it is effectively the same as if it has flowed in one 'bigger' resistance. The total resistance is therefore simply the sum of all the individual resistors

$$R_{series} = R_1 + R_2 + R_3...$$ (1.17)

When resistors are placed in parallel then the voltage across them must be the same, but the current divides up between the different paths, so it is now easier for the total current to pass through. The total resistance must therefore be less than any one resistor alone. It is equal to:

$$R_{parallel} = \frac{1}{\frac{1}{R_1} + \frac{1}{R_2} + \frac{1}{R_3}...}$$ (1.18)

In the special case of just *two* resistors in parallel, this simplifies to:

$$R_{parallel} = \frac{R_1 R_2}{R_1 + R_2}$$ (1.19)

This is called the **product-over-sum rule** and is a very useful formula to remember.

1.9: Capacitance

A capacitor is formed from two conductors or plates separated by an insulator called the **dielectric**. There is, therefore, some capacitance between all metal or conducting objects. The total capacitance between a pair of parallel plates is:

$$C = \varepsilon \frac{A}{d} \quad \text{farads}$$ (1.20)

Where:
ε = Electric permittivity of the dielectric, in farads per metre (8.85×10^{-12}F/m for air).
A = Plate area, in square metres.
d = Distance between plates, in metres.

We will not waste much time with this equation; it is presented here only because it leads to a useful maxim: increasing the distance between conductors and/or reducing their area will *reduce* the capacitance between them. It is worth remembering this when creating a physical layout for a circuit.

The current in a capacitor is proportional to the rate of change of voltage across it. In other words, whenever the voltage across a capacitor changes, the current either increases or decreases in an attempt to keep the voltage at its original value.

Specifically, the current is equal to:

$$i = C \frac{dv}{dt} \quad \text{amps}$$ (1.21)

18

Essential Electronics

If we try to change the voltage very quickly then a capacitor will allow a great deal of current to flow, and if this cannot be supplied then we cannot physically change the voltage as fast as we had expected. This property allows capacitors to be used for smoothing out unwanted ripple voltages in power supplies, for example, but this same property is also what prevents us from amplifying very high frequency signals with ease.

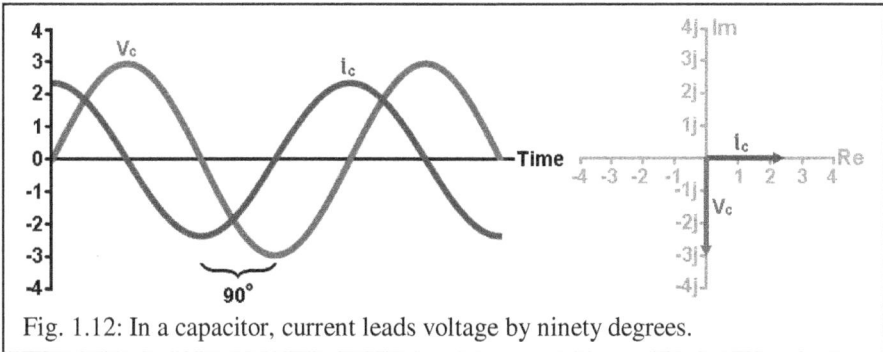

Fig. 1.12: In a capacitor, current leads voltage by ninety degrees.

Fig. 1.12 represents a sinusoidal voltage applied to a capacitor, and the resulting current flowing in it (the units are arbitrary). It is clear that the voltage lags behind the current by ninety degrees or a quarter cycle. Now, beginners may wonder, how does the current 'know' what the voltage is going to be, before it happens? Of course, the only sane answer is that it doesn't know –it's an illusion. The current is not caused by the voltage itself, like in a resistor, but by the *rate* at which the voltage is changing, i.e. the slope of the voltage curve. For a sine wave the rate of change is fastest at the zero crossing so the current is maximum here, and it is not changing at all –albeit momentarily– at the peaks and troughs, so the current is zero there. The result is that the rate of change of a sinusoidal wave is also sinusoidal wave, but offset by ninety degrees. This does create the unfortunate impression that the peak (say) of one wave is somehow 'caused by' the nearest peak of the other, which is not the case. Let us firmly break this illusion by using a triangle wave as the example, as shown in fig. 1.13 (again, the units are arbitrary).

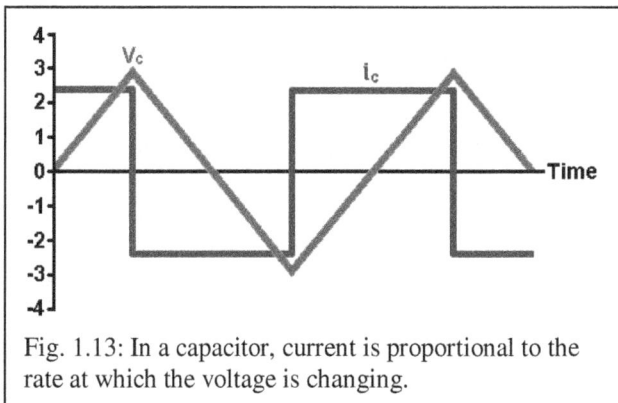

Fig. 1.13: In a capacitor, current is proportional to the rate at which the voltage is changing.

On the rising edge of the triangle wave the rate of change of voltage is positive and constant, so the current is also positive and constant. On the falling edge of the triangle wave the

19

gradient is again constant but this time negative, so the current reverse sign and becomes negative. Thus with each cycle the capacitor charges, discharges, passes through zero and charges with the opposite polarity. But since the waveforms are completely different shapes, how does phase shift come into it? The answer is that it is the individual *harmonics* in the square (current) wave that are shifted forwards by ninety degrees, relative to the harmonics of the triangle (voltage) wave. The higher frequency harmonics are also increased in amplitude due to the capacitor's falling reactance, i.e. more current flows at higher frequencies. The combined effect of phase shifting and altering the amplitude of the harmonics is to create the square wave from the triangle wave.

Since electrons cannot cross the gap between the plates they instead gather on one plate and are repelled in equal measure from the other. The charge collected on either plate of a capacitor is given by:

$$q = Cv \quad \text{coulombs} \tag{1.22}$$

Where C is the capacitance and v is the voltage between the pates. This charge will be negative on the plate where electrons are clustered and positive on the opposite plate from which they have emigrated, and since they cancel each other out the *total* charge collecting in the device is zero. It is therefore not correct to say that a capacitor 'stores' charge. What a capacitor stores is *energy*, in the form of the electric field between the plates. The stored energy is equal to:

$$w = \frac{1}{2}Cv^2 \quad \text{joules} \tag{1.23}$$

Notice that the energy is proportional to the square of the voltage, so doubling the voltage will quadruple the stored energy. This is why high-voltage capacitors must be treated with caution as they can store enough energy to give a painful shock even if the capacitance is quite small.

Knowing that p = iv, the power dissipated in a capacitor is plotted in fig. 1.14 by the line P_c. Notice that it is alternately positive and negative. But what is negative power? It is simply power that is flowing in the opposite direction to our reference, which means it is being

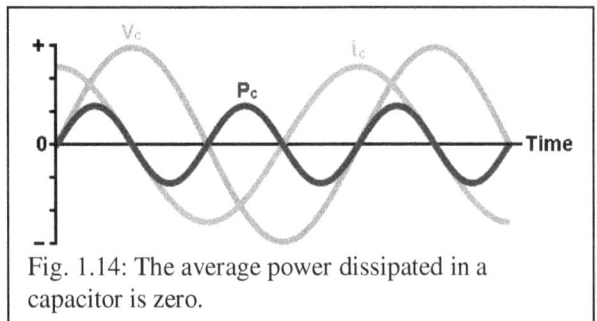

Fig. 1.14: The average power dissipated in a capacitor is zero.

returned to our source, so our supposed power source is not really acting as the source anymore! In other words, during every-other quarter cycle, energy is sent to the capacitor and stored, and during the other quarter cycles that energy is released back to us. Over time the positive and negative power cancels out, so the average power dissipated is zero, and this is called **reactive power**. This means that no energy is converted into heat in a capacitor, at least in theory. Practical capacitors do have some unavoidable series resistance which *will* dissipate a little power, but this is not usually important except in high-current power supplies.

1.9.1: Capacitive Reactance

A capacitor presents a *kind* of resistance to the flow of current, but it is called **reactance** instead. Returning to a sine wave, as the frequency is increased the slopes of the wave get steeper, i.e. the rate of change increases. Therefore, for a given voltage amplitude the resulting current in a capacitor must increase with frequency, which is another way of saying the reactance *decreases*. Specifically, the reactance of a capacitor, X_C, is:

$$X_C = \frac{v_c}{i_c} = \frac{1}{2\pi fC} \quad \text{ohms} \tag{1.24}$$

Where f is the frequency and C is the capacitance. From the above it is clear that as the frequency falls, the reactance increases. At zero hertz the reactance blows up to infinity or, in other words, a capacitor looks like an open circuit to DC. This handy property allows capacitors to be used for coupling, for example, where we want to block a DC voltage but allow AC signals to pass through. Thus also leads to a useful maxim: *the mean average current in a capacitor is zero.*

Reactance, like resistance, is equal to the voltage across the device divided by the current through it. But reactance is different from resistance because it does not dissipate any power, owing to the ninety degree phase difference between voltage and current. Unfortunately, the previous equation does not make this clear because it ignores this phase difference. Therefore, let us jump ahead for a moment and make a small modification to the reactance to turn it into an *impedance* (see section 1.11) which *does* take phase shift into account. The modification is to divide by j, or multiply by –j which amounts to the same thing. The *impedance* of a capacitor, Z_C, is then:

$$Z_C = \frac{1}{j2\pi fC} = -j\frac{1}{2\pi fC} \quad \text{ohms} \tag{1.25}$$

But what does the –j really *mean*? To answer this, let us fall back on Ohm's law:

$$Z_C = -j\frac{1}{2\pi fC} = \frac{-j\,v_c}{i_c}$$

The only difference between this expression and the version of Ohm's law shown in section 1.8 is the presence of the –j. And from section 1.5 we know that multiplying something by –j corresponds to rotating a phasor clockwise by ninety degrees, which is what has happened to the voltage phasor in fig. 1.12. Thus the –j in the above equation is merely a label that indicates the voltage is shifted back by ninety degrees so that it lags the current; it reminds us that we are dealing not just with an

21

impedance but a *capacitive* impedance. By contrast, inductive reactance is indicated by j (no minus sign).

1.9.2: Capacitors in Series and Parallel

When capacitors are placed in series it can be viewed as a single capacitor that has extra plates sandwiched in the middle of its dielectric. These plates do nothing very useful, they simply increase the distance between the two outermost plates, which therefore reduces the total capacitance. The total capacitance is equal to:

$$C_{series} = \frac{1}{\frac{1}{C_1} + \frac{1}{C_2} + \frac{1}{C_3} \cdots}$$
(1.26)

But for just two capacitors in series this simplifies to:

$$C_{series} = \frac{C_1 C_2}{C_1 + C_2}$$
(1.27)

When capacitors are connected in parallel we are effectively creating a new component with more plate area, and therefore more capacitance. The total capacitance is equal to:

$$C_{parallel} = C_1 + C_2 + C_3 \ldots$$
(1.28)

The equations for capacitors in series or parallel are, as it were, the opposite of those for resistors.

1.10: Inductance

An inductor is usually created by winding wire into a coil (although even a straight wire has inductance, in the range of a few nanohenries per centimetre). The inductance of a long, thin coil or solenoid is:

$$L = \mu N^2 \frac{A}{1} \quad \text{henries}$$
(1.29)

Where:
μ = Magnetic permeability of the core material, in henries per metre ($4\pi \times 10^{-7}$ H/m for air).
N = Number of turns.
A = Cross sectional area of the coil, in square metres.
l = Length of the coil, in metres.

Again, we will not dwell on this equation except to note that the smaller the loop area, the smaller the inductance, which is useful to remember when creating physical circuit layouts.

The properties of an inductor are the dual of those of a capacitor. The voltage across an inductor is proportional to the rate of change of current in it. An inductor might therefore be said to resist changes in current through itself, whereas a capacitor resists changes in voltage across itself. In other words, whenever the current in an inductor changes, it develops a voltage across itself in an attempt to keep the current at its original value. Specifically, the voltage across an inductor is equal to:

$$v = L\frac{di}{dt} \quad \text{volts} \tag{1.30}$$

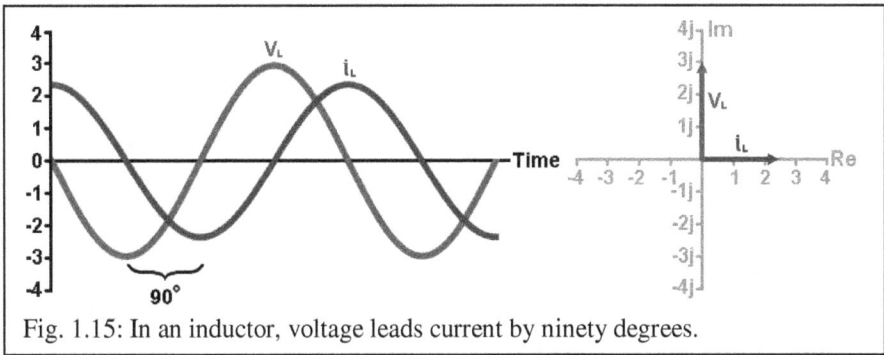

Fig. 1.15: In an inductor, voltage leads current by ninety degrees.

Fig. 1.15 represents a sinusoidal current in an inductor, and the resulting voltage across it (the units are arbitrary). It is clear that the current lags behind the voltage by ninety degrees, or a quarter cycle. Again, it must be emphasised that the inductor voltage does not 'know' what the current is going to be before it happens; the voltage is not caused by the current but by the *rate* at which it is changing, i.e. the slope of the current curve. As with a capacitor this is both a feature and a defect. In particular, if we attempt to change the current very rapidly (by suddenly switching it off, for example) the inductor will generate an enormous voltage in an attempt to keep the current flowing. This **flyback voltage** can be damaging to the inductor itself and to other components. On the other hand, this same phenomenon is exploited to great benefit in transformers.

An inductor stores energy in the form of a magnetic field (a capacitor stores energy in the form of an electric field). The energy stored is equal to:

$$w = \frac{1}{2}Li^2 \quad \text{joules} \tag{1.31}$$

Since voltage and current are ninety degrees out of phase in an inductor, the average power dissipated is theoretically zero, just as for a capacitor. Energy is sent to, or returned from, an inductor during the opposite quarter cycles from those of a capacitor. However, practical inductors are made from wire and therefore have some resistance which will dissipate at least some power.

1.10.1: Inductive Reactance

An inductor has a reactance which is the dual of capacitive reactance. In other words, inductive reactance *increases* with frequency:

$$X_L = \frac{v_L}{i_L} = 2\pi fL \quad \text{ohms} \tag{1.32}$$

Where f is the frequency and L is the inductance. From this equation it is clear that at zero hertz the reactance becomes zero, so an inductor is a short circuit at DC (or at least it would be if it weren't for its own wire resistance).

Fig. 1.16 compares the reactance of an ideal 1H inductor and 100nF capacitor against frequency. Their duality becomes more obvious when the scales are logarithmic, as in b. Notice that at one

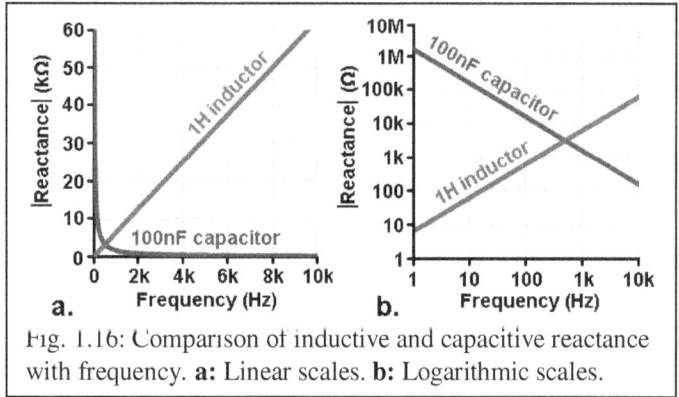

Fig. 1.16: Comparison of inductive and capacitive reactance with frequency. **a:** Linear scales. **b:** Logarithmic scales.

particular frequency the magnitudes of the reactances are equal; the significance of this is explored in section 1.21 on resonance.

To turn inductive reactance into inductive impedance we multiply by j, which indicates that the voltage leads, rather than lags the current, as can be seen in the phasor diagram in fig. 1.15:

$$Z_L = \frac{jv_L}{iL} = j2\pi fL \quad \text{ohms} \tag{1.33}$$

The relationship between current and voltage in an inductor or capacitor is easily remembered with the mnemonic "CIVIL". In a capacitor C, current I comes before (i.e. leads) voltage V, while I comes after (i.e. lags) V in an inductor L.

1.10.2: Inductors in Series and Parallel

When inductors are placed in series we are effectively creating a composite component with more turns, so the inductance increases. The total inductance is equal to:

$$L_{series} = L_1 + L_2 + L_3... \tag{1.34}$$

When inductors are placed in parallel the current divides between them, so the rate of change of current experienced by each individual inductor is reduced, and with it the induced voltage. The total inductance is therefore less than any one inductor alone and is equal to:

$$L_{\text{parallel}} = \frac{1}{\frac{1}{L_1} + \frac{1}{L_2} + \frac{1}{L_3} \cdots}$$ (1.35)

For two inductors this simplifies to:

$$L_{\text{parallel}} = \frac{L_1 L_2}{L_1 + L_2}$$ (1.36)

These equations are similar to those for resistors. However, this assumes the individual magnetic fields do not interact with each other, which is not always a fair assumption. Fortunately, inductors are fairly uncommon in audio circuits and rarely need to be used in series/parallel combinations, so we will not press the matter.

1.11: Impedance

Impedance is the general term for resistance, reactance, or any combination of both. We can therefore re-write Ohm's law in a way that covers all situations, not just resistors:

$$v = iZ$$ (1.37)

Here Z is the impedance in ohms, which may be real, imaginary, or complex. If it is real (purely resistive) then we are back to the simpler version of Ohm's law. But if it imaginary (purely capacitive and/or inductive), or complex (a bit of everything), then we have a phase difference between the v and the i in the equation.

What is the impedance of the network in fig. 1.17? Since the two components are in series we simply add them:

$$Z = R + \left(-j\frac{1}{2\pi fC} \right) = R - j\frac{1}{2\pi fC}$$

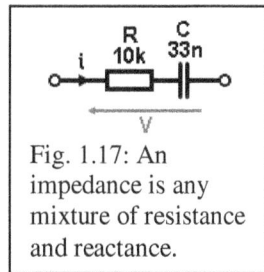

Fig. 1.17: An impedance is any mixture of resistance and reactance.

Now suppose we force a 1mA_{rms} 1kHz sinusoidal current, i, to flow in this impedance; what voltage, v, will appear across it? By applying Ohm's law we could write:

$$v = iZ = iR - j\frac{i}{2\pi fC}$$

But this answer does not help very much since it is in two parts corresponding to the individual voltages across the resistor and capacitor. We cannot simply add these voltages together since one is real and the other imaginary, i.e. ninety degrees out of phase.

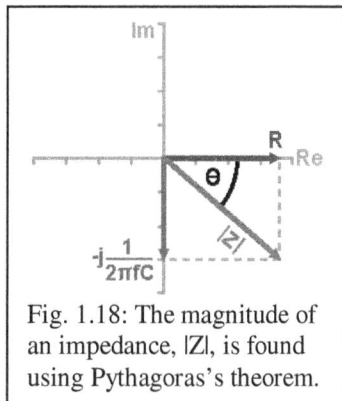

Fig. 1.18: The magnitude of an impedance, |Z|, is found using Pythagoras's theorem.

25

To solve this problem we must instead find the *magnitude* of the impedance. By definition, this is the voltage across the impedance, divided by the current through it, regardless of any phase difference. The two parts of the impedance are shown in fig. 1.18 as vectors[*] (not to scale). Taking them as the two sides of a right-angled triangle, the magnitude is the hypotenuse and can therefore be found using Pythagoras's theorem. In other words, to find the magnitude of an impedance we discard the j, square the real and imaginary parts separately, add them, then take the square root. This is also called geometric addition:

$$|Z| = \sqrt{R^2 + X^2} = \sqrt{R^2 + \left(-\frac{1}{2\pi fC}\right)^2} = \sqrt{10000^2 + \left(-\frac{1}{2\pi \times 1000 \times 33 \times 10^{-9}}\right)^2}$$

$$|Z| = \sqrt{10000^2 + (-4823)^2} = 11102\Omega$$

The voltage across the impedance is therefore:

$$|v| = i|Z| = 1\text{mA} \times 11.102\text{k}\Omega = 11.102\text{V}_{\text{rms}}$$

The angle θ in fig. 1.18 is the angle between the voltage and current, and from trigonometry it is equal to:

$$\theta = \tan^{-1}\left(\frac{X}{R}\right) = \tan^{-1}\left(\frac{-4823}{10000}\right)$$

$$\theta = -25.7°$$

Thus the voltage across the impedance is $11.102\text{V}_{\text{rms}}$, lagging the current by 25.7°. Notice that by putting the resistor in series with the capacitor the phase shift has been reduced compared with what it would have been with the capacitor alone. It is as if the resistor 'wants' the voltage and current to be in phase and so 'pulls' them closer together (or you might say it is the capacitor that wants them to be ninety degrees out of phase and so pulls them apart!). It is also important to remember that these figures only apply at the specified 1kHz spot frequency; at other frequencies the numbers will be different. Another interesting result of geometric addition is that the magnitude of an impedance is always strongly dominated by whichever term is greater, even if it is only a little bit greater.

---

[*] Vectors, not phasors. A phasor represents an alternating quantity whereas here we're representing fixed components.

1.12: Kirchoff's Voltage Law

It was hinted at earlier that when an EMF is applied to a circuit an equal-but-opposite potential difference is set up across the circuit. This leads us to Kirchoff's voltage law (KVL) which states that **the algebraic sum of voltages around any closed loop is zero**. An 'algebraic sum' is one that takes into account the sign (positive or negative) of the things being added together. An alternative version is to say that the sum of EMFs around any closed loop is equal to the sum of potential differences around the loop, i.e. the causes must be equal to the effects.

Fig. 1.19 shows an example circuit loop, also called a **mesh**. If we proceed clockwise around the loop then any voltage arrows that also point clockwise are taken as positive, and those that point against us are taken as negative. The sum of the voltages must be equal to zero:

$$V_1 - V_2 - V_3 + V_4 - V_5 = 0$$

Substituting in the actual values of voltage:

$$20 - 2 - 10 + (-3) - 5 = 0$$

Thus KVL is satisfied and the circuit is not defying the laws of physics, much to everyone's relief.

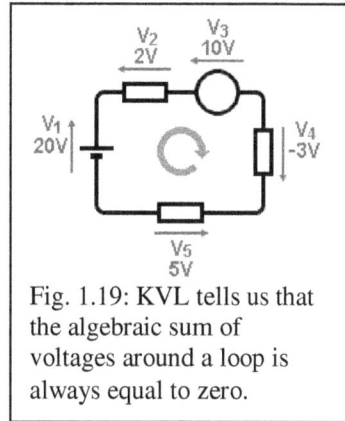

Fig. 1.19: KVL tells us that the algebraic sum of voltages around a loop is always equal to zero.

Alternatively, we could rearrange the equation so all the EMFs (generator voltages) are on one side and the potential differences (voltage drops) are on the other. In this form all the EMFs that point clockwise are taken as positive, while the potential differences that point clockwise are taken as negative since they must oppose the EMFs. Also don't forget that a negative voltage pointing in one direction is the same thing as a positive voltage pointing in the opposite direction:

$$V_1 - V_3 = V_2 - V_4 + V_5$$
$$20 - 10 = 2 - (-3) + 5$$

Both sides of the equation are indeed equal, so KVL is satisfied.

1.13: Kirchoff's Current Law

Current cannot magically spring from nowhere any more than it can vanish into nothingness. Kirchoff's current law (KCL) formalises this by stating that **the algebraic sum of currents entering any node is zero**. Alternatively we could say that the sum of the currents entering a node must be equal to the sum of current leaving the node. A 'node' is simply a point in a circuit, and fig. 1.20 shows an example.

Current arrows that point into the node are taken as positive, and those pointing away are taken as negative. We can therefore write an equation for the sum of currents entering the node, which must be equal to zero:

$$i_1 + i_4 + i_5 - i_2 - i_3 = 0$$

Substituting in the actual values of current:

$$3.5 + (-2.5) + 1 - (-2) - 4 = 0$$

Thus KCL is satisfied.

Alternatively we could rearrange the equation so all the incoming currents are on one side and the outgoing currents are on the other:

$$i_1 + i_4 + i_5 = i_2 + i_3$$

$$3.5 + (-2.5) + 1 = (-2) + 4$$

In this form all the arrows that point into the node appear on one side of the equation, and those that point away are on the other side. Do not forget that a negative current pointing in one direction is the same thing as a positive current pointing in the opposite direction.

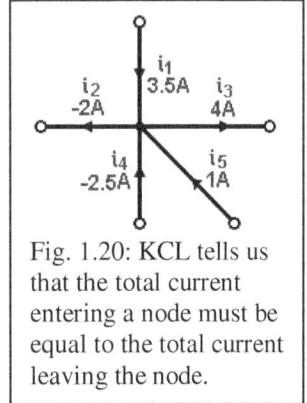

Fig. 1.20: KCL tells us that the total current entering a node must be equal to the total current leaving the node.

1.14: Sources

Real, non-ideal circuits can often be modelled and analysed by thinking of them as being made up of combinations of fictitious but otherwise ideal components. Various practical devices behave as sources of energy, or generators, and such components can therefore be modelled using ideal sources of current or voltage (augmented with some other components to reflect their real-life imperfection, see later).

1.14.1: Voltage and Current Sources

The symbol for a voltage source is a circle with a labelled arrow indicating the voltage it generates, as shown in fig. 1.21. This may be a fixed value or it may be written as an equation. A symbol is often added to the circle to indicate the sort of voltage waveform it produces, such as a sine or square wave. A

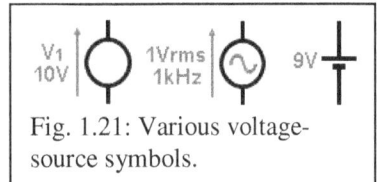

Fig. 1.21: Various voltage-source symbols.

battery symbol is also sometimes used to indicate a fixed (DC) voltage source, instead of a circle, and such symbols can be stacked to give an immediate impression of how large the voltage is. This useful diagrammatic technique is often found in older textbooks but appears to be less common nowadays. Note that the long bar in the battery symbol is always the positive terminal[*] so a voltage arrow is not absolutely necessary here. An ideal voltage source produces its specified voltage and can supply whatever current is needed to maintain that voltage across its terminals, even if that means infinite current. In other words, it has zero internal impedance. A voltage source that produces zero voltage is simply a short circuit.

The symbol for a current source is a pair of overlapping circles with a labelled arrow indicating the current it produces, as shown in fig. 1.21 (usually this arrow is drawn to the side but I think this looks too much like a voltage arrow, so I prefer to place it on the wire itself). A current-source generates a given amount of current and will vary the voltage across itself to

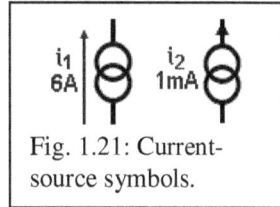

Fig. 1.21: Current-source symbols.

whatever value is necessary to maintain that current, even if that means infinite voltage. In other words, it has infinite internal impedance. A current source that produces zero current is simply an open circuit.

1.14.2 Dependent Sources

A source that generates a predetermined voltage or current is said to be **independent**; it does not know or care what is happening in the rest of the circuit. A battery, for example, produces a rated voltage between its terminals regardless of whether it is in circuit or sitting alone on a shelf, so it behaves like an independent voltage source. A **dependent** source is one whose output depends on some quantity elsewhere in the circuit. For example, an amplifier might be described as a dependent voltage source, since its output voltage depends on the voltage applied to its input terminals.

Of particular interest to us are voltage-controlled voltage sources (VCVS) like the amplifier just described. These have a conversion factor or **gain** of so many output volts per input volt. Fig. 1.21a shows an example of a VCVS which accepts an input voltage v_{in} and multiplies it by a factor of A (its gain).

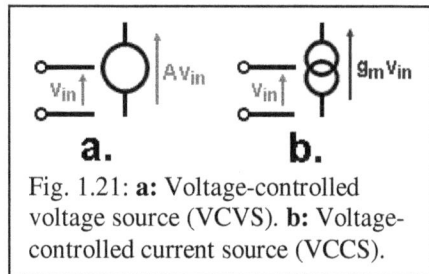

Fig. 1.21: **a**: Voltage-controlled voltage source (VCVS). **b**: Voltage-controlled current source (VCCS).

We will also commonly encounter voltage-controlled current sources (VCCS). These produce a current that depends on some input voltage, and the conversion factor is called the **transconductance**, in so many output amps per input volt. Fig. 1.21b shows an example that accepts an input voltage v_{in} and multiplies it by the factor g_m (its transconductance). There are, of course, current-controlled sources too, but they are rarely needed for audio electronics theory.

---

[*] It is amazing how many publications manage to get this wrong.

1.15: Thévenin's Theorem

Thévenin's theorem states that that **any two-terminal linear network can be modelled as a voltage source equal to the open-circuit voltage, in series with an impedance equal to that measured between the terminals with all independent sources set to zero**.

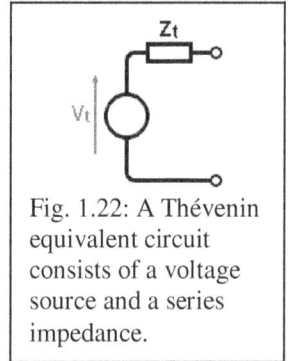

Fig. 1.22: A Thévenin equivalent circuit consists of a voltage source and a series impedance.

A battery is a good –but not perfect– approximation to a voltage source. If its terminals are shorted together an enormous current will flow, but it will not be infinite as it would be with an *ideal* voltage source. It is almost as if the battery really *is* ideal but there is a resistor hidden inside which limits the current to some maximum value when the terminals are shorted. In reality there is no hidden resistor; it is the internal chemical reactions and the resistance of the various physical parts that limit the current, but since these things are hidden inside the casing it hardly matters to us exactly what limits the current. From our point of view on the outside world a single resistor is all we need to model the overall behaviour of the battery. This is what Thevenin's theorem is saying. A practical (linear) device can be modelled as black box containing an ideal voltage source and an impedance which limits the short-circuit current to the value we observe in real life.

For example, suppose we have a battery that gives an open-circuit reading of 12V on an ideal voltmeter as shown in fig. 1.23a. This will be the value of the Thévenin source used to model the battery. Now, suppose we short-circuit the terminals

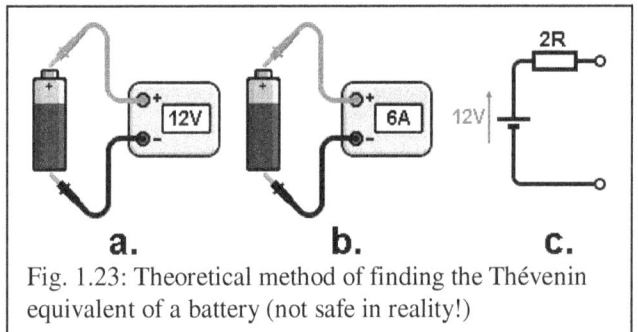

a. **b.** **c.**

Fig. 1.23: Theoretical method of finding the Thévenin equivalent of a battery (not safe in reality!)

with a perfect ammeter (which would be dangerous in practice as the battery would heat up rapidly) and obtain a reading of 6A as shown in b. From Ohm's law the internal resistance must be 12V / 6A = 2Ω. Armed with these figures we can now model the battery using the Thévenin equivalent circuit in c.

In this case the circuit is purely resistive, but the theory applies equally well to complex impedances when dealing with AC sources. The Thévenin impedance may variously be called the **internal impedance**, **output impedance**, or **source impedance**; they all refer to the same thing.

1.16: Norton's Theorem

Norton's theorem states that **any two-terminal linear network can be modelled as a current source equal to the short-circuit current, in parallel with an impedance equal to that measured between the terminals with all independent sources set to zero**.

Using the same battery as in the previous example, the short circuit current was 6A, which will be value of the Norton source used to model it. Since the open circuit voltage was 12V this would require that same 6A to flow in a 2Ω resistor, which is therefore the Norton resistance.

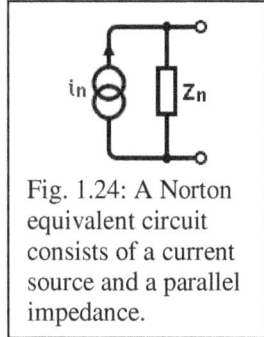

Fig. 1.24: A Norton equivalent circuit consists of a current source and a parallel impedance.

Norton and Thévenin equivalent circuits are duals of one another, as illustrated in fig. 1.25. Provided we are dealing only with resistance and not impedance, it is easy to convert one into the other since the resistance stays the same; we need only use Ohm's law to convert the value of the current source into a voltage source, or *vice versa*.

Which model is used is really a matter of preference, though one will often be much more convenient for the problem at hand. For example, it is nearly always more convenient to model a pentode or transistor using a Norton equivalent, whereas a triode can be modelled as either type with equal ease.

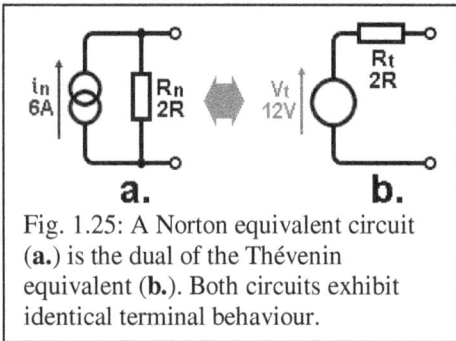

Fig. 1.25: A Norton equivalent circuit (**a.**) is the dual of the Thévenin equivalent (**b.**). Both circuits exhibit identical terminal behaviour.

Note that a Thévenin or Norton equivalent is only equivalent to a real component as far as its terminal behaviour is concerned. The internal behaviour may be quite unrealistic. For example, a Thévenin source dissipates no power under open circuit conditions whereas a Norton source does. Clearly, these models should not be used for power calculations inside a real component.

1.17: Potential Dividers

An enormously useful concept when dealing with voltage amplifiers is the potential divider, which produces an output voltage that is some fraction of the input voltage. In other words, it divides the input voltage up into smaller amounts. The most familiar example is a simple potentiometer or volume control, but even quite complex circuits can usually be distilled down to a simple collection of sources and potential dividers.

Consider the simple potential divider shown in fig. 1.26, which has a voltage V_1 applied to its input and a voltage V_2 taken from its output. We wish to know the fraction of voltage appearing at the output, relative to the input. In other words, we want to know the voltage gain of the circuit, V_2/V_1.

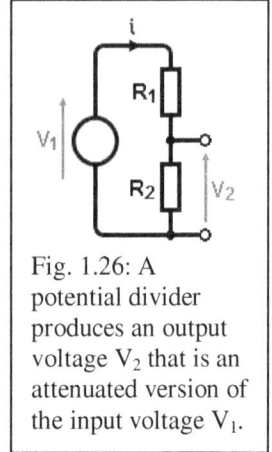

Fig. 1.26: A potential divider produces an output voltage V_2 that is an attenuated version of the input voltage V_1.

To find V_2 we need to know the current in R_2, since Ohm's law will then tell us the voltage across it. Now, as far as the voltage source is concerned the two resistors are in series and form a total resistance of $R_1 + R_2$. Using Ohm's law the current flowing around the loop must therefore be:

$$i = \frac{V_1}{R_1 + R_2}$$

The voltage across R_2 (the output voltage) is therefore:

$$V_2 = iR_2 = \frac{V_1 R_2}{R_1 + R_2}$$

We want the gain of the circuit V_{out}/V_{in}, or V_2/V_1, so dividing both sides by V_1 we finally obtain:

$$B = \frac{V_2}{V_1} = \frac{R_2}{R_1 + R_2}$$

By convention the letter B is used for the gain of passive circuits that usually attenuate the voltage, whereas the letter A is used for the voltage gain of active circuits that usually provide amplification. Since this circuit reduces or attenuates the voltage it may seem a little peculiar to talk about its 'gain', but this is perfectly acceptable and standard terminology. Gain is a just a multiplication factor, so a loss is simply a multiplication or gain of less than one.

This equation is also a simple example of a **transfer function**, that is, an equation that describes the relationship between an output signal and an input signal. Multiplying the input signal by the transfer function (i.e. the gain) produces the output signal; it describes how the signal is 'transferred' from input to output.

Exactly the same formula applies to impedances as well as resistances, but the numbers may be complex when dealing with impedances. The general formula for a potential divider including any number of impedances is:

$$B = \frac{\text{Impedance across which the output voltage is taken}}{\text{Total impedance}} \tag{1.38}$$

This formula is probably used more often than Ohm's law when designing circuits, so it is well worth memorising.

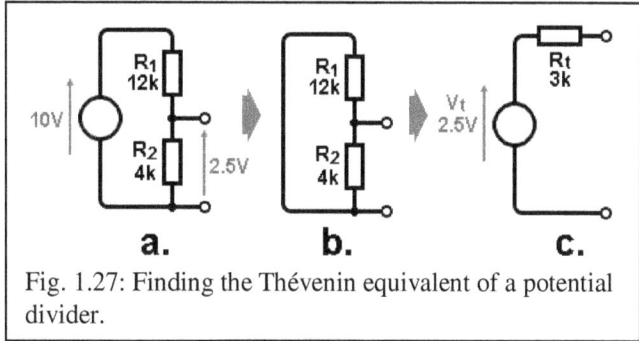

Fig. 1.27: Finding the Thévenin equivalent of a potential divider.

At this point it may also be useful to consider the Thévenin equivalent circuit of a potential divider. Fig. 1.27a shows an example with some known values. To find the Thévenin source voltage, V_t, we need to know the open-circuit voltage, that is, the output voltage with no additional load. In this case the divider has a gain of:

$$B = \frac{R_2}{R_1 + R_2} = \frac{4k}{4k + 12k} = 0.25$$

So the voltage appearing across its unloaded output terminals is $10 \times 0.25 = 2.5V$ (the remaining 7.5V must be across R_1, in accordance with Kirchoff's voltage law).

To find the Thévenin source resistance we need to know the resistance measured between the terminals with all independent sources set to zero. In this case there is only one source, and setting it to zero makes it a short circuit, as shown in b. From this it is obvious that the resistance between the terminals is the parallel combination of the two resistors, so using the product-over-sum rule:

$$R_t = R_1 \parallel R_2 = \frac{R_1 R_2}{R_1 + R_2} = \frac{12k \times 4k}{12k + 4k} = 3k\Omega$$

The completed Thévenin equivalent is shown in fig. 1.27c. It is useful to remember that the source impedance of any two-component potential divider is equal to the two impedances in parallel, as this result is used time and again in audio design.

1.18: Current Dividers

The dual of the potential divider is the current divider. This is a handy rule to know when doing circuit analysis as it can save a lot of algebra. Looking at fig. 1.28 we can see that the resistors are in parallel, so the voltage across each one is the same. A supply current i_1 flows into them and produces a voltage equal to:

$$v = i_1(R_1 \parallel R_2) = i_1 \frac{1}{\dfrac{1}{R_1} + \dfrac{1}{R_2}}$$

But suppose we want to find the current i_2. Again from Ohm's law:

$$v = i_2 R_1$$

Substituting this into the previous expression we get:

$$i_2 R_1 = i_1 \cfrac{1}{\cfrac{1}{R_1} + \cfrac{1}{R_2}}$$

And dividing both sides by R_1 leaves:

$$i_2 = i_1 \cfrac{1}{1 + \cfrac{R_1}{R_2}} = i_1 \frac{R_2}{R_2 + R_1}$$

Or in other words the current gain is:

$$B = \frac{R_2}{R_2 + R_1}$$

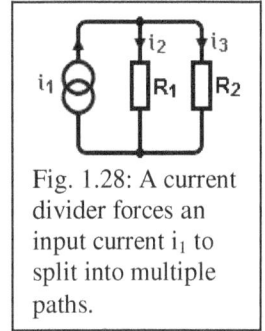

Fig. 1.28: A current divider forces an input current i_1 to split into multiple paths.

This can be extended to a general formula for a current divider with any number of parallel paths:

$$B = \frac{\text{Total parallel impedance through which the current does not flow}}{\text{Total parallel impedance}} \tag{1.39}$$

1.19: The Principle of Superposition

The principle of superposition applies to linear circuits and states that if an input –let's call it x_1– produces an output y_1, and a different input x_2 produces an output y_2, then applying both inputs at the same time will produce an output which is the sum of the individual outputs: $y_1 + y_2$. This applies to any number of inputs and they do not have to be applied to the same point in the circuit.

For example, what is the unknown voltage V_3 in fig. 1.29? It is the sum of voltages that would be caused by V_1 and V_2 independently. By setting the source V_2 to zero (making it a short circuit) we have a potential divider formed by R_1 and the parallel combination $R_2 \| R_3$, as shown in b. Using the formula for a potential divider the voltage is quickly found to be 2.5V.

Now by setting V_1 to zero instead, as shown in c., we are left with a potential divider formed by R_2 and the parallel combination $R_1 \| R_3$, and the voltage is now found to be 3V. The original unknown voltage V_3 is the sum of these two voltages: $2.5 + 3 = 5.5V$.

Superposition also allows us to separate AC and DC conditions when considering circuit operation. We can, for example, work out what all the DC voltages and currents are with one set of calculations and, quite separately, work out what all the AC voltages and currents are with another set, knowing that in the real circuit the AC signals 'ride on top' of the DC signals. Not having to think about both things at once tends to make circuit design easier.

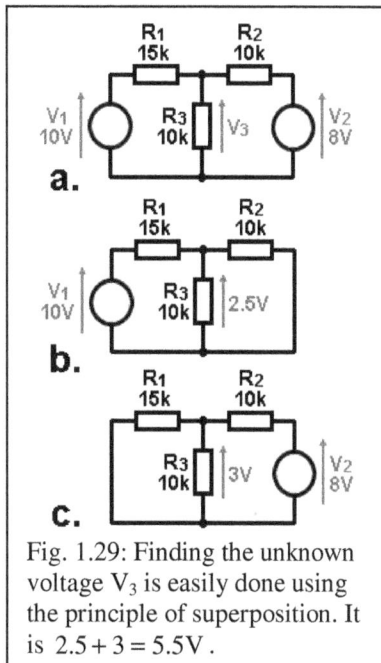

Fig. 1.29: Finding the unknown voltage V_3 is easily done using the principle of superposition. It is $2.5 + 3 = 5.5V$.

1.20: Linear Circuit Analysis

Very often in electronics we are faced with networks of several components in various configurations. Such a huge variety of arrangements may seem intimidating, but most can be simplified by applying the rules of series and parallel components, superposition, and potential dividers.

We already know that several resistors in series can be reduced down to a single 'bigger' resistor that is equal to all the individual resistances, but what is the total resistance of the network in fig. 1.30a? A good first step is to re-draw the circuit in a way that is easier to understand, as shown in b. Then, group the components into simple blocks or impedances, drawn as large rectangles, as in c. Finally, write expressions for each of the impedances. In this case Z_1 contains nothing but R_1:

$$Z_1 = R_1$$

Z_2 contains two parallel resistors, so using the product-over-sum rule:

$$Z_2 = \frac{R_4 R_5}{R_4 + R_5}$$

Z_3 contains two resistors in series:

$$Z_3 = R_2 + R_3$$

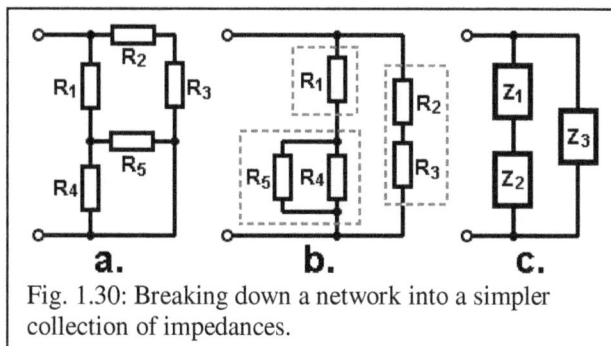

Fig. 1.30: Breaking down a network into a simpler collection of impedances.

35

We could now combine Z_1 and Z_2 into one 'bigger' impedance and re-draw the circuit yet again, although it hardly seems worth it in this case; the total impedance is clearly Z_1+Z_2 in parallel with Z_3. Again, using the product-over-sum rule:

$$Z_{total} = \frac{(Z_1+Z_2)Z_3}{(Z_1+Z_2)+Z_3} = \frac{\left(R_1 + \dfrac{R_4 R_5}{R_4 + R_5}\right)(R_2+R_3)}{R_1 + \dfrac{R_4 R_5}{R_4 + R_5} + R_2 + R_3}$$

This is not exactly a pretty equation, but it is nonetheless a simple matter to enter all the resistance values to find the total.

Networks containing only resistors are all very well, but what about when inductors and capacitors get involved? For example, what is the impedance of the network in fig. 1.31a? The principle is exactly the same as previous, although it is often worth replacing $j2\pi f$ with s. In other words, we write the reactance of a capacitor as $1/(sC)$ and the reactance of an inductor as sL. Technically this is called the Laplace transform and has a deep mathematical meaning, but here we are using it in the simplest possible manner –it's just a change of symbols to make the equations easier to handle.

The circuit in fig. 1.31 looks particularly intractable, but re-drawing it reveals it to be a simple arrangement of three parallel impedances. In this case Z_1 contains nothing but R_2:

$$Z_1 = R_2$$

Z_2 contains R_1 in series with C:

$$Z_2 = R_1 + \frac{1}{sC}$$

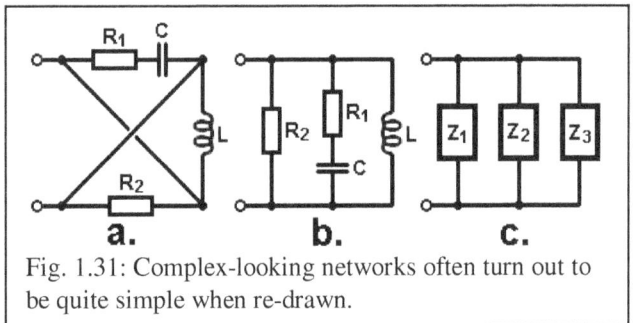

Z_3 contains nothing but L:

$$Z_3 = sL$$

Fig. 1.31: Complex-looking networks often turn out to be quite simple when re-drawn.

To find the total impedance is easier to convert to conductances (we cannot use the product-over-sum rule since we are dealing with more than two impedances):

$$\frac{1}{Z_{total}} = \frac{1}{Z_1} + \frac{1}{Z_2} + \frac{1}{Z_3} = \frac{1}{R_2} + \frac{1}{R_1 + \dfrac{1}{sC}} + \frac{1}{sL} = \frac{s^2 LC(R_1+R_2) + s(L + CR_1 R_2) + R_2}{sLR_2(1+sCR_1)}$$

Re-inverting conductance to get back to impedance:

$$Z_{total} = \frac{sLR_2(1+sCR_1)}{s^2LC(R_1+R_2)+s(L+CR_1R_2)+R_2} = \frac{j\omega LR_2 - \omega^2 LCR_1R_2}{-\omega^2 LC(R_1+R_2)+j\omega(L+CR_1R_2)+R_2}$$

This may look intimidating, but after entering component values it will reduce considerably, and can be broken into real and imaginary parts from which the magnitude and phase could be found.

Sometimes the principles described above are not enough to break down particularly complicated networks. Unfortunately, there is not space here to go into more examples, but they can be found in proper circuit-analysis textbooks. These will also cover more formal (albeit tedious) techniques such as mesh and nodal analysis, and the star-delta transform, among others. These days, however, if a problem can't be solved on the back of an envelope then we usually turn to computer simulation.

1.21: Resonance

In section 1.11 on impedance an example was given using a resistor and capacitor. But when capacitors and *inductors* are mixed in the same circuit something interesting happens. It will be remembered that capacitors have negative impedance while inductors have positive impedance, so under certain conditions these two will cancel each other out , or **resonate**.

For example, fig. 1.32 shows a series RLC network, and since the components are in series the total impedance is the sum of the individual impedances:

$$Z = R + j\omega L + \left(-j\frac{1}{\omega C}\right)$$

(ω is a shorthand for $2\pi f$, remember). Since j is a common factor we can collapse the above expression into a plain old complex number:

$$Z = R + j\left(\omega L - \frac{1}{\omega C}\right)$$

Fig. 1.32: A series resonant circuit reaches a minimum impedance of R at the resonant frequency.

Note that the capacitive reactance is subtracted from the inductive reactance, so instinctively we should guess that with a particular set of variables we can make them cancel out completely, leaving only the resistance. This will occur at a particular frequency, called the **resonant frequency**, which is equal to:

$$f_0 = \frac{1}{2\pi\sqrt{LC}} \quad \text{hertz} \tag{1.40}$$

Note that for this circuit the resonant frequency is unaffected by the resistance.

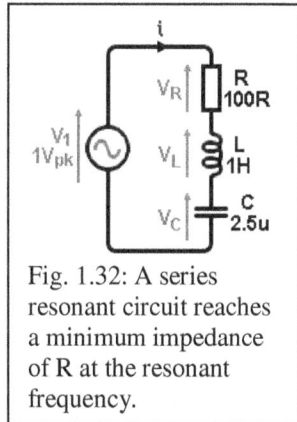

When the network is connected to a voltage source and the current is measured as the frequency is increased, we will find that the current rises to a maximum at the resonant frequency. This is shown in fig. 1.33 which plots the variation in current with a $1V_{pk}$ sine wave applied, for three different resistor values. At the resonant frequency there is nothing to limit the current except the resistor, i.e. the total impedance reaches its minimum value here.

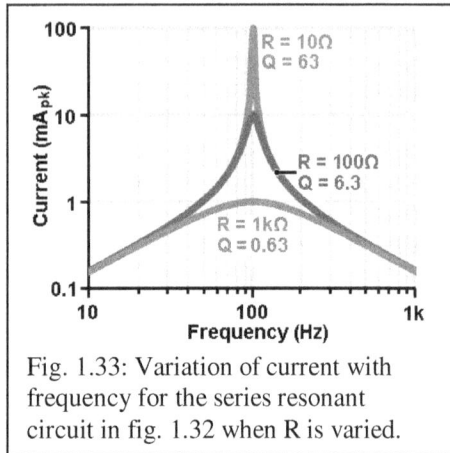

Fig. 1.33: Variation of current with frequency for the series resonant circuit in fig. 1.32 when R is varied.

The 'peakiness' of the response is defined as the **Q** of the circuit, and is equal to:

$$Q = \frac{1}{R}\sqrt{\frac{L}{C}} \quad \text{(no units)} \tag{1.41}$$

Increasing the resistance therefore subdues or damps the resonance, which is clear from fig. 1.33. The amount of damping may alternatively be expressed in terms of the **damping ratio** or **damping factor**[*];

$$\zeta = \frac{1}{2Q} \quad \text{(no units)} \tag{1.42}$$

If we *want* a circuit to resonate then we will usually talk about its Q, whereas if resonance is something we want to avoid then it is common to work with ζ, as will be seen later.[†] This is simply a matter of convenience and convention.

Fig. 1.34 shows the voltages in the circuit at resonance when R is 100Ω. A striking feature is that the voltages across the inductor and capacitor greatly exceed the source voltage; we have a sort of amplification, more usually called **magnification**

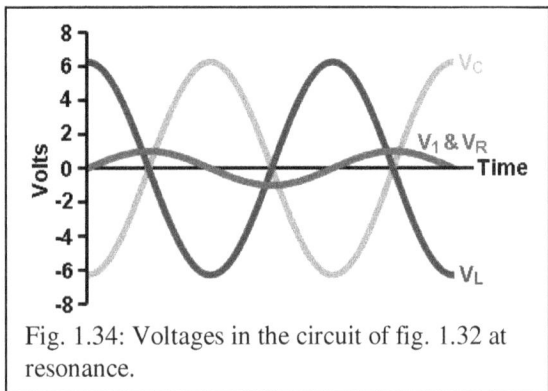

Fig. 1.34: Voltages in the circuit of fig. 1.32 at resonance.

---

[*] This is not the same as the 'damping factor' often quoted for audio power amplifiers, which really ought to be called the damping *figure*.
[†] ζ is the Greek letter zeta.

in this situation. This does not defy Kirchoff's voltage law, however, because the relative phases of the voltages are such that, at any particular instant, the sum of the voltages is still equal to the source voltage. In this case (i.e. at resonance) v_C and v_L are exactly equal but out of phase with one another and so cancel completely, leaving v_R equal to the supply voltage v_1.

A typical parallel resonant circuit is shown in fig. 1.35. Here the resistance often represents the unavoidable resistance of a practical inductor. For this circuit the resonant frequency is:

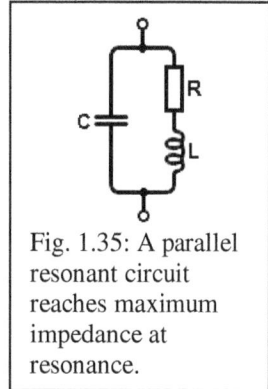

Fig. 1.35: A parallel resonant circuit reaches maximum impedance at resonance.

$$f_0 = \frac{1}{2\pi}\sqrt{\frac{1}{LC} - \left(\frac{R}{L}\right)^2} \qquad (1.43)$$

Note that with this arrangement the resistor does affect the resonant frequency to some extent; a higher resistance (more damping) results in a lower f_0. A parallel resonant circuit reaches *maximum* impedance at resonance. In most cases resonance is something we wish to avoid in audio circuits since it results in peaks in the frequency response and rapid phase changes, as will be seen shortly.

1.22: Filters

In section 1.17 we examined the potential divider, using only resistors. But if we replace one or both of those resistors with a capacitor or inductor then we have a potential divider whose gain changes with frequency due to the varying reactance. In other words, we have a filter. Filters may be intentional or unavoidable, as there are stray capacitances and inductances everywhere.

Fig. 1.36 shows an RC filter, also called a **low-pass** filter for reasons that will become obvious. Using the formula for a potential divider (equation (1.38)), the gain of the circuit is:

$$B = \frac{V_{out}}{V_{in}} = \frac{-j\dfrac{1}{\omega C}}{R + \left(-j\dfrac{1}{\omega C}\right)}$$

Multiplying top and bottom by $j\omega C$:

$$B = \frac{1}{1 + j\omega RC}$$

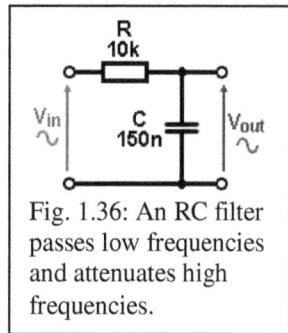

Fig. 1.36: An RC filter passes low frequencies and attenuates high frequencies.

This is the gain or transfer function of the filter, in complex notation. We are usually more interested in the magnitude, which is found by squaring the real and imaginary parts and taking the square root of their sum:

$$|B| = \frac{1}{\sqrt{1 + (\omega RC)^2}} \qquad (1.44)$$

The phase is:

$$\theta = \tan^{-1}(-\omega RC) \quad \text{degrees} \qquad (1.45)$$

Note that both these expressions depend on frequency.

When plotting graphs it is usual to convert the voltage gain into decibels using:

$$dB = 20 \cdot \log\left(\frac{v_{out}}{v_{in}}\right) \quad \text{decibels} \qquad (1.46)$$

The magnitude of the gain –better known as the **frequency response**– is plotted in fig. 1.37 along with the phase response. At very low frequencies the capacitor is effectively an open circuit so the input signal is not attenuated, i.e. the gain is 0dB and there is no phase shift. This region is called the **passband**. At high frequencies the reactance of the capacitor falls and so shunts the signal, attenuating it and shifting its phase. This region is called the **stopband**. The phase shift approaches −90° at the highest frequencies, that is, a sine wave output will lag the input.

Somewhere in the middle, the reactance of the capacitor becomes equal to the resistance, and at this point the voltage gain is −3dB and the phase shift is halfway between minimum and maximum, i.e. −45°. This point is known variously as the **cut-off frequency**, roll-off point, pole, corner, turnover, or −3dB frequency.

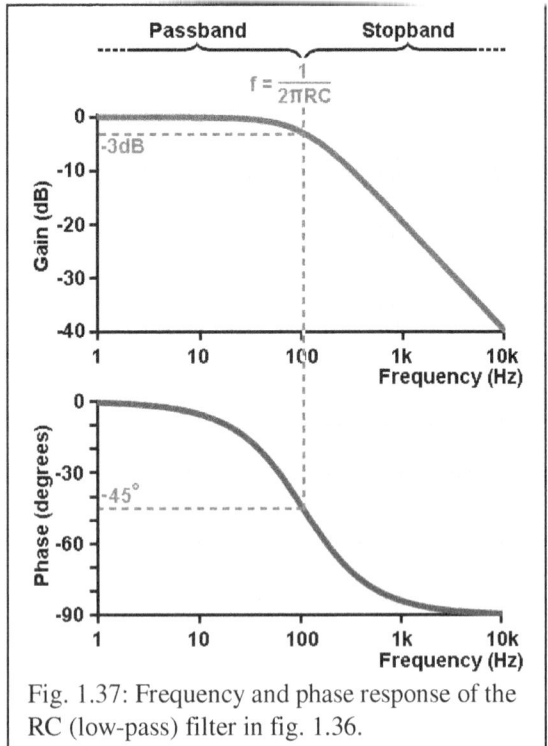

Fig. 1.37: Frequency and phase response of the RC (low-pass) filter in fig. 1.36.

At this point the frequency is equal to:

$$f_c = \frac{1}{2\pi RC} \quad \text{hertz} \tag{1.47}$$

This formula must certainly be memorised. Frequencies above this point are progressively attenuated while those below pass more-or-less unhindered, which is why it is called a low-pass filter, or sometimes a high-cut filter. The difference between the upper and lower −3dB points of a filter is called the **bandwidth**. In this case there is no lower −3dB point, so the bandwidth is equal to the upper (and only) cut-off frequency.

By substituting equation (1.47) into (1.44) we get another useful way to express the gain of an RC filter:

$$B = \frac{1}{\sqrt{1 + \left(\frac{f}{f_c}\right)^2}} \tag{1.48}$$

Where f_c is the cut-off frequency and f is the frequency in question. For example, if we wanted the filter to produce a particular amount of loss B, at a particular frequency f, this expression could be rearranged to find the necessary cut-off frequency:

$$f_c = \frac{Bf}{\sqrt{1 - B^2}} \tag{1.49}$$

By swapping the positions of the resistor and capacitor we create a CR or **high-pass** filter, as in fig. 1.38. This is the dual of the filter just shown, and the amplitude and phase are given by:

$$|B| = \frac{\omega RC}{\sqrt{1 + (\omega RC)^2}} \tag{1.50}$$

$$\theta = \tan^{-1}\left(\frac{1}{\omega RC}\right) \tag{1.51}$$

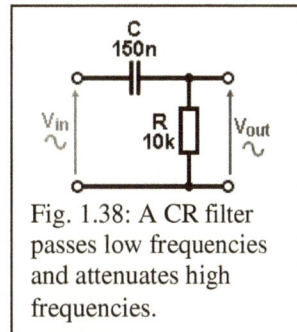

Fig. 1.38: A CR filter passes low frequencies and attenuates high frequencies.

These are plotted in fig. 1.39, and the symmetry with fig. 1.37 is obvious. At very low frequencies the capacitor approaches an open circuit and so 'blocks' these frequencies and shifts their phase, while the highest frequencies pass through largely unaffected. The phase shift approaches +90° at low frequencies, so an output sine wave will lead the input. The cut-off frequency is again $1/(2\pi RC)$ and at this point the phase shift is again halfway between maximum and minimum.

41

Substituting this into equation (1.50) also produces:

$$B = \frac{f/f_c}{\sqrt{1+\left(f/f_c\right)^2}} \qquad (1.52)$$

The above rearranges into the useful expression:

$$f_c = \frac{f\sqrt{1-B^2}}{B} \qquad (1.53)$$

Both the previous filters contain only one reactive component and are called **first-order** filters. Such filters have some universal characteristics: Beyond the cut-off frequency the attenuation approaches a rate of −20dB/decade, which is the same as −6dB/octave. It never *quite* achieves this rate, although from the previous figures it is clear that it gets pretty close once you are much beyond the cut-off frequency. A first order filter can only introduce a maximum of ±90° phase shift, and this figure is also never quite reached except at infinity or zero hertz. A first-order filter can also be recognised by its step response – see section 1.23.

A **second-order** filter contains two reactive components. One way to achieve this is to cascade two first-order filters as in fig. 1.40. Since each filter section provides −20dB/decade of attenuation, the total rate of attenuation for a second-order filter approaches −40dB/decade or 12dB/octave. Similarly, since each filter section can contribute up to −90° phase shift, the total shift for a second-order low-pass filter approaches −180°. More filter sections could be cascaded to create even higher-order filters, each one adding a further −20dB/decade and −90° phase shift. It is an essential bit of electronics knowledge that filter

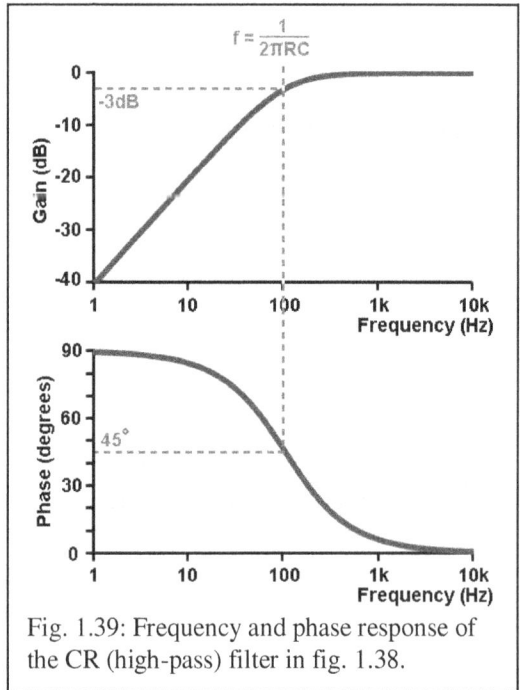

Fig. 1.39: Frequency and phase response of the CR (high-pass) filter in fig. 1.38.

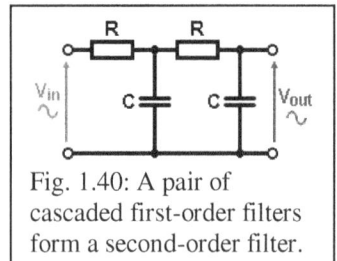

Fig. 1.40: A pair of cascaded first-order filters form a second-order filter.

slopes *always* come in multiples of −20dB/decade.

Fig. 1.41 shows the frequency response up to a fifth-order filter, where in each case all the resistors are the same value (R), and likewise the capacitors (C). The cut-off frequency for such an arrangement is still given by equation (1.47), but the total attenuation at this frequency increases by about −6.5dB for each extra filter section (not −3dB as might be expected, because each filter section loads the previous one and alters its behaviour[*]). The transfer functions for cascaded filters can again be derived using the principles of potential dividers, but it is much quicker to turn to circuit simulation for such designs.

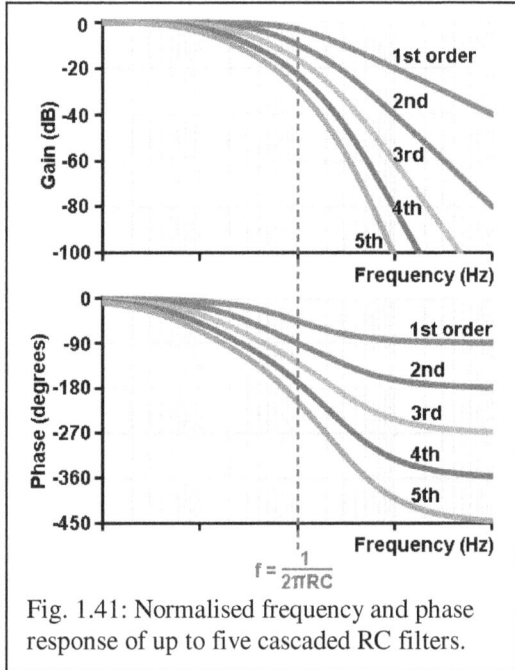

$$f = \frac{1}{2\pi RC}$$

Fig. 1.41: Normalised frequency and phase response of up to five cascaded RC filters.

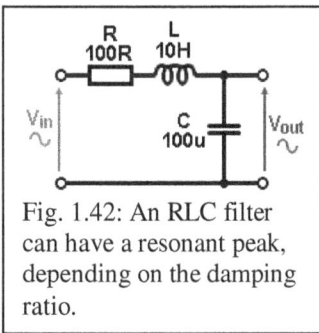

Fig. 1.42: An RLC filter can have a resonant peak, depending on the damping ratio.

Another type of second-order filter which is commonly encountered is an RLC filter, shown in fig. 1.42 (actually all we really need is L and C, but the resistance is an unavoidable part of a practical inductor and its effect on the circuit is worth knowing about). For variety, component values have been added which might represent a smoothing filter in a power supply. Having already encountered resonance in section 1.21 we should expect it to have some interesting effects on the frequency and phase response.

It will be remembered that the resistance in a resonant circuit controls the damping ratio, or Q, depending on what you're interested in. In this case the damping ratio is:

$$\zeta = \frac{R}{2}\sqrt{\frac{C}{L}} = \frac{R}{4\pi f_0 L} \tag{1.54}$$

---

[*] For this reason it is unwise to call a cut-off frequency a '−3dB frequency', because it is not always −3dB.

43

Where f_0 is the resonant
frequency. If the damping ratio
is less than 1 then the circuit is
said to be **underdamped**. A
damping ratio of exactly 1 is
called **critical damping** while
larger values are said to be
overdamped.

The frequency and phase
response are shown in fig 1.43,
with R varied from 100Ω to
3kΩ. With a small value of
resistance the damping ratio is
low (Q is high), and the current
that flows at resonance is
therefore quite large. The
voltage across the capacitor
(and inductor) is magnified, i.e.
we get a resonant peak in the
frequency response, which is
obvious in the figure. The
phase snaps rapidly towards
$-180°$ around the resonant
frequency.

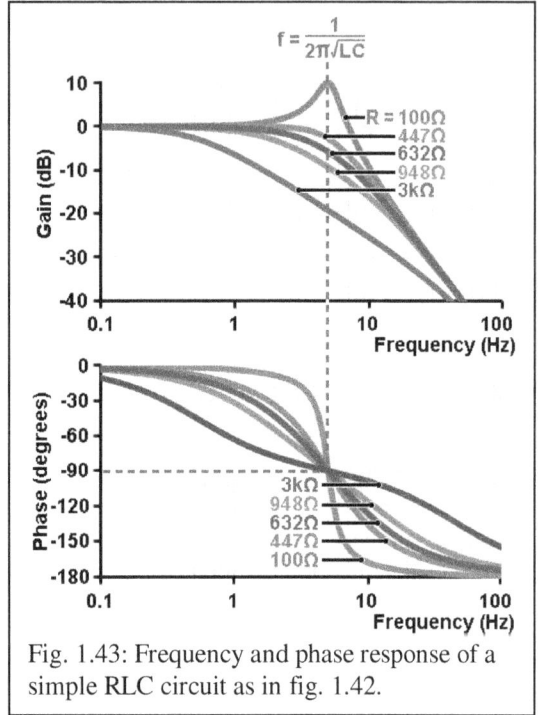

Fig. 1.43: Frequency and phase response of a
simple RLC circuit as in fig. 1.42.

If the resistance is increased then the damping increases, the peak is progressively
squashed, and the phase transitions become gentler. Increasing the resistance to
447Ω gives a damping ratio of $1/\sqrt{2}$ in this case. This is a special value of damping
as it marks the point where the resonant peak is completely suppressed, resulting in
the flattest, widest possible frequency response. This is called a **maximally flat** or
Butterworth response and is achieved when the damping resistance is equal to $\sqrt{2}$
times the reactance of the inductor (or capacitor) at the resonant frequency.[*]

A resistance of 632Ω results in a damping ratio of one, or critical damping. This is
also a special value as it marks the point where overshoot of the step response is
completely suppressed –see section 1.23. The resistance is now equal to twice the
reactance of inductor (or capacitor) at the resonant frequency.

A resistance of 948Ω results in a damping ratio of 1.5, since the resistance is now 1.5
times the previous value. This is a special value as it gives an identical response to a
pair of cascaded RC filters as in fig. 1.40.

---

[*] Incidentally, a damping factor of $\sqrt{3}/2$ is called a Bessel response and results when
$R = \sqrt{(3L/C)}$. Such a response is sometimes used in audio because it gives the most
constant phase delay.

44

Increasing the resistance to 3kΩ results in a damping ratio of 4.7 (there is nothing special about this value). The circuit is now very overdamped and this has the effect of creating two different cut-off frequencies or poles at $f = 1/(2\pi RC)$ and $R/(2\pi L)$. The lower pole causes the frequency response to start falling at a first-order rate of −20dB/decade, increasing to −40dB/decade once the second pole is reached, just beyond the limit of the graph. This effect of making a circuit behave like two quite different cascaded filters is sometimes called **pole splitting**.

1.22.1: Bode Plots

Filters form a large and wearisome discipline in electronics, and many long and boring books have been filled with their theory. 'Proper' filter theory is beyond the scope of this book, but fortunately it is not really necessary for practical work since the filters we meet in hi-fi preamps tend to be simple and relatively easy to analyse on paper. Nowadays it is easier still to model circuit behaviour on a computer, but it is worth being aware of a few old-fashioned tricks too.

The frequency and phase response of a filter can be estimated using a **Bode plot**. A Bode plot ignores the awkward curves in the real-life response and instead approximates everything using straight lines called **asymptotes**. This allows us to plot the response just by knowing a few key points on the graph.

For example, fig. 1.44 shows the Bode plot of a generic RC filter (compare with fig. 1.37). The response is assumed to be perfectly flat up to the cut-off frequency (pole), after which is it assumed to fall at a constant rate of −20dB/decade. The true response is shown faint, and it is clear that the Bode approximation

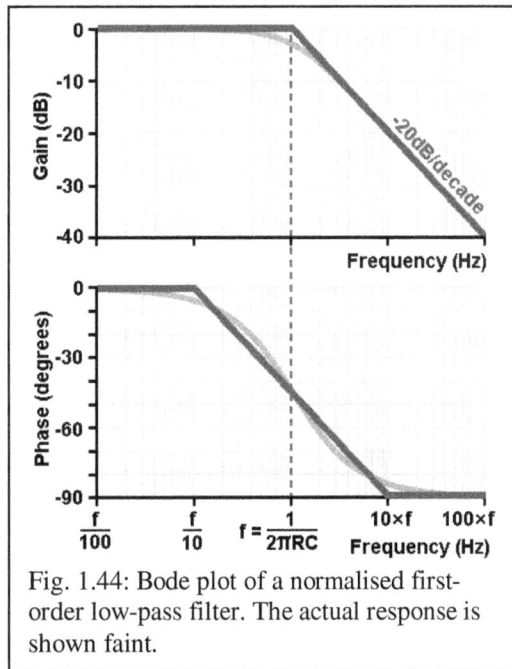

Fig. 1.44: Bode plot of a normalised first-order low-pass filter. The actual response is shown faint.

is quite good, with the greatest error occurring at the cut-off frequency where the true response is actually down by 3dB.

To plot the phase response it is assumed that it starts diving from zero at a constant rate, beginning at a frequency ten times lower than the cut-off frequency and reaching −90° at a frequency ten times higher than the cut-off frequency. The shift at

45

the cut-off frequency itself then comes out at −45° which is correct. The true response is again shown faint, and it is clear that the phase approximation is not as good as the amplitude approximation. Fortunately, we are usually much less interested in the phase response, and it is common not to bother with it.

A great convenience of Bode plots is that if we have several cascaded filters then we can estimate the overall frequency and phase response simply by adding individual Bode plots together, *provided each filter does not greatly load the previous one*. This is briefly illustrated in fig. 1.45:

- Fig. 1.45a shows the frequency response of a simple RC filter with a cut-off frequency of 48Hz. Above this point the response falls at the usual rate of −20dB/decade.
- Fig. 1.45b is a shelving filter, which is basically a potential divider with a capacitor added to allow high frequencies to bypass the divider. The loss at low frequencies is −12dB, but above 530Hz the response rises at a first order rate, reaching 0dB at 2122Hz. Also note that the impedances in this filter are much larger than in the first, so loading effects will be negligible.
- Fig. 1.45c shows the result of adding the two plots together, which is the

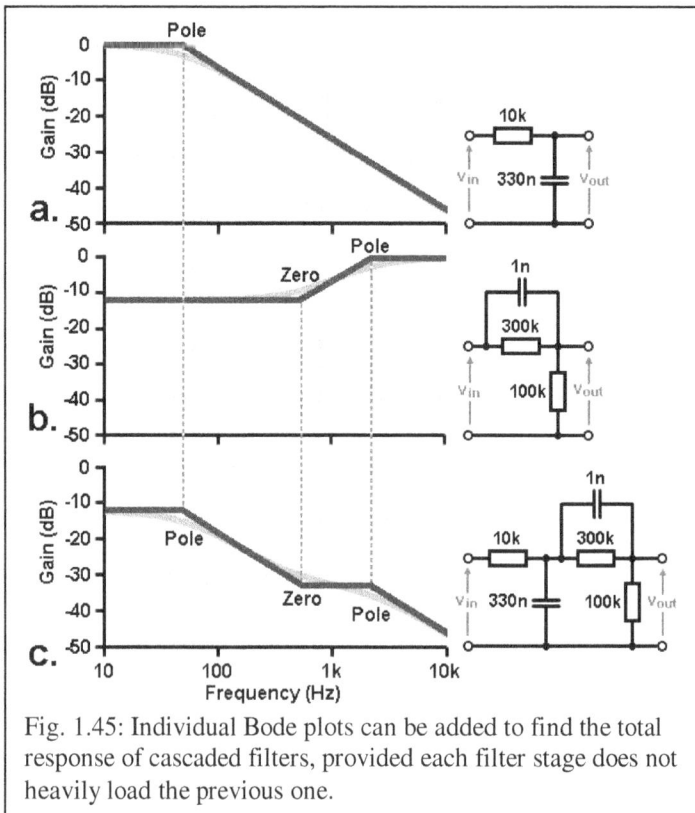

Fig. 1.45: Individual Bode plots can be added to find the total response of cascaded filters, provided each filter stage does not heavily load the previous one.

response of the two filters when cascaded. At low frequencies the first filter contributes no loss while the second causes -12dB loss, so the total is -12dB. At 48Hz the first filter starts to contribute -20dB/decade loss, so the total response then falls at this rate, and this continues until 530Hz at which point the rising response of the second filter cancels out the falling response of the first, resulting in a flat response. Until, that is, the second filter runs out of boost at 2122Hz, whereupon the first filter takes over again and the response continues falling. The actual response of the circuit is shown faint and it is clear that the approximation is pretty good.

The cut-off frequencies on a Bode plot where the response bends down are called **poles** –a name which has been surreptitiously dropped into earlier discussion. The points where it bends up are called **zeros**. In fig. 1.45c, for example, the circuit contains two poles and one zero. These names are used most often when talking about RIAA equalisers (chapter 10) so it is worth remembering which is which.

1.22.2: Phase Delay

When an electrical signal is shifted in phase it is also shifted in time. The amount of time delay (or advance) caused by a certain phase lag (or lead) is called the **phase delay**, t_{pd}. The relationship between phase shift, frequency, and time or phase delay is simple:

$$t_{pd} = -\frac{\theta_{(radians)}}{\omega} = -\frac{\theta_{(degrees)}}{360° \times f} \qquad \text{seconds} \qquad (1.55)$$

Where:
θ = phase shift in radians or degrees
ω = angular frequency in radians per second
f = frequency in hertz

For example, fig. 1.46a shows two sine waves of different frequencies each of which has been phase shifted by $-90°$ or one quarter cycle (the minus sign indicates phase *lag*). It is plain to see that because the two sine waves are of different frequencies they have not been shifted by the same amount of *time*, i.e. they experience the same phase *shift* but different phase *delays*. When they are added together it is also obvious that the envelope of the resultant signal has changed as a result of the dissimilar phase delays. This shape-change is sometimes called **phase distortion**, and a similar example was shown earlier in fig. 1.9.

Fig. 1.46b shows the other side of the coin: starting with the same two sine waves but shifting them each by the same amount of *time*, i.e. the same phase delay. It is now the phase-shifts which are different, but notice the envelope of the resultant signal is not changed or distorted, it is merely delayed.

In radio transmission, information is often sent in waveform 'packets' whose envelope has a specific and important shape, such as a pulse train of digital information or the vaguely stair-case waveforms use in analog television. Any

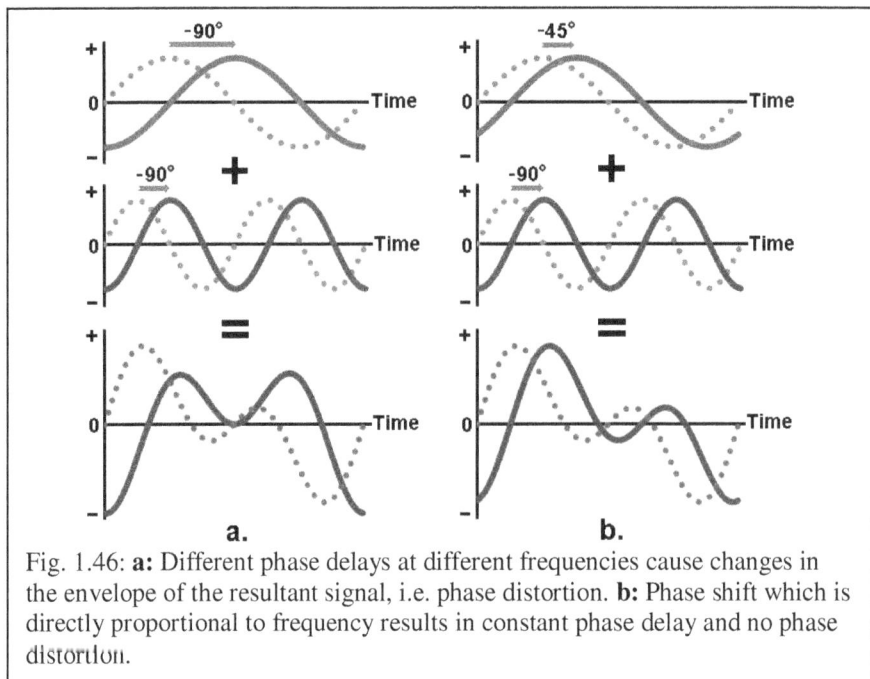

Fig. 1.46: **a:** Different phase delays at different frequencies cause changes in the envelope of the resultant signal, i.e. phase distortion. **b:** Phase shift which is directly proportional to frequency results in constant phase delay and no phase distortion.

distortion of the shape of these packets can lead to reception problems. Radio and TV designers therefore try to design systems with constant phase delay over the frequency range of interest.[1] This is much less important with audio signals, but there can be special conditions under which phase distortion may be audible, particularly with low-frequency transients like bass drums which may become 'stretched in time' or 'smeared' by phase distortion.

Filters unavoidably cause phase shift, but how much phase *distortion* does this lead to? An easy way to get a feel for this is to plot the phase response on *linear* axes. For example, fig. 1.47 shows the phase response of an RC (low-pass) filter with a cut-off frequency of $f_c = 20kHz$. Ideally we would want the phase to be directly proportional to frequency as this would result in constant phase delay. This is represented by the straight, dotted line. If this could be achieved then we would have a **linear-phase filter** with no

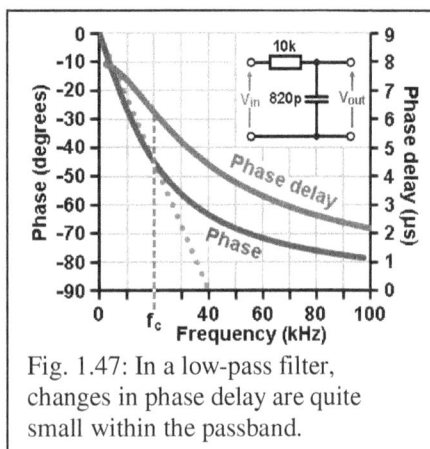

Fig. 1.47: In a low-pass filter, changes in phase delay are quite small within the passband.

[1] Amos, S. W & Birkinshaw, D. C. (1958). *Television, Principles and Practice Volume 2*. Illife & Sons Ltd., London. pp30-5.

phase distortion. In practice this is only possible with digital rather than analog filters, but from the figure we can see that below f_c the actual phase response is a fairly good approximation to the straight line.

The phase delay is plotted against the right-hand axis, and it can be seen that low frequencies experience more delay than high frequencies as they pass through the circuit. For example, a signal down near 20Hz is delayed by about 2µs more than one at 20kHz. Fortunately this is a very small difference (it's the same as the time period of a 500kHz signal) so it is not likely to be audible. Frequencies are attenuated above f_c and so contribute less to the overall resultant signal, so we don't much care about phase delay in the stopband. In other words, simple low-pass filters do not cause serious phase distortion.[2]

By contrast, fig. 1.48 shows a similar plot for a CR (high-pass) filter with a cut-off frequency of 20Hz. The phase delay of this filter is completely different from the previous example. Here a 20Hz signal is delayed by about six *milli*seconds more than one of much higher frequency within the passband. In other words, high-pass filters introduce far more phase distortion than low-pass filters. It is also worth pointing out that since the phase shift is positive (lead) the phase delay is now negative, implying a time *advance*. However, the signal does not really travel into the future, it only looks that way when a steady-state sine wave is viewed on an oscilloscope.

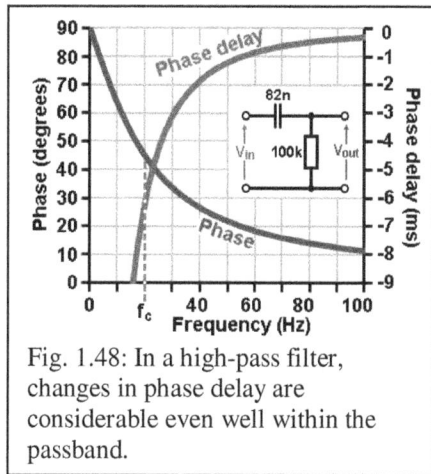

Fig. 1.48: In a high-pass filter, changes in phase delay are considerable even well within the passband.

Higher-order and resonant filters tend to introduce even more rapid phase changes, leading to even more significant phase distortion. This is sometimes expressed mathematically as **group delay**, t_{gr}, which is the (negative) *rate of change* of phase:

$$t_{gd} = -\frac{d\theta_{(radians)}}{d\omega} = -\frac{1}{360°}\frac{d\theta_{(degrees)}}{df} \quad \text{seconds} \quad (1.56)$$

Group delay is a more abstract idea than phase delay but it can be thought of loosely as the time delay experienced by the signal envelope itself –which is composed of a collection or 'group' of frequencies– when passing through a bandpass filter. However, the concept of group delay is better suited to radio theory where information is transmitted with predetermined envelope shapes, so we will not meet it again in this book.

[2] 'Cathode Ray' (1956). Distortion – Audible and Visible, *Wireless World*, April, pp188-93.

1.23: Step Response

Until now we have been mainly concerned with signals that are continuous and predictable, with no sudden changes in their behaviour. Any sine waves we have looked at were assumed to have been going on long before we arrived to look at them, and were assumed to continue long after we lost interest in them. Such nice, unvarying conditions are said to be **steady state**.

In reality, of course, a signal must have a beginning and an end, and probably all sorts of changes in between. A great deal of information about how a circuit behaves in the face of such changing conditions can be obtained by looking at its **step response**, also sometimes referred to lazily as its time response. In other words, we apply a sudden jump or 'step' to the input and examine what happens at the output. The step response of a circuit is intimately linked with its frequency response, and by examining one we can draw conclusions about the other.

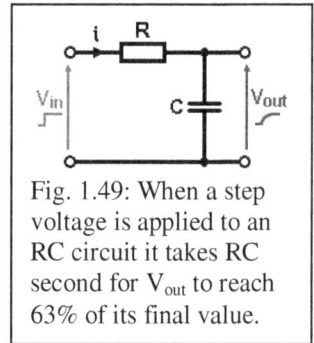

Fig. 1.49: When a step voltage is applied to an RC circuit it takes RC second for V_{out} to reach 63% of its final value.

Fig. 1.49 shows an RC (low-pass) filter to which a voltage step is applied. The output cannot follow the step precisely, because it takes time for the capacitor to charge up. A step represents an instantaneous change, i.e. infinite *rate of change*, so at the very moment the step is applied the capacitor behaves like a short circuit and takes a lot of current, limited only by the resistor. As it charges up, the voltage across it

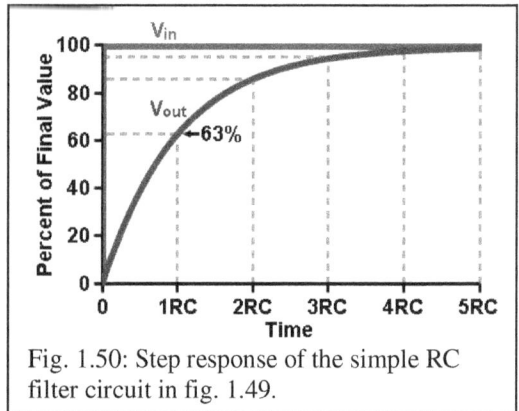

Fig. 1.50: Step response of the simple RC filter circuit in fig. 1.49.

counteracts the input voltage and the current gradually reduces. When the capacitor voltage equals the input voltage, current can no longer flow and the capacitor is fully charged; after this it behaves as an open circuit since the input signal is no longer changing and is therefore DC, so we have reached the steady state.

The voltage across the capacitor at any time (t) after the step is first applied is given by the equation for a charging capacitor:

$$v_{out} = V_{in}\left(1 - e^{-\frac{t}{RC}}\right) \tag{1.57}$$

50

Essential Electronics

Fig. 1.50 shows a generalised graph of the input and output voltages after the step is applied. The time axis happens to be divided into amounts of time equal to R×C seconds, as this value of time has special significance. When the time is equal to RC the previous equation becomes:

$$v_{out} = V_{in}\left(1 - e^{-\frac{RC}{RC}}\right) = V_{in}\left(1 - e^{-1}\right) = V_{in}\left(1 - 0.37\right)$$

$$v_{out} = 0.63 V_{in}$$

This special value of time is called the **time constant**, τ = RC.* It is the time taken for the capacitor voltage to reach 63% of its final value and it *also* happens to be the time period of the cut-off frequency of the filter when expressed in radians per second rather than hertz:

$$\frac{1}{\tau} = \omega_c = 2\pi f_c \qquad (1.58)$$

In theory it takes an infinite amount of time for the capacitor to become fully charged. However, for a simple RC filter it takes four time constants to reach 98% of the final value, which is close enough for most purposes, so this figure of 4τ is called the **settling time**. The time taken to reach 90% of the final value is called the **rise time** and is equal to 2.2τ.

In fig. 1.51 the positions of the resistor and capacitor are swapped to form a CR (high-pass) filter. When a step input is applied, the capacitor

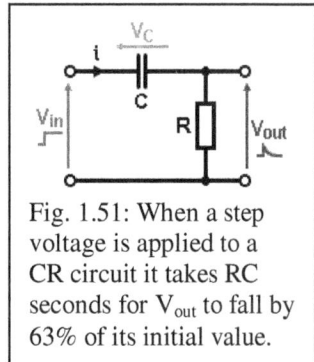

Fig. 1.51: When a step voltage is applied to a CR circuit it takes RC seconds for V_{out} to fall by 63% of its initial value.

again objects to the change and behaves momentarily like a short circuit. The output voltage therefore jumps to the same value as the input, and current flows in the resistor. Since this current has to flow through the capacitor too, it charges up and the voltage across it increases in exactly the same way as described previously. Since the capacitor voltage rises, the voltage across the resistor (v_{out}) must fall by the

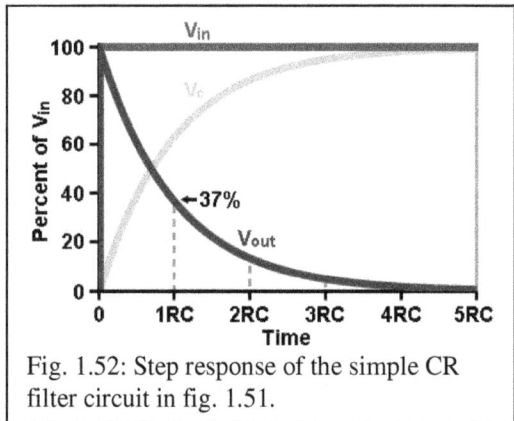

Fig. 1.52: Step response of the simple CR filter circuit in fig. 1.51.

---

* τ is the Greek letter tau. Curiously, most English speakers pronounce this 'tow' to rhyme with 'now', while a few renegades pronounce it 'torr', but the Greeks themselves pronounce it 'taff'.

same proportion (remember KVL). When the capacitor is fully charged the current stops flowing and we have reached the steady state; the output voltage drops to zero and the capacitor is effectively an open circuit. Another way of looking at it is to say that the sharp edge of the step contains high frequency Fourier components and it is those that are initially passed by the capacitor while the 0Hz average component is blocked.

Fig. 1.52 plots the relative voltages in the CR circuit. Again, it takes RC seconds for the voltage across the *capacitor* to rise to 63% of its final value, and this is therefore the time taken for the voltage across the resistor, v_{out}, to *fall* to 37% of its final value.

Fig. 1.53 shows the same LCR second-order circuit as in section 1.22 earlier. By applying a step input we are in fact applying a whole spectrum of frequencies, one of which must inevitably be the resonant frequency. If the circuit is underdamped then this frequency will be magnified, causing the output voltage to exceed the input voltage, or overshoot, and it will continue to oscillate or **ring** until this energy is finally burned off as heat in the resistor. The larger the resistance, the greater the damping ratio, the faster the energy dissipates, and the faster the ringing decays.

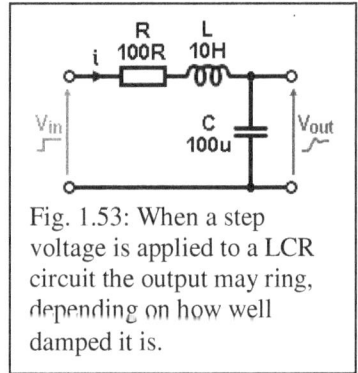

Fig. 1.53: When a step voltage is applied to a LCR circuit the output may ring, depending on how well damped it is.

The step responses with different values of resistance are shown in fig. 1.54. A value of 100Ω results in a damping ratio of 0.16 (underdamped) so the output rings, in this case taking about one second to die away.

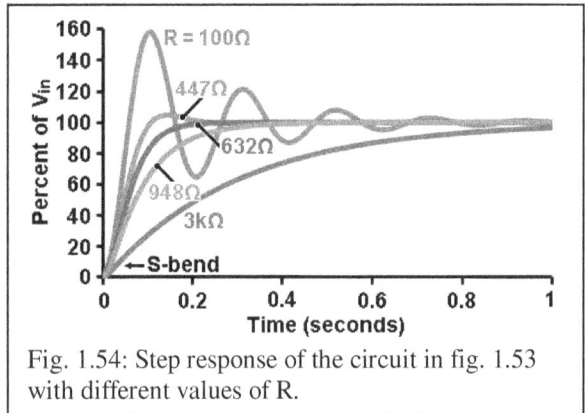

Fig. 1.54: Step response of the circuit in fig. 1.53 with different values of R.

A value of 447Ω gives a damping ratio of $1/\sqrt{2}$. This is a special value as it is the lowest value of damping that is defined as producing an overshoot but no undershoot (actually there *is* a small undershoot but it is less than 2% of the final value. The overshoot is 4.3%). This corresponds to a maximally flat or Butterworth frequency response.

A value of 632Ω gives a damping ratio of unity or critical damping. This is also a special value as it is gives the fastest possible rise time with no overshoot/ringing at all.

A value of 948Ω gives a damping ratio of 1.5 and makes the circuit behave identically to a pair of cascaded RC filters.

A value of 3kΩ results in a damping ratio of 4.7 (well overdamped). The step response is now quite sluggish as it takes a long time to reach its final value.

By damping a circuit we trade rise time for less overshoot and gentler changes in the frequency and phase response. The more we increase the damping, the more the circuit begins to behave like a first order RC filter. A first-order system can always be spotted by the fact that when a step is applied, the output voltage starts to rise immediately at its fastest rate, and slows thereafter. But with second- and higher-order systems the output starts to change slowly at first, then gets faster, then slows again, creating an S-bend in the step response. This bend is labelled in fig. 1.54, and it becomes less conspicuous as the damping is increased since the circuit begins to look more like a first-order filter.

The step response of an amplifier can reveal important information about its internal behaviour and frequency response, which is especially useful when it is tiresome to measure the frequency response directly. In practice we don't use a single step for testing amplifiers but a square-wave or pulse train whose frequency is sufficiently low for the system to approach the steady state before each new step comes along.

1.24: Negative Feedback

Feedback, as the name suggests, involves feeding a portion of the output signal back to the input of an amplifier.[*] If a signal is fed back so that it serves to increase or add to the input signal then it is **positive feedback** (older texts may use the term **regeneration**). This can cause the amplifier to self-oscillate, which is not usually what we want for audio. Conversely, if the signal fed back serves to decrease or subtract from the input voltage then it is **negative feedback** or **NFB**, also referred to in special cases as **degeneration**. This is the kind of feedback we are most interested in as it provides some useful results, and from now on the term 'feedback' is assumed to mean *negative* feedback unless stated otherwise.

---

[*] Some books claim rather lazily that negative feedback was invented by Harold S. Black in 1927. This is of course quite silly; negative feedback occurs throughout nature so one can hardly 'invent' it. Engineers have been using negative feedback for centuries, including in early radio receivers. However, those early receivers used positive feedback to maximise the circuit gain and then added some adjustable negative feedback to cancel out some of the positive, to prevent the circuit from oscillating. Black's insight was to realise that using *nothing but* negative feedback could bring about all sorts of improvements in circuit behaviour, *if* you are willing to sacrifice gain to get them. In other words, he did not invent negative feedback, he invented the negative-feedback amplifier.

The topic of feedback can grow very complicated very quickly, as anyone who has studied control-theory will know, so we will only touch upon some important elementary features here. Fortunately, most valve amplifiers use such simple feedback circuits, and apply so little feedback, that we do not need to drown in theory to understand them.[†]

1.24.1: The Universal Feedback Equation

Fig. 1.55a shows a block diagram of a non- inverting voltage amplifier with some gain, A_o. This might represent a single gain stage or a much larger circuit or, indeed, an opamp. Now, it must be understood that what the amplifier actually amplifies is the voltage appearing *between* its two input terminals, that is, the voltage difference between them, v_2. In a. the voltage between the two terminals is simply equal to the raw source signal v_1. Because this system has no feedback it is said to be running **open loop** and the overall gain is called the **open-loop gain**, which in this case is A_o, so the voltage appearing at the output is quite simply A_o times v_1. A potential divider has also been added to the output of the circuit but it is not doing anything very useful just yet; we simply get an attenuated version of the output signal appearing at the junction of the two resistors, as shown. This smaller signal v_3 is therefore some **fraction**, B, of the output signal, so it is $A_o B$ times the input signal, and this is the signal we will now feed back.

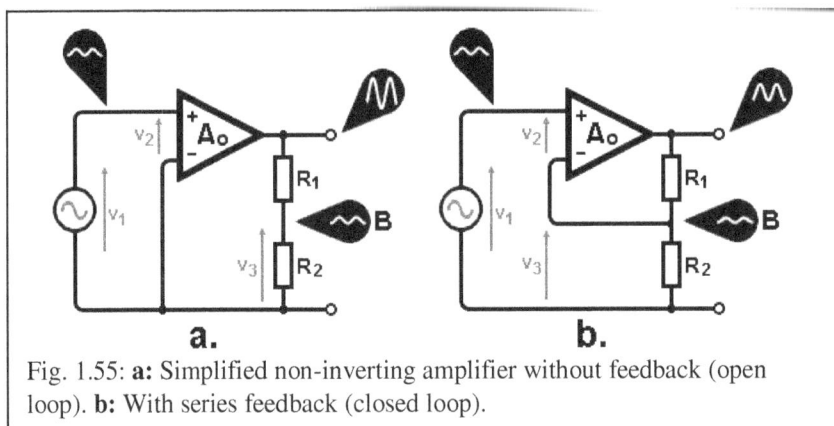

Fig. 1.55: **a:** Simplified non-inverting amplifier without feedback (open loop). **b:** With series feedback (closed loop).

[†] Most readers will be aware that feedback is something of a contentious issue to some audiophiles who object to it on a more-or-less religious basis. They insist that it is somehow bad for audio, perhaps because they imagine the signals re-circulate around the amplifier like an echo (they don't). This vague mistrust is often ascribed to early transistor amplifiers that used a lot of feedback without proper attention to stability and slew rate, resulting in genuinely poor quality sound. But that should be blamed on bad design, not feedback. When applied properly, feedback improves practically everything. Nevertheless, valves naturally lend themselves to zero-feedback designs if you like that sort of thing.

Moving to the circuit shown in b. the feedback voltage is applied to the amplifier's input terminal labelled '−', so it is subtracted from the raw source voltage applied to the '+' terminal. The voltage *difference* between the two terminals is therefore reduced somewhat since the feedback signal cancels out a little of the source signal or, in other words, we have negative feedback. Since the feedback forms a loop around the amplifier the system it is now said to be running **closed loop**, and the overall gain is called the **closed-loop gain**, which we will label A.

Suppose the voltage between the input terminals, v_2, is 1V. This will be amplified by A_o and the voltage being fed back is A_oB volts. Using KVL, the total source voltage that must be present for us to get into this situation in the first place is therefore $1+A_oB$ volts,[*] and from this simple argument we obtain the **universal feedback equation** for closed loop gain:

$$A = \frac{A_o}{1 + A_o B} \tag{1.59}$$

Where:
A = closed-loop gain
A_o = open-loop gain
B = feedback fraction

The term A_oB is called the **loop gain** because it is the total gain 'around the loop', that is, if we broke the feedback loop at a carefully-chosen point and measured the gain between the two broken ends.

The expression $1+A_oB$ is the input needed to get the same output as we had before the feedback was applied. In other words, it is the factor by which the overall gain of the system has been reduced; it is the amount of feedback being used. This expression crops up a lot in feedback theory so it too has a special name: the **feedback factor** (not to be confused with the feedback *fraction*, B). It is common to express the feedback factor in decibels and to talk about an amplifier as using so-many dBs of feedback –see later.

The voltage v_2 appearing between the amplifier's input terminals is called the **error voltage** because it is the difference between the input and feedback voltages, it being assumed that we would like the output signal –and therefore the feedback signal– to be identical to the input signal (or an amplified version of it). The more feedback is applied, i.e. the greater we make v_3, the smaller this error becomes and the more closely the output will resemble the input. The error can never be reduced to zero, however, because then there would be nothing for the amplifier to amplify in the first place!

---

[*] Feedback problems often require us to start at the output and work backwards to figure out what the input must have been.

Looking at equation (1.59), if A_o and/or B were very large or infinite then the loop gain A_oB would likewise be very large. So large, in fact, that the '1' could be ignored and the equation would simplify to:

$$A \approx \frac{1}{B} \qquad (1.60)$$

In this special case we find A_o has vanished from the equation. Since B is simply the gain of the potential divider we could rewrite this as:

$$A \approx \frac{R_1 + R_2}{R_2} = 1 + \frac{R_1}{R_2} \qquad (1.61)$$

This equation will be familiar to anyone who has used opamps before. It is a powerful result because it means that if we have enough loop gain then we can control the characteristics of the overall system entirely with the feedback components which set the value of B, and these components (e.g. resistors and capacitors) will usually be much more linear than the amplifying devices themselves. The performance of the system should then be just as reliable as the feedback components, even though the amplifying devices may drift or be replaced with ones of slightly different characteristics.[*]

Unfortunately, this idyll rarely applies to valve circuits as they almost never have enough loop gain to make A_o disappear from the equation. Usually we must look to opamps/solid-state circuits to achieve this. The cathode follower (chapter 7), however, is a special case where there is so much feedback that performance does become largely independent of the actual valve being used.

The amplifier in fig. 1.55b was non-inverting and subtracted the feedback voltage from the source voltage. An alternative possibility is to use an *inverting* amplifier and to *add* the feedback voltage to the source voltage. Since the feedback signal will be 180° out of phase with the input, summation causes it to cancel some of the input, so we would still have negative feedback. Fig. 1.56a shows a system similar to fig. 1.55a except the raw input is now being applied to the '−' or inverting input, so the output will be inverted. The potential divider does exactly the same thing as before, but it has been redrawn to make the next step seem less radical.

We cannot simply connect the feedback signal to the inverting input since it would then be shorted by the voltage source. Instead, we must apply the source voltage to the other end of the potential divider, as shown in b. The current being fed in from

---

[*] This is why some (usually older) textbooks say that negative feedback makes a system more 'stable'. This is a poor choice of word, however, because applying negative feedback usually makes an amplifier more likely to oscillate, i.e. *un*stable, and it is not very helpful when one book says that feedback improves 'stability' while another says exactly the opposite. It would be more correct to say that negative feedback can make a system more *consistent*.

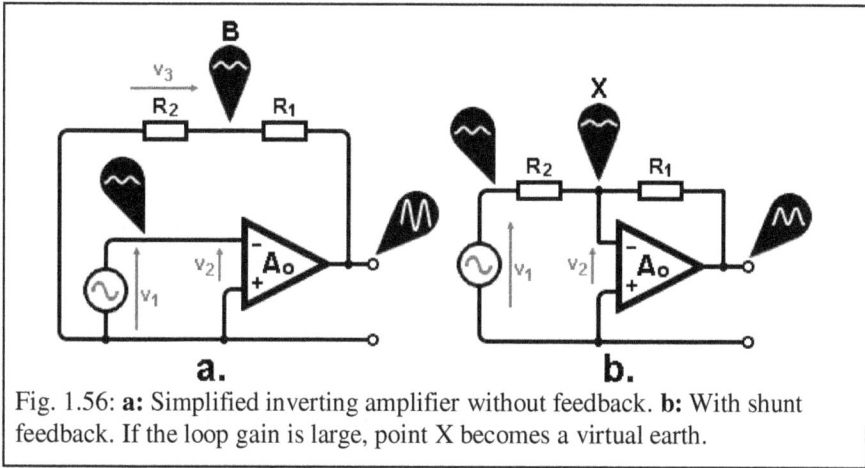

Fig. 1.56: **a:** Simplified inverting amplifier without feedback. **b:** With shunt feedback. If the loop gain is large, point X becomes a virtual earth.

the source then adds to the current being fed back, producing a summed voltage at the junction of the two resistors –the principle of superposition at work.

But what is the closed-loop gain of this circuit? At first glance we might expect it to be given by equation (1.61) again, but that is not the case. Because we had to move the point where the source signal is applied it is now attenuated by the potential divider before it reaches the '–' input. In other words, while the feedback signal is attenuated by a factor of $B = R_2/(R_1+R_2)$, the source signal is *also* attenuated but 'in the opposite direction', by a factor of $R_1/(R_1+R_2)$,[†] which is the same as $1-B$. The closed-loop gain is therefore:

$$A = (1-B)\frac{A_o}{1+A_oB} = \frac{A_oR_1}{A_oR_2 + R_1 + R_2}$$

But if the loop gain is large then this simplifies to:

$$A = \frac{R_1}{R_2} \tag{1.62}$$

This too will be familiar to anyone who has used opamps before. It also emphasises a general rule that if the feedback and source signals are not both applied directly to the amplifier's input terminals then we must be very careful about how we apply the universal feedback equation.

1.24.2: Virtual Earth

Staying with fig. 1.56b, if the loop gain is very large then something else interesting happens. Suppose the amplifier has an open-loop gain of $A_o = 1000$ but we add enough feedback to reduce the closed-loop gain to 1. If the source signal is 1V, say, then we will receive a 1V signal at the output (actually –1V since the

---

[†] This assumes the amplifier's output impedance is very small.

amplifier is inverting). But we know that the true gain of the amplifier *inside* the loop is 1000, so the error voltage present between its input terminals, v_2, must be $1V / 1000 = 1mV$. This is much smaller than the source signal. In fact, it is so small that we might shrug our shoulders and call it zero volts, so node 'X' looks more-or-less like it is grounded or earthed, at least as far as AC signals are concerned. For this reason it is called a **virtual earth**. Whatever the source voltage is, nearly all of it appears across the input resistor R_2, and input current flows into the virtual earth (and out again through R_1). This can be useful when mixing signals together because we can have several input resistors feeding in different signals without having to worry about them interacting with each another –they appear to be isolated from one another by the virtual earth. Such circuits are known as **virtual-earth amplifiers** or **virtual-earth mixers**.

1.24.3: General Properties of Negative Feedback

The main motivation for using feedback is that it improves an amplifier's linearity, i.e. it reduces distortion by the feedback factor. If an amplifier produces second harmonic distortion, for example, then this will be fed back so the error signal is effectively 'pre-distorted' in a complementary way, so the distortion cancels itself out when it reaches the output again.[*] On the other hand, the distortion that was fed back is effectively distorted again by the amplifier, so we get distortion of the distortion. Feedback therefore tends to smear the distortion spectrum so that it contains more high-order (and possibly more noticeable) Fourier components, even though the *total* distortion is reduced. This is one of the more rational objections raised by anti-feedback protestors.

It must also be understood that the reason an amplifier produces distortion in the first place is that its open-loop gain is not constant, and distortion is only reduced by whatever the feedback factor happens to be at that particular moment. In an amplifier which produces crossover distortion, for example, the gain drops to practically nothing during the crossover region, so the feedback factor also drops to nothing during this brief period. Feedback is therefore not very effective at correcting this sort of problem. Feedback also reduces the self-generated noise appearing at the output of an amplifier, but since it reduces the gain by the same amount, the signal-to-noise ratio is unchanged by feedback, all else being equal.

Another important motivation for using feedback is that it extends the bandwidth of the system. For example, fig. 1.57 shows the Bode plot of a fictional amplifier with a gain of 100 (40dB) at higher frequencies but only 50 (34dB) at lower frequencies (the actual response is shown faint). The second plot shows what happens when a feedback fraction of 0.02 is used. This results in a feedback factor $(1+A_oB)$ of 3 (10dB) at the higher frequencies but only 2 (6dB) at lower frequencies where there is

---

[*] This explanation makes it sound as if the signal passes around the loop again and again like an echo, but in fact the only thing that is ever amplified is the error voltage, and that happens on and instant-by-instant basis.

less loop gain. Notice that this is also the vertical distance between the two plots. The effect of feedback is therefore to flatten and widen the frequency response. In fact, the poles in the open-loop response are always raised or lowered by the

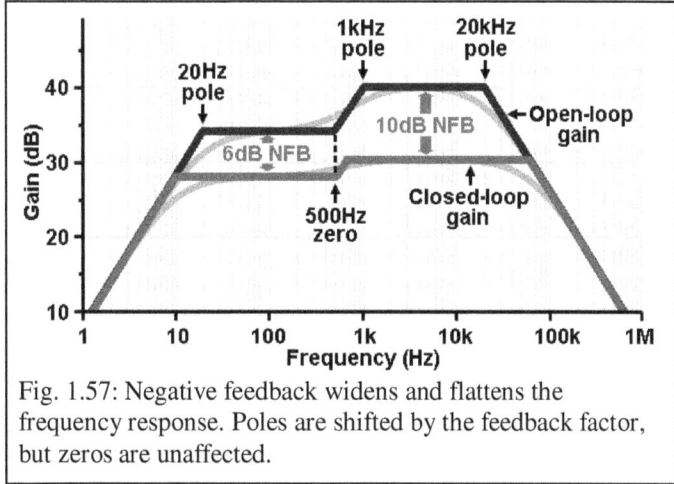

Fig. 1.57: Negative feedback widens and flattens the frequency response. Poles are shifted by the feedback factor, but zeros are unaffected.

feedback factor. Zeros, however, remain unmoved by feedback. This is a fairly extreme example meant to illustrate the point; we would not usually start with such an uneven frequency response and then try to fix it with feedback. It is better to make the amplifier good to start with, then use feedback to make it even better.

When feedback is applied over a single amplifying device or gain stage it is called **local feedback**, whereas feedback that encloses multiple stages is called **global feedback**. As a general rule, global feedback is more powerful than local feedback; in other words, several gain stages with global feedback will produce less distortion than if the same stages each have enough local feedback to produce the same total gain. On the other hand, it is more difficult to prevent a global feedback loop from oscillating. It is of course possible to have a stage with local feedback that is itself inside a larger system that has global feedback; such a system is said to have **nested** feedback loops.

The way in which feedback is derived affects the system's output impedance. If the feedback voltage is proportional to the voltage across the load, regardless of the current in the load, then it is called **voltage feedback**, and is illustrated in fig. 1.58a. Voltage feedback tends to make the whole system behave more like an ideal voltage source as far as the load is concerned. In other words, it will try to supply whatever load current is necessary to maintain the prescribed output terminal voltage.

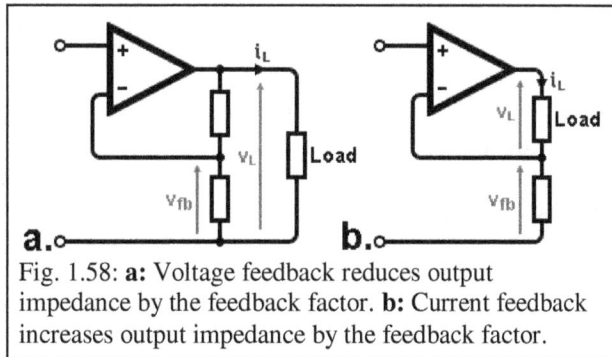

Fig. 1.58: **a**: Voltage feedback reduces output impedance by the feedback factor. **b**: Current feedback increases output impedance by the feedback factor.

59

This is a round-about way of saying that voltage feedback reduces the output impedance of the amplifier. For example, if we add enough voltage feedback to reduce the gain by half (a feedback factor of 2) then the output impedance will also be halved. This is the sort of feedback most often used in audio amplifiers.

If the feedback voltage is proportional to the current in the load, regardless of the voltage across it, then it is called **current feedback**. The load is then itself part of the feedback divider and therefore controls the feedback fraction, as illustrated in fig. 1.58b, and this has just the opposite effect of voltage feedback. It tends to make the whole system behave more like an ideal current source as far as the load is concerned. In other words, it will try to supply whatever output voltage is necessary to maintain the prescribed output current into the load, which means the output impedance of the amplifier is increased. For example, if we add enough current feedback to reduce the gain by half then the output impedance will be doubled.

The way in which feedback is applied affects the system's *input* impedance. If the feedback voltage is added in series with the source voltage as in fig. 1.55b earlier, then the input impedance is increased by the feedback factor. This is often called a series feedback amplifier. If the feedback voltage is applied in parallel with the source voltage as in fig. 1.56b, then the input impedance is reduced by the feedback factor. This arrangement is often called a shunt feedback amplifier. The distinction between series and shunt feedback is not always obvious, but the circuits will become familiar with experience.

1.24.4: Stability

For hi-fi what we usually want is negative feedback. Unfortunately, there are always unavoidable phase shifts in the open-loop circuit, and they are bound to amount to ±180° at some frequency. At this point the negative feedback will become positive. Looking at the universal feedback equation, if the loop gain A_oB happens to equal to −1 then the closed-loop gain blows up to infinity and the amplifier will oscillate. This leads us to the **stability criterion** which tells us how to avoid oscillation:

- At the frequency where the loop gain equals 1 (0dB), the phase shift must be less than 180°.
- At the frequency where the phase shift reaches 180°, the loop gain must be less than 1 (<0dB).

One way to check that the stability criterion is met is to examine the frequency and phase responses using Bode plots. Fig. 1.59 shows a Bode plot of a fictitious amplifier that has an open-loop gain of 100 (40dB) at low frequencies and, for whatever reason, three poles. Each pole introduces −20dB/decade attenuation and up to −90° phase shift. At some frequency, therefore, the phase shift *will* reach −180° (and beyond). The true phase response is shown here, rather than a Bode approximation. Only the upper end of the amplifier's bandwidth is shown in this case, but exactly the same treatment would apply to the low-frequency end, except the phase shift will tend towards +180° rather than −180°.

The loop gain A_oB can also be plotted. In this case half the output voltage has been fed back at all frequencies, so the feedback fraction B is 0.5 (−6dB). The loop gain is therefore a line that is 6dB lower down than the open-loop gain line. The true response is shown in this case, rather than a Bode approximation. The closed-loop gain is also plotted, and at low frequencies it is equal to:

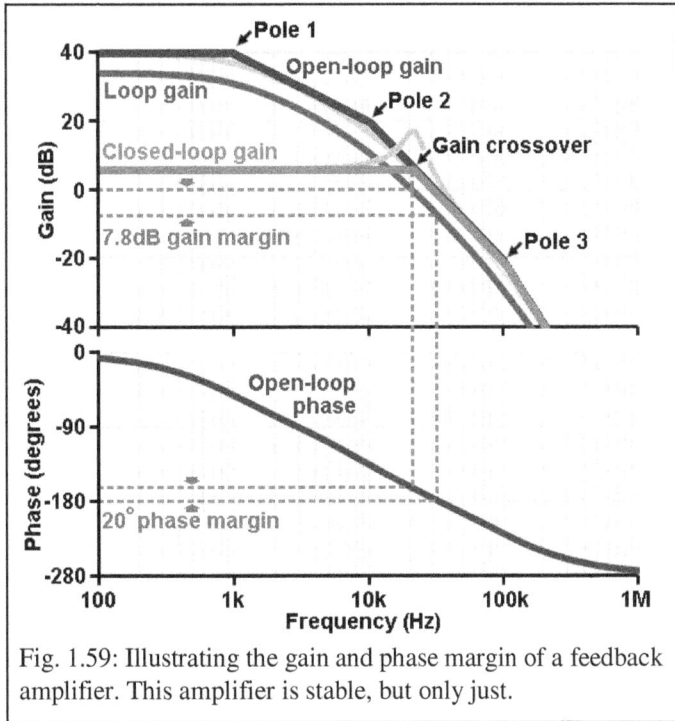

Fig. 1.59: Illustrating the gain and phase margin of a feedback amplifier. This amplifier is stable, but only just.

$$A = \frac{A_o}{1 + A_oB} = \frac{100}{1 + 100 \times 0.5} = 1.96 \quad (5.8dB)$$

Looking at the graph, at the frequency where the loop gain falls to unity (0dB) the phase shift has reached −160°. The difference between this and −180° is 20°, which is called the **phase margin**. The phase margin is a measure of stability as it indicates how much extra phase shift would be needed to reach the dangerous figure of −180° where the loop gain is unity. This amplifier has too little phase margin, causing a resonant peak to appear in its closed-loop response (shown faint). A phase margin of at least 66° or more is necessary to avoid such peaking completely (Butterworth response), while 76° or more is needed to prevent overshoot in the step response (critically damped).

Now look at the frequency where the phase shift really *does* hit −180°; the loop gain is −7.8dB. The difference between this and 0dB is 7.8dB and is called the **gain margin**. This is another measure of stability as it indicates how much more open-loop gain would be needed to raise the loop gain to the dangerous figure of unity (0dB) where the phase shift is −180°.

61

Now let us make an observation: the frequency where the loop gain drops to 0dB is the same as where the closed-loop gain line intersects with the open-loop gain line, which is called the **gain-crossover frequency**. The rate at which the two (idealised) lines converge is called their **rate of closure**. For example, in fig. 1.59 the open-loop gain line is falling at −40dB/decade and the closed-loop gain line is horizontal, so their rate of closure is 40dB/decade. By looking at the rate of closure it is possible to (almost) guarantee that the stability criterion is met without having to draw the loop-gain line or find the exact gain and phase margins. The following rules apply:

- If the rate of closure is more than 40dB/decade (18dB/octave) then the phase shift will reach 180° and oscillation is virtually guaranteed, unless the loop gain happens to be much less than unity. This is not a practical way to design an amplifier.
- If the rate of closure is 40dB/decade (12dB/octave) then the system probably won't produce sustained oscillations since the phase shift is likely to be somewhat less than 180°. It may get pretty close, however, so will develop a resonant peak in the frequency response and produce unsustained oscillations (ringing) instead. This is what has happened to the amplifier in fig. 1.59.
- For good stability the rate of closure should be 20dB/decade (6dB/octave). The phase shift will then remain well below 180° and the system will be well behaved. If this cannot be achieved then it will be necessary to modify the design by applying less feedback and/or moving the open-loop poles and zeros around. There are many ways this might be done, but they will be application-specific, so this is not the place to discuss them.

A special condition exists if we can arrange for the phase shift not to reach −180° before the open-loop gain A_o falls to unity (0dB). If this is achieved then we could apply any amount of feedback and the stability criterion would always be met; the system would never oscillate even if we fed back *all* of the output signal (B = 1). Such a system is said to be **unconditionally stable**.

Fig. 1.60 shows an example of such a system. The phase margin is a healthy 80° when B = 1 (all the output voltage is fed back) despite there being three open-loop poles. This is achieved because two of the poles are far removed from the lowest or **dominant pole**, so they don't add much phase shift until well beyond the 0dB gain point. Notice that the rate of closure is 20dB/decade and there is no resonant peaking. The closed-loop phase response is also shown for interest. It tends to be much easier to achieve this situation when the amplifier is a simple one, containing perhaps just one amplifying device, i.e. local feedback. This relative freedom from instability makes local feedback more attractive than global feedback in some cases.[*]

---

[*] Some audio commentators posit that global feedback is bad whereas local feedback is just about tolerable (perhaps because they know that local feedback is impossible to avoid –even Ohm's law is an expression of local feedback). This is only slightly less silly than saying all feedback is bad.

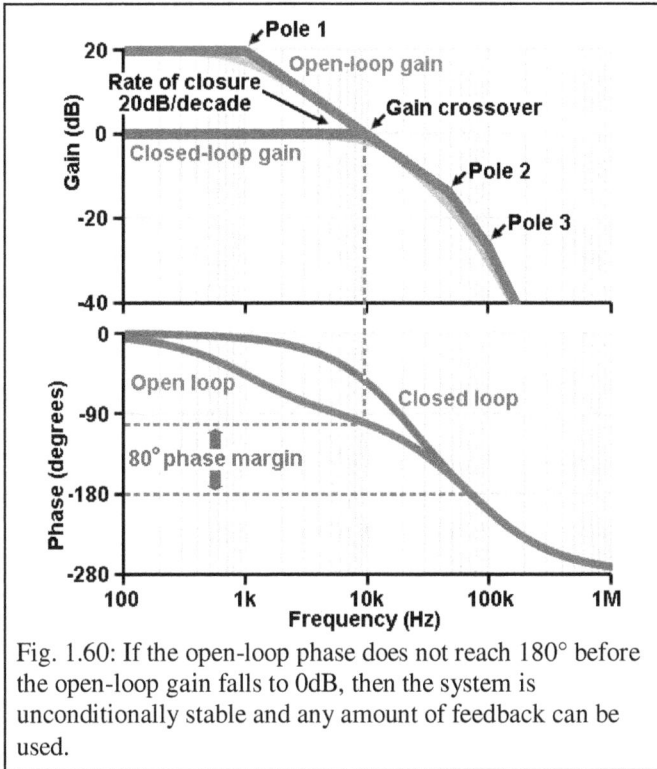

Fig. 1.60: If the open-loop phase does not reach 180° before the open-loop gain falls to 0dB, then the system is unconditionally stable and any amount of feedback can be used.

Chapter 2: Practical Components

In chapter 1 we troubled ourselves only with ideal components –the kind we find in circuit simulators but not in suppliers' catalogues. Real components come in various types with various limitations and imperfections. It is essential to have some general knowledge of these types and limitations so that appropriate devices can be chosen to ensure proper and reliable circuit operation, within the desired budget. As for the imperfections, they are countless, and a full treatment could take forever. This chapter will therefore discuss only the most significant defects and those which are pertinent to valve audio design. For the fine detail you should always check the manufacturer's datasheets and application notes. We will also restrict the treatment to through-hole components only, as surface-mount types are not nearly as useful for valve amp construction.

2.1: Resistors

Resistors are by far the most nearly-perfect of electronic components. They are readily available with values from milliohms to a gigohm or so, although valve circuit design rarely calls for anything outside the range of 1Ω to 10MΩ. They are categorised according to the material used to make the resistive element, and for modern design work the two principle types are film (i.e. carbon film and metal film) and wirewound.

2.1.1: Resistance

The most important thing about a resistor is of course its nominal resistance. Resistors are made in sets of 'preferred' or 'standard' values called the **E-series**, ranging from E6 to E192. For example, the E12 series contains twelve logarithmically-spaced preferred values, per decade, as listed in table 2.1. Higher or lower values are found by multiplying or dividing these preferred numbers by factors of ten. The advertised resistance is the 'nominal' value, that is, the resistance it is *supposed* to have. In reality we are at the mercy of manufacturing processes, so there is a tolerance band associated with the advertised resistance. Traditionally, the lower E-series had poorer tolerance and were therefore cheaper than the upper series, but nowadays resistors from the E24 series are about as cheap as anything else. Valve circuits tend to be quite non-critical so there is usually no need to venture above the E24 series.

E6	E12	E24
10	10	10
		11
	12	12
		13
15	15	15
		16
	18	18
		20
22	22	22
		24
	27	27
		30
33	33	33
		36
	39	39
		43
47	47	47
		51
	56	56
		62
68	68	68
		75
	82	82
		91

Table 2.1: Standard resistor values covering one decade.

2.1.2: Tolerance

The tolerance of a resistor is given as a percentage of the nominal resistance value. Experience suggests that most devices have a roughly Gaussian or 'normal' distribution, so finding a resistor whose value is right at the extreme end of its tolerance band is less likely than finding one close to the nominal value. Resistance variation is usually distributed equally above and below the nominal value, so a '±' can be assumed when talking about resistor (and other component) tolerances. Vintage resistors might have tolerances as bad as 10%, so a nominally 100kΩ resistor could be expected to fall anywhere between 90kΩ and 110kΩ. Manufacturing is much better these days and 1% resistors are now more-or-less standard. In general, you pay more for tighter tolerance, although 1% devices are often no more expensive than 5% resistors, at least for low-power (≤½W) devices. Anything better than 1% will be considerably more expensive, however.

2.1.3: Power Rating

Often of more significance to us than the tolerance is the resistor's power rating. This is the maximum average power that it can dissipate continuously while still meeting the manufacturer's guarantees. Datasheets may also quote pulsed power or short-term overload ratings, but these are of less interest to us. The average continuous power rating usually applies for ambient (i.e. air) temperatures of up to 70°C. The temperature inside a valve chassis could easily reach 40°C and may even exceed 50°C in a hot country, but is not likely to reach 70°C. Nevertheless, for good long-term reliability it is standard engineering practice always to choose a resistor that is rated for at least twice the power we expect it to endure during continuous operation.

2.1.4: Voltage Rating

It is not always appreciated that resistors have a voltage rating. This is the maximum voltage that can be allowed between the terminals; higher voltages may cause flashover of the resistive element. Some typical ratings for film resistors are:

⅛W, 200V
¼W, 200V
½W, 350V
1W, 500V

Always check the manufacturer's datasheet, however. The voltage rating is of particular importance in the high-voltage environment of a valve amp, and it often requires the use of ½W devices or better, even though the expected power dissipation may be much less. For this reason the resistors in all the circuits in this book may be assumed to be ½W unless otherwise indicated. Many builders choose to use 1W or even 2W devices as standard, which pushes such concerns to the bottom of the list (they also have nice visual appeal). If the necessary voltage rating isn't available then it is a simple matter to connect a number of resistors in series to share the burden, since the voltage drop across each one is proportional to its resistance.

2.1.5: Excess Noise

All resistances generate Johnson (thermal) noise, but practical resistors also generate **excess noise** caused by spontaneous fluctuations in the conductivity of the resistive material. Excess noise is inversely proportional to frequency (1/f or pink noise) and is proportional to the current in the resistor.[*] But since the current is

Resistor type	Excess noise (µV/V/decade)
Wirewound	~0
Metal film	0.01 – 0.3
Metal oxide	0.1 – 1.0
Carbon film	0.01 – 0.4
Carbon composition	0.1 – >1.0

Table 2.2: Resistor excess noise indices.

proportional to the voltage across the resistor, excess noise is usually quoted in the form of a 'noise index', NI, in microvolts (RMS) of noise per volt DC applied to the device, per decade of bandwidth, or µV/V/decade. This is the value of a fictitious noise voltage generator in series with the resistor –see chapter 5. It tends to decrease with higher power ratings and with smaller resistance values. Table 2.2 shows some typical noise index figures.

The 'per decade' part of the noise index is not always mentioned on datasheets, leading users to underestimate excess noise in calculations. The number of decades in a given bandwidth f_{hi}–f_{lo} is equal to $\log(f_{hi}/f_{lo})$. But just to make things more complicated, the total noise in each decade must be added together geometrically, so the total noise in a given bandwidth, per volt DC, is the noise index multiplied by the *square root* of the number of decades.[1] For the audio band this works out to be:

$$v_{N(excess)} = NI \times V_{dc} \times 1.7 \quad \text{microvolts} \tag{2.1}$$

2.1.6: Temperature Coefficient

Usually of lesser importance to the valve amp designer is the **temperature coefficient of resistance**, or simply **tempco**. This figure indicates by how much the resistance can be expected to change (usually in parts per million) per degree Celsius change in temperature. Carbon film resistors have tempcos up to 700PPM/°C, so a 20°C rise could cause more than 700×20 = 14000PPM change in resistance, which is 1.4%. This will be part of the reason why the sound of certain amplifiers seems to change as they warm up. Metal film resistors have much better tempcos.

2.1.7: Parasitics

A practical resistor is not a pure resistance but has some unavoidable or 'parasitic' self inductance and shunt capacitance, as modelled in fig. 2.1. Self inductance increases with resistance and power rating. The inductance of low power

---

[*] For this reason it is also sometimes called current noise, but this name is deprecated as it could be confused with the shot noise of current which is quite different.
[1] Motchenbacher, C. D. & Fichen, F. C. (1973). *Low Noise Electronics Design*. John Wiley & Sons Inc., New York.

(<1W) film resistors will be in the region of a few nanohenries, rising to tens of nanohenries for more powerful devices. Ordinary wirewound resistors have a few microhenries of inductance. The component leads will also add about ten nanohenries per centimetre, and hand-wired valve equipment often leaves the leads quite long. Even so, these inductances are so small that they really

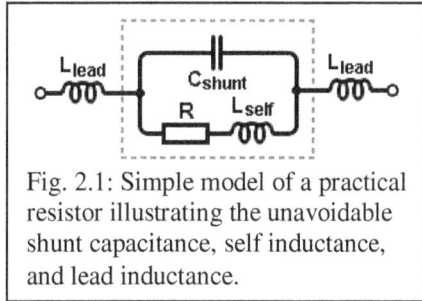

Fig. 2.1: Simple model of a practical resistor illustrating the unavoidable shunt capacitance, self inductance, and lead inductance.

only become important at radio frequencies, and even then only when the resistance value is small so there is little circuit damping (a whole $1\mu H$ has a reactance of only 1Ω at 159kHz, which is easily swamped). The shunt capacitance of through-hole film resistors tends to hover around 0.3pF which is also small enough not to be a concern in valve amp design, but wirewounds may have several picofarads which could be more troublesome.

2.1.8: Metal Film

The most common type of resistor is metal film. These are made from a ceramic rod onto which a film of metal alloy is deposited. A laser then burns away material to leave behind a helical path of metal film with the desired resistance. A coat of protective epoxy is then applied and is *usually* coloured blue or purple, although there are plenty of exceptions. These days they are about the same size and cost as carbon-film types but are more stable (tempco around 25 to 100PPM/°C), and they generate essentially no distortion –see fig. 2.2. Their excess noise is usually slightly lower or at least no worse than for carbon film resistors, at around 0.01 to 0.3μV/V/decade. They are the resistor of choice for nearly all audio applications.

2.1.9: Metal Oxide

Metal Oxide resistors are a particular type of metal film. They are commonly used for moderate power ratings (>1W) and tend to have poorer noise, distortion, and tempco (≈350PPM/°C) than ordinary metal films. They are not normally used in critical audio positions but may be found in power supplies.

2.1.10: Carbon Film

Until a few years ago the most common type of resistor was the carbon film, but is has since been usurped by the metal film. Carbon film resistors are constructed in the same manner as metal films but using carbon instead of metal alloy. They are *usually* coloured beige or brown, but there are plenty of exceptions. Their tempco is unimpressive at around 300 to 700PPM/°C while their excess noise is usually around 0.01 to 0.4μV/V/decade, which is not much worse than for metal films.

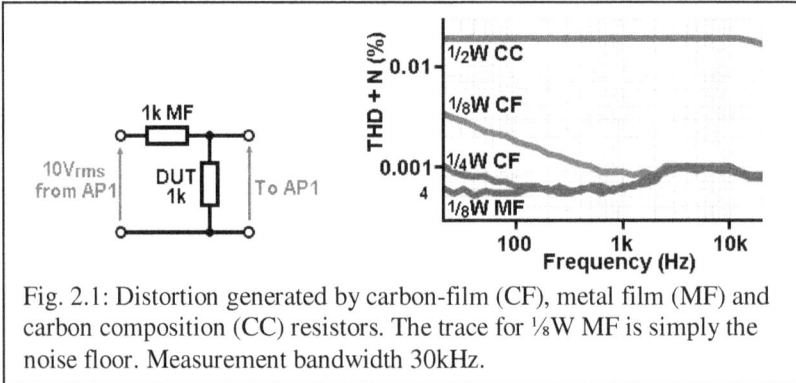

Fig. 2.1: Distortion generated by carbon-film (CF), metal film (MF) and carbon composition (CC) resistors. The trace for ⅛W MF is simply the noise floor. Measurement bandwidth 30kHz.

Carbon film resistors are known to produce distortion that increases at low frequencies and with increasing power dissipation, presumably due to thermal cycling. Fig. 2.1 shows the distortion generated by an ⅛W carbon-film when dissipating 25mW; inspection of the distortion residual showed it to be pure third harmonic. A ¼W device produces less distortion, probably because it is physically larger and therefore has a longer thermal lag and lower bulk temperature. A metal film resistor, however, generates no measurable distortion even when it is an ⅛W device. Some commentators have claimed carbon film resistors have a warm sound whereas metal films sound somehow 'clinical', and although the distortion generated by an individual resistor will be far below that of any valve, it is just possible that the cumulative distortion of *many* carbon resistors might have some noticeably 'warming' effect on the sound.

2.1.11: Carbon Composition

The traditional resistor type found in vintage equipment is carbon composition. These are made from ceramic and carbon dust bonded together in resin, and the whole device is made from a solid lump of this material, so it may be called a 'bulk' resistor. The tempco of carbon composition resistors is dire and therefore

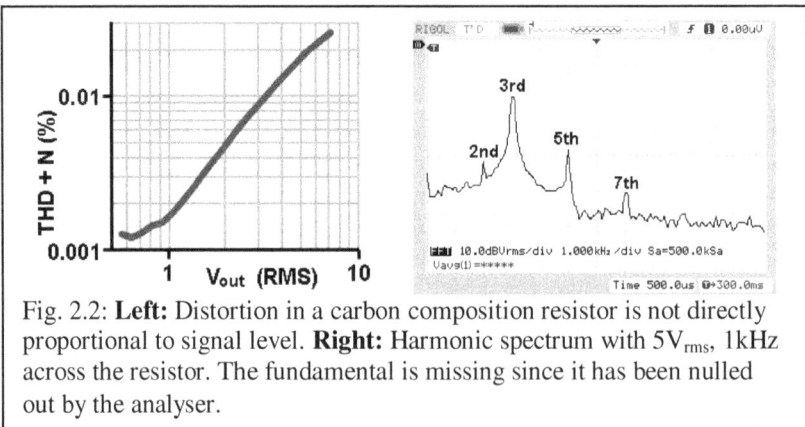

Fig. 2.2: **Left:** Distortion in a carbon composition resistor is not directly proportional to signal level. **Right:** Harmonic spectrum with 5V_rms, 1kHz across the resistor. The fundamental is missing since it has been nulled out by the analyser.

not quoted on datasheets; it will certainly be much worse than 1000PPM/°C. Devices found in vintage amps will often have drifted in value by an alarming amount. They are the noisiest of all the resistor types with up to and above 1μV/V/decade excess noise. Vintage samples will sometimes develop crackling or popcorn noise too.

Carbon comp's are still available as they are sometimes useful for their high pulsed-power capability and low inductance, but they are a very poor choice for modern audio circuits. Not only are they noisy, they are highly nonlinear too, owing to their very poor voltage coefficient of resistance. In other words, their resistance changes with applied voltage. A typical value is 0.035%/V for a ½W device, so a voltage swing of 200V could result in an alarming 7% change in resistance. High-power devices tend to be even worse.

Fig. 2.1 shows that with 5V$_{rms}$ across a ½W carbon comp', 0.019% distortion is generated at all frequencies (the droop above 10kHz is due to the measurement bandwidth limit). Fig. 2.2 shows distortion versus the voltage across the resistor for the same test circuit, and it can be seen that the distortion increases at a faster rate than the voltage (roughly to the power of 1.3). The harmonic spectrum with 5V$_{rms}$ across the resistor is also shown, and is dominated by a decaying series of odd harmonics, but with some second harmonic visible too.

2.1.12: Wirewound

The most conceptually simple type of resistor is the wirewound, which is nothing more than a coil of wire wound on a ceramic rod. Their construction makes them capable of handling high power levels, and devices rated for less than 3W are rare. The smaller types (which are still fairly bulky)[*] may be epoxy or ceramic coated, while the larger types may be housed in a metal casing that must be bolted to a heat sink to achieve the full power rating. They do not come in a wide range of values (often restricted to the E6 range) and are rarely larger than 100kΩ. Because they are made of nothing but pure metal they generate no excess noise or distortion. Their tempco is usually in the range of 100 to 300PPM/°C. However, the value of new (non-precision) wirewound resistors may drift by as much as 1% over time as the tensioned wire relaxes, but they can be stabilised or 'aged' by baking at 135°C for a day.[2] Precision types with tempcos under 10PPM/°C are also available.

A common complaint about wirewounds is that their coiled nature leads to relatively high self inductance. Special low-inductance devices can be bought (they use elaborate winding techniques) but, as noted earlier, even a few microhenries is not likely to pose a problem in valve amp design. It is actually the shunt capacitance which is more concerning, as it can be as high as 10pF.

---

[*] It is now possible to buy surface-mount wire wounds, but they're not exactly tiny.
[2] Scroggie, M. G. (1971). *Radio and Electronic Laboratory Handbook*, 8th edition. London, Illife. p198.

2.1.13: Fusible / Flameproof Resistors

Fusible resistors are designed to withstand their rated power dissipation indefinitely but they will fuse (i.e. go open-circuit) within a few seconds if the dissipation is several times greater. They are also called flameproof resistors because they are designed to fail without producing a flame. They are used when it is desirable to limit current as well as providing a fusing action, since ordinary fuses may allow damaging transients of tens of amps to flow before blowing.

2.1.14: Potentiometers

A potentiometer or 'pot' consists of a track of resistive material across which a moving wiper can be swept. Pots usually have poor tolerance –10 to 20%– and the range of resistance values is normally restricted to multiples of 10, 22 or 47 in Europe, or 10, 25 and 50 in the US. In practice the two families are identical since the tolerances overlap; they are simply labelled differently for the different markets, for historical reasons.

If the track resistance is evenly distributed along its length then it is said to have a **linear taper**, and these are often denoted with the letter B. When the wiper is in the centre of a linear pot, half the total track resistance lies on either side of it. Pots are also available with more of the resistance 'concentrated' towards one end of track, and the most common example is the **logarithmic taper**, also called audio taper, often denoted by the letter A. Log pots are normally used for volume controls because they cause the voltage division to vary in a way that complements the response of the human ear, giving a smoother variation in perceived loudness with knob rotation. Practical log pots are not truly logarithmic,

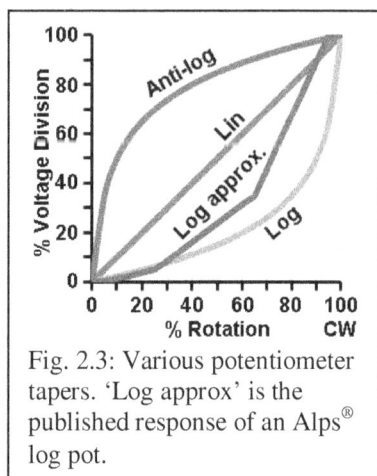

Fig. 2.3: Various potentiometer tapers. 'Log approx' is the published response of an Alps® log pot.

however; they merely approximate it using two or three different pieces of linear track. In fig. 2.3 the published response of an Alps® log pot is shown for comparison with the idealised response. Other tapers such as reverse-log or anti-log (often denoted by the letter C) are available for special applications, but they are less common and more expensive.

Pots may be of the larger and more robust type intended to have a control knob attached, or they may be very small trimpots intended to be adjusted infrequently with a screwdriver (set and forget). Both types are available with single turn or multi-turn action, the latter allowing slower and more controlled adjustment. Multiple pots may be attached to the same shaft in which case they are said to be 'ganged'. Dual-gang pots are quite common –for stereo volume controls, for example– and larger gangs are available at a higher price. There are also slider pots,

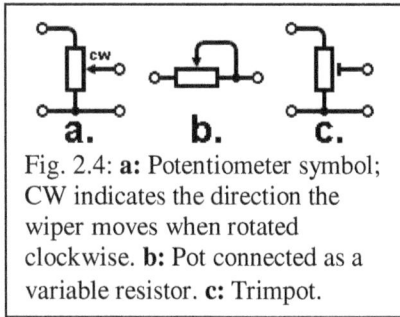

Fig. 2.4: **a:** Potentiometer symbol; CW indicates the direction the wiper moves when rotated clockwise. **b:** Pot connected as a variable resistor. **c:** Trimpot.

familiar to anyone who has ever used a fader on a mixing console. Pots may be used in the strictest sense of a potential divider, e.g. volume control, or as simple variable resistor, as shown in fig. 2.4. In this book 'CW' is used on diagrams to indicate the end of the pot that the wiper moves towards when turned clockwise, as viewed from the front, i.e. the side with the knob attached!

The most common type of pots have a carbon track and are usually rated for ½W or less for linear taper, and less still for log taper because the dissipation is not evenly distributed. They have poor tempcos so are not suitable for precision applications when used as variable resistors. However, when used as potential dividers they are in fact very stable since both sections of the track vary by the same proportion. Either way, the author was unable to detect distortion in any carbon pots or trimpots; presumably their bulk is too great to exhibit the same problem as small carbon-film resistors. More sophisticated (i.e. expensive) pots have tracks made from cermet (ceramic-metal composite), or conductive plastic. They are common for power ratings up to a couple of watts and are superior to carbon in most ways. For higher-power applications wirewound pots are available, more usually called **rheostats**. They are expensive and are rarely used these days since transistors offer cheaper and more flexible methods of power control.

Beginners often fall into the trap of applying a perfectly safe voltage to a pot wired as a variable resistor, but when they turn down the resistance they discover the pot burns out. This happens because the power rating of a pot applies to the *whole* resistive track. Trying to dissipate the full rated power in only a small portion of the track will inevitably lead to smoke. It is better to think of the power rating as a current rating instead, since the maximum current that can be allowed to flow in the track is equal to $\sqrt{P/R}$, where R is the *total* track resistance. If this current can be sustained through the whole track then it can be sustained in a small portion of the track too, regardless of the pot setting.

2.2: Capacitors

Capacitors receive a great deal of criticism from audio subjectivists, not all of which is deserved. In fact, they seem to attract more vilification than any other type of component, which is certainly unfair. If resistors can be considered to be the most nearly-perfect of all electronic components, then capacitors come in an easy second, well ahead of inductors, transformers, valves, opamps, or anything else. I personally suspect that much of the mistrust stems from non-technical enthusiasts being deeply suspicious of the fact that capacitors contain an insulator, making them look like a 'break' in the signal path, whereas inductors at least have a nice solid bit of wire connecting one terminal to the other (anything else is presumably too magical even to comprehend). This attitude of course betrays a lack of understanding of how electricity actually works. When capacitors are chosen properly any undesirable effects they may have will be buried deeply below the noise and distortion created by other components, and quite inaccessible even to golden ears. This does not require resorting to expensive 'audiophile grade' devices, either.

Capacitors are categorised according to their **dielectric**,[*] which is the insulating material between their plates. If we have a pair of fixed metal plates then the capacitance between them will be smallest when the dielectric is a vacuum (air is practically the same). Inserting *any* other material into the space between the plates will cause the capacitance to increase. Exactly how much it increases depends on the material's electric permittivity, ε, which is a measure of how much energy can be packed into the material in the form of an electric field. Broadly speaking, materials with higher ε tend to have greater defects too, and devices which pack a lot of capacitance into a small volume are more likely to be mistrusted by audiophiles, with some justification.

Functionally, capacitors are divided into polar and non-polar types. Polar capacitors are sensitive to the voltage polarity applied across them, and connecting them the wrong way round may lead to catastrophic failure. Polar capacitors are basically synonymous with aluminium and tantalum electrolytic capacitors. Conversely, non-polars do not care whether the applied voltage is positive or negative. The vast majority of non-polar capacitors are ceramic or plastic.

2.2.1: Capacitance

The most important property of a capacitor is of course its capacitance. In valve audio applications we are likely to use values ranging from about 10pF to 10000µF or so, making capacitors the widest ranging of components (nine decades!). Many high-value capacitors are available only in the E6 range while most others come in the E12 range. In the US/Canada it is not unusual to find nominal values in multiples of 2.0, 3.0 and 5.0 instead of 2.2, 3.3 and 4.7. However, capacitors usually have poor tolerance so the difference between these historical and standard values is seldom of any significance.

2.2.2: Tolerance

Capacitors have notoriously poor tolerance. Ordinary plastic capacitors are usually 10% tolerance, while most electrolytic capacitors are 20%. Vintage devices are likely to be much worse. Precision capacitors can be bought with tolerances as good as 1% but are limited to values less than about 10nF, and they are five to ten time more expensive than standard devices. The alternative is to measure and select components by hand, or better still to design circuits so they are immune to ordinary variation in capacitance.

---

[*] From *dia* meaning 'through', and *electric* which originally meant 'insulator'.

2.2.3: Voltage Rating

Of particular importance to valve amp design is a capacitor's voltage rating. This is the maximum working voltage than can be sustained across the device before dielectric breakdown becomes a possibility. This value may be specified as the maximum *DC* working voltage, but this is a somewhat redundant statement; the voltage rating is the same regardless of whether it is DC or a repeating peak value. Some datasheets may also quote pulsed or short-term overload ratings, but we are unlikely to need to operate so far off the map.

Plastic and ceramic capacitors are readily available with voltage ratings up to 1kV and beyond, which is more than enough to accommodate most valve circuits. Electrolytic capacitors are easy to obtain up to 450V, beyond which the availability drops like a stone. It is therefore often in our interest to design circuits which operate at less than 450V.

2.2.4: Leakage

No practical dielectric material is a perfect insulator; really it is just a very large resistance. Therefore, the device can be represented as a perfect capacitor with a parallel resistance, called the leakage resistance since it allows direct current to leak past the capacitor. In general, leakage current can be assumed to be negligibly small for plastic and ceramic capacitors but *not* negligible for electrolytic capacitors, where it ranges from a few microamps to a few hundred microamps. Leakage is worst when voltage is first applied and decays to normal running levels after a few minutes –see also section 2.2.14.

2.2.5: ESR

Perfect capacitors are completely reactive but practical capacitors will always convert a little bit of power into heat. We do not need to know exactly what causes this; the important thing is that from our point of view on the outside world it is as if there is a small resistance in series with the capacitor, which is called the **effective series resistance**, or **ESR**. This typically ranges from a small fraction of an ohm to a few ohms, though it varies somewhat with frequency and temperature. ESR is a problem in large power supplies because it limits the rate at which current can be drained out of charged capacitors, and the internal heat dissipated by the ESR may damage the device. However, the exact value of ESR is seldom needed for valve amp work, even in the power supply, as the ripple current rating is usually of more direct relevance (section 2.2.7).

2.2.6: Loss Tangent / Dissipation Factor

The unavoidable parallel (leakage) and equivalent series resistance within a practical capacitor are collectively referred to as **dielectric loss** (not to be confused with dielectric absorption), and are encapsulated in a rating called the **loss tangent**: tan(δ), or tan(d).

The figure δ (delta) is equal to 90 − φ , where φ is the phase angle between the current and voltage. Ideally φ would be ninety degrees, so both δ and tan(δ) would be zero. Since capacitive reactance and ESR both vary with frequency, tan(δ) must also vary with frequency. It is often quoted at 120Hz for electrolytics, whose tan(δ) then ranges from about 0.1 to 0.5. Plastic capacitors may quote figures at higher frequencies.

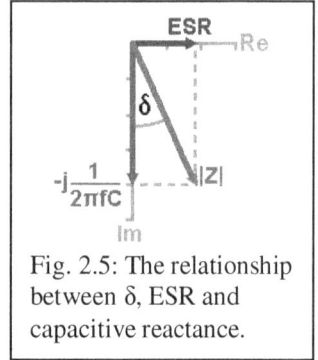

Fig. 2.5: The relationship between δ, ESR and capacitive reactance.

Because tan(δ) is only a number it does not tell us whether we are dealing with parallel or series resistance, or both, so for calculation purposes we assume it all comes from ESR. The device can then be modelled with the vector diagram in fig. 2.5 where tan(δ) is the ratio of the real and reactive parts of the impedance. Therefore, it is *also* the ratio of the *real* power being dissipated as heat in the resistance, and the *reactive* power flowing back and forth in the capacitance. The loss tangent is therefore also sometimes called the **dissipation factor**.

$$\tan(\delta) = \frac{ESR}{|X_C|} = 2\pi fC \times ESR \qquad (2.2)$$

Annoyingly, some manufacturers prefer to ignore the rules of common sense and basic mathematics by quoting the loss tangent as a percentage. Apparently, a dissipation factor of 10% is supposed to mean tan(δ) = 0.1, for example. Others don't even add the percent symbol, making their figures look one hundred times worse than they actually are. Fortunately, the loss tangent is rarely a consideration in ordinary valve amp design, even in the power supply.

2.2.7: Ripple Current

An ideal capacitor does not dissipate power, it only stores it temporarily and then returns it (reactive power). Practical capacitors have some ESR which *will* dissipate some energy as heat. Heat is particularly bad news for electrolytic capacitors because it causes the electrolyte to evaporate. For this reason a maximum ripple current value may be specified. This is the maximum RMS value of current that can be allowed to flow in the device, and exceeding it will reduce its lifetime and may cause it to leak electrolyte or even explode. The ripple current rating is usually only a concern in power supply circuits so it is typically quoted at 100Hz or 120Hz, corresponding to full-wave rectified mains. Naturally, a high ripple-current rating goes hand in hand with low ESR and a small loss tangent.

2.2.8: Dielectric Absorption

If a capacitor is initially charged up to a voltage and the terminals are then briefly shorted together, it should immediately discharge and we would find the voltage is zero after the short is removed. In practice, however, the voltage is found to recover slightly. This effect is called dielectric absorption. It is a molecular effect rather beyond the scope of this book, but it can be appreciated by imagining the capacitor to consist of a series of RC stages as shown in fig. 2.6. When the terminals are briefly shorted, only the nearest capacitor is actually discharged fully. When the short is removed the other capacitors are able to recharge the first one, so the terminal voltage rises again. Dielectric absorption is often blamed for 'capacitor sound', and various claims have been made about its perceived

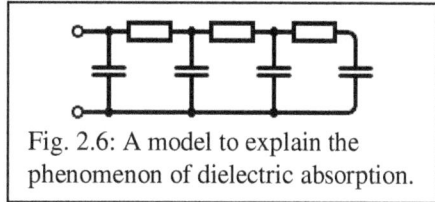

Fig. 2.6: A model to explain the phenomenon of dielectric absorption.

subjective effect ('smearing' is a popular choice), but it is a red herring. It is true that many capacitors which generate high levels of distortion also tend to have high dielectric absorption, but there is no evidence to confirm that the one actually *causes* the other. Silvered mica capacitors, for example, have unusually high dielectric absorption yet produce no distortion in ordinary tests.

2.2.9: Parasitics

As already mentioned, practical capacitors have parallel leakage resistance and effective series resistance. They also have self inductance, which tends to increase with capacitance, ranging from perhaps 10nH for plastic capacitors in the nanofarad range to tens of nanohenries for large electrolytics. Lead inductance will add a further ~10nH per centimetre. All this creates a series resonant circuit as shown by the model in fig. 2.7. At low frequencies the capacitive reactance dominates, so the device behaves more-or-less the way we would hope, i.e. like a

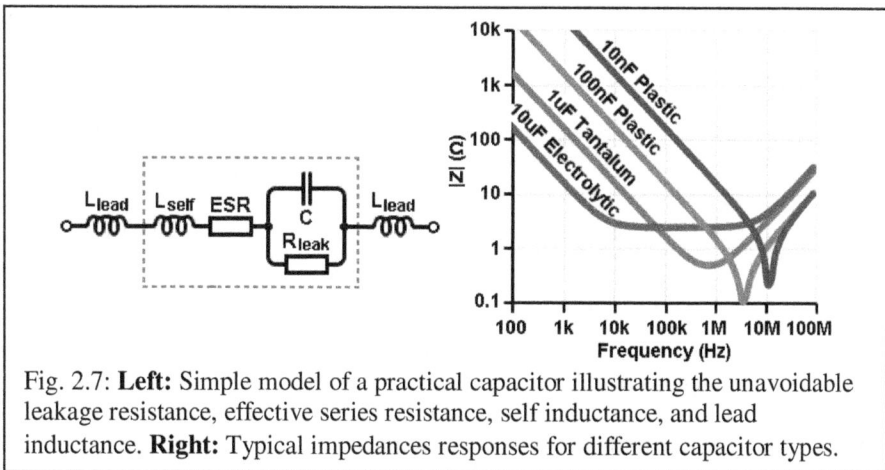

Fig. 2.7: **Left:** Simple model of a practical capacitor illustrating the unavoidable leakage resistance, effective series resistance, self inductance, and lead inductance. **Right:** Typical impedances responses for different capacitor types.

75

capacitor. But at very high frequencies the inductance dominates, so the impedance *rises* with frequency –the exact opposite of what we want from a capacitor. In the middle is the **self resonant frequency** where the total impedance drops to a minimum value determined by the ESR. We can therefore only rely on a capacitor to behave like a capacitor below its self-resonant frequency. Plastic and ceramic capacitors have resonant frequencies of several megahertz making them practically perfect as far as audio applications are concerned, whereas electrolytic capacitors are rather worse. Fig. 2.7 shows how the impedance might be expected to vary in some different capacitor types. Note, however, that just because the impedance starts rising does not mean the device becomes completely ineffective as a capacitor above resonance; the total impedance is what matters in circuit.

2.2.10: Air

The most theoretically perfect capacitor is one with a vacuum or air dielectric. Unfortunately, to get more than a few picofards of capacitance the plate area must be huge, which would lead to physically enormous and impractical components for audio work. Air capacitors are therefore restricted to the small (<50pF) variable or 'trimmer' capacitors used in radio-frequency circuits. The big tuning capacitors found in valve radios manage a few hundred picofarads.

2.2.11. Silvered Mica

Silvered-mica capacitors are made from silver deposited onto mica, which is a natural mineral. They are usually limited to values up to a few nanofarads and are relatively expensive. They have low ESR, are very stable even at high temperatures, and produce no detectable distortion. However, they suffer from disappointingly high dielectric absorption and are rarely used for audio since polystyrene or ceramic (C0G/NP0) capacitors can do the job much more cheaply.

2.2.12: Ceramic

Ceramic capacitors –now often called 'multilayer' capacitors– are usually the cheapest type for values ranging from picofarads to tens of nanofarads. They are often recognisable as orange- or blue-coloured discs or beads. There are many different types of ceramic capacitor, plus a bewildering array of specifications associated with each, but for audio purposes we can divide them all into two families:
- Type-I which includes C0G/NP0 types;
- Type-II which covers everything else.

The type-II 'everything else' includes Z5U and X7R ceramic dielectrics, as well as many others; the different names indicate different working temperature ranges and tempcos. They are the common, low-cost type of ceramic capacitor, and are ubiquitous in digital decoupling circuits. However, they have very poor characteristics for audio –too many to be worth listing them all– but the most obvious criticism is that their capacitances changes considerably with applied voltage and temperature, leading to high levels of distortion. They are also

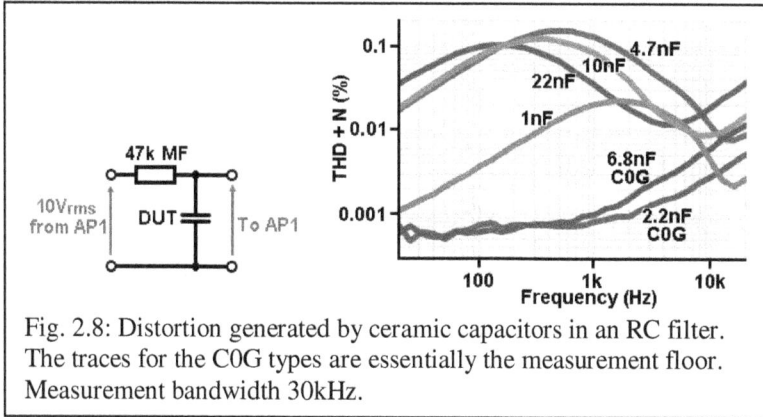

Fig. 2.8: Distortion generated by ceramic capacitors in an RC filter. The traces for the C0G types are essentially the measurement floor. Measurement bandwidth 30kHz.

piezoelectric meaning they produce a voltage when subjected to pressure or shock (i.e. they are microphonic), which can only make things worse.

Fig. 2.8 gives some idea of the scale of the problem; a random selection of ceramic capacitors of unknown origin tested in a simple RC filter. For all type-II devices the distortion was severe. Worse still, it was strongly temperature dependent. At room temperature it was found to be almost pure third harmonic but by warming the device up with a hot air gun the second harmonic would rise out of the noise floor and sink again as it cooled down (the 22nF device, however, produced equal second and third harmonic regardless). When arranged as a CR filter as in fig. 2.9 the picture becomes truly tragic with distortion reaching 1% at low frequencies. Clearly, these capacitors are unsuitable for audio-path applications and it is no surprise that they are universally hated by audiophiles.

Fortunately, C0G/NP0 capacitors are a completely different story. 'NP0' stands for 'negative-positive-zero' meaning they have near-zero temperature coefficient of capacitance (<30PPM/°C). NP0 is the old name for what is now called C0G, so they are affectionately referred to as 'cogs'. Their loss tangent, dielectric absorption and all-round stability are far superior to other ceramics, and they are as good as most

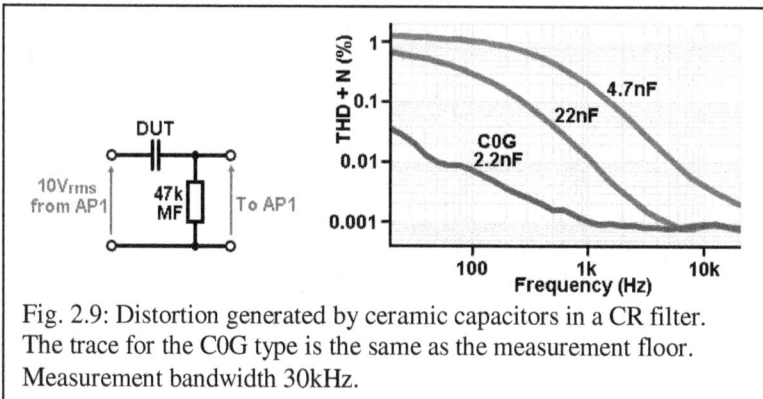

Fig. 2.9: Distortion generated by ceramic capacitors in a CR filter. The trace for the C0G type is the same as the measurement floor. Measurement bandwidth 30kHz.

plastic capacitors. This can be seen from fig. 2.8 and 2.9 where the C0G capacitors generate no measureable distortion –the rise in the stop-band is not distortion but the increasing noise floor of the filter. This sort of ceramic can be used with confidence in audio circuits.

2.2.13: Plastic / Film

There is a whole family of capacitors that can be lumped under the title of 'plastic' or 'film' types. They are the most common type for values in the range of nanofarads to a couple of microfarads with voltage ratings up to 1kV and beyond. Plastic capacitors are subdivided into two constructional types:
- Metal film, also called film-and-foil;
- Metallised.

Metal film capacitors are the best quality. They are made from a Swiss roll (US: jelly roll) of metal foils and a thin film of the dielectric material. Some film capacitors have a band marked on one end; this does not mean they are polarised but indicates which end is connected to the outer foil. This end can then be connected to the lower-impedance part of the circuit so that it acts like a shield around the rest of the capacitor, which reduces sensitivity to hum.

Metallised capacitors are made from a film of dielectric onto which metal is 'sputtered' to form the plates, rather than using separate foils. This achieves greater capacitance in a given volume but poorer performance in most other respects. Metallised capacitors are sometimes advertised as being self-healing, meaning that if a voltage overload causes a spark between the plates, a neat hole will be punched through the dielectric without shorting or melting the plates together, so the device will continue working afterwards. This is particularly important for capacitors intended for use on the mains supply, which are designated Class-X or Class-Y. These capacitors can also be used as signal capacitors of course, and their thick dielectric gives them acceptable distortion performance.

Plastic capacitors have extremely low leakage, low dielectric absorption and good tempcos, making them the capacitor of choice for audio circuits. The exact properties vary somewhat between the different types of plastic dielectric, and of the many different types in use the most significant are (in rough order from best to worst):
- PTFE (Teflon);
- Polystyrene (KS);
- Polypropylene (KP);
- Polyphenylene Sulphide;
- Polyester (PET / Mylar / KT).

(If the letter M appears before KS, KP, or KT, it indicates the capacitor is a *M*etallised type.)

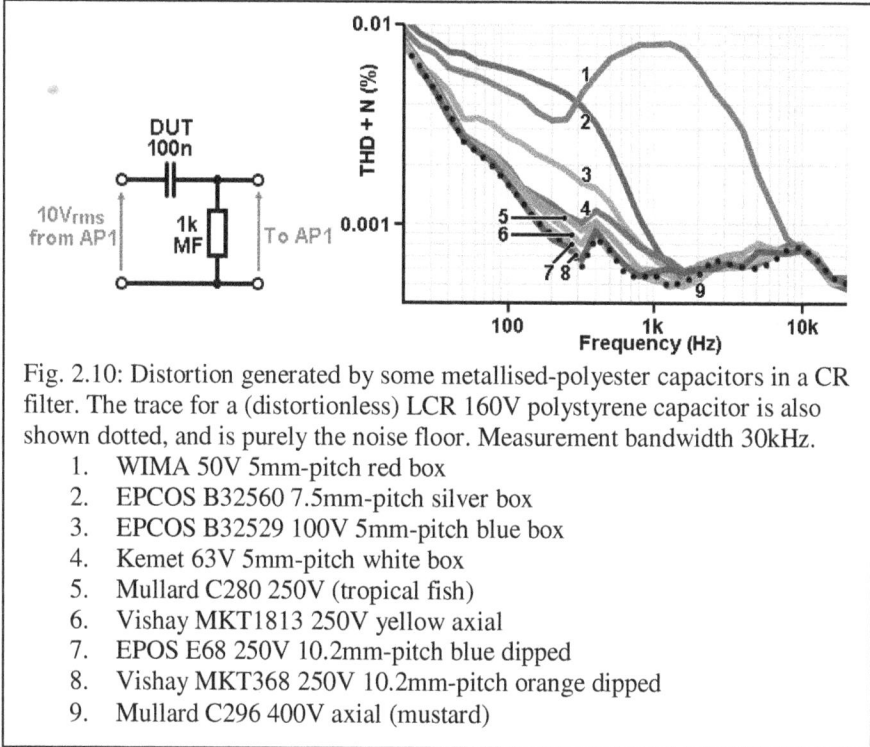

Fig. 2.10: Distortion generated by some metallised-polyester capacitors in a CR filter. The trace for a (distortionless) LCR 160V polystyrene capacitor is also shown dotted, and is purely the noise floor. Measurement bandwidth 30kHz.

1. WIMA 50V 5mm-pitch red box
2. EPCOS B32560 7.5mm-pitch silver box
3. EPCOS B32529 100V 5mm-pitch blue box
4. Kemet 63V 5mm-pitch white box
5. Mullard C280 250V (tropical fish)
6. Vishay MKT1813 250V yellow axial
7. EPOS E68 250V 10.2mm-pitch blue dipped
8. Vishay MKT368 250V 10.2mm-pitch orange dipped
9. Mullard C296 400V axial (mustard)

All of these are essentially distortionless and perfectly suitable for audio with the possible exception of polyester –the cheapest and most common type. There is much variation in the linearity of polyester capacitors, as demonstrated by fig. 2.10. Small-volume box capacitors perform poorly, but at the same signal levels, higher-voltage physically-large devices show little or no measurable distortion. The devices which do produce distortion display a decaying series of odd harmonics. The worst performer in fig. 2.10 was a WIMA red box that apparently suffers from two sources of distortion: the first is the same as that exhibited by all capacitors while the second is proportional to the power flow in the device, producing the pronounced hump in the graph. Its spectrum is particularly rich (and variable), including some second harmonic, as shown in fig. 2.11. The way forward

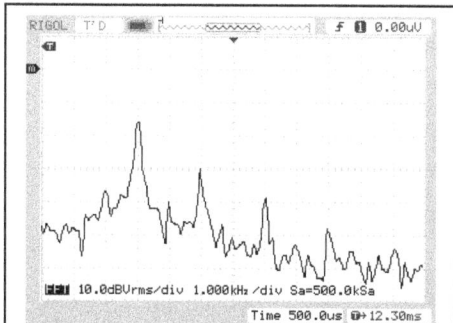

Fig. 2.11: Distortion spectrum of the WIMA red box capacitor tested in fig. 2.10 at 1kHz. THD+N was 0.0081%. The highest peak is the third harmonic; the fundamental is missing since it has been nulled out by the analyser.

79

is clear: if you have to use polyester it is best to use devices with high voltage ratings, i.e. thick dielectrics.

Relatively high-value (tens of microfarads), high-voltage polypropylene capacitors are often available at fairly low cost in the form of 'motor run' capacitors normally intended for industrial applications. They often have a screw mounting ideal for hand-wired amplifiers and are quite popular for power-supply decoupling. They should not be confused with 'motor start' capacitors, which are electrolytic and not really designed for continuous use.

2.2.14: Aluminium Electrolytic

Aluminium electrolytic capacitors are the most common type for values above a few microfarads. They are universally hated by subjectivists but, when a lot of capacitance is needed, it is impossible (or at least wincingly expensive) to avoid them. It is also unnecessary, because by following a few basic principles it is easy to use electrolytics in a way that has no detrimental effect on audio fidelity.

An aluminium electrolytic capacitor is made from two aluminium foil 'plates', one of which is anodised on one side to form an insulating oxide layer, which is the dielectric. The second foil would not make very good contact with the rough oxide dielectric so it is coated with a conductive gel, which is the electrolyte. The electrolyte squeezes into all the microscopic pits in the oxide layer, which amount to a huge plate area, and this is what gives these devices such high capacitance in a small package.

There is a limit to how thick the aluminium-oxide layer can be grown or formed, and this sets a practical upper limit for the voltage rating of around 450V. It is therefore advantageous to design circuits that use power supply voltages less than this. Devices boasting higher voltage ratings are occasionally encountered but should be regarded with some suspicion. If a higher voltage rating is needed then capacitors can be connected in series to share the total drop. However, the leakage current is unlikely to be the same in multiple devices and this will cause unfair voltage sharing. To swamp this effect it is necessary to connect a resistor in parallel with each capacitor, and a rule of thumb is $R = 50 / C$ as shown in fig. 2.12. The total useful capacitance is of course reduced since the capacitors are in series. It must

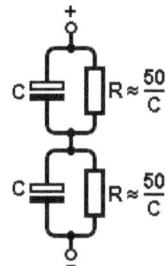

Fig. 2.12: Electrolytic capacitors can be connected in series to increase the voltage handling. To ensure equal voltage sharing, a resistor should be connected in parallel with each.

also be pointed out that the metal body of a capacitor is usually connected to the negative terminal, so when two devices are connected in series the body of one of them will rest at half the total voltage and could present a shock hazard in a high-voltage circuit.

80

The oxide dielectric is initially formed by passing current through the device during manufacture. However, if the device is left unused for a long time then the oxide will deteriorate, reducing the insulation between the plates. When a voltage is eventually applied, an abnormally large leakage current will flow until the oxide has reformed. In extreme cases –such as vintage equipment that has been unused for years– the sudden high level of leakage current may cause the electrolyte to vaporize and violently explode the device. To avoid this it is necessary to power up old equipment in a way that limits the current to a few milliamps so the oxide can reform slowly (a rule of thumb is to allow one minute of reforming for every month that the capacitor has been unused). A common method is to use a variable autotransformer (Variac™) to increase the voltages gradually, but any method of limiting the leakage current will do just as well.

For modern, healthy devices the worst-case operational leakage current (i.e. with the maximum permissible voltage applied) is frequently quoted as $I = 0.01CV_{max}$, where C is the capacitance and V_{max} is the voltage rating. It will therefore range from a few microamps for small capacitors to several hundred microamps for large, high-voltage ones. By dividing the working voltage by the leakage current we can alternatively find the parallel leakage resistance:

$$R_{leak} = \frac{V}{I} \approx \frac{V_{max}}{0.01CV_{max}} = \frac{100}{C}$$

It would be wise not to take this equation too literally, however, as the leakage resistance may not be constant with applied voltage. Nevertheless, this should be quite sufficient for design work. The leakage current is highest when voltage is first applied and over a few minutes will fall to whatever value is necessary to maintain the dielectric (which is still high compared with plastic and ceramic capacitors). This constant variation in the formation of the dielectric is what gives electrolytics such poor tolerance and variation with time, temperature, and even frequency.

The chemistry of an electrolytic creates an implicit diode in parallel with the expected capacitance. When the voltage is of the correct polarity the diode is reverse-biased and all is well. But if the wrong polarity is applied, and it exceeds a few hundred millivolts, the diode becomes forward biased and a large reverse-leakage current will flow. This causes the electrolyte to vaporise, and the growing pressure may rupture or explode the device. Modern capacitors have a rubber bung designed to vent gas in a controlled manner,[*] or a creased metal top that will burst in a vaguely controlled way, but explosions can still occur, so you should never stand over new equipment when it is being powered up. Whatever happens, reverse voltages of more than a few tens of millivolts must be avoided. Polarity is implied by the circuit symbol; for the IEC international symbol the white bar is the positive end,

---

[*] I vividly remember reverse-biasing a small electrolytic capacitor by accident which, after half a minute or so, suddenly began spewing a seemingly endless cloud of white, acrid smoke that filled the whole room and lingered for hours.

and for the US/Canadian symbol the flat bar is the positive end, as shown in fig. 2.13.

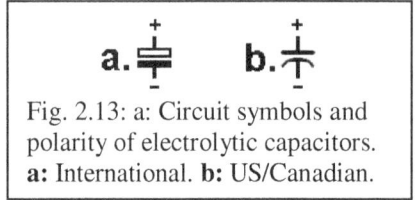

Fig. 2.13: a: Circuit symbols and polarity of electrolytic capacitors. **a:** International. **b:** US/Canadian.

There is a common myth that electrolytics *must* have some DC polarising voltage to work properly. Indeed, before the 1980s, mixing consoles usually employed unipolar power supplies as there was still a question as to whether the electrolytic coupling capacitors would be reliable without a firm polarising voltage. It has since been determined that they are, in fact, perfectly reliable with no polarising voltage (at least, modern devices are), and modern mixing consoles use them this way by the thousand (with bipolar supplies), without problems. Another myth is that a polarising voltage of 70% of the maximum rated voltage –or some other magic number– will minimise distortion and maximise lifetime. It is true that a polarising voltage will sometimes reduce distortion (see fig. 2.17), but the magic value is usually just a few volts. It is also entirely inconsistent and unpredictable.

The nominal lifetime of most electrolytics is quoted as 1000 hours, although more and more product lines are now boasting longer lifetimes. This still does not sound much, but it must be understood that this is the guaranteed life when running on the edge of their maximum ratings. By being kinder to them, lifetime is extended. A commonly quoted rule of thumb is to expect the working life to double for every 10°C lower in temperature that they are operated.[*] Cheap electrolytics are rated for a maximum ambient temperature of 85°C, assuming free air flow. However, in the claustrophobic interior of a valve chassis it is well worth considering 105°C rated devices which are not much more expensive, or even 125°C rated devices, which are a lot more expensive.

Electrolytics have high ESR compared with plastic capacitors, and it tends to be worse in devices with very low voltage ratings (e.g. <25V). It increases significantly at low temperatures, though this is not likely to be a problem in a valve amp.[†] At normal operating temperatures (i.e. above room temperature) the ESR is fairly constant, and if necessary can be estimated from the loss tangent / dissipation factor quoted on the datasheet (section 2.2.6). The loss tangent usually ranges from about 0.1 to 0.5 at 120Hz, corresponding to an ESR of a few hundred milliohms to a few ohms, but special low-ESR devices are also available. However, this figure is seldom needed for valve amp work, even in the power supply, as the ripple current rating is usually of more direct relevance.

---

[*] Though it is perhaps unrealistic to expect more than ten years of reliable operation.
[†] Electrolytics, like Goldilocks, prefer not to be too hot nor too cold.

All capacitors have some unavoidable self inductance which will resonate with the capacitance at a particular frequency, and for electrolytics this lies in the region of tens of kilohertz. Above this the impedance starts rising as the inductive

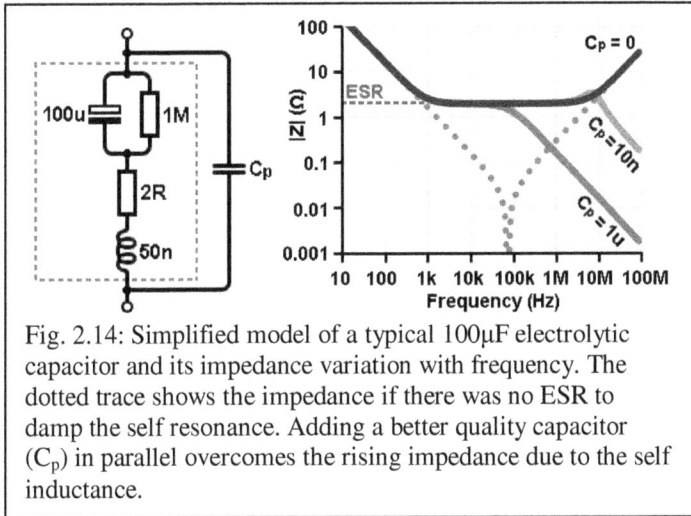

Fig. 2.14: Simplified model of a typical 100µF electrolytic capacitor and its impedance variation with frequency. The dotted trace shows the impedance if there was no ESR to damp the self resonance. Adding a better quality capacitor (C_p) in parallel overcomes the rising impedance due to the self inductance.

reactance takes over, so the device becomes less and less effective at passing high-frequencies. If the electrolytic had nothing *but* capacitance and inductance then at the resonant frequency the impedance would drop to nothing, but in practice there is always the ESR to damp the circuit, which also unfortunately causes the impedance to bottom out right in the audio range. To alleviate this effect it is common to place a plastic or ceramic capacitor in parallel with an electrolytic so that at high frequencies, where the electrolytic starts to give up, the smaller capacitor takes over and maintains a low impedance path. The effect on total impedance is illustrated in fig. 2.14. There is no particular rule governing the size of the second capacitor –a few tens of nanofarads or more will do. However, this technique does rather rely on the second capacitor itself having low inductance, which may not be the case in a hand-wired amp where the leads are not cut short, so in some designs one sees several parallel capacitors of ever diminishing value in an attempt to 'smooth out' the impedance response. Whether this has any beneficial effect on actual audio performance is questionable, but it is unlikely to do any harm.[†]

Of greater concern to us as audio designers is the fact that electrolytics generate significant distortion, just like cheap ceramics. Fig. 2.15 shows the effect of using an electrolytic as a coupling capacitor. The distortion is mainly second harmonic with some third, and it increases with the signal voltage across the capacitor. However, the most important thing to realise is that this distortion is significant *well above the cut-off frequency*. For example, a 10µF device gives a pole at 16Hz –already below the audio range– yet its distortion reaches 0.01% at 40Hz. Fortunately this also means the solution is a simple one: make the capacitance so large that the cut-off frequency is pushed into the sub-1Hz region, so the signal voltage across the

---

[†] However, in digital circuits >50MHz this can make things worse. In such cases it is better to use several identical capacitors in parallel to avoid multiple resonant effects.

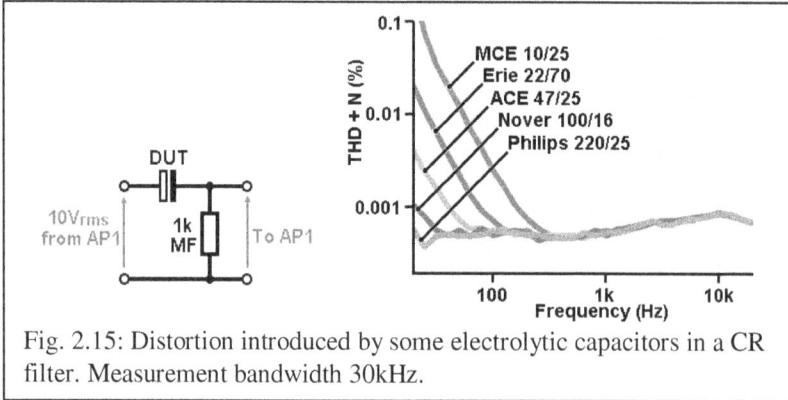

Fig. 2.15: Distortion introduced by some electrolytic capacitors in a CR filter. Measurement bandwidth 30kHz.

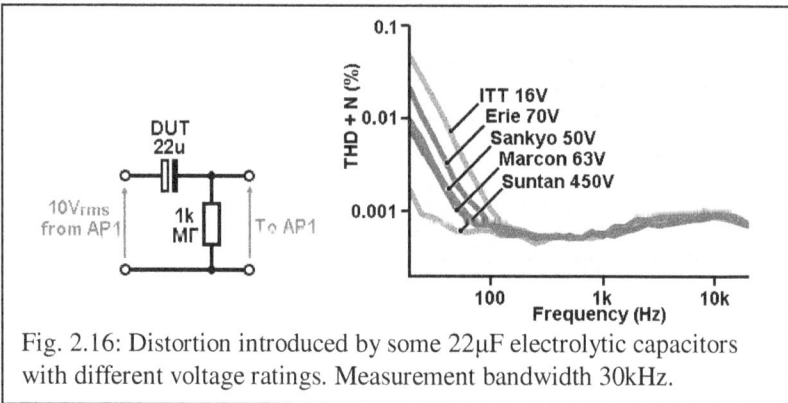

Fig. 2.16: Distortion introduced by some 22µF electrolytic capacitors with different voltage ratings. Measurement bandwidth 30kHz.

capacitor becomes negligible. As with other sorts of capacitor, distortion improves with higher voltage ratings, as illustrated by fig. 2.16.

Also, the second harmonic distortion is strongly affected by the DC bias voltage applied to the capacitor. For many (but not all) a bias of somewhere between zero and 10V will reduce the second harmonic, so total distortion improves[3] (which is good news for cathode bypass capacitors). Much larger bias voltages nearly always cause distortion to increase. This effect can be seen in fig. 2.17 where a 1µF device was tested with various bias voltages (provided by batteries to minimise noise); notice that there is almost an order of magnitude difference between the best and worst cases. Some audio designers add plastic capacitors in parallel with any electrolytic coupling capacitors in an attempt to improve distortion. Unfortunately, this has no effect unless the added capacitor is large enough to reduce the signal voltage across the electrolytic by a significant amount, i.e. the plastic device needs to have almost as much capacitance as the electrolytic, which presumably defeats the object.

---

[3] Bateman, C. (2002). Capacitor Sound 5, *Electronics World* (December), pp44-51.

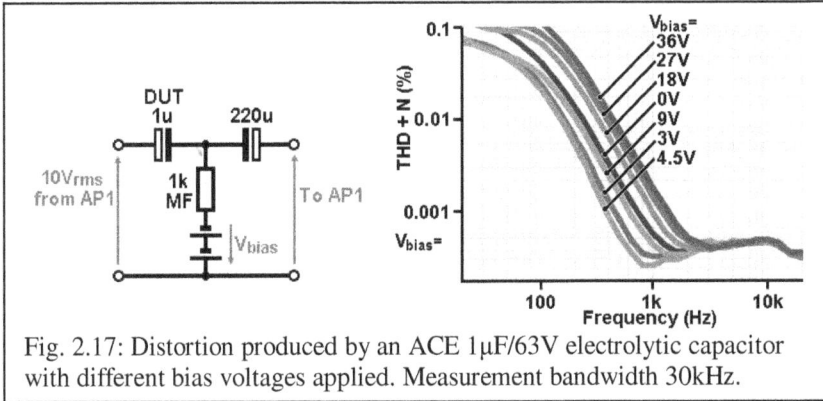

Fig. 2.17: Distortion produced by an ACE 1μF/63V electrolytic capacitor with different bias voltages applied. Measurement bandwidth 30kHz.

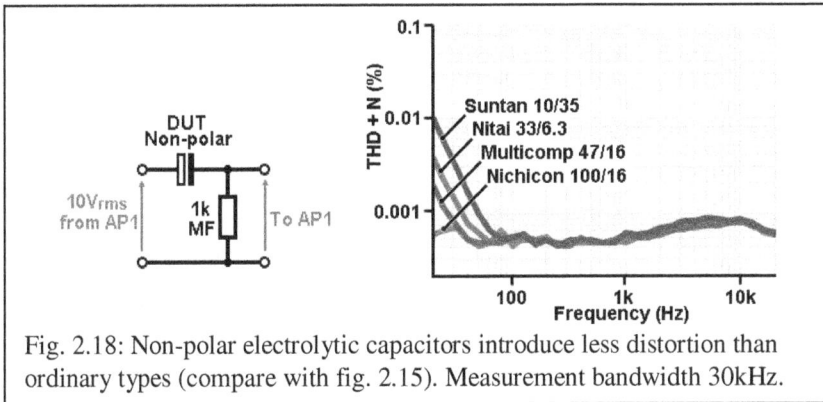

Fig. 2.18: Non-polar electrolytic capacitors introduce less distortion than ordinary types (compare with fig. 2.15). Measurement bandwidth 30kHz.

It is possible to buy non-polar electrolytic capacitors, which are effectively two ordinary electrolytics connected back-to-back in the same can. Whatever the voltage applied, one of the capacitors is always incorrectly biased, but since its reverse leakage current is limited by other one it cannot come to harm. They are less common and more expensive than ordinary electrolytics, but their distortion is about an order of magnitude lower, as illustrated by fig. 2.18. On the other hand, it is so easy to avoid electrolytic distortion by using a sufficiently large amount of capacitance that there are few cases where a non-polar is much of an advantage.

The rules for using electrolytic capacitors are simple:

- They must not be reverse biased;
- They must not be allowed to get too hot;
- They must never be used in circuits where the exact value of capacitance is a critical factor, e.g. time constants in filters (because electrolytics have poor tolerance);
- The cut-off frequency must be well below the audio range if distortion is to be avoided;
- If convenient, apply a polarising voltage of <10V.

85

2.2.15: Tantalum

Tantalum capacitors are a type of electrolytic capacitor (although when anyone talks of electrolytics they nearly always means the aluminium kind). Tantalum capacitors are made in a similar way to aluminiums, except the oxidised plate is made from tantalum. This rare metal forms a much thinner oxide layer than aluminium, allowing greater capacitance per unit volume. They are commonly used as power-supply decoupling capacitors in low-voltage circuits where their advantage is that they tend to have lower ESR and leakage current than aluminium types (though perhaps not as low as is sometimes implied). They also tend to have better tolerance and do not dry out with age. They are most common in values of a few hundred nanofarads to a few microfarads, and are restricted to fairly low voltage ratings. Anything larger than about 22μF and/or more than 50V is likely to be disproportionately expensive.

Tantalums generate serious distortion –about an order of magnitude worse than aluminium electrolytics– as shown in fig. 2.19. The distortion residual proved to be a decaying series of second, third and fourth harmonics. They also have a nasty habit of exploding violently at the slightest sign of improper use, whereas aluminium electrolytics will usually survive a brief voltage overloads or even reverse polarity. These caveats make tantalums of limited use in valve amplifiers.

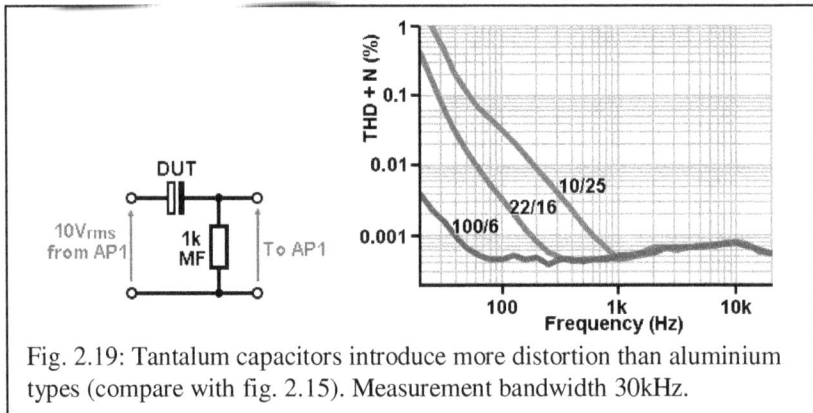

Fig. 2.19: Tantalum capacitors introduce more distortion than aluminium types (compare with fig. 2.15). Measurement bandwidth 30kHz.

2.3: Inductors

Together with resistors and capacitors, inductors complete the holy trinity of passive components. However, inductors are far less common in low frequency circuits since the inductance must be fairly large to be useful, making them physically bulky and expensive –often they must be custom made. They are also exceedingly sensitive to picking up hum, and highly non-linear. Yet for reasons that are rather inscrutable, inductors and transformers have a cult following of audiophiles while capacitors seem to have no such disciples. This is quite irrational, as inductors are tragically inferior to capacitors in terms of linearity and just about everything else for that matter. Perhaps it is their general rareness that gives them exclusive appeal.

With capacitors, *any* material inserted between the plates will result in greater capacitance than with air/vacuum alone, so there is a huge variety of practical dielectric materials to choose from. With inductors cores the story is quite different; most materials have no significant effect on inductance compared with air/vacuum. The only substances that really make a difference are nickel, cobalt, and iron.

The most common materials used for inductor cores are:
- Air (plastic and other non-iron materials also count as air); practical up to a few millihenries although larger ones may be found in loudspeaker crossovers, at great expense;
- Ferrite (ceramic and iron composite); very common in values up to about 10mH while larger values are usually specialist items;
- Silicon steel; used in low-frequency transformers and power chokes.

Since inductors are not common components it is difficult to make universal statements about their ratings and uses; you really must consult the specific datasheet. The one rating that will certainly be given is the maximum DC current. This may be limited by the wire gauge (i.e. fusing) or by the need to keep the inductance within a certain tolerance. Graphs may also be provided showing the variation of inductance with DC current. Inductors are often used in valve amplifiers in the form of power supply chokes, and such chokes will also have a voltage rating which is the maximum that can safely be sustained between the conductor and the core/frame. It is difficult to make high inductances in the presence of large DC currents, so chokes also tend to have very poor tolerances, perhaps as bad as ±50%.

2.3.1: Distortion

To give an idea of the amount of distortion that might be encountered in an inductive circuit, three similar inductors were tested:
- A ferrite-core inductor used in the passive equaliser of an AMS-Neve 1073 microphone preamplifier. DC resistance 113Ω.
- A small, E/I-laminated power choke (with air gap). DC resistance 35Ω.
- A very old interstage coupling transformer (ungapped), using the secondary winding for the test, with the primary open circuit. DC resistance 30Ω.

There is nothing very special about these components; they were chosen simply because they all gave readings close to 1 henry on a digital inductance meter.

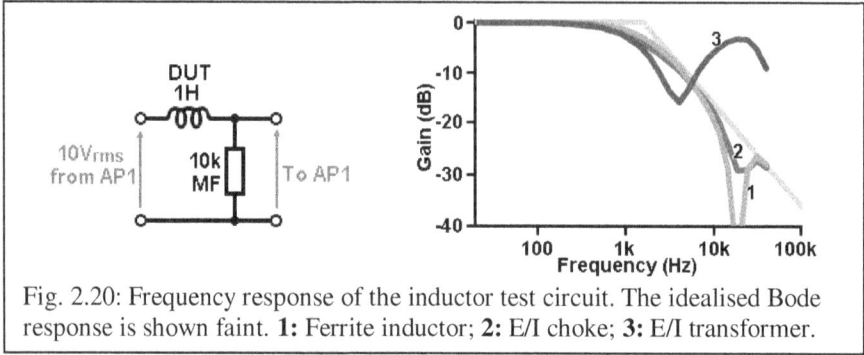

Fig. 2.20: Frequency response of the inductor test circuit. The idealised Bode response is shown faint. **1:** Ferrite inductor; **2:** E/I choke; **3:** E/I transformer.

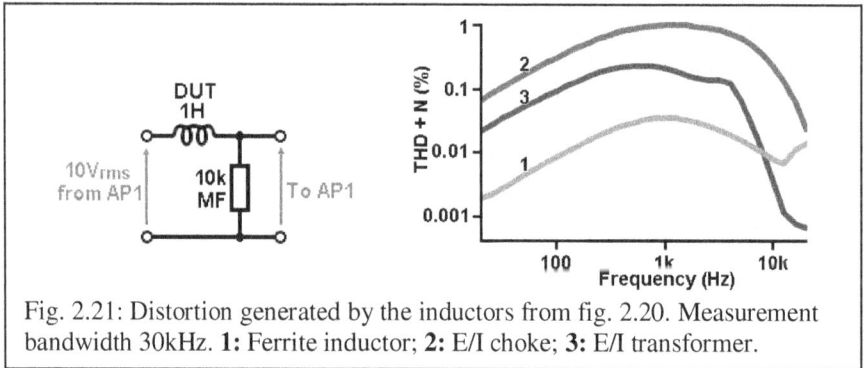

Fig. 2.21: Distortion generated by the inductors from fig. 2.20. Measurement bandwidth 30kHz. **1:** Ferrite inductor; **2:** E/I choke; **3:** E/I transformer.

Fig. 2.20 shows the frequency response of the three devices when used in a simple LR filter, indicating they are indeed all close to 1H (the notch is caused by shunt capacitance resonating with the inductance). The notch for the interstage transformer is much lower in frequency since the shunt capacitance on the unused winding is reflected back through the device, resulting in a much larger effective capacitance on the used winding.

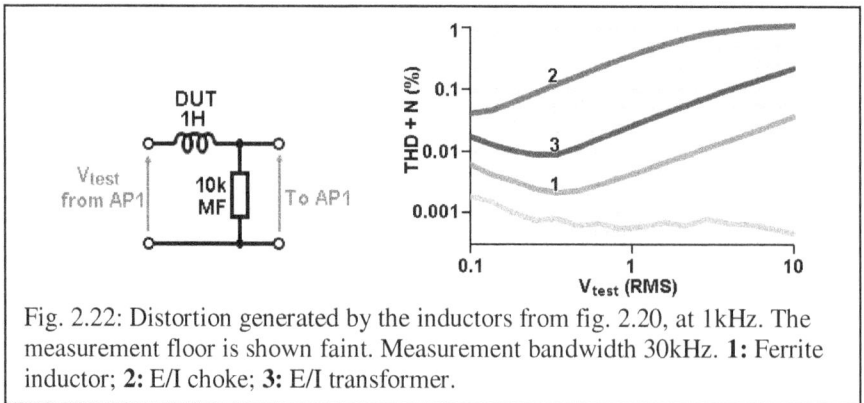

Fig. 2.22: Distortion generated by the inductors from fig. 2.20, at 1kHz. The measurement floor is shown faint. Measurement bandwidth 30kHz. **1:** Ferrite inductor; **2:** E/I choke; **3:** E/I transformer.

Fig. 2.21 shows distortion versus frequency. It is unimpressive in all devices, although it is reassuring to see that the ferrite inductor is superior, since it was designed for small-signal use. Curiously, the gapped-choke produced the worst distortion. This and the absence of a deep notch in its frequency response suggest it suffers from unusually high core loss. In general, distortion in inductors is directly proportional to level, as exhibited in fig. 2.22 (the rise at low frequencies is due to the rising noise floor of the system), although the power choke shows signs of levelling off, for reasons unknown. In all cases the distortion was dominated by the third harmonic followed by a decaying series of the other odd harmonics at much lower level.

2.3.2: Parasitics

An inductor can be modelled accurately enough for most purposes by including a series resistance representing the winding resistance, and a shunt capacitance representing the various distributed capacitances between the turns and across the whole device, as in fig. 2.23. The power lost in the winding resistance is usually referred to as the 'copper loss' or 'I^2R loss', and the power wasted through eddy currents/hysteresis as the 'core loss' or 'iron loss'. Core loss can be

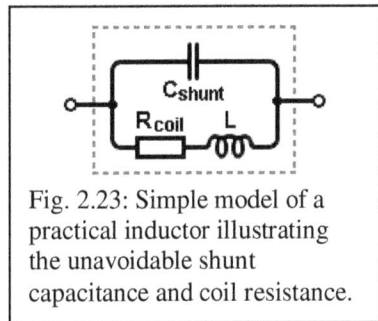

Fig. 2.23: Simple model of a practical inductor illustrating the unavoidable shunt capacitance and coil resistance.

modelled as a resistance in parallel with the device, but can often be ignored for audio-frequency work.

2.3.3: Ferrite Beads

In some circuits it can be beneficial to pass a wire through a ferrite bead or wrap it a few times around a ferrite ring, to suppress radio frequency oscillation or interference. The ferrite is designed so that at low frequencies it behaves as an inductance of a few hundred nanohenries, but at high frequencies (several megahertz) it behaves as a resistance of a few tens of ohms and so damps resonances by absorbing RF energy and converting it into heat. Since the inductance is very small it will have no effect on audio distortion, despite having a non-linear core.

2.4: Power Transformers

Power transformers have a relatively simple job to perform and there is very little that the circuit designer needs to know in order to use them. The principle concern is that they provide the desired secondary voltages with sufficient current capability for the task. This is discussed in more detail in chapter 11.

Power transformers are invariably of the laminated E/I type or doughnut-shaped toroidal type. Fig. 2.24 shows some typical examples. Smaller E/I transformers are usually PCB or U-clamp mounted while larger ones will often have end frames allowing them to be mounted in any orientation. Some come with metal shrouds for complete electric shielding and added fire proofing (common in the US/Canada). Toroidal transformers are more ideal than E/I types; they have less leakage flux (i.e. less radiated noise and higher efficiency), less winding resistance, and are smaller for a given power rating, so unsurprisingly they are often (but not

Fig. 2.24. Various types of transformer.
a: E/I with shroud (foot mounting);
b: E/I with end frames;
c: E/I with U-clamp;
d: Toroidal.

always) more expensive. On the other hand they tend to have higher capacitance between primary and secondary than E/I types, which may allow noise from the mains to leak into the audio circuit more readily.

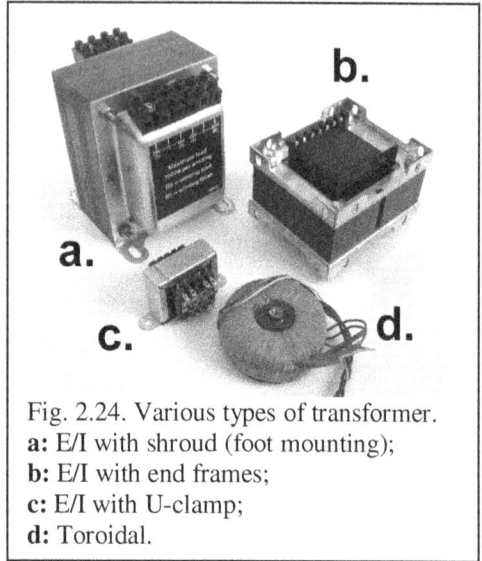

2.4.1: VA (Power) Rating

The VA rating of a transformer is equal to the full-load RMS secondary voltage multiplied by the permitted full-load RMS secondary current, and is therefore the average power that that can be safely demanded from the transformer. It is given in volt-amps rather than watts because it takes into account the fact that the load may be reactive rather than purely resistive. The most familiar example of this is a capacitor-input rectifier. The total power (in VA) that the transformer has to handle is then the magnitude of the real power dissipated in the load, combined with the reactive power flowing back and forth in the capacitance. This is explained in more detail in chapter 11.

The overall VA rating of a transformer can be reliably estimated from the size of its core. Modern transformers, especially, come in a limited range of standard core sizes, and devices of the same physical size will nearly always have the same total VA rating. However, it should perhaps be pointed out that the power limitations are mainly due to the thickness and resistance of the copper wire (I^2R losses) and not to the core itself as is sometimes supposed. Powerful transformers have big cores not because they intrinsically need so much iron, but because they need to use thick wire which takes up a lot of winding space.

If the device has only one secondary or several identical secondaries then the total VA rating is divided equally among them, so the current capacity of each winding is equal to the apportioned VA divided by the voltage of the winding in question.

Transformers with several different secondary windings might give the rating of each winding in terms of VA, or as separate RMS voltage and current ratings. But just to confuse matters, power transformers intended specially for valve amps often use a more old fashioned ratings system. Typically, AC RMS figures are quoted for the heater winding/s as usual, since they are assumed not to use a rectifier, but the manufacturer then second-guesses the user by quoting the maximum *DC* load current for the high-voltage winding, assuming a standard full-wave rectifier and a 'typical' (but unstated) power factor. If this is in any doubt then the manufacturer should be contacted.

2.4.2: Regulation

Some transformer datasheets will quote a 'regulation' figure (something of a misnomer). This is a measure of the change in secondary voltage between no load and full load, the difference being caused mainly by the voltage drop across the winding resistance as the load current increases. In Britain the regulation is given as a percentage according to:

$$\%_{regulation} = 100 \times \frac{V_{no\ load} - V_{full\ load}}{V_{no\ load}} \tag{2.3}$$

Fig. 2.25 shows the range of regulation to be expected from ordinary mains transformers. Why this regulation figure became the standard metric rather than the simple ratio of the no-load and off-load voltages is anyone's guess. Perhaps it was handy in the days before pocket calculators and circuit simulators. These days it would really be more helpful if manufacturers simply quoted the winding resistances and the no-load and full-load voltages. In theory, the secondary-referred winding impedance

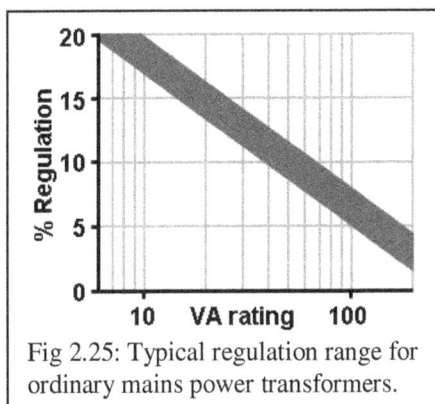

Fig 2.25: Typical regulation range for ordinary mains power transformers.

can be calculated from the regulation percentage:

$$Z_{sec} = \frac{\%_{regulation} \times V_{full\ load}}{(100 - \%_{regulation}) \times I_{full\ load}} \tag{2.4}$$

But since regulation percentages is always quoted in suspiciously round numbers, any calculations using them must be assumed to be approximate only.

2.4.3: Phasing

When windings are connected in series it is important to get their phasing correct so the voltages add in phase. If the phasing is incorrect then the two voltages will cancel out uselessly (actually there are some special cases where this *is* useful – see section 11.10). Similarly, when windings are connected in parallel they *must* be identical and in phase. Getting this wrong will effectively short-circuit the transformer with disastrous results. Windings of different voltages must not be connected in parallel under any circumstances. The relative phasings of windings is usually indicated by dots on the transformer diagram; each dot indicates the end of the winding that is in phase with all the other dots. In other words, if the voltage happens to be increasing at one of the dots, then it is increasing at all of the other dots too.

Many off-the-shelf devices have dual primary windings which can be connected in series for 230/240V mains or in parallel for 115/120V mains, as shown in fig. 2.26. The most common, modern transformers also have dual (identical) secondary windings which can be left entirely separate, or be connected in series to obtain double the voltage or to generate a bipolar supply, or be connected in parallel to obtain a single voltage with maximum current. Again, it is essential to observe the correct phasing when doing this.

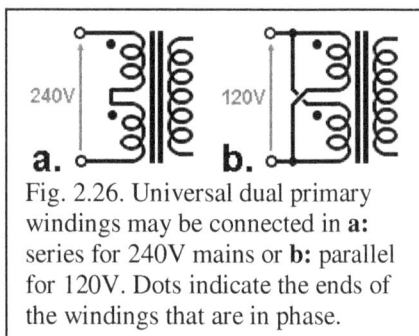

Fig. 2.26. Universal dual primary windings may be connected in **a:** series for 240V mains or **b:** parallel for 120V. Dots indicate the ends of the windings that are in phase.

2.5: Small-Signal Transformers

Small-signal transformers are common in valve circuits for converting low impedances such as speakers and headphones into high impedances that can be driven more easily by the valves. They are also the traditional method of boosting tiny voltage signals such as those from a microphone of phono pickup, with minimal noise degradation. It is now possible to perform this task equally well and at a fraction of the cost using solid-state circuits, but transformers are still used when total galvanic isolation is needed, and of course when they are deemed to contribute pleasing aberrations to the sound (there is still a healthy market for recording equipment using transformers, for exactly this reason). Nevertheless, at the low voltages –i.e. low flux densities– used with small-signal transformers, linearity tends to be quite good, although it is essential that no DC current flows in the windings as this will quickly lead to saturation unless the device has been specifically designed for it, i.e. the core is gapped. Bandwidth gets worse as the turns ratio increases, and is also a strong function of the load impedance, winding geometry and general component quality. Any statement of bandwidth that does not specify the source and load impedances is entirely meaningless.

Small-signal transformers are very sensitive to picking up hum, especially from the magnetic fields of nearby power transformers. A high-quality device will be enclosed in a mu-metal can which provides magnetic shielding (sometimes this costs as much as the transformer itself), and it should also be grounded to provide electric shielding. Other metals –including steel– are fairly useless for magnetic shielding and really only provide electric shielding.

For small-signal applications we are rarely interested in power handling, although a limit may be specified in the datasheet nonetheless; what matters is impedance transformation and turns ratios, that is, how much boost or attenuation the device provides.

2.5.1: Turns and Impedance Ratios

At the very least, the manufacturer will provide the turns ratio –which may also be called the voltage ratio– and a winding diagram of the device. The turns ratio is the voltage gain that the transformer provides. In many cases there will be dual primaries or secondaries so different turns ratios can be obtained by series or parallel connections.

The impedance ratio is the square of the turns ratio. Thus if a transformer has a step-up turns ratio of 4, it has an impedance ratio of $4^2 = 16$. In other words, impedances on the primary will be reflected across to the secondary and appear magnified by 16 times when looking back into the secondary. Likewise the impedances on the secondary will be reflected back to the primary and appear 16 times smaller when looking into the primary. Conversely a step-*down* transformer will make impedances on the secondary appear larger on the primary, and *vice versa*. Although a transformer may be primarily advertised as step-up or step-down there is no reason why it can't also be used in reverse, as the application demands.

2.5.2: Source and Load Impedances

A good datasheet will specify a source impedance. This is the maximum recommended source impedance when used in an audio system and should also be the value used by the manufacturer when obtaining other performance data. If the device is used with a different source impedance then it may or may not meet the quoted specifications. In particular, increasing the source impedance will result in less low-frequency bandwidth and worse distortion (see fig. 2.30). The winding resistances may also be quoted, and a very good datasheet will also provide the primary inductance (measured with all other windings open circuit) and the primary and secondary leakage inductances (measured with all other windings shorted).

The datasheet will also quote the input impedance of the primary when the secondary is loaded with a recommended load resistance. The input impedance of the primary is the combination of primary inductance and wire resistance, and the secondary load which is reflected back to the primary (just to confuse matters this primary input impedance is sometimes ambiguously called the 'load impedance' of the transformer). A good transformer will be designed so that the reflected impedance dominates the total, i.e. the primary inductance will be large enough and the wire resistance small enough that they have little effect on the primary impedance within the audio band.

With only a resistive secondary load a transformer will often produce a resonant peak in the frequency response at high frequencies. To counter this, an RC Zobel network will often be recommended in parallel with the secondary load resistance. This is usually designed to give a near-Butterworth frequency response from the whole circuit (see fig. 2.29), and this in turn makes the primary impedance more constant with frequency. These loading recommendations do not *have* to be obeyed but it may be difficult to improve on the manufacturer's recommendations. However, if the transformer is unknown then it is worth experimenting with RC values while feeding in a 1kHz square wave from a representative source resistance, and monitoring the secondary waveform on an oscilloscope. The Zobel network can then be tuned to obtain an overshoot but no undershoot (no ringing), or perhaps no overshoot at all. In general, the greater the turns ratio and/or the higher the impedances the transformer is designed for, the more heavy-handed the Zobel network will need to be to get a flat frequency response.

2.5.3: Parasitics

Looking in textbooks you soon discover that there are several variations on the equivalent circuit for a transformer. The model in fig. 2.27 is probably the most useful as it includes enough elements to represent most of a transformer's defects. Each winding includes wire resistance and shunt capacitance. In reality these are distributed across each and every part of the windings, but for reasons of practicality they are lumped into single components in the transformer model. The position of C_{pri} varies in different textbooks,[4] but I find the model matches reality most usefully when it is in the position shown. There is also interwinding capacitance between primary and secondary, but it is normally so small compared with the other winding capacitances that it can be lumped in with them, so it is shown faint in fig. 2.27. Many transformers include an inter-winding screen which can be grounded to eliminate this capacitance.

There is also the primary inductance, and primary and secondary leakage inductances. Leakage inductance is that measured across a winding with all other windings shorted. It results from the small portion of magnetic flux generated by the driven winding which fails to link with any other windings, because it leaks out into the air or otherwise manages to skip them on its circuitous journey around the core.

Core loss can be modelled by adding a resistor in parallel with the primary inductance, but small-signal transformers are so efficient that this is normally ignored. The primary and secondary equivalent circuits are linked by a theoretically *ideal* transformer which converts voltages on its primary to voltages on its secondary according to its turns ratio N_2/N_1. It has been drawn faint to emphasise the fact that, being ideal, it has *infinite* impedance and therefore does not load the rest of the circuit. It is well worth playing with this model in a circuit simulator to get a feel for how the various components affect circuit behaviour. The model works for power transformers too, of course.

---

[4] For example: Langford-Smith, F. (1953). *Radio Designer's Handbook*, 4th ed. Iliffe & Sons Ltd., London, p204.

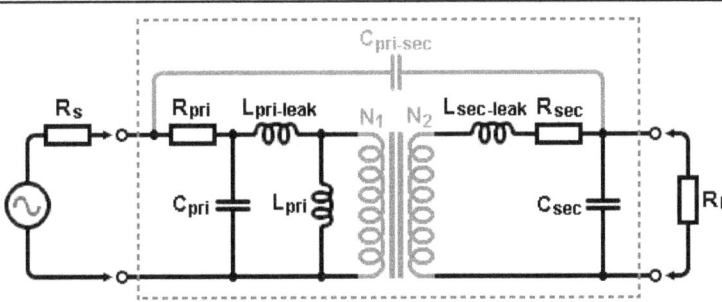

Fig. 2.27: Practical transformer model.
R_s = External source resistance
R_{pri} = Primary winding resistance
C_{pri} = Primary capacitance
$L_{pri\text{-}leak}$ = Primary leakage inductance
L_{pri} = Primary (self) inductance
$L_{sec\text{-}leak}$ = Secondary leakage inductance
R_{sec} = Secondary winding resistance
C_{sec} = Secondary capacitance
$C_{pri\text{-}sec}$ = Primary-to-secondary capacitance
R_l = External load resistance

This equivalent circuit is visually intuitive because it shows all of the parasitic elements separately and in their logical places. It is, however, not so intuitive from the point of view of circuit analysis. It can be simplified by referring all of the impedances on one side to the other side, eliminating the ideal transformer in the process. This allows us more easily to determine the frequency/step response of the system. It is most common to refer the secondary impedances to the primary. This is done by dividing them by the impedance ratio, which is the square of the turns ratio: $(N_2/N_1)^2$. Also note that dividing the *impedance* of a capacitor by $(N_2/N_1)^2$ amounts to multiplying its *capacitance* by $(N_2/N_1)^2$. This leads to the circuit in fig. 2.28, and the system can now be viewed broadly as two RLC filters in cascade.

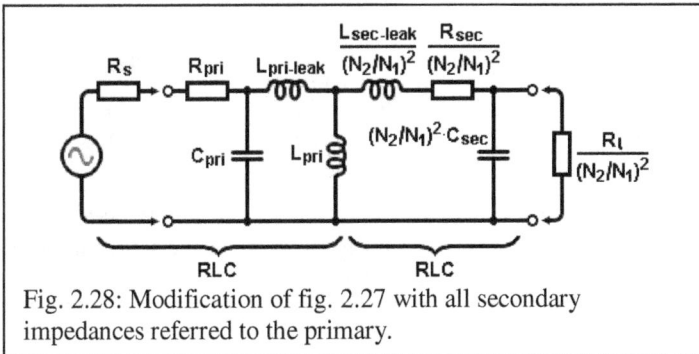

Fig. 2.28: Modification of fig. 2.27 with all secondary impedances referred to the primary.

95

2.5.4: Distortion

To give some idea of the distortion behaviour of signal transformers, a microphone input transformer was tested. This device was made for AMS Neve by Vigortronix and has a step-up turns ratio of 4 (with its two secondaries connected in series) and is designed for a maximum nominal source impedance of 200Ω with a 20kΩ secondary load. Fig. 2.29 shows the frequency response of the test circuit, and it is clear that the manufacturer's recommended Zobel network does a good job of eliminating the resonant peak at 60kHz. Since the turns ratio is 4, the voltage gain should be ×16 or 12dB, but in this case 2dB is lost across the source impedance and winding resistances.

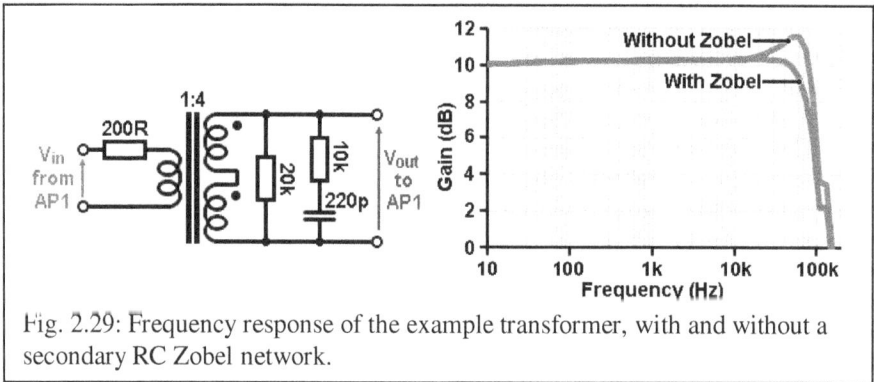

Fig. 2.29: Frequency response of the example transformer, with and without a secondary RC Zobel network.

Fig. 2.30 shows the distortion performance with a constant output voltage of 3.8V$_{rms}$ (the voltage directly across the primary winding was 1V$_{rms}$). Distortional falls as frequency is increased, eventually disappearing into the noise floor around 0.001%. However, if the source impedance is increased then the distortion suffers greatly. This is characteristic of all transformers.

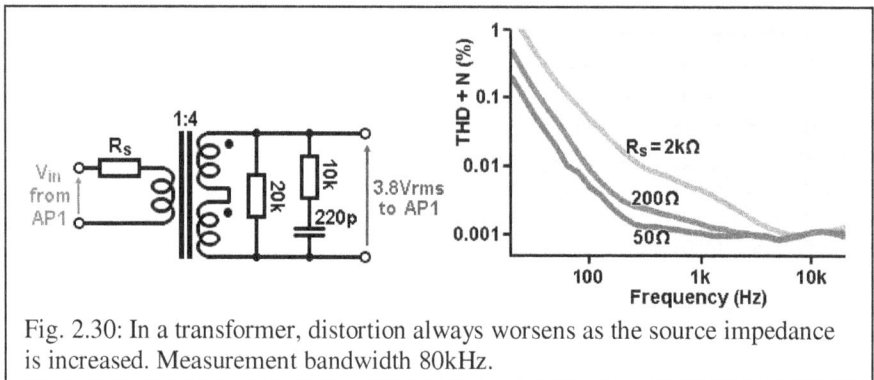

Fig. 2.30: In a transformer, distortion always worsens as the source impedance is increased. Measurement bandwidth 80kHz.

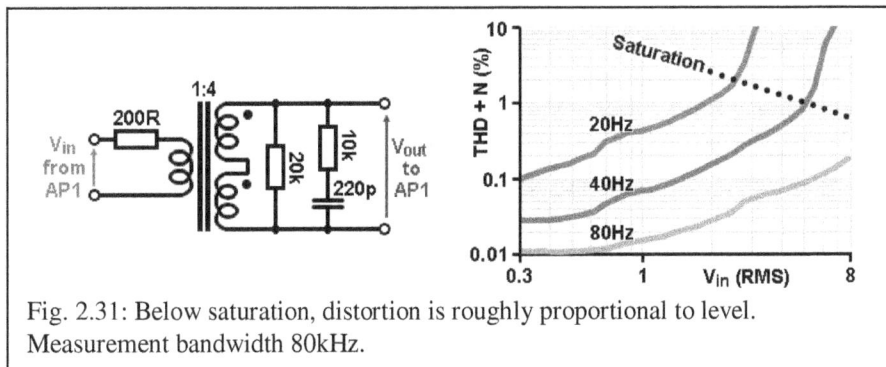

Fig. 2.31: Below saturation, distortion is roughly proportional to level. Measurement bandwidth 80kHz.

Fig. 2.31 shows how distortion varies with level. Under normal working conditions the distortion is proportional to level, but with a sufficiently large primary voltage the core will begin to saturate and distortion skyrockets. Also note that if the frequency is doubled then it takes twice the primary voltage to cause saturation. Below saturation the distortion was mainly third harmonic with some second harmonic varying with level. Close to saturation the third harmonic increased so much that any second harmonic bacame insignificant by comparison. At even higher levels the other odd harmonics rose out of the noise floor. The distortion in these figures may seem excessive but it should be noted that a typical microphone output is of the order of a few millivolts rather than volts, so in normal use the distortion will be proportionately lower, probably below 0.01% even at 20Hz.

2.6: Valves

The sort of valves used for small-signal and audio work are classically referred to as **receiving valves** because they were used mostly in radio receivers. The alternative type, **transmitting valves**, are naturally designed for radio transmitting duty and are typically very high power but not very linear, so are not used for audio preamp work. Receiving valves can be broadly divided into two groups: preamp types intended mainly for voltage amplification, and power-amplifier types intended to deliver several watts of output power. However, there is no strict definition of either group, and no clear division between them. In general, valves with ≤2.5W heaters are considered to be preamp types, but either type can in theory be used for either purpose, though it may not be economical to do so!

Critics will sometimes deride certain valve types, claiming that they are not really 'audio valves', presumably implying that they are not inherently linear enough to satisfy their tastes. In truth very few valves were ever designed specifically for audio use, and those that were are often not much more linear than the general-purpose types. Indeed, most 'audio valves' appear to have been designed mainly with low microphonics and heater hum in mind, rather than exceptional linearity. Also, manufacturers were fairly inconsistent with their advertising. Many valves were variously recommended for UHF, RF, AF, general purpose or any mixture of uses

97

depending on the market being targeted at the time, and it is often questionable what the 'original' intended purpose was. Audio engineers may not like to admit it, but dealing with low frequencies is far easier than dealing with radio and higher frequencies, and the majority of

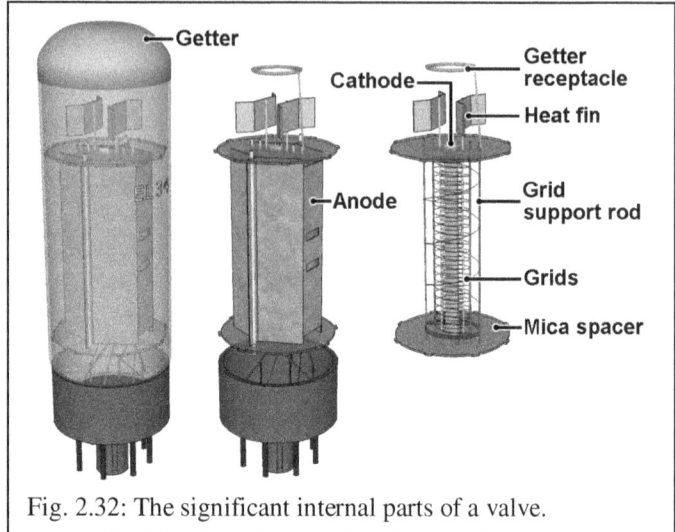

Fig. 2.32: The significant internal parts of a valve.

devices that work well for radio will by extension work well for audio too.

Valves designed specifically for radio frequencies are usually optimised for low interelectrode capacitance. The anode connection, in particular, may be brought out through a top cap. Thick lead wires and multiple electrode connections also reduce lead inductance. RF pentodes may also be optimised for low screen current to minimise partition noise and maximise efficiency. None of these things are detrimental for hi-fi work either. However, RF valves may have poorer microphonics, heater hum, and flicker noise, as these are low-frequency effects of little interest to radio engineers. Nevertheless, almost any valve can be pressed into audio service by the competent circuit designer, and these days a more healthy attitude is to say that there is no such thing as an 'audio valve'; ingenuity and necessity will make a successful amplifier circuit out of any bottle. If good linearity is desirable then sensible choice of devices, operating points and negative feedback are required, while if the warm 'valve sound' is what is desired, anything goes.

Fig. 2.32 shows the internal construction of a typical valve (in this case an EL34 power pentode, but preamp valves are constructed in the same way), and the important elements will now be described.

2.6.1: The Anode

The anode is usually made from sheet nickel or nickel-coated steel, stamped and pressed into a box or roughly cylindrical shape. It may also be blackened with graphite to reduce secondary emission and to improve heat radiation. Cooling fins are often added to the anodes of power valves, while really powerful valves may have anodes of solid carbon.

Most of the heat in a valve comes from power dissipated by the heater and anode, which can only escape by conduction through the pins and by radiation to the glass envelope, and then by radiation and convection to the outside air. Low-current preamp valves will run quite cool, but those dissipating more than a couple of watts may reach an envelope temperature of around 200°C.

Cooling is most effective when the valve is mounted vertically, especially if there is free air flow from the base. If the valves are mounted on the top of a chassis then a ring of holes drilled around the valve base can be beneficial in creating a natural up-draught or chimney effect. Corrugated valve coolers like that in fig. 2.33 work similarly and are remarkably effective. Some datasheets (especially for rectifiers) will forbid certain mounting positions as it can lead to the internal parts sagging under their own weight, but this is unlikely for preamp valves. When positioning valves, a rule of thumb is to separate them by a distance at least equal to their own diameter, to prevent them from heating each other up.[5] If closer spacing is required then forced-air cooling may be necessary.

Fig. 2.33: Valve with corrugated valve cooler fitted. Holes around the socket encourage updraught as well as allowing multiple mounting positions.

2.6.2: The Grids

A grid is made by winding a helix of wire concentrically around the cathode. The control grid is the one closest to the cathode and, to a first approximation, the closer it is to the cathode (relative to the anode) the higher the μ of the valve, and the closer the pitch of its winding, the higher the g_m. A similar principle holds for the screen and other grids if present.

The grid is usually wound from fine molybdenum wire which has a fairly high work function and therefore does not emit electrons easily; this is important for minimising grid current. In special-quality devices it may be gold-plated to increase the work function still further. In most valves the grid is wound as an elliptical helix on a pair of copper or nickel support rods (which also act as heat sinks) which slot into mica spacers at each end. In power valves, extra fins may be welded to the rods to improve heat radiation. Most grid heating is due to radiation from the cathode and anode.

---

[5] Voorhoeve, N. A. J. (1953) *Low-Frequency Amplification*, Philips Technical Library, Eindhoven. p88.

Towards the end of the valve era more and more valves (mainly television types) adopted 'frame grids'. A frame grid has a pair of metal supports welded across the ends of the usual grid support rods to create a rigid, rectangular frame. This allows extraordinarily fine grid wire to be wound tightly across the frame, creating a flat or planar grid with better tolerances, which in turn allows the grid to be placed much closer to the cathode.[6] Frame-grid valves therefore tend to have high μ and much higher g_m than traditional types, and they can be more linear too, due to the more ideal parallel construction (the ECC88/6DJ8 is a popular example). However, despite having a more rigid frame, these valves suffer acutely from microphonics since even the tiniest movement of the grid amounts to a significant change in relative electrode spacings.

2.6.3: The Cathode

In early valves the heater and cathode are one and the same, so they are called **directly heated** valves. In the very earliest types the cathode –more properly called the **filament** in such devices– was a simple tungsten wire which had to be heated to around 2500K for satisfactory electron emission. These valves glowed as bright as light-bulbs and were therefore also called **bright emitters**.

Later valves used **thoriated-tungsten** filaments. These had thorium oxide mixed with the tungsten to reduce the work function of the metal, so it only had to be heated to about 2000K. Thoriated-tungsten cathodes are quite fragile and such valves must be handled with care; they also benefit from soft-start heating circuits. Power triodes such as the 2A3, 300B, 845 and many transmitting valves fall into this category and are still popular among those valve aficionados who can afford them.

The third stage of development was the **indirectly heated** cathode in which the cathode is heated by an entirely separate heating element. The cathode is made from a nickel tube which is coated with a proprietary mixture of barium and strontium carbonates. This mixture has a very low work function and only needs to operate at about 1000 to 1100K (the heater in turn must operate at about 1500K to achieve this). The heater is again a tungsten wire, insulated with aluminium oxide and inserted into the cathode tube. This has many advantages over direct heating, including reduction of hum and the freedom to operate many valves from the same heater supply. Not surprisingly, indirectly-heated valves became the standard receiving type from the 1930s onwards (except for some rectifiers), and are the only valve type considered in this book.

The life of the cathode is ultimately limited by the evaporation of barium from the coating, which is an unavoidable consequence of actually using the valve. However, this lifetime is potentially hundreds of thousands of hours, far longer than many valves last in practice. This is because there are other processes besides evaporation which cause performance to degrade.

---

[6] Mullard Ltd. (1960). Frame Grid Valves. *Wireless and Electrical Trader*, 8 October.

Once heated, the cathode emits electrons. The maximum amount of current that can be 'sucked' from the cathode is called the **saturation current** or **emission current** and is dependent on the cathode temperature. At normal operating temperature the saturation level is far in excess of the normal working currents, so not all the electrons emitted by the cathode are immediately 'used' for anode current; they instead build up in a cloud around the cathode. This cloud is called the **space charge** and acts like a reservoir from which the varying electron stream (anode current) flows, while the cathode keeps the reservoir filled.

During use, residual gas molecules that happen to be floating around may cross the path of the electron stream and electrons may be knocked off them, creating positive ions. These ions will themselves be repelled by the positive anode and accelerate towards the grid and cathode. If they reach the grid they will constitute a reverse grid current (section 2.6.17), while if they reach the cathode they will attack its surface in two ways:
- By chemical reaction; this reduces the work function of the oxide coating and is called **cathode poisoning**;
- By brute-force impact; this destroys the coating and is known as **cathode sputtering**, or more loosely as **cathode stripping**.[*]

Not surprisingly, this is bad news for the cathode, whose emissive properties will be degraded. Fortunately the space charge is negative and repels such ions, acting like a protective blanket around the cathode. Therefore, a receiving valve must not be operated at full saturation since there will then be no protective space charge. The only time when saturation is allowable is at power-on when the cathode is just warming up, at which point the cathode is fairly immune to chemical attack since it is not yet very hot, and because there will be very little anode current to ionise the gas in the first place. A very aged or poisoned cathode can sometimes be 'rejuvenated' to some degree by running the heaters at roughly 150% of rated voltage while simultaneously drawing a heavy anode current for a few seconds. The improvement may not be permanent, however.

Some users prefer to pre-heat the cathode for a few seconds, allowing the space charge to build up before applying the anode voltage. This is a double-edged sword, however, as it can inadvertently lead to the growth of **interface resistance** if the valve happens to be left in this 'standby' mode for long periods. If the cathode is heated but no anode current is drawn for some hours, a pernicious layer of barium-orthosilicate may grow between the nickel tube and the oxide coating. This interface resistance acts like an unbypassed cathode resistor, reducing the available g_m of the device even though it may test normal for emission. It also greatly increases flicker

---

[*] True cathode stripping occurs when the anode voltage is high enough to tear negatively-charged pieces of the oxide coating away from the cathode. However, this requires a field strength in excess of 4MV/m, well above anything found in receiving tubes. Herrmann, G. & Wagener, S. (1951). *The Oxide-Coated Cathode*. Chapman & Hall Ltd., London. p111

noise.[7] Worse still, once formed this layer cannot be removed. It was once said that the useful life of a valve could be almost completely defined by its rate of growth,[8] so late generation computer valves which *had* to spend much of their time in cut-off were built with high-purity cathodes which virtually eliminated the problem, but this is not likely to apply to popular audio types.

The safest way to avoid this alternative form of cathode poisoning is simply to ensure that at least *some* anode current flows whenever the cathode is heated. A reduction of heater power also provides a considerable improvement.[9] If you have no choice but to leave the valves in standby then reduce the heater voltage to perhaps 80% of its normal value, and allow a trickle of anode current to flow at all times.

2.6.4: The Heater

In indirectly-heated valves the heater is made from tungsten wire which is folded or coiled, and coated with aluminium oxide to insulate it from the cathode. The heater may further be twisted to create a double helix which improves the cancellation of its magnetic field and so reduces hum.[10] The thickness of the insulation determines how much voltage can be tolerated between heater and cathode before too much current leaks between them, leading to excessive hum and noise –see section 2.6.13.

The heater is a very simple element and can be likened to an ordinary light bulb, although it operates at much lower temperature. For interest, fig. 2.34 shows

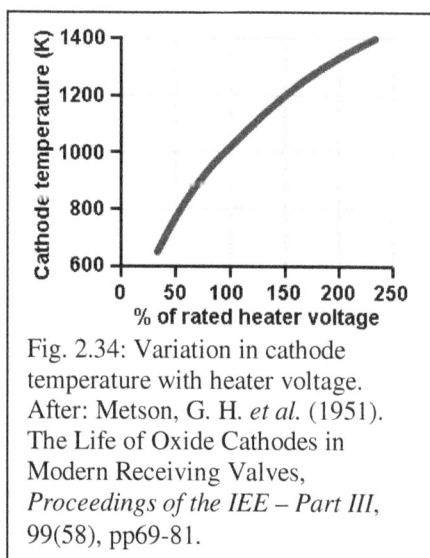

Fig. 2.34: Variation in cathode temperature with heater voltage. After: Metson, G. H. *et al.* (1951). The Life of Oxide Cathodes in Modern Receiving Valves, *Proceedings of the IEE – Part III*, 99(58), pp69-81.

how the cathode temperature varies with applied heater voltage. Usually the heater is omitted from the valve symbol used on circuit diagrams since it is assumed that the reader will provide the necessary heater power without having to be told. After all, the valve won't do very much without it!

---

[7] Lindemann, W. W. & Van der Ziel, A. (1952). On the Flicker Noise Caused by an Interface Layer, *Journal of Applied Physics,* 23, pp1410-1411.

[8] Eaglesfield, C. C. (1951). Life of Valves with Oxide-Coated Cathodes, *Electrical Communication*, June, pp95-102.

[9] Metson, G. H. (1956). On the Electrical Life of an Oxide-Cathode Receiving Tube, *Advances in Electronics and Electron Physics*, 8, pp403-46.

[10] Hasset, W. A. (1961). The Materials and Shapes of Vacuum Tube Heaters, *Electronic Industries*, December, pp118-22.

2.6.5: Electrostatic Shields

Some devices contain an electrostatic shield separating multiple valves or enclosing the whole valve. This may be a mesh or solid sheet, and is sometimes mistaken for the anode. In some devices it is internally connected to the cathode while in others it is brought out to a pin which should be grounded to achieve complete electrostatic screening.

2.6.6: The Getter

During manufacture, when the valve is being evacuated, it is placed inside an induction coil to heat the internal parts and release trapped gas. It is the job of the getter to 'get' this gas and remove it from the atmosphere of the valve. A metal receptacle is coated with getter material (often a barium compound) which, when rapidly heated by induction, explodes and immediately combines with other gases before condensing on the inside of the glass envelope, creating the familiar mirrored surface which is the getter proper.[*] Once sealed, the vacuum in a valve is around 13 to 130μPa (10^{-7} to 10^{-6}mmHg). The getter then continues to absorb gas throughout the life of the device, particularly when it is above 150°C,[11] and a very old and weak valve can often be recognised by the fact that the getter has become almost transparent. If air gets into the valve then the getter will quickly turn white, indicating the valve is useless.

The getter receptacle is usually positioned at the top of the valve and is carefully orientated so the getter material is directed away from other electrodes when fired. In many valves the receptacle is only supported by a single wire –often welded to the anode– and can vibrate freely; in other words, it is mechanically underdamped. In such valves this structure may be the greatest source of microphonics.[12] In better designs the receptacle is supporting on two sides –often with wide strips of metal rather than thin wires– and connected only to the mica spacer, which greatly reduces this problem.

It is possible that a valve which has been left unused for many years may have a soft (i.e. poor) vacuum owing to the ingress of a tiny amount of air around the glass/pin seals. If the device is then powered up and the cathode heated, it will be at greater risk of cathode poisoning or sputtering. It may therefore be beneficial to bake old valves in an oven at about 100°C for a few hours to activate the getter and restore the vacuum (higher temperatures may spoil phenolic bases, however).[13]

---

[*] A brownish-coloured getter is also quite normal and results when the flashing is done very rapidly.

[11] Wolfe, G. (1962). Getters. In: RCA Electron Tube Division, *Electron Tube Design*, pp519-25.

[12] Dagpunar, S. S. *et al.* (1960). Microphony in Electron Tubes, *Philips Technical Review,* 22(3), pp71-88.

[13] Jones, M. (2000). New Life for Old Valves, *Electronics World* (November), pp863-7.

2.6.7: The Mica Spacers

The various parts of a valve are held in place by punched mica spacers that fit snugly into the bottle. Unexpected leakage currents between electrodes may arise due to contamination of the mica spacers, and slots are often cut into them to increase creepage distances. Multiple spacers may be used to shield others from getter material, and also to increase rigidity and reduce microphonics.[14]

2.6.8: Variable-Mu / Remote Cut-off Valves

So-called variable-mu valves have the control grid wound with variable pitch. The closely-wound sections have greater control over their portion of the electron stream and will reach cut-off sooner than the coarsely-wound bits, as the bias voltage is increased. The effect is like several parallel-connected valves each with different μ. However, the effect on the device as a *whole* is that the *transconductance* varies strongly with bias voltage, which allows these valves to be used as voltage-controlled amplifiers. The amplification factor μ does not, in fact, vary by nearly as much, since this is determined mainly by grid-cathode distance which is fixed.

These valves were originally developed in America by Dalanthe and Snow[15] who went to great lengths to explain why variable transconductance was a desirable property in radio modulation, yet they then chose to call the device 'variable-mu' rather than the more obvious 'variable-g_m'. This mixed message in the original name has been a source of puzzlement ever since. Over time, British authors reinterpreted the name to mean 'variable mutual conductance' (the old term for transconductance)[16] which makes some sense, while in America the name was partially dropped in favour of 'remote cut-off tube' since the mutual characteristic graph reaches cut-off relatively far from the vertical axis (i.e. the grid base is large). Other sources referred to them as 'super-control valves' which is as vague as it is glamorous. These days it is better to call them 'variable-g_m' since that is exactly what they are. Such valves are by definition highly non-linear and therefore unsuitable for most hi-fi applications, the main exception being audio compressors / limiters.

2.6.9: Valve Ratings

Depending on which datasheet you read, different valve ratings systems may be encountered. Older datasheets often quote **design centre** figures. These

---

[14] Bondy, M. & Jinetopulos, M. (1962). Design of Mica Spacers. In: RCA Electron Tube Division, *Electron Tube Design*, pp465-7.

[15] Stuart, B. & Snow, H. A. (1930). Reduction of Distortion and Cross-Talk in Radio Receivers by Mean of Variable-Mu Tetrodes, *Proceedings of the IRE*, 18(12), pp2102-27.

[16] Bainbridge-Bell, L. *et al.* (1947). Letters to the Editor, *Wireless World*, February, p74

ratings make assumptions about how the circuit designer is going to use the valve and are really a set of maximum recommended *operating* conditions designed to keep the valve within its true absolute maximum limits, allowing for variations in mains voltage. Such ratings may be misleading if we intend to use the valve in a somewhat different manner from that assumed by the datasheet.

Later (and modern) datasheets normally state the **absolute maximum** ratings. These are exactly what they sound like: absolute maximum values that should not be exceeded under any circumstance. These sorts of ratings are rather more useful since they indicate explicitly what the acceptable range of usage is, and it is left to the circuit to designer to make sure he stays within the limits, which is as it should be. The datasheet may then go on to list some recommended operating conditions too, of course, but at least the engineer knows where he stands.

2.6.10: Anode Voltage Rating

The absolute maximum voltage that the anode can withstand, before it is at risk of flashover or arcing, is given on the datasheet as V_{a0}, i.e. the anode voltage with zero anode current. The anode voltage must *never* exceed this value, so this is usually taken as the maximum allowable supply voltage (once an arc has occurred it will often leave a carbonised trail that will arc again and again, even if the voltages are reduced). There is general speculation that many modern-production valves cannot withstand the same peak anode voltages as their vintage ancestors, although admittedly there is no systematic evidence for this. Nevertheless, in the interests of general reliability it is prudent to stay well below V_{a0}.

There is also a maximum allowable *average* anode voltage, given on the datasheet as $V_{a(max)}$. This figure indicates the maximum allowable quiescent anode voltage. While amplifying a signal the *instantaneous* anode voltage is allowed to swing above $V_{a(max)}$ provided it does not exceed Va_0.

2.6.11: Anode Dissipation Rating

The maximum average power dissipation that the anode can tolerate will be given on the datasheet as $P_{a(max)}$ or $W_{a(max)}$. This can be plotted as a curve on the anode characteristics –see section 3.3.5. Dissipating too much power will cause the anode to glow red hot, which is referred to as **red-plating**. If left unchecked the heat may cause the internal parts to soften and deform out of their normal positions, and in extreme cases may even melt the anode or glass envelope. The *peak* power is allowed to exceed $P_{a(max)}$ during use, provided the average remains below the limit, and this is not uncommon in power output circuits. However, there is seldom any need to treat preamp valves so harshly.

When there are multiple valves in the same envelope, dissipation limits may be quoted for each anode independently, and for the device in total. For example, the 6SN7GT dual-triode quotes 5W for either triode alone but only 7.5W for both combined.

2.6.12: Heater Voltage and Current Rating

This first thing listed on the datasheet is usually the required heater voltage and current. The most popular audio valves have 6.3V heaters with the current (and therefore heater power) depending on the size of the device. Power valves must handle large cathode currents and therefore have large cathodes and need commensurately higher heater power. In fact, the heater voltage/current is often quite unique to certain valve types, and when faced with an anonymous valve from which the printing has rubbed off, measuring the heater current can provide a useful clue about what type it may be.

The valve characteristics are normally expected to be within tolerance when the heater voltage is within ±10% of its nominal value. This accommodates normal mains voltage variations (although Mullard specified ±7% for the heater voltage, and some special-quality valves expected ±5%). Operating the heater *above* its allowed range will lead to premature ageing without bringing many significant advantages and is therefore to be avoided. Conversely, many designers prefer to operate at the lower end of the allowed voltage range as it encourages longevity while still meeting the datasheet specifications (voltage regulators also come in round numbers like 12V rather than 12.6V). The lower temperature reduces cathode evaporation and interface-resistance growth, and improves heater reliability.[*] Potential lifetime therefore increases, and under certain conditions noise is reduced too (see section 5.3.2).

It is sometimes argued that reduced heater power will lead to reduced lifetime, but this is only half true. When the heater power is reduced, the cathode emission reduces, and therefore the saturation current is also reduced. If the heater voltage is *excessively* low then it is possible that an operating current that would normally be considered quite safe will in fact result in saturation, and this will indeed lead to reduced lifetime. Also, reducing the heater voltage too much causes the anode characteristics to 'slump', increasing r_a and reducing g_m (μ remains largely unchanged and may even increase slightly[17]), thereby reducing the valve's electrical usefulness. However, operating the heater only 10%-low results in minimal slumping of the characteristics and tends to maintain them for longer too, i.e. it increases the useful lifetime.[†] For a 6.3V heater this means operating at not less than 5.7V.

---

[*] According to RCA Electron Tube Division, (1962), *Electron Tube Design,* p91, heater failure rate is approximately proportional to the 12.5[th] power of heater voltage. Reducing the heater voltage by only 10% may therefore improve heater reliability by almost 400%.
[17] Winter, A. J. (1953). The Effect of Filament Voltage Upon Vacuum Tube Characteristics, *Trans. IRE* (January), pp47-59.
[†] According to Martin, A. V. J. (1967), Factors Determining Tube Life, *Radio & TV News* (July), p111, cathode life is inversely proportional to the 9th power of heater voltage. A 10% reduction may therefore increase cathode life by some 260%.

When the heater is cold its resistance is very low, typically about one sixth of its normal hot resistance. When a voltage is first applied there will be an inrush current that decays exponentially as the wire heats up and the resistance rises, until the steady state is reached. Fig. 2.35 shows the typical current variation after the rated voltage is applied. This inrush current may require the use of unexpectedly large fuses in the power supply. It also causes some

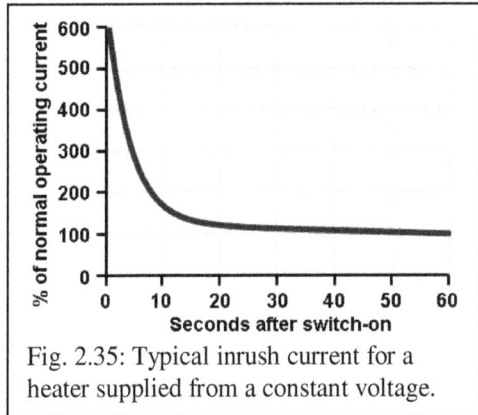

Fig. 2.35: Typical inrush current for a heater supplied from a constant voltage.

heaters to flash bright white at power on –a frequent cause for alarm among beginners– but this is perfectly normal and is simply a matter of the specific valve design. Heater failure is extremely rare, and soft-start circuits are quite unnecessary for indirectly heated valves (although they won't hurt either).

The datasheet will sometimes claim that the heaters are intended to be supplied from a constant voltage, i.e. parallel connected, the current rating being only approximate. Alternatively the specification may be for a constant-current supply, i.e. series connection, the voltage rating being only approximate. Valves advertised for series connection were often part of a manufacturer's family of devices such as the Philips/Mullard 'U' series, which have matching warm-up characteristics so no individual heater is over stressed during the warm-up phase. Other valves such as the Philips/Mullard 'E' and 'P' series were advertised as being suitable for both series and parallel connection, including the popular ECC81/2/3 dual-triodes which have three available heater pins so their two heaters can be connected either in parallel for 6.3V/300mA operation or in series for 12.6V/150mA.

However, these differences in usage appear to be mainly historical, and even the manufacturers' databooks are inconsistent with their advice.[†] Heaters are trivially simple devices and are so consistent and predictable that they can be operated in series or parallel as desired, whatever the datasheet implies. The so-called 'approximate' voltage or current ratings are usually quite accurate, and it is unlikely that the values in practice will come out more than 10% off the quoted values, whether constant voltage or constant current operation is used. It is only necessary to ensure that no heater is over stressed during the warm-up phase, which is easy with modern electronics, so there is no longer a pressing need for heaters with matching

---

[†] For example, manufacturers couldn't always agree about what the heater voltage was supposed to be! RCA quoted the 7F7 as having a 6.3V/300mA heater whereas Sylvania quoted it as a 7V/320mA (but only provided application data for 6.3V use). Similarly, Philips quoted the heater of the DAF96 as 1.4V 25mA for parallel operation but 1.3V for series operation.

warm-up times. However, it is worth pointing out that if valves have previously been tested and carefully matched, with the heaters connected in parallel, then connecting them in series may cause them to become unmatched, due to the small differences in heater power in each case

2.6.13: Heater-Cathode Voltage Rating

The datasheet will normally quote a maximum permissible voltage allowed between heater and cathode, $V_{hk(max)}$, below which heater-to-cathode leakage current is guaranteed not to exceed a certain value. Most valves have a guaranteed heater-cathode insulation resistance of $>10M\Omega$, and under normal conditions it actually reaches hundreds of megohms, so leakage current can be expected to be less than a microamp in either direction.

Heater-cathode leakage is mainly due to the emission of electrons from the cathode or the exposed ends of the heater, which are then attracted to whichever electrode is more positive. Some leakage is also due to migration of ions directly through the heater insulation. Leakage is usually worse when the heater is positive with respect to the cathode than when it is negative,[18] and a few valves quote different V_{hk} limits for positive and negative polarities.

However, $V_{hk(max)}$ is more of an advised limit than a true maximum, and exceeding it may not cause immediate problems; leakage will instead increase gradually over time, leading to excessive hum or intermittent pops and crackles. Admittedly, the heater insulation *will* break down completely if enough voltage appears across it, but this limit is well beyond the number quoted on the datasheet. For many valves the stated value is ±90V (when in doubt this is a safe value to assume) but some go much higher, although whether this is always because of a genuine difference in heater insulation or simply a relaxing of the allowed leakage current limits remains to be seen.

Leakage between heater and cathode is most annoying if the heater is AC powered since an AC current leaking into the cathode circuit will develop an AC hum voltage across any impedance it encounters, which will be amplified. Leakage is not a problem if the heater is powered with regulated DC, unless it becomes so excessive that the valve develops intermitted crackle; this is a known problem in certain amplifier circuits using high voltages and cathode followers with no heater elevation (see below). However, if a valve does develop this problem then it may still be perfectly usable if it is substituted into in a circuit where it is no longer stressed.

Obviously, the heater-cathode voltage rating is of particular importance in circuits where the cathode is at a very different voltage from the heater, such as cathode followers, SRPPs, μ-followers and so on. In such cases it may be necessary to elevate or 'float' the heater supply on a voltage that is close to the cathode voltage

---

[18] Dingwall, A. G. F. (1962) Heaters. In: RCA Electron Tube Division, (1962). *Electron Tube Design*, pp232-43.

(see section 11.8). When many valves share the same heater supply it will be necessary to elevate it to a voltage that accommodates the $V_{hk(max)}$ ratings of all the devices. This may require a compromise value that satisfies all the valves at idle, but may be exceeded when signal voltages are present, e.g. when the cathode voltage of a cathode follower swings up to its peak value.[*] If heaters are connected in series then they can be arranged so the valves with the highest cathode voltages are supplied from the higher-voltage end of the heater chain. In more extreme cases it becomes necessary to use more than one heater supply for different parts of a circuit, each elevated to a different voltage.

At the points where the heater actually touches the inside of the cathode tube a kind of semiconductor is created. At these spots the leakage becomes quite non-linear with applied voltage. It is not necessarily zero when V_{hk} is zero, and may level off or saturate if V_{hk} is large enough. Consequently, it is often found that elevating the heater by a few tens of volts (up to ±50V, say) will reduce hum –either by minimising leakage or by saturating it– even if other factors don't require it. Usually, though, we have little choice about the elevation voltage because it is dictated by a cathode follower or some other amplifier stage with a high cathode voltage.

2.6.14: Heater-Cathode Resistance Rating

Many datasheets quote a maximum allowable resistance between cathode and heater, $R_{hk(max)}$, which often takes the form of a bias resistance. A limit of around 150kΩ is typical. This limit is determined by leakage current between heater and cathode; the larger the resistance between the two, the greater the noise voltage that will develop across it due to leakage current. Again, this is more of an advised limit than an absolute one, and exceeding it does not usually cause problems provided $V_{hk(max)}$ is not also exceeded.

2.6.15: Grid Voltage Ratings

The screen-grid in a pentode has a maximum allowable voltage rating, V_{g20}, beyond which leakage or flashover across the mica spacers may occur. This limit should be strictly obeyed and, again, there is speculation that many modern-production valves cannot be relied upon to meet the specification found on vintage datasheets. A maximum average voltage will also be given as $V_{g2(max)}$.

Logically there must also be a maximum voltage that can be sustained between the control-grid and cathode, but it is seldom stated on the datasheet. This is, however, very significant for DC-coupled amplifiers when first switched on, as the grid may be pulled to a high voltage while the cathode is still cold and not conducting (see

---

[*] Curiously, the presence of signal voltage on top of the average heater-cathode voltage may actually improve matters. See: Gentry, C. H. R *et al.* (1965). Cathode/Heater Insulation Failure in Oxide-Cathode Valves. *Proc. IEE*, 112 (8), pp1501-8.

section 3.13.5). Having destroyed a few valves this way the author suggests ±100V (when cold) as a sensible maximum.

2.6.16: Grid Dissipation Ratings

Every grid has a maximum allowable dissipation limit before it will deform or melt. For the control grid (and suppressor grid if present) it is rarely stated since it is assumed that no significant grid current flows during normal use, so dissipation is nil. The screen grid in pentodes does pass current and will certainly have a stated dissipation limit, $P_{g2(max)}$, which should be strictly observed. This is not likely to be a problem in a preamp, but power valves are often operated closer to their limits and over-dissipation of the screen grid is a leading cause of catastrophic failure in power pentodes.

2.6.17: Grid Current

Ideally we would like the control grid to have infinite input impedance. In other words, when we apply a voltage to the grid, no current flows in it. Unfortunately this is quite impossible to achieve in practice, i.e. some grid current *does* flow. This current is undesirable because it introduces noise into low-level circuits, causes distortion of the signal voltage on the grid, and in extreme cases may cause the bias to runaway and give rise to red-plating (next section).

Grid current tends to be worst when the valve is new and falls to a lower value after several hours of burn-in, finally degrading again towards the end of the valve's life. However, even after burn-in it may take 15 to 30 minutes after each switch-on for it to settle at a steady level. Variation between individual valve samples is also wide, and specific information on it is rarely provided by the manufacturer, so if valves with unusually low grid-current are needed then they must be selected by hand. In general, low-g_m valves exhibit less grid current than high-g_m types.

There are several individual sources of grid current that contribute to the total:
- **Forward grid current:** If the grid is positive with respect to the cathode then it will act like an anode, collecting electrons. In other words, the grid and cathode together act like a small valve diode. This gives rise to positive or forward grid current (i.e. flowing into the valve) which increases rapidly as the grid is made positive. In fact, forward grid current begins to flow even when the grid is still a couple of volts negative, and may become significant around $V_{gk} = -1V$;
- **Grid primary emission:** During manufacture the grid may be contaminated with cathode material. Barium which evaporates from the cathode will also deposit onto the grid during the lifetime of the valve. Since the grid is close to the cathode it is also hot, so the contamination will emit electrons despite the high work function of the grid wire itself. This constitutes a small negative or reverse grid current (i.e. flowing out of the valve);

- **Ion current:** Gas ions that have become positively charged by collision with electrons may be collected by the grid. This results in negative or reverse grid current. The rate of gas ionisation increases with anode current, so ion current increases as the valve is biased hotter;
- **Photoelectric emission:** As the anode is bombarded by the electron stream, ultraviolet light and (harmlessly) weak X-ray radiation is generated. The grid is exposed to this radiation, causing electrons to be liberated from it by photoelectric emission. This results in reverse grid current which increases at higher anode voltages and currents. In valves with good vacuums this effect is even greater than ion current;[19]
- **Direct leakage:** Ordinary resistive leakage between the grid and other electrodes will contribute grid current that is proportional to grid voltage. It is most severe between grid and anode since the voltage difference is greatest there.

All these sources of grid current combine to produce a characteristic like that in fig. 2.36. Over the normal region of operation, grid current can be expected to be of the order of tens of nanoamps, and the main contributor is direct grid-anode leakage. A dirty valve socket will only make this worse. Nevertheless, the input resistance of the grid is still many megohms and can usually be considered infinite. But as the grid voltage is increased towards −1V, forward grid current takes over and rapidly increases; the grid is now acting like an anode, and its input resistance quickly drops to a few kilohms. If the source cannot easily

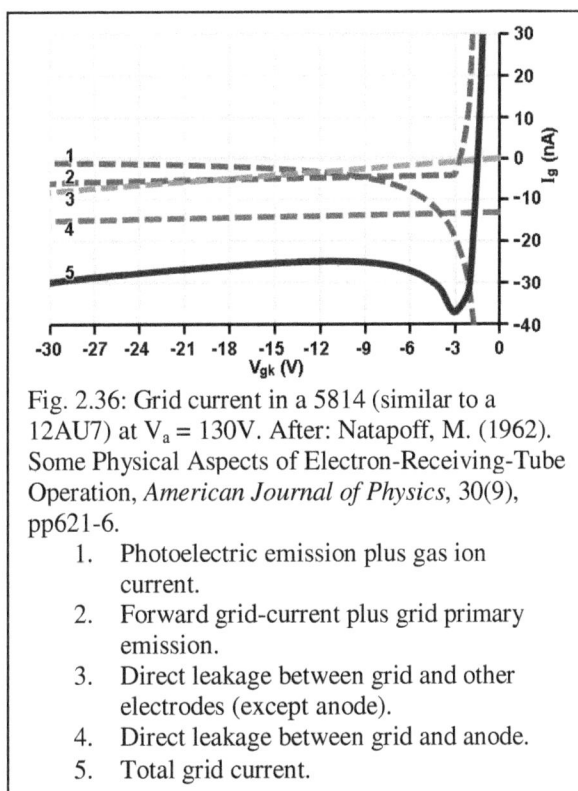

Fig. 2.36: Grid current in a 5814 (similar to a 12AU7) at $V_a = 130V$. After: Natapoff, M. (1962). Some Physical Aspects of Electron-Receiving-Tube Operation, *American Journal of Physics*, 30(9), pp621-6.

1. Photoelectric emission plus gas ion current.
2. Forward grid-current plus grid primary emission.
3. Direct leakage between grid and other electrodes (except anode).
4. Direct leakage between grid and anode.
5. Total grid current.

---

[19] Fairstein, E. (1958). Grid Current in Electron Tubes, *Review of Scientific Instruments*, 29(6), pp524-6.

supply this current then it will lead to distortion, so it is standard practice always to bias a valve colder than −1V. The datasheet may even specify the bias voltage below which the grid current may exceed some stated limit, often 300nA.

2.6.18: Grid-Leak Resistance Rating

The datasheet will quote a maximum allowable grid-leak resistance, $R_{gk(max)}$. This is the maximum resistance (at DC) allowed between grid and cathode. In modern parlance it would be called a 'pull-down' resistance. This resistance must be small enough that the grid current which flows in it does not influence the bias voltage much. In other words, it provides a path for charge collected by the grid to 'leak' away (see section 3.3.2).

Most grid current is negative, flowing out of the grid and down the grid leak resistance, so a positive voltage will therefore develop across it, between grid and cathode. Since this will bias the valve hotter, the grid will heat up further and grid primary emission will increase, exacerbating the effect. This problem is greatest when there is very little load resistance in series with the valve to limit the anode current to safe levels. For example, in power output stages where the main load is an output transformer, a grid-leak resistor which is too large may cause the valve to go into thermal runaway and red-plate until it burns itself out or a fuse blows. The $R_{gk(max)}$ limit for power valves may therefore be quite low. Cathode-biased circuits are more resistant to runaway than fixed-biased ones since any increase in anode current will cause an increase in the voltage across the bias resistor which counteracts the increase in current. Therefore, two values for $R_{gk(max)}$ may be quoted, the larger one being for cathode bias (a method for calculating one from the other is provided by Langford-Smith).[20] Most preamp circuits, on the other hand, have high-resistance anode loads, so it is usually impossible for over dissipation to occur, whatever the bias may be. The $R_{gk(max)}$ limit is therefore much less troublesome in a preamp and values as high as 10MΩ are often allowed at small anode currents. If in doubt, a value of 1MΩ is a safe assumption.

---

[20] Langford-Smith, F. (1953). *Radio Designer's Handbook*, 4th ed. Iliffe & Sons Ltd., London, p83

112

Chapter 3: Fundamentals of Amplification

This chapter deals with the design and analysis of the basic triode gain stage –the main building block of a valve preamp. Here we will focus on 'traditional' circuit techniques using only resistors and capacitors, plus the valve itself of course. Later chapters will cover more modern refinements such as current-source loading and diode biasing. But to appreciate whether such refinements are indeed helpful, or simply a waste of money, we first need to understand simpler circuit variations.

Electronic components work together to form an overall system with certain performance characteristics. Each individual component cannot usually be considered solely in isolation, so in this sense analog circuits are gestalt, that is, they amount to more than the sum of their individual parts. Inevitably, therefore, designing a circuit to achieve a certain specification requires us to consider almost every part of the system almost simultaneously. When choosing even a single component we really have to think about what pre-existing circuit chunk it is going to affect, and what chunks are going to affect *it*, even before those chunks exist. This makes analog design an unavoidably iterative process, beginning with the broad brush strokes like general circuit topology, then reworking the picture several times with more and finer detail until the final compromise (it is always a compromise, never a perfect solution) is settled upon. This global consideration of the system marks the difference between the competent designer and the amateur tinkerer.

3.1: Basic Theory of Valves

A thermionic valve or vacuum tube contains two or more electrodes suspended inside a glass bulb, which may charmingly be called the bottle. Any air inside the bulb is thoroughly removed during manufacture, leaving only a vacuum. The first electrode is called the **cathode**[*] and gives off electrons very easily when heated. Inside the cathode tube is the heater. Operation could hardly be simpler: pass current through the heater so that it gets hot and the cathode will in turn also get hot. However, since the heater is entirely unconnected from the cathode and plays no special part in the audio circuit, it is nearly always omitted from circuit diagrams and is not considered to be a 'working' electrode.

When the cathode reaches its normal working temperature of about 1000 kelvin or 727°C, electrons boil off it and drift around near its surface, forming an electronic cloud called the **space charge**. Without some other influence the space charge would build up to the point where it is so dense that it repels any further electrons from boiling off the cathode, which by itself would not be very useful.

---

[*] From the Greek *kata hodos* meaning 'down way'. This is why the letter 'k' is still used to indicate the cathode.

3.1.1: Valve Diodes

To make a useful valve, another electrode must be introduced, called the **anode**[‡] (or **plate** in America). Since we now have two electrodes the whole device is called a diode and is illustrated in fig. 3.1. If the anode is made positive with respect to the cathode (electrode voltages are always measured relative to the cathode) it will attract electrons from the space charge. These electrons accelerate towards the anode and crash into it. In fact, if the anode is at +100V then the electrons will be travelling at about 5927km per second when they impact![*] New electrons continue to boil off the cathode to keep the space charge topped up.

Since electrons carry negative charge from cathode to anode, the current reckoned in the same direction must also be negative. But by convention everyone prefers to work with positive numbers where possible, so we usually reckon the current as flowing from anode to cathode so that it comes out as a positive number. This current is called the anode current, I_a. If the anode voltage is negative, however, then electrons will not be attracted to it and no current will flow. A valve is therefore a one-way device, hence the name.

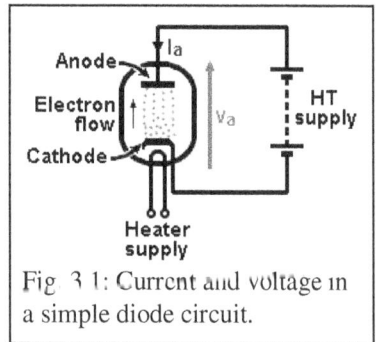

Fig. 3.1: Current and voltage in a simple diode circuit.

The voltage measured between anode and cathode is called the anode voltage V_a, or more accurately V_{ak}. The supply voltage is called the **HT** in this book. This is a historical term that stands for *high tension* which can be taken to mean *high voltage*. In America the notation **B+** is often used to mean the same thing. In modern transistor circuits the positive supply voltage is variously referred to as V_{CC}, V_{DD} or simply V+.

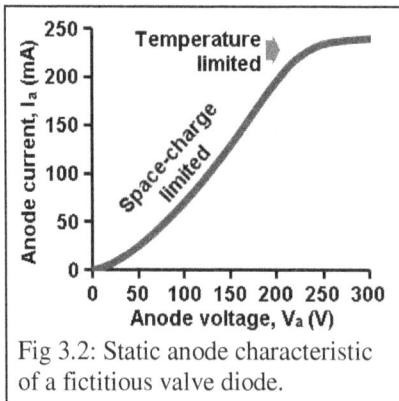

Fig 3.2: Static anode characteristic of a fictitious valve diode.

If the HT is gradually increased then current in the diode will also increase, so for every value of anode voltage (V_a) there is a corresponding value of anode current (I_a). If these points are plotted on a graph we obtain something like that in fig. 3.2, which is called the **static anode characteristic**, or sometimes just the **I/V characteristic**. The characteristic is curved near the bottom when the anode voltage is low because here the anode has only a weak influence over the electrons,

---

[‡] Originally spelled 'anhode', from the Greek *ana hodos* meaning 'up way'.
[*] Interestingly, it does not matter how far they have to travel; only the voltage through which the electrons accelerate determines their velocity.

114

compared to the repulsion of the space charge. As the voltage increases and its electric field intensifies it becomes more effective at drawing electrons out of the space charge, so they become faster and more numerous and the curve bends upwards. If the anode voltage is made really high then it will attract all the electrons that the cathode is capable of emitting, so eventually the curve levels off, and this is called **saturation**. Since this saturation region depends on the emission of the cathode which in turn depends on its temperature, the valve is said to be **temperature limited**. The lower part of the graph is then termed **space-charge limited**. Valves are not normally designed to be operated at saturation as this can be damaging to the cathode's surface, so audio circuits always work in the space-charge limited region.

For a geometrically perfect diode the curvature of the graph in the space-charge limited region follows the **three-halves power law**:

$$I_a = GV_a^{3/2} \qquad\qquad (3.1)$$

Here G is called the *perveance* of the device –a constant related to its physical geometry. It is this law that gives triodes their characteristic distortion spectrum, which we will revisit later. However, real valves are not geometrically perfect and so only approximately follow this law, leading to subtle differences in the behaviour of different valves.

3.1.2: Triodes

By interposing a third electrode between the anode and cathode we can exert some additional control over the anode current, so this new electrode is called the **control grid** or g_1. Since we now have three electrodes (the heater doesn't count, remember) the device is called a triode. If a small positive voltage is applied between grid and cathode, electrons in the space charge will be drawn towards the grid, but because the grid is wound with very fine wire most of them will fly straight through the gaps and be captured by the electric field of the anode.

Conversely if the grid is made negative it will repel electrons, encouraging them to stay in the space charge, so anode current will be restricted. If it is made *very* negative then the anode current can be choked off altogether even though the anode may be hundreds of volts more positive. This condition is called **cut-off**, and the grid voltage required to put the valve into cut-off is sometimes called the **grid base**.[*] Because the grid is very much closer to the cathode than the anode, a change in grid voltage has a more powerful effect on anode current than a similar change in anode voltage. This is related to a property called µ, discussed in the next section.

---

[*] 'Grid base' was a common term in older literature but seems to have fallen out of fashion. It is, however, a very useful concept and well worth resurrecting.

115

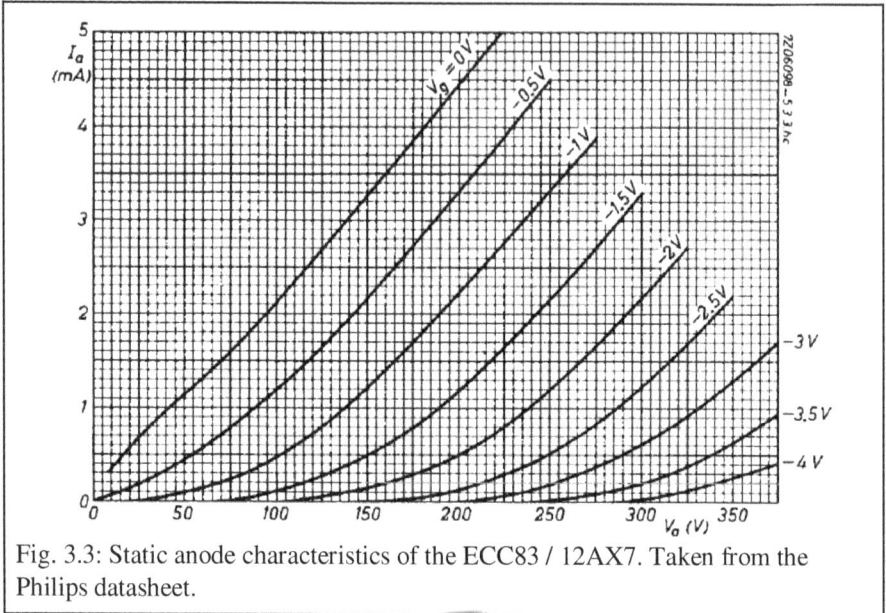

Fig. 3.3: Static anode characteristics of the ECC83 / 12AX7. Taken from the Philips datasheet.

Thanks to the grid, for every value of anode voltage there is now a range of possible anode currents, depending on the grid voltage. The static anode characteristics of a triode can therefore be drawn as a family of curves, each corresponding to a different grid voltage, so they are called the **grid curves**. Fig. 3.3 shows the characteristics of an ECC83 / 12AX7 triode. Notice that the saturation region doesn't even appear on this graph as it is far above the working range.

Another way to present exactly the same information is shown in fig. 3.4. Here the curves correspond to different anode voltages, with grid voltage now on the horizontal axis. This graph is called the **mutual characteristics**, or somewhat misleadingly the **transfer**

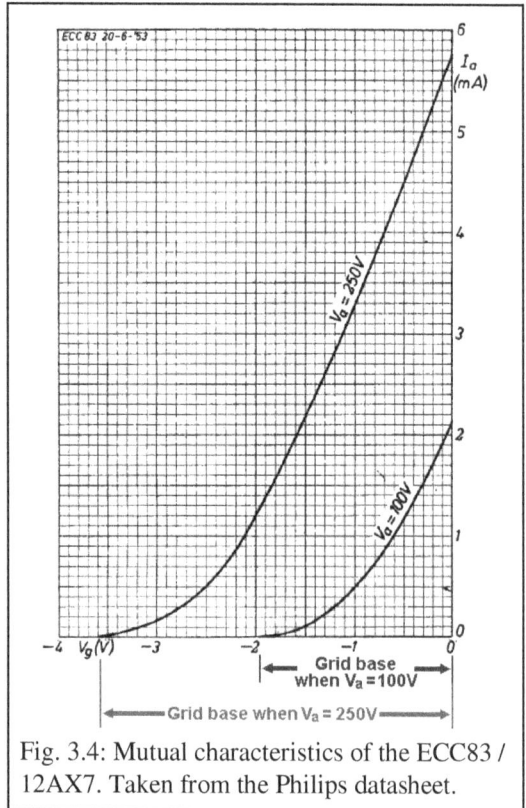

Fig. 3.4: Mutual characteristics of the ECC83 / 12AX7. Taken from the Philips datasheet.

116

characteristics. The static anode characteristics are generally the more useful form, but given either graph it is possible to draw the other. Notice that in fig. 3.4 each curve forms a sort of triangle with the axes; the grid voltage required for cut-off, at any particular anode voltage, corresponds to the width of the base of each triangle, hence the term *grid base*.

3.2: The Valve Constants

There are three important characteristics for any triode, called (somewhat optimistically) the **valve constants**. These are:

- Anode resistance (r_a);
- Amplification factor (μ);
- Transconductance (g_m).

These constants are useful for understanding and analysing circuit behaviour, and for comparing the relative merits of different kinds of valve. All three can be derived from the static anode characteristics or mutual characteristics graphs, as demonstrated in the following sections.

At any given point on the anode characteristics graph the three valve constants are related by van der Bijl's equation:[1]

$$\mu = g_m \times r_a \tag{3.2}$$

This means that only two constants need to be found graphically –usually g_m and μ because they're the easiest to determine with reasonable accuracy– while the third can be calculated.

None of the constants are, in fact, constant. For example, fig. 3.5 shows how they vary with anode current in an ECC83 / 12AX7 at a constant anode voltage. At other anode voltages the figures would be slightly different. Furthermore, as the valve ages, r_a tends to increase and g_m decreases, but μ remains more-or-less unchanged, falling only slightly. It should also be pointed out that valves have fairly poor tolerances and always have. Too many enthusiasts get agitated after finding that their specimens, when measured, do not match the published values

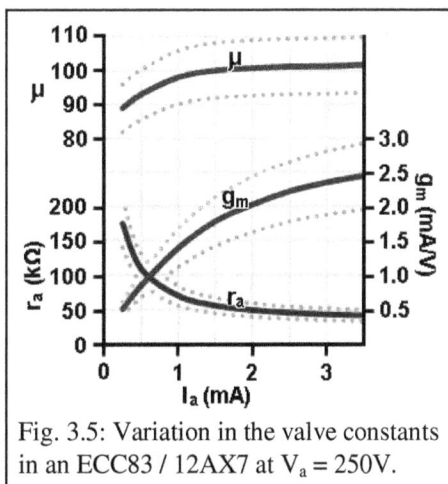

Fig. 3.5: Variation in the valve constants in an ECC83 / 12AX7 at $V_a = 250V$.

---

[1] van der Bijl, H. J. (1919) Theory and Operating Characteristics of the Thermionic Amplifier. *Proceedings of the IRE,* 7(2), p109. This equation is sometimes attributed to Barkhausen.

precisely. This has led to some unfounded remarks about certain manufacturers producing low quality valves. In reality, the tolerances on the valve constants were always understood to be ±20% for r_a and g_m, and ±10% for μ. These ranges are depicted by the dotted lines in fig. 3.5. It is therefore a waste of time to inspect the published graphs with a fine magnifying glass in an attempt to pin down a particular value to the nth decimal place. Simply getting within ±5% of the real-world value should be considered fortunate. Analog design can rarely be done entirely on paper anyway; some practical experimentation, measurement, and adjustment-on-test (AOT) is inevitable.

3.2.1: Anode Resistance, r_a

Fig. 3.6 shows the static anode characteristics of the ECC83 again. At point 'A' on the graph we have 150V across the valve and 2.1mA anode current flowing in it, so it could be imagined to be equivalent to a 150V / 2.1mA = 71kΩ resistor. However, at point 'B' it looks like a 75V / 0.8mA = 94kΩ resistor, so clearly this device is nothing like a real fixed resistor: it is non-linear. Any of these values of resistance taken at spot points on the graph are called **beam resistance**. Beam resistance is only a DC value and is virtually useless to the circuit designer.

There is a much more useful way to figure the resistance of a valve. If we choose a point on the curve and then look at a small portion of the curve either side of that point it actually looks fairly straight, i.e. linear. For example, around point A, changing the anode voltage by 50V causes the anode current to change by about 1.1mA. Over this small section of the graph the resistance appears to be about 50V / 1mA = 50kΩ. This is the resistance that the valve presents to small variations in anode voltage (albeit superimposed on a steady DC voltage), so it is an AC property and is called the internal **anode resistance**. It is given the symbol r_a (or r_p for **plate resistance** in America), and corresponds to the inverse of the slope of the grid curve. The steeper the slope, the lower this resistance, and wherever the slope of the curves is the same r_a must also be the same. We only need to glance at the graph to see that each grid curve has roughly the same shape as the next, so for any particular anode current r_a is roughly

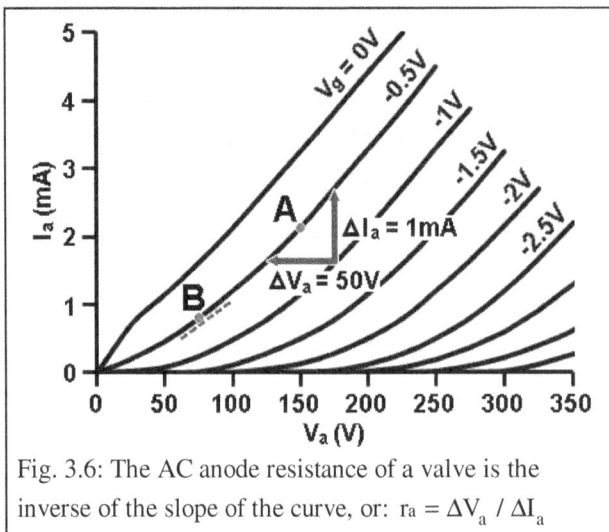

Fig. 3.6: The AC anode resistance of a valve is the inverse of the slope of the curve, or: $r_a = \Delta V_a / \Delta I_a$

the same across the board. Anode resistance is, therefore, a 'constant', roughly speaking. However, it is more constant at higher anode currents where the grid curves are fairly straight and parallel.

At point B the lines are more curved than at A, but by drawing a tangent to the grid curve (shown dashed) we can estimate the anode resistance to be around 70kΩ. It is quite obvious, therefore, that at lower anode currents the anode resistance increases due to the **bottom bend** or 'bunching' at the foot of the grid curves.

3.2.2: Amplification Factor, μ

Looking at the static anode characteristics it is obvious that the grid curves have more-or-less the same shape as one another. Making the grid more negative simply shifts the curve to the right, which reduces the anode current for a given anode voltage. But the initial anode current could then be restored by increasing the anode voltage. For example, in fig. 3.7 an anode current of 2mA is achieved at an anode voltage of 90V when the grid is at 0V (point A). Reducing the grid voltage to −1V would reduce the current to 0.4mA, but it could be restored to 2mA by increasing the anode voltage to 190V (point B). The distance from A to B is 100V on the V_a scale but only −1V on the V_g scale. The ratio of these two is called the **amplification factor** of the valve, and is given the symbol μ (mu). In this case μ = 100V / 1V = 100 (the minus sign is ignored). This figure is also the maximum possible voltage gain that the triode can achieve in any circuit.

We can see that μ is related to the horizontal separation of the grid curves, and over most of the graph the separation is about the same, so μ is another valve constant; indeed it is the *most* constant of the three constants. What's more, μ does not degrade much as the valve ages, unlike the other constants, which makes it a very useful figure for circuit designers.

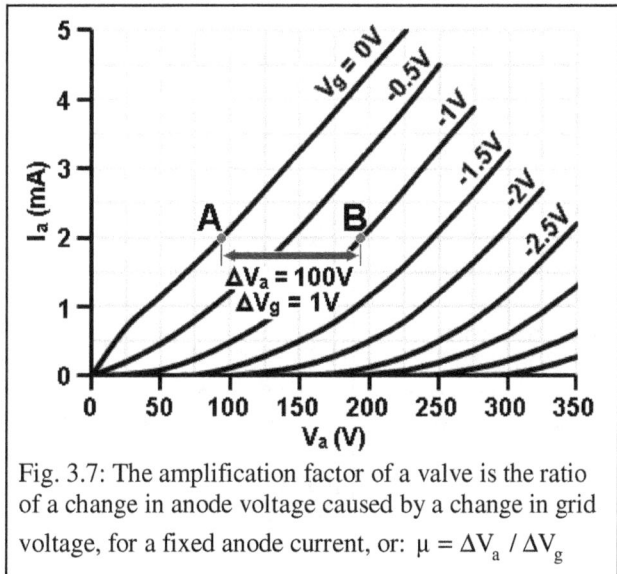

Fig. 3.7: The amplification factor of a valve is the ratio of a change in anode voltage caused by a change in grid voltage, for a fixed anode current, or: $\mu = \Delta V_a / \Delta V_g$

119

Another way to describe μ is to say that it is the measure of how much more effective the grid is than the anode at controlling anode current. For a triode, the three-halves power law can then be written:

$$I_a = G\left(V_g + \frac{V_a}{\mu}\right)^{3/2} \tag{3.3}$$

The term V_a/μ is equal to the (magnitude of the) cut-off voltage, i.e. the grid base, and G is again the perveance, which is about 1.6×10^{-3} for the ECC83. In practice neither μ nor G are completely constant. Fortunately, this equation is not something we need to consider for actual circuit design; it is presented here mainly for academic interest.

3.2.3: Transconductance, g_m

      The first two valve constants were derived by holding the grid voltage constant (to find r_a) or by holding the anode current constant (to find μ). The final important parameter is found by holding the anode voltage constant. From fig. 3.8 we can see that if V_a is 150V then changing V_g from $-1V$ to $0V$ causes anode current to increase by about 2.2mA. This control that the grid voltage has over the anode current is called the **transconductance**, or in older books the **mutual conductance**, and is given the symbol g_m. Transconductance is related to the vertical separation of the grid curves, which is again moderately constant over much of the graph.[*]

In this case the transconductance is $2mA / 1V = 2mA/V$. The units indicate that we get a 2mA change in anode current per volt change at the grid. The modern SI unit of conductance is the siemens (S) where $1S = 1A/V$. Some older or American texts may use the mho (ohm spelled backwards) which is equivalent to the siemens. In this case we could therefore write the transconductance as 2mA/V, 2mS (milli-siemens), 2000μmho

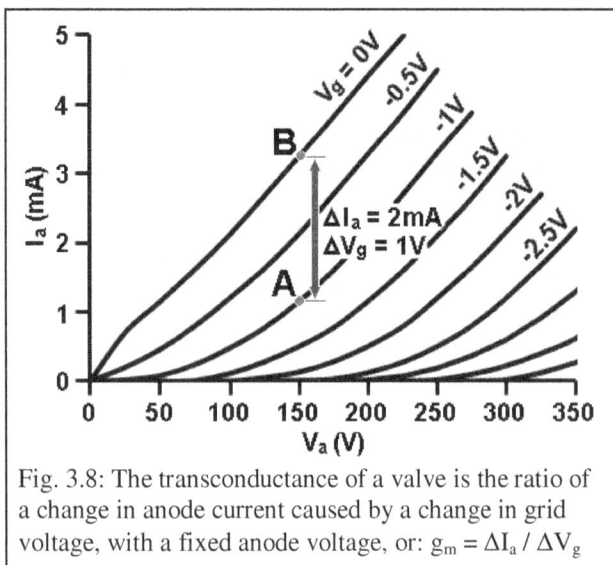

Fig. 3.8: The transconductance of a valve is the ratio of a change in anode current caused by a change in grid voltage, with a fixed anode voltage, or: $g_m = \Delta I_a / \Delta V_g$

---

[*] Incidentally, the transconductance also corresponds to the *gradient* of the curves on the *mutual* characteristics graph. When older texts talk about the 'slope' of a valve they are referring to its transconductance, not its anode resistance.

(micro-mho), or 2000μA/V. However, it is more sensible to stick to milliamps-per-volt in order to distinguish between *trans*conductance which is a property of active components like valves, and pure conductance which is simply the inverse of resistance. These are also the units used on most valve datasheets.

3.3: Amplification

From the anode characteristic graphs shown earlier it is obvious that if we apply an audio signal voltage to the grid, i.e. vary the grid voltage, then current in the valve is bound to vary too. We can put this current variation to good use by putting a resistor in series with the valve so that a corresponding audio voltage is generated across the resistor, which is also called the **load**. Fig. 3.9 shows this in essence. R_a is the **anode resistor** (not to be confused with the anode resist*ance* r_a) and provides a DC current feed from the HT as well as serving as the load. The exact choice of this component is dealt with later. A battery is shown in series with the signal generator so the AC audio voltage at the grid is superimposed on a small negative DC voltage; more on this in section 3.3.2. Notice also that the cathode is shared with both the input *and* output ports of the circuit, i.e. it is 'common' to both and assumed to be grounded or at zero volts. This circuit arrangement is therefore called a **common-cathode** amplifier.

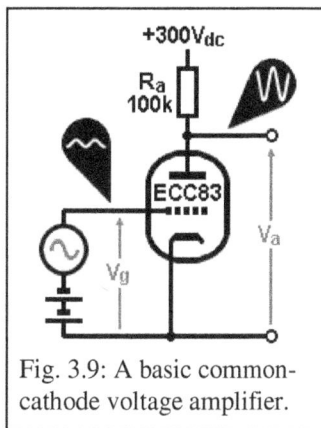

Fig. 3.9: A basic common-cathode voltage amplifier.

An obvious question is: what is the voltage gain of this amplifier? If we knew the value of AC anode current caused by a given AC grid voltage then we could apply Ohm's law, multiplying the AC current by the value of R_a to find the AC output voltage (remember, these AC values are actually superimposed on steady DC values but we can treat the AC and DC parts separately). At first glance we might try to multiply the AC grid voltage by the transconductance to find the AC anode current: $i_a = v_g \times g_m$, but this will not work because g_m only applies when the anode voltage is constant, whereas here it is changing. There are two approaches we can take to solve this problem: a graphical method or a more mathematical method. We will deal with the graphical method first and the mathematical model later in this chapter.

3.3.1: The Load Line

The graphical method of analysing the circuit is to draw a **load line** on the static anode characteristics. This will shows us any and all of the possible voltages and currents in the circuit, and is the most powerful design tool in our armoury. Referring to fig. 3.9, imagine the valve is completely cut off so that no anode current flows. Hence there can be no voltage dropped across R_a, so the full HT reaches the anode. This point can therefore be plotted on the anode characteristics at:

$V_a = 300V$, $I_a = 0mA$ (point A in fig. 3.10).

Now suppose the valve is a short circuit (which is impossible in reality but needn't hinder our imagination). The full HT is now dropped across R_a with none across the valve. From Ohm's law the anode current would have to be $300V/100k\Omega = 3mA$. This point can also be plotted at:

$V_a = 0V$, $I_a = 3mA$ (point B in fig. 3.10).

We could plot some intermediate values too, but because Ohm's law is a linear equation and the axes on the graph are linear too, there is no need; we can simply draw a straight line between A and B, to complete the load line as shown in fig. 3.10. Examining the load line we see that it is intersected at various points by the grid curves. Each intersection shows us what V_a and I_a will be for any given value of

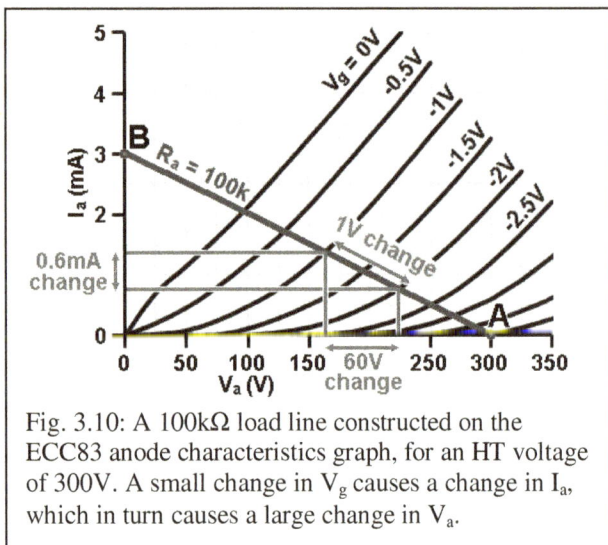

Fig. 3.10: A 100kΩ load line constructed on the ECC83 anode characteristics graph, for an HT voltage of 300V. A small change in V_g causes a change in I_a, which in turn causes a large change in V_a.

V_g, and these are the only combinations that are possible in this circuit. Load lines for other values of anode resistor could also be plotted for comparison (e.g section 3.6).

Having drawn the load line it is now possible to work out the voltage gain of the circuit. Suppose we made the grid two volts negative. We can deduce from the graph that 0.8mA would flow and the anode voltage would be 224V. If we then raised the grid voltage to one volt negative, current would increase to 1.4mA and the anode voltage would be pulled down to 164V. We have only changed the grid voltage by $2-1=1V = 1V$, yet the anode voltage has changed by $224 - 164 = 60V$. The gain is therefore:

$$A = \frac{V_{out}}{V_{in}} = \frac{60}{1} = 60$$

Note that this is less than the µ of the valve (about two thirds in fact, a ratio which will appear again in section 3.6.1). Also notice that when the grid voltage swings up the anode voltage swings down, so the output signal will be up-side down or 180°

out of phase compared with the input signal. This circuit is therefore said to be **inverting**. The gain may be written with a minus sign to indicate the inversion (–60 in this case), although this is often ignored.

However, there is something suspicious about this gain calculation. The grid curves are not evenly spaced along the load line, so if we did our calculation near the top where the curves are more stretched apart we would get a larger figure for gain than if we did it near the bottom where they are more bunched up. So which bit of the load line do we use? To answer this question we first need to understand how to bias the valve, and from there discover how the choice of bias influences the distortion generated in the circuit.

3.3.2: Biasing

Imagine a perfect sine wave. The wave has high peaks and low troughs. In between there must be some average level, so the wave could be said to be both rising above and falling below this average level. Now to speak in terms of voltage we have a sine wave that swings both positive and negative *relative* to some average voltage. This average voltage is, in a manner of speaking, our starting point, and it is up to us to decide where it should be on the load line so the valve can amplify both up-going and down-going parts of the signal. This is called **biasing** the valve.

Suppose we fix the grid voltage at −1.5V and call this our **bias point**, as indicated in fig. 3.11. We are not yet inputting any signal, the valve is simply at rest. The valve is said to be in a state of **quiescence**, and we can see from the graph that the quiescent bias voltage of −1.5V causes a quiescent anode current of just over 1mA and a quiescent anode voltage of about 195V. These are the steady DC values onto which the audio voltage will be superimposed.

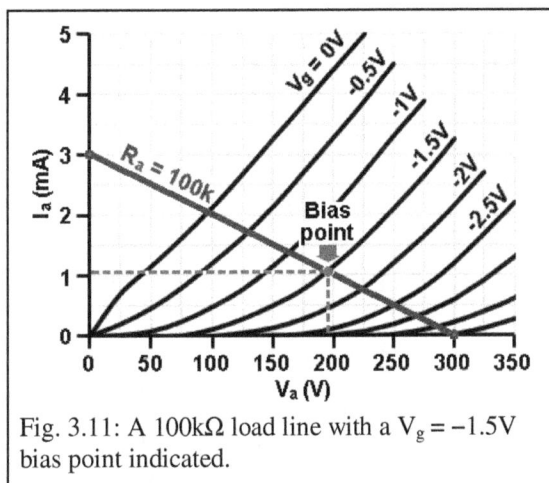

Fig. 3.11: A 100kΩ load line with a $V_g = -1.5V$ bias point indicated.

There are several ways to apply bias in practice. A very direct method is shown in fig. 3.12. Here a bias battery is shown, but this could be substituted with a negative voltage supply from some other source. The negative voltage is applied to the grid via R_g which is the **grid-leak** resistor. This resistor feeds the DC voltage reference to the grid but prevents the audio signal from being shorted to ground through the low internal impedance of the bias battery. R_g might typically be 1MΩ in value. The **coupling capacitor**, C, blocks DC and so prevents the negative voltage on the grid

123

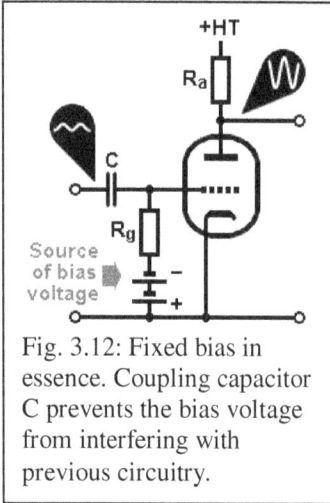

Fig. 3.12: Fixed bias in essence. Coupling capacitor C prevents the bias voltage from interfering with previous circuitry.

from interfering with previous circuitry, so it is sometimes also called a **blocking capacitor**. C and R_g together form a high-pass filter, so they must be suitably chosen to allow the desired audio frequencies to reach the grid without undue attenuation. This type of biasing is called **fixed bias** because the bias voltage is fixed and cannot change during normal circuit operation. Of course, a disadvantage of using a bias battery is that batteries need periodic replacement, although we might expect several years of life from a lithium battery since the grid current is in the range of nanoamps. On the whole, however, fixed bias is rarely used for preamp valves because cathode bias is more convenient, as we shall see.

Fig. 3.13 shows another form of biasing called **grid-leak bias** or **contact bias**. This relies on the small but ever-present grid current to generate a negative voltage at the grid. Since the grid current must be positive (i.e. *into* the grid) for this to work, and grid current is only positive when the grid is no more than a couple of volts negative (section 2.6.17), this sort of biasing cannot be used to generate large bias voltages. Also, the grid current is normally much less than a microamp so R_g must be very large to generate any useful bias. When a signal is applied, the average grid voltage will be

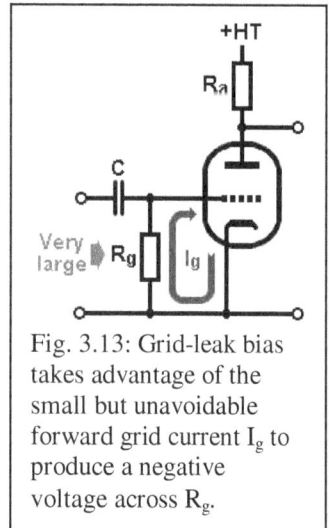

Fig. 3.13: Grid-leak bias takes advantage of the small but unavoidable forward grid current I_g to produce a negative voltage across R_g.

'pumped' more negative due to the charging of capacitor C, so the bias will vary with the applied signal. This technique is therefore not suitable for linear (e.g. hi-fi) amplifiers as it is not very predictable and results in excessive noise and distortion.

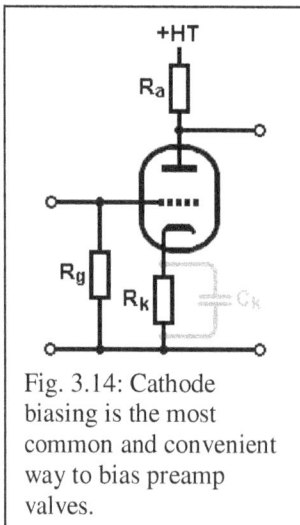

Fig. 3.14: Cathode biasing is the most common and convenient way to bias preamp valves.

Let us now reverse our way of thinking. Instead of making the grid negative with respect to the cathode, make the cathode *positive* with respect to the grid. This is a perfectly reasonable thing to do since the valve only reacts to the *difference* between the grid and cathode voltages; the *absolute* measured voltages are not important. It does not matter whether the grid is actually at a negative voltage

124

while the cathode is at zero volts, or if the cathode is positive and the grid is at some lower voltage, the result is the same from the valve's point of view. Either way, a negative voltage has been applied *between* grid and cathode.

If we put a resistor in series with the cathode then any current that flows through the valve will also flow through the resistor and will cause a voltage drop across it. This will raise the cathode voltage above the grid, so the valve provides its own bias. This method of biasing is shown in fig. 3.14 and is variously called **cathode bias, self bias**, or **automatic bias**. R_k is called the **cathode resistor** or simply **bias resistor**. Cathode biasing has several advantages:

- It is self adjusting; if the average current through the valve increases, so does the bias voltage, which opposes the increase in anode current. This means the bias will adjust itself naturally to changes in HT, component tolerance and ageing. There is also less chance of the valve going into uncontrolled runaway and overdissipating in the event of some failure, which is useful when dealing with power valves.
- The grid-leak resistor tethers the grid directly to ground or zero volts, so a coupling capacitor is not essential (unless there is an *incoming* DC voltage that we need to block).
- A bypass capacitor C_k (shown faint) can be added in parallel with R_k to control the gain, frequency response, and linearity of the circuit. This is explained in more detail in section 3.7.

A sort of hybrid between fixed and cathode biasing can be achieved by using a diode rather than a resistor to provide the bias voltage. This is explored in more detail in section 6.3, but for now let us stick to the traditional way of doing things.

3.3.3: The Cathode Load Line

When designing a triode gain stage the usual procedure is to choose the triode, then choose a suitable anode load and draw the load line. Then choose the bias point and (if we are using cathode bias) find the necessary value of cathode resistor. More details of this whole process will emerge throughout this book, but for now let us concentrate on finding the cathode resistor. Fig. 3.15 shows a load line, repeated from fig. 3.11 earlier. Since we are now about to

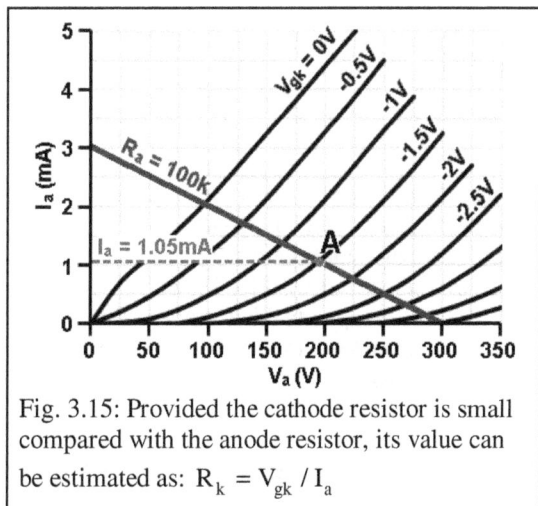

Fig. 3.15: Provided the cathode resistor is small compared with the anode resistor, its value can be estimated as: $R_k = V_{gk} / I_a$

125

interfere with the cathode voltage we should be careful to avoid any ambiguity when talking about grid voltage, so we will be explicit and talk about the grid-to-cathode voltage V_{gk}.

If our chosen bias point happens to be $V_{gk} = -1.5V$ (point A in fig. 3.15) then this is the same as saying we want the cathode voltage to be raised to $+1.5V$, assuming the grid is tethered to ground. The graph indicates that the quiescent anode current will be 1.05mA. We now know the voltage across and current through the bias resistor, so Ohm's law suggests a value of:

$$R_k = \frac{V_k}{I_a} = \frac{1.5V}{1.05mA} = 1.4k\Omega$$

This is only a standard value in the E48 range and higher, so we might use 1.5kΩ instead, which is in the E12 range. This is really just a first approximation, however, because as soon as we add R_k to the circuit it will alter the load line slightly, since R_k is now in series with the valve together with R_a. Fortunately, we can double check our first approximation by redrawing the load line to correspond to the *total* resistance now in

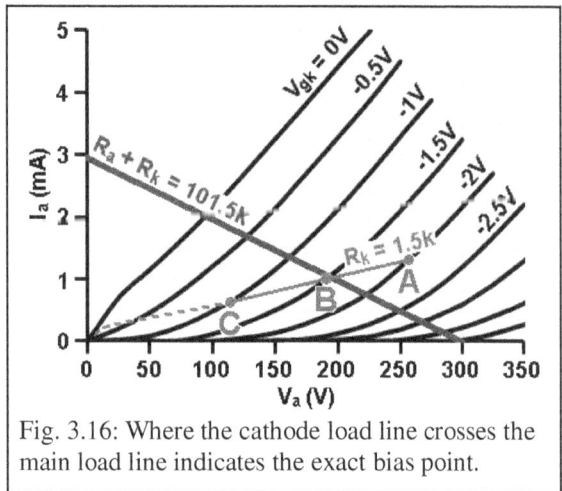

Fig. 3.16: Where the cathode load line crosses the main load line indicates the exact bias point.

series with the valve, which is 101.5kΩ. This has been done in fig. 3.16 where it is obvious that because R_k is very small compared with R_a it has made practically no difference to the slope of the load line, but we shall persevere for the sake of demonstration.

The main load line takes into account the total resistance in series with the valve and therefore shows the total range over which the valve can operate, but we can also draw a **cathode load line** to show exactly where the bias point will fall when part of that total resistance is used for R_k. This line will not be perfectly straight so we may need to plot a few points. First suppose that the cathode voltage were somehow 2V, what must the anode current be? Ohm's law guarantees:

$$I_a = \frac{V_k}{R_k} = \frac{2V}{1.5k\Omega} = 1.3mA$$

This point is plotted in fig. 3.16 at $V_{gk} = -2V$, $I_a = 1.3mA$ (point A).

Now repeat the process for some other possible cathode voltages, say 1.5V and 1V:

$$I_a = \frac{V_k}{R_k} = \frac{1.5V}{1.5k\Omega} = 1.0mA \quad \text{(point B)}$$

$$I_a = \frac{V_k}{R_k} = \frac{1V}{1.5k\Omega} = 0.7mA \quad \text{(point C)}$$

Joining these points up produces the cathode load line. Other points could be added to extend it, as shown dashed. Where the two load lines cross indicates the actual bias point, and in this case it is extremely close to our initial choice. In fact, with medium- and high-μ preamp valves we rarely need to go to the trouble of drawing a cathode load line, because R_k will always end up small compared with R_a, so a first approximation is all that is needed. With power valves, however, the bias is often more critical, so drawing cathode load lines can be useful.

3.3.4: Resistor Ratings

Now that we have selected a bias point we know the average current flowing in the anode and cathode resistors, so we can work out the power dissipated in each component using $P = I^2R$. From fig. 3.16 the quiescent current looks to be about 1mA, so the power in the anode resistor will be:

$$P = I^2R = 0.001^2 \times 100000 = 0.1W$$

It is good practice to double the required power rating of a resistor to ensure reliability, so in this case we might be fooled into using a ¼W device. However, in chapter 2 it was pointed out that most ¼W resistors are only rated to withstand 200V, and judging from fig. 3.16 the voltage across R_a could just about reach this value (i.e. if $v_{gk} = 0V$). Furthermore, we should also allow for current surges at start-up, or for a faulty valve with an anode-to-cathode short. What we need is a resistor that can withstand the full supply voltage, so a ½W (or better) device is called for.

The power in the cathode resistor is:

$$P = 0.001^2 \times 1500 = 0.0015W \quad \text{or } 1.5mW$$

This is next to nothing, so even an ⅛W device would do, although there is no reason why we couldn't use something larger. It does not need to be rated for the full supply voltage because, even with an anode-to-cathode short, nearly all of the voltage would be dropped across R_a leaving very little across R_k.

3.3.5: The Safe Operating Area

Each valve type has certain limitations with regard to applied voltages, current capability, and power dissipation. By plotting the limits on the static anode characteristics we can see clearly which areas are out-of-bounds and which are safe to work in; this is called the **safe operating area** or **SOA**.

The maximum allowable peak anode voltage, V_{a0}, is quoted as 550V for the ECC83. The anode voltage must *never* exceed this value so it is usually taken as the maximum allowable supply voltage. This is indicated in fig. 3.17 by shading everything to the right of $V_a = 550V$ in black. The maximum allowable average anode voltage, $V_{a(max)}$, is quoted as 350V. The bias point must not lie to the right of this value, so this area is shaded grey. However, while amplifying a signal the *instantaneous* anode voltage is allowed to swing above $V_{a(max)}$ provided it does not exceed V_{a0}.

The upper limit of the SOA is determined by the maximum power that can be dissipated by the anode,[*] $P_{a(max)}$. For the ECC83 the limit is 1W, and this can be plotted by taking some example anode voltages and then calculating what maximum anode current is allowed to flow, using $I = P / V$:

1W / 400V = 0.0025A
1W / 350V = 0.0029A
1W / 300V = 0.0033A
1W / 250V = 0.004A
1W / 200V = 0.005A

A curve has been drawn through these points in fig. 3.17. Anywhere above this curve the anode dissipation will exceed 1W, so it too is shaded

Having dealt with the 'hard limits' quoted on the datasheet we can also shade in some areas that can be considered out of bounds as far as useful operation, i.e. reasonable fidelity, is concerned. First, at very low anode currents the grid curves

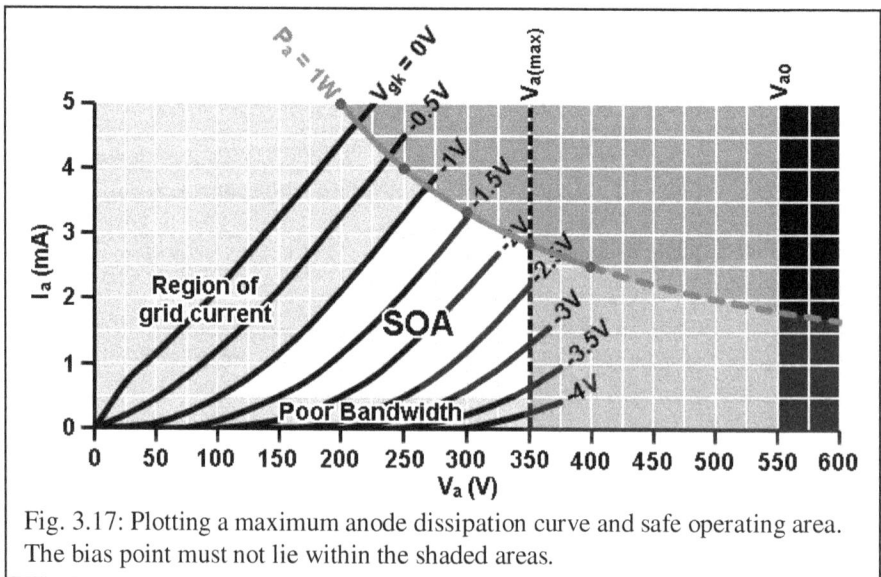

Fig. 3.17: Plotting a maximum anode dissipation curve and safe operating area. The bias point must not lie within the shaded areas.

---

[*] Some datasheets may quote a maximum allowable anode current, which will put an additional upper limit on the SOA, but this is uncommon for preamp valves.

become very bunched up and shallow. Operation in this region can be unpredictable, the output resistance becomes large, and bandwidth and slew rate will suffer (see later). Therefore, everything below $I_a = 0.5mA$ has been shaded. This is a fairly arbitrary figure chosen entirely by eye and will vary with different valve types, depending on how discerning the circuit designer is.

The second 'soft limit' is the region where forward grid current flows, as this causes severe distortion and clipping (section 3.5). As noted in section 2.6.17, this usually becomes significant when V_{gk} is less than about $-1V$, so everything to the left of this grid curve is also shaded. This figure can be assumed to be the same for any valve type.

The unshaded white area is now the SOA and the bias point must lie somewhere within it. With preamp design it is fairly easy to stay within the SOA and in most cases the supply voltage will be less than any of the voltage ratings of the valve, so we can forget about them completely. Power valves, on the other hand, often operate much closer to their maximum ratings so special attention must be paid to their SOAs.

In this case the SOA is frustratingly small. This is mainly because this valve has a very small grid base, so excluding everything to the left of the $-1V$ grid curve wipes out a lot of potential operating area. Getting good performance from an ECC83 / 12AX7 requires a fairly high supply voltage, a high-impedance load, and a carefully selected bias point. Failure to make the right choices perhaps explains why some audiophiles have reported poor results from this valve, when in fact it is one of the lowest distortion triodes available –when used properly.

3.4: Distortion and the Transfer Function

It will be remembered from chapter 1 that an equation which relates the output of a circuit to its input is called a transfer function. A transfer function can also be represented graphically with the 'output quantity' on the vertical axis and the 'input quantity' on the horizontal axis. In this case, rather than starting with a neat equation and plotting it, we can extract the transfer function from the load line. For the amplifier circuit we are considering, the input quantity is the grid voltage, and the output quantity is the anode voltage. But since the anode voltage is proportional to the anode current flowing in R_a we can use the current just as well; the overall shape of the function will be the same, and it is really just the shape we're interested in.

The graphical construction is shown in fig. 3.18. At each point where a grid curve intersects the load line a corresponding point has been plotted at the same I_a and V_{gk} on a new set of I_a/V_{gk} axes, and the transfer function emerges. A distortionless (i.e. linear) circuit would produce a perfectly straight line, whereas this one gets progressively more curved near the bottom due to the bunching of the grid curves, although the upper end is not too bad. For minimum distortion we would therefore

129

try to keep the operating point on the straighter portion of the line. However, we also do not want to get too close to the region where forward grid current starts to flow, around $V_{gk} = -1V$, and in this case that doesn't leave us much room to manoeuvre. In other words, this particular circuit can reliably handle only small input signals, not more than $1V_{pp}$ say, centred on a bias of $-1.5V$.

During normal design work there is actually no need to go to the trouble of drawing the transfer function; it is presented here simply to illustrate how the linearity (or lack thereof) of a common-

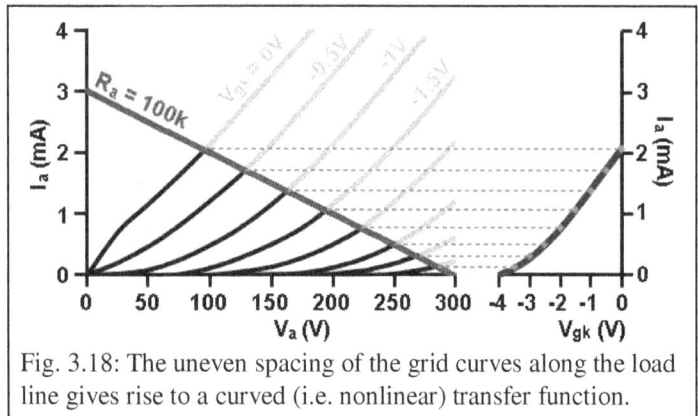

Fig. 3.18: The uneven spacing of the grid curves along the load line gives rise to a curved (i.e. nonlinear) transfer function.

cathode gain stage is related to the spacing of the grid curves along the load line. Choosing a load and bias point is a simple matter of inspecting the anode characteristics and judging which combination will result in pleasingly equal grid-curve spacings, achieve the desired headroom (section 3.5), and avoid any forbidden areas as defined in the previous section. Note that this does not require any special tools or calculations, it can all be done by eye.

3.4.1: Harmonic Distortion

Now that we understand what it means to bias the valve and how the grid-curve spacing dictates linearity, we can see what effect the choice of bias point has on the operation of the circuit. Consider fig. 3.19 which shows a 100kΩ load line again, this time with a bias point at −2V (B). Notice that this is mid-way between the 0V grid curve and the −4V grid curve where the valve reaches cut-off. Biasing in this area is called **centre biasing** and offers maximum headroom before the signal is clipped, all else being equal.

Now suppose we apply a $4V_{pp}$ sine wave to the grid. When the grid voltage swings 2V positive the operating point will move up the load line and reach the 0V grid curve at point A. The anode current peaks at just over 2mA and the anode voltage is pulled down to 95V. When the grid voltage swings 2V below the bias point during the other half of the waveform, the valve just reaches cut-off at point C and the anode voltage hits 300V. Notice that we have swung the grid voltage symmetrically around the bias point but the anode voltage has swung *asymmetrically* around its quiescent point, because the grid curves are unequally spaced. The output waveform, which has been plotted below the graph, is bent out of shape and is no longer a perfect sine wave; it is distorted.

130

It is obvious from fig. 3.19 that the output waveform is stretched on the down-going half and squashed on the up-going half (don't forget that it is inverted). Since the waveform is now asymmetrical it must contain even-order harmonics (2nd, 4th etc.). In fact, triode distortion is nearly always characterised by a dominance of second harmonic, followed by a decaying series of all other harmonics. The larger the audio signal is, the more of the load line / transfer function the

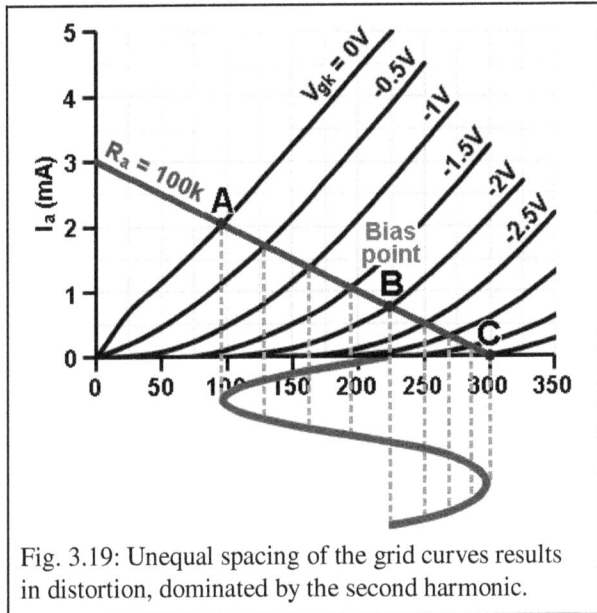

Fig. 3.19: Unequal spacing of the grid curves results in distortion, dominated by the second harmonic.

operating point moves over, and so the greater the distortion. In most cases it is directly proportional to signal level up until clipping. Fig. 3.20 shows the distortion spectrum measured in a real circuit at 1kHz when driven as in fig. 3.19. The fundamental has been nulled out by the distortion analyser so the remaining distortion products can be resolved more clearly, and the dominance of the second harmonic is plain to see. The series would continue indefinitely, but in this case harmonics higher than the fourth are below the noise floor.

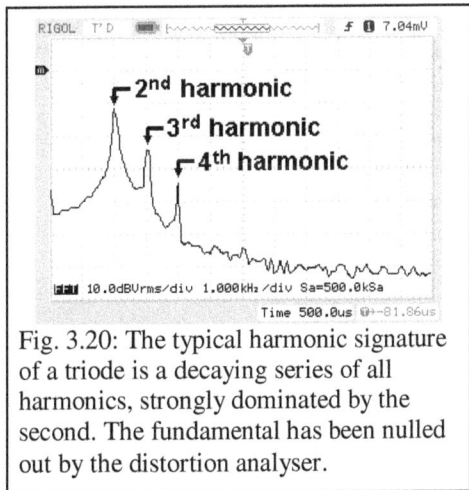

Fig. 3.20: The typical harmonic signature of a triode is a decaying series of all harmonics, strongly dominated by the second. The fundamental has been nulled out by the distortion analyser.

In theory, the amount of distortion can be quantified using the load line, although it should be taken as a rough guide only since real valves never conform exactly to their datasheets. Referring to fig. 3.19 again, the anode voltage swung about 128V more negative but only about 77V more positive of its quiescent value.

The percentage of second harmonic distortion is given by:[2]

$$H_2\% = \frac{AB - BC}{2(AB + BC)} \times 100 \qquad (3.4)$$

$$H_2\% = \frac{128 - 77}{2 \times (128 + 77)} \times 100 = 12.4\% \text{ second harmonic distortion.}$$

This is an alarmingly high figure and an unavoidable result of trying to take advantage of the whole transfer function. In section 3.6 we will see how the choice of load resistor affects this figure.

3.4.2: Intermodulation Distortion

When a single frequency is applied to a non-linear circuit, the output will be distorted and will contain harmonics. But when more than one frequency is applied at the same time to a non-linear circuit, the frequencies will interact with one another to produce new frequencies that are the *sum and difference* of the originals, and also the sum and difference of all the new frequencies produced. This is called **intermodulation distortion** or **IMD**.

Fig. 3.21 shows the result of applying a pair of sine waves, 250Hz and 1kHz, both 1V$_{rms}$ each, to the same ECC83 circuit as used previously. The output signal (time domain) is shown above its Fourier spectrum (frequency domain). The two fundamental components are clearly visible as the two large peaks, but there is a collection of smaller peaks too, some of which are harmonics and some are intermodulation products. Note, however, that the input signals in this case are inordinately large to illustrate the effect clearly –they are in fact overdriving the valve.

It is argued by some audiophiles that harmonic distortion tests do not reveal enough information about an amplifier and that intermodulation tests should be used instead. This is a redundant statement, however, because IMD is proportional to harmonic distortion; an amplifier which produces more harmonic distortion than another will produce more IMD too. The trouble is that the ear is

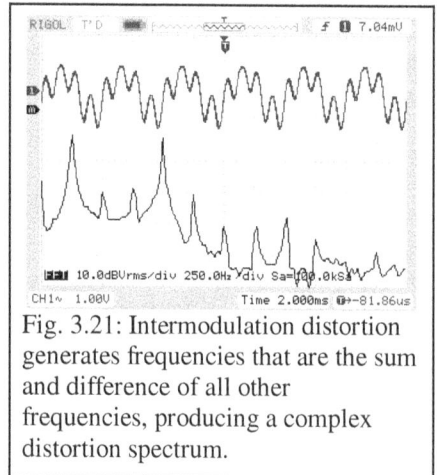

Fig. 3.21: Intermodulation distortion generates frequencies that are the sum and difference of all other frequencies, producing a complex distortion spectrum.

---

[2] Langford-Smith, F. (1957). *Radio Designer's Handbook* (4th ed.), Iliffe and Sons ltd., London, p491. This formula is an approximation as it assumes *only* second harmonic distortion is produced. In reality other distortion products are present, though to much lesser degrees.

much more tolerant of low-order harmonics than high-order ones. An amplifier producing 1% second harmonic might sound acceptable whereas 1% fifth harmonic would be painfully unpleasant, but neither THD *or* IMD (single figure) tests indicate what harmonics are actually present. What is really needed is for any distortion measurement to be accompanied by a table or graph of the harmonic amplitudes.

Harmonic distortion is at least a 'one number' figure of merit that everyone can understand. It is a fairly easy test to perform, and results for different amplifiers are directly comparable. By contrast, coming up with a universal standard IMD test is notoriously difficult because there are any number of different test frequencies and relative amplitudes to choose from, and the resulting figures can be greatly skewed by the order of the distortion present.[*] Even if everyone agrees always to use the same standard IMD test, the problem of distortion order still makes it extremely difficult to compare one manufacturer's figures with another, unless they additionally provided a table or graph of the distortion spectrum for a single tone – which is of course a THD test!

3.5: Clipping and Headroom

Having looked at what happens for small signals (and some not so small) when amplified by a non-linear circuit like a triode, we will briefly push things beyond the normal hi-fi realm. Driving the valve beyond its limits of 'clean' amplification is called **overdrive** –a word that will be all too familiar to electric guitarists. One way to overdrive the valve is to drive the grid very negative, towards the bottom of the load line where the grid curves become bunched up and gain begins to fall. If we continue to drive the grid negative then the valve will eventually reach cut-off and stop conducting completely, so the output waveform will level off. In other words, the circuit runs out of headroom and the output wave will be clipped on its positive side (since the stage is inverting). The grid voltage may continue to swing more negative but the valve can do nothing but remain in cut-off.

Fig. 3.19 suggested that cut-off ought to occur around $V_{gk} = -4V$, although the precise point is difficult to determine, owing to the increasing bunching of the grid curves. The transition into cut-off is therefore fairly gentle, causing a rounding of the edges of the clip. Fig. 3.22 shows a test circuit that has been biased deliberately close to cut-off, and then overdriven. In the accompanying oscillogram the grid voltage has been inverted and scaled to allow easy comparison with the anode waveform. Biasing a valve close to cut-off like this results in less quiescent power dissipation in the valve, so it is loosely referred to as **cool biasing** or **cold biasing**. Solid-state circuits, by contrast, tend to clip rather harder, and remembering that sharp corners indicate the presence of high-frequency components, solid-state circuits tend to

---

[*] A refreshingly simple explanation of this effect can be found in: Crowhurst, N. H. (1959). *Basic Audio Vol. 2*, F Rider Publisher Inc., New York, pp108-11. See also: 'Cathode Ray' (1955). Measurement of Non-Linearity Distortion, *Wireless World*, July. pp317-23.

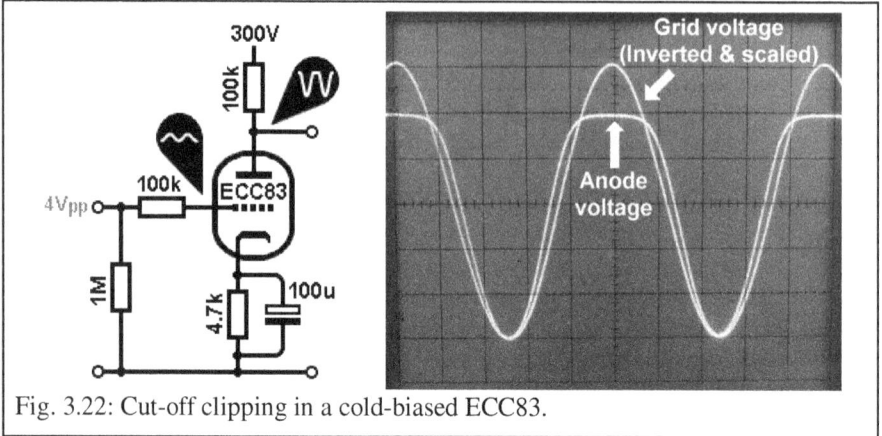

Fig. 3.22: Cut-off clipping in a cold-biased ECC83.

generate higher levels of high-order harmonics when clipped, making it more immediately audible than in a valve circuit.

There is an obvious question that some readers have probably been asking since the start: why are there no more grid curves shown to the left of the $V_{gk} = 0V$ curve? Curves for positive V_{gk} values do indeed exist, but they are not usually included on the graph because it is not practical to operate the valve in this region. The reason is that when the grid voltage approaches the cathode voltage, electrons being drawn from the space charge get attracted to the grid rather than to the anode, since the grid is very much closer, rather as if we had placed a diode between grid and cathode. Positive current therefore flows into the grid and down through the cathode. In other words, we get forward grid current. This current causes a voltage drop across whatever resistance happens to be in series with the grid, making it harder to drive the grid beyond $V_{gk} = 0V$. To put it crudely, less of our input voltage actually makes it to the grid. The more we attempt to make the grid positive the more current flows into it to prevent us from doing so, as if some invisible volume control is being suddenly turned down. In more technical terms, the input impedance of the valve suddenly falls from many megohms to a few kilohms or less, and this effect is known as **grid-current limiting**.[*] The grid signal will therefore be clipped on the positive side, but since the stage is inverting the anode signal will appear clipped on the negative side. But it must be fully understood that it is actually the *grid signal* which is being clipped; the valve continues to amplify what appears on its grid quite normally.

---

[*] Some modern texts may misleadingly refer to this as **saturation** because it resembles the clipping effect produced in a saturating transistor. However, saturation has a more specific meaning for valves. As explained earlier in section 3.1.1, it is the absolute maximum current that can be drawn from the cathode. This is well beyond the current levels ever found in a normal triode circuit.

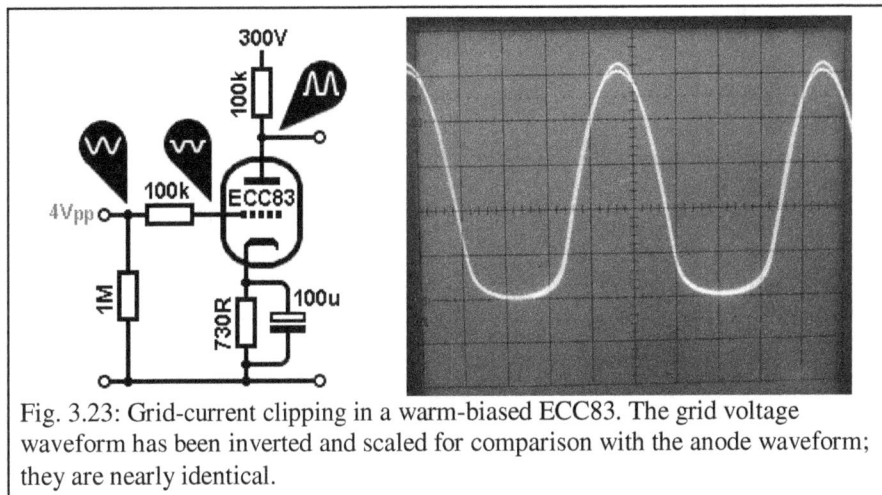

Fig. 3.23: Grid-current clipping in a warm-biased ECC83. The grid voltage waveform has been inverted and scaled for comparison with the anode waveform; they are nearly identical.

Fig. 3.23 shows another test circuit along with the waveforms viewed on the grid and anode during overdrive. Again, the grid signal has been inverted and scaled for comparison and it can be seen that both signals are nearly identical – the valve is amplifying just what it sees on the grid. Since biasing a valve close to the region of grid-current results in more quiescent power dissipation, it is called **warm biasing** or **hot biasing**.

In a similar manner as for cut-off clipping, the transition into grid-current limiting is relatively soft. Significant forward grid current begins to flow even before the grid reaches $V_{gk} = 0V$, and the ECC83 datasheet warns that it may reach 0.3µA at $V_{gk} = -0.9V$. The greater the source resistance, the greater the voltage drop caused by this grid current, and therefore the more abruptly the signal will be clipped. In other words, as clipping is approached, distortion will begin to increase sooner if the source impedance is large. The 100kΩ **grid stopper** (section 3.10) used in this example is deliberately excessive to illustrate clipping more clearly; in practice we would avoid such a large value in a hi-fi amp.

From the foregoing we can see that the ultimate headroom limits of a valve stage are set by cut-off at one end, and grid-current limiting at the other. However, we usually build in a little margin for error by defining our own 'soft' headroom limits, with the $V_{gk} = -1V$ grid curve at one end and the visible bunching of the curves at the other, as described in section 3.3.5 previously. Fortunately, the relatively soft nature of valve clipping means they can be operated quite close to the ultimate limits before the sound becomes really objectionable. No doubt this is part of 'the valve sound'. Many enthusiasts are quite happy with valve power amplifiers that deliver just a couple of watts and probably clip quite often during a listening session, whereas with transistor amplifiers most listeners find they need tens or hundreds of watts, not so much for extra loudness but for extra headroom, to avoid their hard clipping.

135

3.6: The Effect of Load on Distortion

It has been hinted at already that the value of anode load resistance has some influence over the distortion generated by the stage. To illustrate this, fig. 3.24 shows load lines for three possible resistances: 220kΩ, 100kΩ and 47kΩ. Clearly, increasing the load resistance causes the load line to rotate down and approach the horizontal axis. A bias of −1.5V has been used for this example, as indicated by the dot on each load line. This is slightly warmer than centre, so the valve will hit grid-current limiting before cut-off.

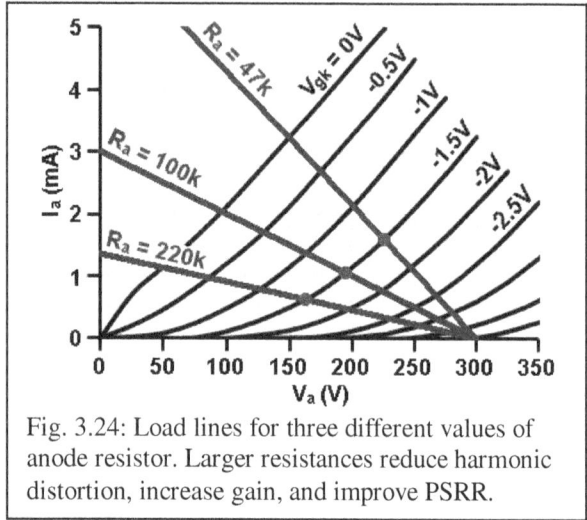

Fig. 3.24: Load lines for three different values of anode resistor. Larger resistances reduce harmonic distortion, increase gain, and improve PSRR.

Examining the figure we see that the greater the load resistance the larger the possible output voltage swing. In

Load resistance	Voltage gain	Maximum output swing
47kΩ	46	130V$_{pp}$
100kΩ	60	180V$_{pp}$
220kΩ	68	205V$_{pp}$

Table 1.1: Figures relating to fig. 3.24.

each case the *range* of possible bias voltages –i.e. the grid base– is exactly the same since the HT is the same for all of them, from which we deduce that the voltage gain must be greater when using larger load resistances. Table 1.1 summarises the figures.

The grid curves are more evenly spaced along the 220kΩ load line than along the 47kΩ load line, and more equal spacing means less distortion. Fig. 3.25 shows the distortion measured in the circuit with Sovtek and Mullard valve samples. Where the curves begin to bend upwards indicates the onset of clipping. Between R$_a$ = 47kΩ and 100kΩ there is a clear improvement in distortion, but between 100kΩ and 220kΩ the change is rather less impressive as we encounter approach the law of diminishing returns. It is also interesting that the highly-revered Mullard actually performs less well than the cheap Sovtek, yet both looked equally healthy on a curve tracer. The moral here is that real-world device variation is significant, so don't expect to be able calculate accurate distortion figures on paper alone (e.g. equation (3.4)). Nevertheless, the general truth is that the larger the load resistance, the greater the gain and available output swing, and the lower the harmonic distortion. The limiting case would be an infinite load impedance, resulting in a horizontal load line and minimum distortion. This is not possible using passive resistors since an infinite resistance would be an open circuit, so the circuit would not function at all. But it *is*

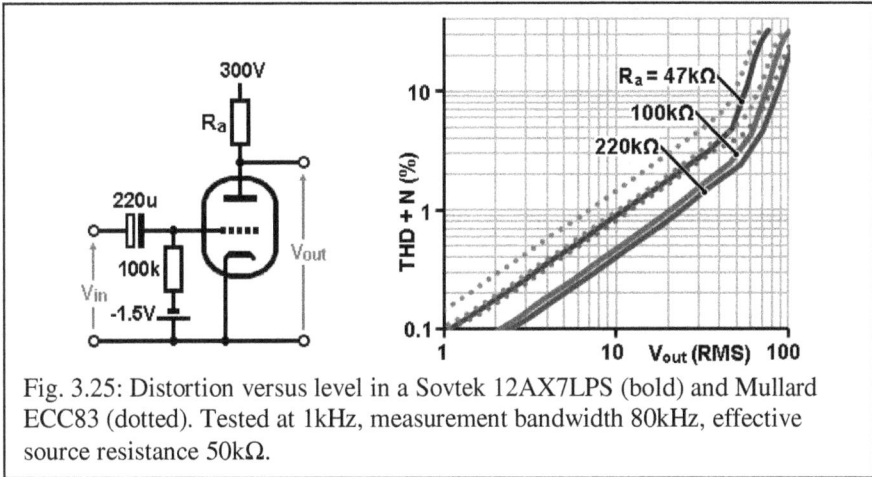

Fig. 3.25: Distortion versus level in a Sovtek 12AX7LPS (bold) and Mullard ECC83 (dotted). Tested at 1kHz, measurement bandwidth 80kHz, effective source resistance 50kΩ.

possible to use an active circuit which presents a near infinite impedance to AC signals while still allowing DC to flow. This is examined in chapter 6.

3.6.1: The Golden Ratio

When the anode load resistor is made equal to twice the internal anode resistance r_a, something special happens. Consider fig. 3.26 which shows the anode characteristics of a 'perfect' triode, and a load line corresponding to $R_a = 2r_a$. As indicated on the graph, the peak anode current that can flow (when $V_{gk} = 0V$) is exactly:

$$I_{peak} = \frac{HT}{(R_a + r_a)} = \frac{2}{3}I_{max}$$

The centre-bias point which gives maximum headroom is therefore half this value or $\frac{1}{3}I_{max}$ and results in a quiescent anode voltage of $\frac{2}{3}HT$, while the voltage gain rather neatly becomes equal to $\frac{2}{3}\mu$. It also offers close to the best compromise between output power and

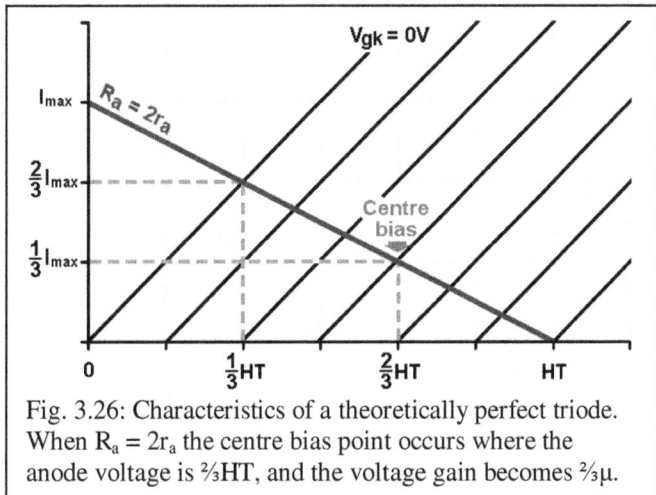

Fig. 3.26: Characteristics of a theoretically perfect triode. When $R_a = 2r_a$ the centre bias point occurs where the anode voltage is $\frac{2}{3}HT$, and the voltage gain becomes $\frac{2}{3}\mu$.

137

efficiency, although this last point is of little significance in a voltage amplifier stage.

A real triode is not perfectly linear so in practice this rule does not hold exactly, but in the days when a designer might have only a volt meter and no access to a full datasheet he could still use this 'golden ratio' of $R_a = 2r_a$ to build a fairly respectable hi-fi circuit. After choosing R_a the bias could be adjusted until the anode voltage was $\frac{2}{3}$HT at which point he could be reasonably sure of maximum headroom, and that the gain was two-thirds μ (the valve constants could be found in quick-reference tables that were published more widely than full datasheets). Of course, we do not have to apply the golden ratio today, but it is mentioned here as a historical reason why you often find anode resistors of roughly twice r_a in classic circuits, and why old textbooks often advise that triodes should be biased to an anode voltage of around $\frac{2}{3}$HT.[*]

3.6.2: The AC Load Line

Our gain stage has so far existed in isolation, but presumably it will be used to drive something else. In other words, the load impedance presented to the valve will not be R_a alone but a combination of R_a and whatever comes next in the audio chain. We therefore need to consider what effect this extra loading may have on the circuit, because it may cause us to reconsider our choice of anode resistor and bias point. Indeed, trying to optimise an analog circuit invariably requires us to consider simultaneously what also comes before and after it. It becomes an iterative process, possibly involving multiple readjustments along the way. This might seem like a lot to think about at once, but with practice –and a few rules of thumb– it becomes second nature.

Let us return to the circuit we had earlier and discover what happens if it is asked to drive a 38kΩ load, R_l (the reason for choosing this unlikely value is explained later in section 3.9.1). The signal voltage appearing at the anode is superimposed on the average anode voltage, so the output signal is really a varying DC voltage. But in most cases what we actually want is a pure AC signal to feed to the next

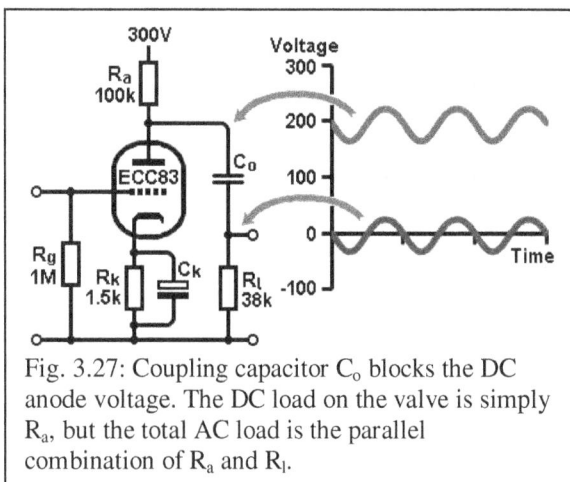

Fig. 3.27: Coupling capacitor C_o blocks the DC anode voltage. The DC load on the valve is simply R_a, but the total AC load is the parallel combination of R_a and R_l.

---

[*] This is in contrast to pentodes and transistors which can swing almost rail-to-rail, so they are often biased to an anode voltage of *half* the supply voltage.

stage of the amp, so an output coupling capacitor C_o is added to block the DC voltage while letting the AC signal pass through, as shown in fig. 3.27. This is called **CR coupling, capacitor coupling**, or usually **AC coupling**. C_o also forms a high-pass filter with R_l, so signals below the cut-off frequency will be attenuated. For now we will assume C_o is large enough[†] to pass all audio frequencies.

As far as DC currents are concerned, nothing has changed; the anode load is simply R_a. AC currents, however, can now also flow in R_l via C_o, so the total load as far as AC signals are concerned is the parallel combination of R_a and R_l, or:

$$R_{ac} = R_a \parallel R_l = \frac{R_a R_l}{R_a + R_l} = \frac{100k \times 38k}{100k + 38k} = 28k\Omega$$

We can show this on the static anode characteristics by first drawing the ordinary DC load line and choosing a bias point somewhere on it, and then drawing an **AC load line**. Since the coupling capacitor does not affect the DC conditions, the AC load line must pass through the bias point. Fig. 3.28 shows the DC load line and a bias point of -1.5V. One way to find the AC load line is to suppose the anode voltage changes by some arbitrary amount, say 100V. From Ohm's law we know the current in the AC load must change by: $100V / 28k = 3.6mA$. We can therefore plot

a point that is 100V lower than the quiescent anode voltage, and 3.6mA higher than the quiescent anode current. This is labelled A, and the AC load line is then drawn through A and the bias point.

The AC load line only applies while an AC signal is being amplified, that is, while the valve is working under dynamic rather than static conditions. It is therefore sometimes called the **working load line**. The DC

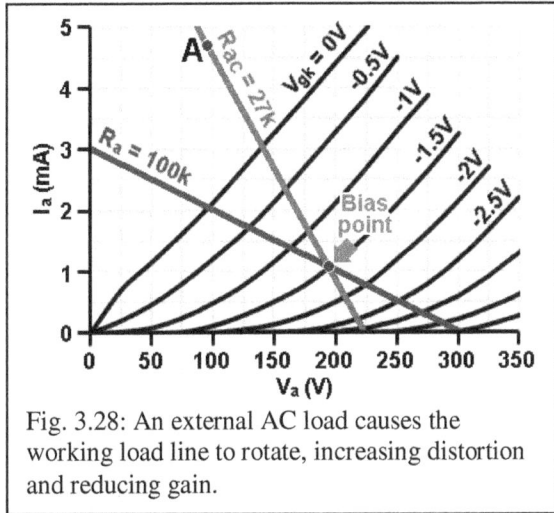

Fig. 3.28: An external AC load causes the working load line to rotate, increasing distortion and reducing gain.

load line has a slope of $-1/R_a$, but the addition of the AC load causes the working load line to rotate clockwise around the bias

Load line	Voltage gain	Maximum output swing	Centre bias point
DC (100kΩ)	60	180V_{pp}	-2V
AC (28kΩ)	30	85V_{pp}	-1.5V

Table 1.2: Figures relating to fig. 3.27.

[†] By 'large' I mean high *capacitance*, not physically large.

point, so it has a steeper slope of $-1/(R_a\|R_l)$.

Astute readers may wonder how the peak current in the AC load can be greater than the current actually supplied by R_a. This is possible because when the circuit is first switched on C_o charges up and stores energy. When the anode voltage swings low the 'extra' current is sourced from the coupling capacitor, and when the anode swings high again, current in R_a is diverted back into C_o to recharge it.

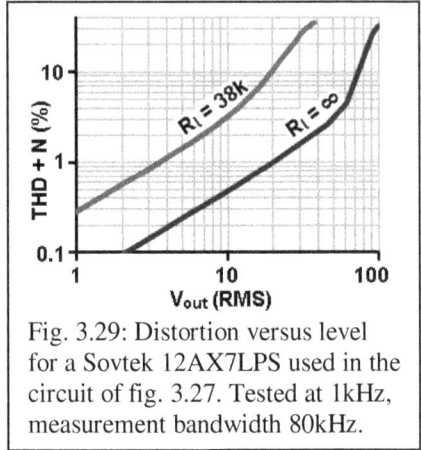

Fig. 3.29: Distortion versus level for a Sovtek 12AX7LPS used in the circuit of fig. 3.27. Tested at 1kHz, measurement bandwidth 80kHz.

The extra loading presented by R_l affects operation in several ways. The maximum output swing and gain are both reduced. In this case the gain is reduced by exactly half for reasons explained later in section 3.9. A further important change is that the point of centre-bias has moved. Although the bias appears warm when looking at the DC load line it is actually about central on the AC load line, because the anode voltage cannot swing so high any more. Table 1.2 outlines these figures.

The steeper load line also causes the distortion to suffer. Fig. 3.29 indicates roughly a 6-fold difference between the unloaded and loaded cases, and fig. 3.30 shows the distortion spectrum at $10V_{rms}$ output in each case. With the AC load the third and fourth harmonics are about 5dB larger, relative to the second harmonic (again, the fundamental has been nulled out). In this case an extreme value for R_l was chosen to illustrate the changes in operation clearly. In practice we would be rather kinder, avoiding such heavy loading by making sure R_l is much larger than R_a, so the working load line will not rotate so much.

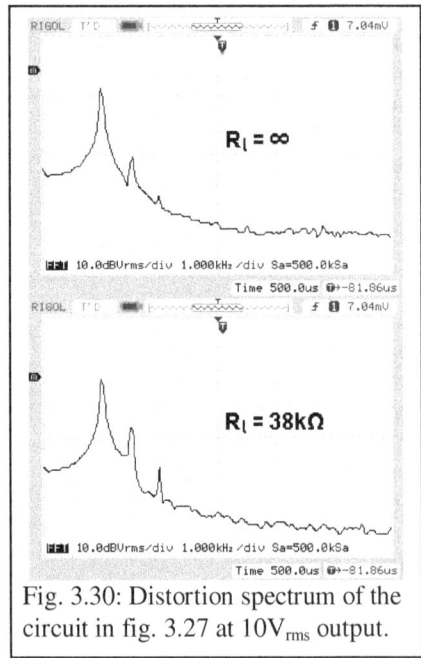

Fig. 3.30: Distortion spectrum of the circuit in fig. 3.27 at $10V_{rms}$ output.

3.7: The Cathode Bypass Capacitor

Now that we have a better understanding of how a triode amplifier works we can turn our attention back to the cathode bias circuit. In several of the previous circuits a capacitor has been added in parallel with the cathode resistor, and an explanation is deserved. To appreciate the function of this capacitor, first imagine

140

that it is *not* fitted; only the bias resistor is used. We know that when the grid voltage increases the anode current also attempts to increase, and this in turn is used to develop a larger voltage signal at the anode. But since the anode current also flows in the cathode resistor R_k, the voltage across R_k will vary in sympathy with the anode current, i.e. it will attempt to follow the grid voltage. This in turn means that the voltage difference *between* grid and cathode will be less than the applied signal voltage alone. But what the valve actually amplifies *is* the difference between grid and cathode, so by itself R_k causes the valve to see a smaller input signal, so the apparent gain of the circuit will be less than the load line suggests. In other words, the cathode resistor introduces local negative feedback, referred to as **cathode current feedback** or **cathode degeneration**. This reduces distortion[*] and increases headroom, but it increases the output impedance and worsens PSRR too (section 3.12).

Less distortion and more headroom may be just what we need in some circuits, but not always. To eliminate the effect of cathode degeneration we need to stop the cathode voltage from following the grid, but we don't want to interfere with the DC bias voltage. This is easily done by adding a **cathode bypass capacitor** C_k, in parallel with R_k (fig. 3.32). Any rapid rise in cathode current will now be diverted into charging the capacitor and, when cathode current falls, the capacitor will supply the deficit from its own charge. Another way of looking at it is to say that the capacitor **decouples** or bypasses to ground any AC signals on the cathode, so signal current does not flow in the cathode resistor, but the DC bias voltage is blocked and so remains untouched. With either explanation the result is the same; the cathode bypass capacitor smoothes out changes in cathode voltage, helping to hold the cathode voltage constant and thereby preventing cathode degeneration.

Of course, at low frequencies the capacitor's reactance rises and it eventually becomes powerless to prevent degeneration, so the gain drops to some minimum value. We therefore get a step or 'shelf' in the frequency response, also called a **pole-zero pair**. This is illustrated in the Bode plot in fig. 3.31 (the actual response is shown faint). Ordinarily we would use a sufficiently large capacitance that this step is pushed far below the audible range, in which case the

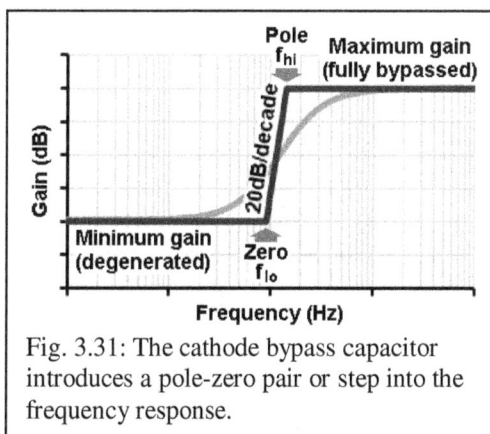

Fig. 3.31: The cathode bypass capacitor introduces a pole-zero pair or step into the frequency response.

---

[*] For some very detailed discussion of this see: Roddam, T. (1951), How to Choose a Valve, *Wireless World*, June pp221-3 & October, pp409-11.
Also the suspiciously similar article: Cooper, J. F. (1957). Which Tube Shall I Use?, *Audio*, November, pp30-7.

141

stage can be said to be **fully bypassed**. The pole frequency (where the gain starts to fall from its upper level) is given by:

$$f_{hi} = \frac{1}{2\pi C_k (R_k \parallel r_k)}$$ (3.5)

Where r_k is the **internal cathode resistance**. This is the internal resistance as seen looking into the cathode, which is equal to $(R_a + r_a)/\mu$. The frequency response then falls at a rate of 20dB/decade until it reaches the zero and levels off at the minimum or degenerated gain of the circuit. The zero occurs at a frequency of:

$$f_{lo} = \frac{1}{2\pi R_k C_k}$$ (3.6)

In order to keep the overall preamp's frequency response flat down to 20Hz, the pole needs to be much lower than this, since other stages will add their own roll-offs and the effects are cumulative. Also, because the bypass capacitor is likely to be an electrolytic it will have poor tolerance, so this is a further incentive to push the pole into subterranean regions. Fortunately, the cathode voltage is small so only a low voltage rating is required, so there is little trouble in using an oversized value of capacitance. Equation (3.6) (the zero) is a lot quicker and more convenient to use than equation (3.5) (the pole) so it is usually only necessary to choose a capacitor that pushes the zero below 1Hz, since the pole is rarely far behind.

Rearranging equation (3.6) and using our existing example of an ECC83 with a 1.5kΩ bias resistor, pushing the zero below 1Hz requires a bypass capacitor larger than:

$$C_k = \frac{1}{2\pi f_{lo} R_k} = \frac{1}{2\pi \times 1 \times 1500} = 106 \times 10^{-6} F \quad or \quad 106\mu F$$

We might settle for 100µF as a convenient value, or use 220µF to be on the safe side. A 10V-rated part would do, but a 25V-rated part is not very bulky and capacitor

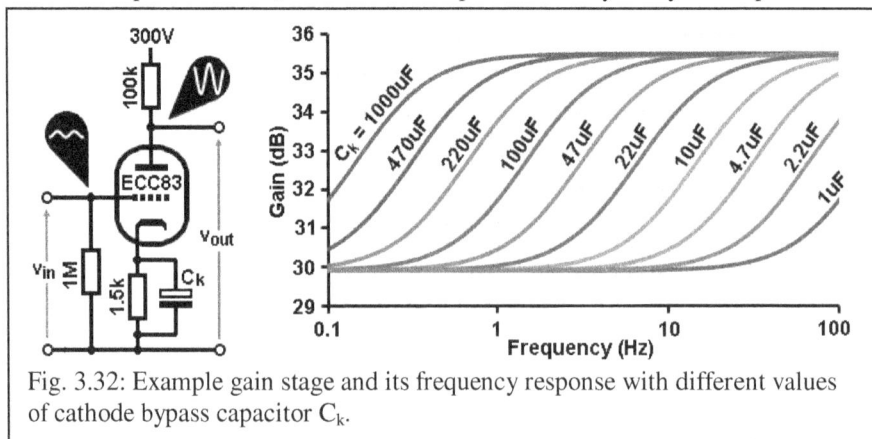

Fig. 3.32: Example gain stage and its frequency response with different values of cathode bypass capacitor C_k.

142

defects tend to fall with higher voltage ratings. Fig. 3.32 shows how the frequency response changes with different values of C_k. Halving or doubling its value shifts the whole response up or down by exactly one octave. The degenerated gain depends on R_k and a formula for finding it is presented in section 3.8.1. Other types of triode will give different values of maximum and minimum gain, depending on their own anode and cathode resistors, but the *shape* of the frequency response is always the same.

3.8: Equivalent Circuits

Using graphical methods to determine the gain and general circuit operation is highly instructive, and with practice it is possible to estimate how a circuit will behave simply by visualising load lines in your head. However, it is also handy to have some more universal equations that can be used for quick calculations, or even a general method for generating equations that apply to our own particular designs. This can be done using equivalent circuits, in which the valves are replaced by Thévenin or Norton sources. However, remember from chapter 1 that these theorems are only suitable for modelling *linear* circuits. We therefore make the assumption that the signal levels are so small that the valves do, in fact, operate more-or-less linearly (i.e. nowhere near clipping), so this approach is also called **small-signal analysis**. At this stage we are only interested in AC signals rather than DC conditions, so any imaginary generators can be purely AC signal generators; we assume the DC voltages and bias points have already been taken care of.

- For the **Thévenin equivalent** the valve is modelled as a *voltage* generator equal to $-\mu \cdot v_{gk}$, in series with a resistance equal to r_a, as in fig. 3.33a.
- For the **Norton equivalent** the valve is modelled as a *current* generator equal to $g_m \cdot v_{gk}$, in parallel with a resistance equal to r_a, as in fig. 3.33b.

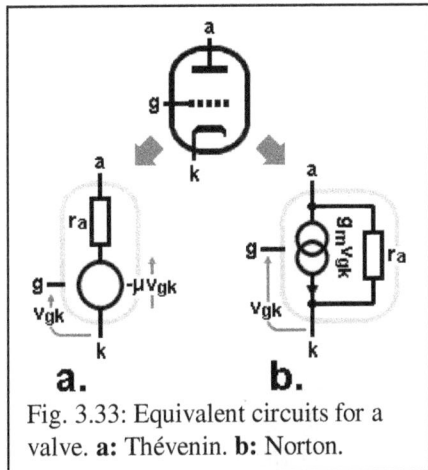

Fig. 3.33: Equivalent circuits for a valve. **a:** Thévenin. **b:** Norton.

The two models are duals of one another and ultimately give the same results, but sometimes one is easier to handle than the other, leading to simpler algebra. It is also important to observe the direction of the voltage/current arrows. This explains the presence of the minus sign in the Thévenin model; the arrow head is towards the anode so it represents the voltage between anode and cathode (the familiar way to talk about voltages), which we know is inverted relative to the voltage between grid and cathode. On the other hand, it is easy to make silly algebraic mistakes with minus signs, so it is also common to see the arrow reversed and the minus sign removed; the result is the same either way.

143

3.8.1: Gain Equations

Fig. 3.34 shows an example of the steps we might take to generate a whole equivalent circuit. The image in a. is a simple gain stage. A power-supply decoupling capacitor C_s has been included for reasons that will emerge. The first step is to convert the valve into either its Thévenin or Norton equivalent, and the Thévenin model is chosen here, which takes us to b. Now, we are only interested in AC signals, so if we imagine the frequency to be high enough that all capacitors can be treated as short circuits then we arrive at c. Notice that by replacing C_k with a short circuit we have also eliminated R_k. The last step is really just tidying up, except for one small modification: since R_k has been eliminated we find that v_{in} is in fact the same as the grid-to-cathode voltage, so it has been relabelled v_{gk} in d.

The final equivalent circuit is a lot simpler than what we started with. The valve now looks like a voltage generator that takes whatever signal voltage appears between grid and cathode (v_{gk}) and multiplies it by $-\mu$. Because the power supply capacitor C_s looks like a short circuit we find that R_a –which is in reality connected between HT and the anode– appears to be connected between anode and ground. In other words, as far as AC is concerned the power supply and ground are *one and the same*, so r_a and R_a form a potential divider. The final output signal is therefore the generator voltage multiplied by the gain of this potential divider.

$$v_{out} = -\mu v_{gk} \frac{R_a}{R_a + r_a}$$

In this case v_{gk} is exactly equal to the input voltage, so we can also write:

$$v_{out} = -\mu v_{in} \frac{R_a}{R_a + r_a}$$

The voltage gain of a circuit is v_{out}/v_{in}, so by dividing both sides by v_{in} we quickly obtain the gain of the fully bypassed gain stage:

$$A_{(bypassed)} = -\mu \frac{R_a}{R_a + r_a} \tag{3.7}$$

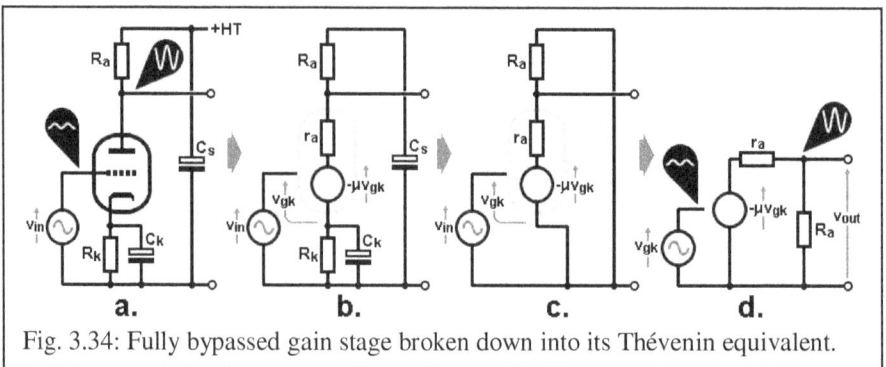

Fig. 3.34: Fully bypassed gain stage broken down into its Thévenin equivalent.

Using the ECC83 / 12AX7 circuit from earlier, where $R_a = 100k\Omega$, $r_a = 60k\Omega$ and $\mu = 95$ (as determined from the anode characteristics *at the bias point*), the gain can be predicted as:

$$A_{(bypassed)} = -95 \times \frac{100k}{100k + 60k} = -59.4 \quad (35.5dB)$$

This is very close to the figure of -60 obtained from the load line in section 3.3.1. Measurement of a Sovtek 12AX7LPS yielded a value of 59.1 (35.4dB), which is reassuring.

The previous model assumed that R_k was fully bypassed. If no bypass capacitor is used, or if the frequency is low enough for it to act like an open circuit, then R_k will degenerate the stage and the gain will be reduced. For this we need a new equivalent circuit, as shown in fig. 3.35. From Ohm's law the current flowing around the circuit must be:

$$i = \frac{v}{R} = \frac{-\mu v_{gk}}{R_a + r_a + R_k}$$

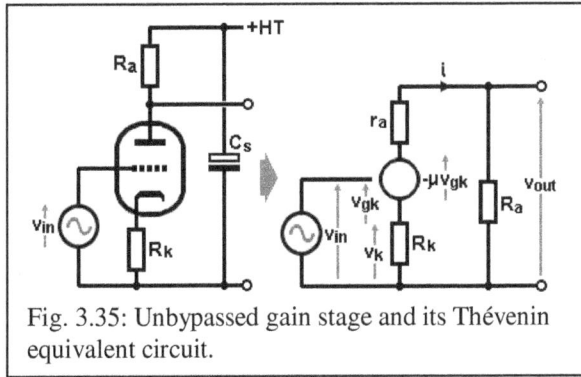

Fig. 3.35: Unbypassed gain stage and its Thévenin equivalent circuit.

But unlike the previous circuit v_{in} is not equal to v_{gk} because some of it is cancelled out by the opposing voltage appearing across the cathode resistor, or in other words:

$$v_{gk} = v_{in} - v_k = v_{in} - (-iR_k)$$

Substituting this into the previous equation and simplifying gives:

$$i = \frac{-\mu(v_{in} + iR_k)}{R_a + r_a + R_k} = \frac{-\mu v_{in}}{R_a + r_a + R_k + \mu R_k}$$

Notice that degeneration has an effect like adding an extra resistor 'μR_k' to the output circuit. By multiplying this current by R_a we get the output voltage, and then also dividing both sides by v_{in} we find the overall gain of the circuit:[*]

$$A_{(degenerated)} = -\mu \frac{R_a}{R_a + r_a + R_k(\mu + 1)} \quad\quad (3.8)$$

---

[*] The degenerated gain can also be derived directly from the universal feedback equation, although you need to be careful to enter the right expressions. The open-loop gain is $\mu R_a/(R_a+r_a+R_k)$ and the feedback fraction is R_a/R_k.

Again using our ECC83 / 12AX7 gain stage, with $R_k = 1.5\text{k}\Omega$:

$$A_{(degenerated)} = -95 \times \frac{100\text{k}}{100\text{k} + 60\text{k} + 1.5\text{k} \times (95+1)} = -31.3 \quad (29.9\text{dB})$$

This is exactly what was shown in fig. 3.32 where the gain bottoms out as the cathode bypass capacitor becomes ineffective. Measurement of a Sovtek 12AX7LPS yielded a degenerated gain of 30.7 (29.7dB), so the calculation is pretty good. But in case readers are now worrying that they will have to go through all this circuit analysis every time they design a preamp, rest assured that it is not essential. Although it is a valuable skill, it is possible to avoid it by using approximations, general equations given in this book, computer simulation, and a little real-life experimentation. Equations (3.7) and (3.8) were derived in full simply to show the usefulness of equivalent circuits.

3.9: Impedance Matching and Bridging

Whenever one circuit is coupled to another it is important to know what effect this will have on the circuit being driven, or on the circuit doing the driving, and we have already touched on this in section 3.6.2. The **output impedance** of the driving circuit, and the **input impedance** of the next circuit, form a potential divider or attenuator, as illustrated in fig. 3.36. Depending on the context these impedances may also be called the **source impedance** and **load impedance** respectively. Like any impedance they may consist of a mixture of resistance, capacitance, and inductance, so we may also talk specifically about the source *resistance* or load *capacitance*, for example. This emphasises that the unavoidable potential divider created between them may also act as a filter.

If the source and load resistances happen to be equal to each other then the greatest amount of *power* is passed from source to load, and this is called **impedance matching**. However, in an audio preamp we are usually more interested in signal *voltages*, and for maximum *voltage* transfer the input impedance must be as large as possible with respect

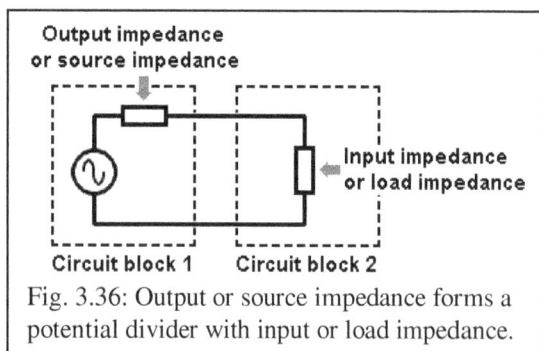

Fig. 3.36: Output or source impedance forms a potential divider with input or load impedance.

to the source impedance, which is called **impedance bridging**. In other words, we usually do our best to ensure the load impedance does not grossly 'load down' the source impedance, or the signal will be attenuated too much.

Although it wasn't stated at the time, we have already seen a prime example of this in fig. 3.34. There we saw that the output resistance *of the valve itself*, r_a, formed a potential divider with the anode load resistor R_a. Therefore, the smaller we make R_a, the smaller the voltage transfer between the valve and R_a becomes, i.e. the smaller the voltage gain. Going further, we can perceptually 'zoom out' and talk about the input and output impedances of the gain stage *as a whole*. These impedances will in turn form potential dividers with the input/output impedances of any other circuit blocks they are connected to. These too can be broken down into equivalent circuits.[*]

3.9.1: Output Impedance

Fig. 3.37a shows the Thévenin equivalent circuit of a fully bypassed gain stage (repeated from fig. 3.34) ready for analysis. Only resistances are considered here, so what we are about to calculate is strictly the output *resistance*. This is rather simplistic since the valve does not contain only the resistance r_a, but also capacitance between anode and cathode, and between anode and grid, as well as lead inductances. Fortunately these components are small enough that they can often be ignored at audio frequencies. For example, the anode-cathode capacitance of the ECC83 is given on the datasheet as about 0.3pF. Even if we increase this to 1.3pF to allow for the additional capacitance of the valve socket and associated wiring, its reactance is over 6MΩ at 20kHz, which is practically infinite compared to the resistances in circuit. These extra 'parasitic' components are, however, important in other ways, as shown later.

To find the output resistance we first set any independent sources to zero. In this case the only 'fixed' or independent source is v_{gk}, and setting it to zero volts makes it a short circuit, taking us to b. This leaves the dependent source $-\mu v_{gk}$, but since there is now nothing for it to amplify, this source also becomes zero, as in c. Finally we can tidy up the drawing as in d., and find the total resistance between the output terminals. In this case it is obviously r_a and R_a in parallel, so we immediately know from the product-over-sum rule (equation (1.19)) that the output resistance of a fully bypassed gain stage is:

$$R_{o(bypassed)} = \frac{R_a \times r_a}{R_a + r_a} \tag{3.9}$$

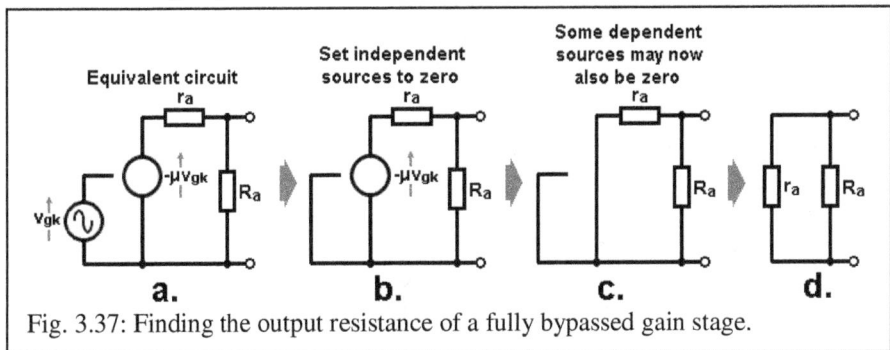

Fig. 3.37: Finding the output resistance of a fully bypassed gain stage.

[*]One way to determine output impedance is to imagine applying a $1V_{ac}$ signal to the *output* of the circuit; how much current would flow into it? The ratio of this enforced voltage and the resulting current is the output impedance.

As an example, let us find the output resistance of our ECC83 gain stage. If $R_a = 100k\Omega$ and $r_a = 60k\Omega$ (at the bias point) then the output resistance becomes:

$$R_{o(bypassed)} = \frac{100k \times 60k}{100k + 60k} = 38k\Omega$$

This means that if the stage is connected to one with a 38kΩ input impedance we will have a 1:1 potential divider, so half the possible signal voltage will be lost across the output resistance or, in other words, the gain of the overall circuit would appear halved. This is exactly what the AC load line in fig. 3.28 showed, since it represented an extra 38kΩ load, which is exactly why it was chosen for the example. Indeed, a quick and dirty way to measure the output resistance of a real-world circuit is to feed the input with a signal generator so that it produces a small output signal, which is measured and noted. Then add an external load resistance and adjust it until the output signal is reduced to half its initial value; at this point the external load is equal to the output resistance.

For a degenerated gain stage the outcome is a little different because one of the properties of feedback is that it alters the input and output impedances. Fig. 3.38 shows the equivalent circuit, and setting v_{in} to zero takes us to b. However, we cannot immediately eliminate the dependent source because v_{gk} is not zero; it is in fact equal to the voltage across R_k. Without going through the full treatment, we already saw in the previous section that R_k has the effect of adding another resistance to the output circuit, μ times larger than R_k, as shown in c. In other words, impedances in the cathode circuit appear multiplied by μ+1 when looking into the anode, as shown in d., where the two resistors have been collapsed into one. The output resistance is therefore $R_a \parallel (r_a + R_k(\mu+1))$ or:

$$R_{o(degenerated)} = \frac{R_a(r_a + R_k(\mu+1))}{R_a + r_a + R_k(\mu+1)} \tag{3.10}$$

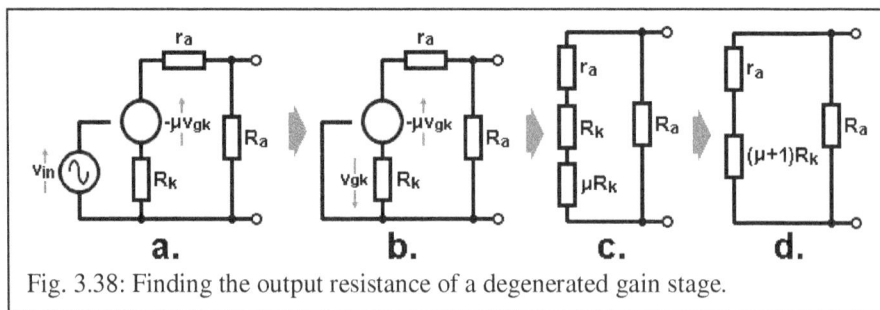

Fig. 3.38: Finding the output resistance of a degenerated gain stage.

For our example circuit, leaving the cathode unbypassed would raise the output resistance to:

$$R_{o(degenerated)} = \frac{100k \times \left(60k + 1.5k(95+1)\right)}{100k + 60k + 1.5k(95+1)} = 67k\Omega$$

3.9.2: Input Impedance and the Miller Effect

Assuming the valve is biased sufficiently negative that grid current does not flow, the input resistance of the grid is practically infinite; it is basically an open circuit. However, a grid leak resistance of some form is always required, so this sets the input resistance of the stage as a whole, as shown in fig. 3.39. This is commonly 1MΩ which is large enough that it should not cause undue attenuation with most ordinary source impedances.

The input *capacitance* is another matter, and one that is not small enough to ignore this time. Any two conductors separated by an insulator form a capacitor, so there are unavoidable parasitic capacitances inside the valve, called the **interelectrode capacitances**. The important ones are

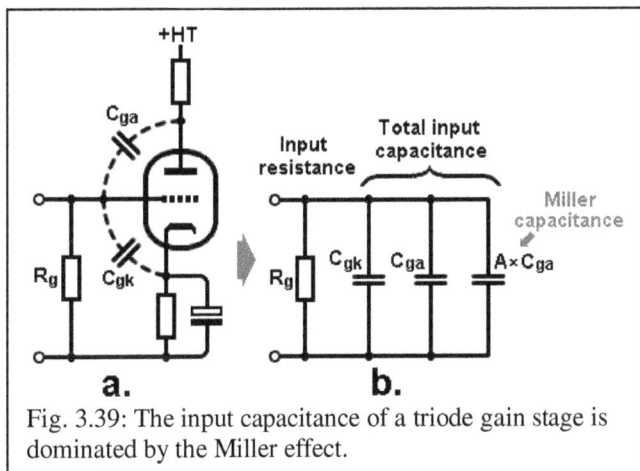

Fig. 3.39: The input capacitance of a triode gain stage is dominated by the Miller effect.

between grid and anode (C_{ga}) and between grid and cathode (C_{gk})[*] as represented in fig. 3.39a. Input current will flow in these capacitances just as if they were actual components connected between the grid and ground, which is how they appear in the equivalent circuit in b. However, because the anode voltage signal is an amplified version of the signal on the grid, the voltage across C_{ga} will be much larger than if the anode voltage were fixed, so the resulting input current must also be larger. For example, if the gain of the stage is −60 and we input a +1V signal, the anode voltage will fall by 60V, so the total signal voltage appearing across C_{ga} will be 61V. As far as the input is concerned it looks like we have an additional capacitive load that is 60 times larger than the C_{ga} we already had. The total input capacitance seen between the input terminals is therefore:

$$C_{in} = C_{gk} + C_{ga}(A+1) \tag{3.11}$$

---

[*] Some datasheets do not provide C_{gk} but instead give the "grid to all except anode" capacitance which can be used instead.

Where A is the magnitude of the voltage gain (i.e. ignoring any minus sign). This multiplication[†] of the input capacitance is called the **Miller effect**, and the additional fictional capacitor is called the **Miller capacitance**. But since this component nearly always dominates the total, engineers casually refer to the whole lot simply as the 'Miller capacitance' of the circuit.

Returning to our example gain stage, the ECC83 datasheet quotes 1.6pF for both C_{ga} and C_{gk}. However, the stray capacitance between the valve socket pins and associated wiring will add to these figures. The ECC83 is unfortunate in having the grid and anode pins right next to each other, which will typically add 1.5pF or so; the same goes for the grid and cathode pins, making an estimated 3.1pF for both. We calculated the voltage gain to be 60 (ignoring any minus sign), so the total Miller capacitance should be:

$$C_{in} = 3.1 + 3.1 \times (60 + 1) = 192.2pF$$

Measurement with a Sovtek 12AX7LPS produced a figure of about 190pF. This input capacitance will form an RC low-pass filter with the source resistance, attenuating high frequencies and possibly causing distortion due to slew rate limiting (section 3.11). Achieving sufficient bandwidth and maintaining low distortion therefore requires careful attention to the Miller capacitance of the following stage. If the stage is degenerated then its gain will be lower and with it the Miller capacitance. The input impedance is therefore increased (a smaller capacitor has greater reactance) which is exactly what we expect from series-applied feedback. From section 3.8.1 the degenerated gain was calculated to be about 31 so we would expect the total input capacitance then to be 102pF. The same Sovtek 12AX7LPS was tested and found to have a degenerated gain of 30.7 and a resulting Miller capacitance of 115pF. The difficulty in predicting strays is an inevitable source of error when trying to predict Miller capacitance.

3.10: Oscillation and Stoppers

In all the previous equivalent circuits we were only concerned with performance at audio frequencies, so most of the parasitic components could be omitted as they are very small. However, these parasitics can still cause problems, specifically unwanted high-frequency oscillation. This is likely to occur in the range of tens to hundreds of megahertz and is fiendishly difficult to detect by normal means (although it may be possible to pick it up on a nearby AM radio). Despite being inaudible, parasitic oscillation will wreak havoc with the audio performance since it consumes headroom and could damage loudspeaker tweeters. Symptoms include

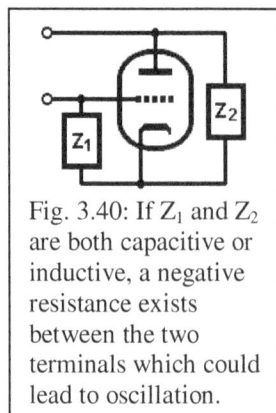

Fig. 3.40: If Z_1 and Z_2 are both capacitive or inductive, a negative resistance exists between the two terminals which could lead to oscillation.

[†] The general term for electronic impedance multiplication or division is **bootstrapping**.

unusually high distortion, rustling sounds, inexplicable voltages or voltages which seem to meander around and refuse to settle when measured. In some cases the oscillation is only triggered while amplifying audio signals of a particular amplitude, making it even more difficult to detect. On an oscilloscope it may appear as a thickening or fuzzing on particular parts of the audio waveform.

Oscillation can occur due to positive feedback via the stray capacitance between early and late stages of an amplifier, in which case better layout and/or shielded cables may be necessary. But even a single stage by itself can oscillate. For example, it can be shown[3] that the impedance existing between the grid and anode of fig. 3.40 is:

$$Z_{ga} = g_m Z_1 Z_2 + Z_1 + Z_2 \qquad (3.12)$$

If Z_1 and Z_2 are both inductive or both capacitive they will combine to form a negative resistance. The worst situation occurs when they are both inductive, because the above equation then becomes:

$$Z_{ga} = -g_m \omega^2 L_1 L_2 + j\omega(L_1 + L_2)$$

The first term is a negative resistance and the second term (in series with it) is an inductance. If a capacitor is then connected in parallel with this combination, i.e. between anode and grid, then we have an oscillator. Just such a capacitor is already built into the valve in the form of C_{ga}, and the inductances are also

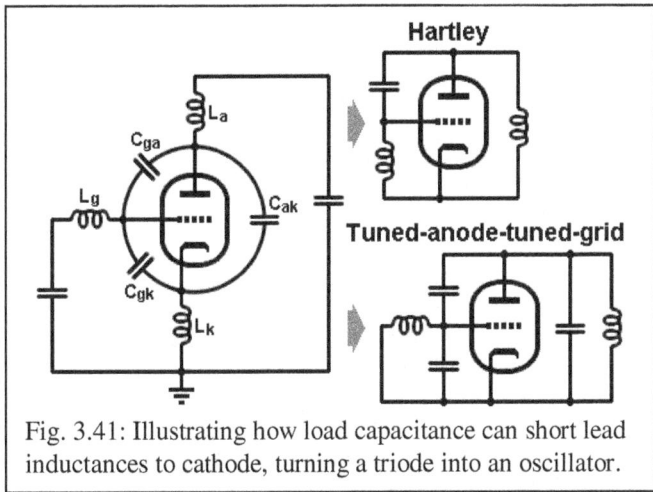

Fig. 3.41: Illustrating how load capacitance can short lead inductances to cathode, turning a triode into an oscillator.

built in, in the form of the wires leading to the grid and anode. In the megahertz range these inductances may effectively become shorted to cathode (ground) by load and stray capacitances, completing a Hartley oscillator or tuned-anode-tuned-grid oscillator, as shown in fig. 3.41.

The name of the oscillator is of little consequence to us; all we really need to know is that the oscillation is caused by inductance in the grid and anode circuits (coupling

---

[3] Amos, S. W. & Birkinshaw, D. C. (1969). *Television Engineering, Principles and Practice, Vol. 3* (2nd ed). Illife Books Ltd, London, p253

transformers therefore increase the risks), and that the negative input resistance is proportional to g_m, so high-g_m valves are more likely to oscillate than low-g_m types. Good layout with short leads can reduce the possibility of oscillation, but is not a guaranteed cure. Reliably stopping oscillation requires the addition of a damping resistor to one or both offending areas, and such resistors are therefore called the **grid stopper** and **anode stopper**, as shown in fig. 3.42. The anode stopper may alternatively be placed in series with the output as in fig. 3.42b, in which case it may be called a **build-out resistor**.

The anode stopper must be small compared with r_a otherwise it will cause excessive increase of output resistance and undue loss of gain. Something between 10Ω and 100Ω is typical. Ideally, the positive resistance of the grid stopper cancels out any negative input resistance, and a value of about $2/g_m$ or 100Ω to 2kΩ is generally sufficient. Larger values can be used for

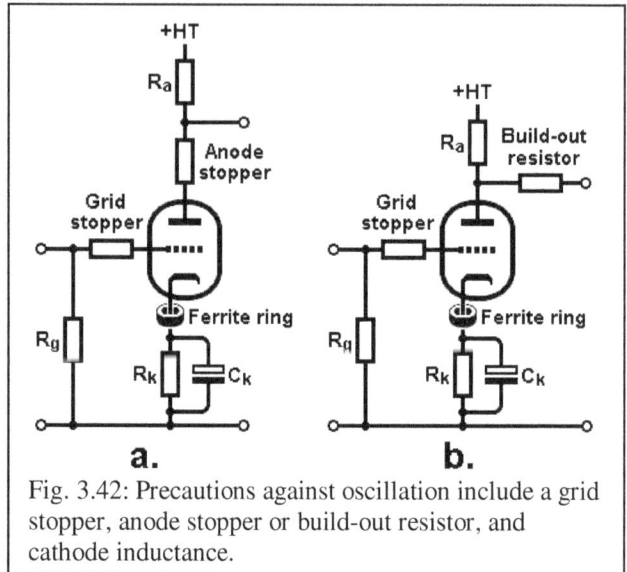

Fig. 3.42: Precautions against oscillation include a grid stopper, anode stopper or build-out resistor, and cathode inductance.

more robust protection, although the added noise may be objectionable in sensitive applications like the front end of a phono stage. Also, do not forget that the grid stopper creates a low-pass filter with the Miller capacitance, so too much resistance could lead to loss of treble.

Stopper resistors must be mounted as close to the valve socket as possible in order to minimise lead inductance. Inductance in the cathode circuit is not usually a problem and may in fact stabilise an otherwise troublesome circuit. A cathode bypass capacitor naturally behaves like an inductor at the problem frequencies, though a ferrite ring around the cathode lead may also help. These precautionary components are very cheap insurance and a tell-tale sign of a knowledgeable circuit designer.

3.11: Slew Rate and Reactive Loads

The slew rate of an amplifier is the maximum rate at which the output voltage can change or 'slew' from one value to another. 'Speed' is sometimes used as a vague sort of synonym for slew rate, though it is not a technical term. The required slew rate depends on the highest frequency we want to amplify, and how much we need to amplify it, i.e. the voltage swing. This is important when driving capacitive loads because the current demanded by a capacitor depends on how fast we try to change the voltage across it. Stating the slew rate of a circuit is a roundabout way of saying what maximum frequency it can deliver into a reactive impedance, *at a given amplitude*. Beyond this limit **slewing distortion** occurs.

It is important to understand that slew rate has nothing to do with bandwidth (a fact which is almost never grasped by the popular audio press). Bandwidth is a different limit that we may or may not hit before we run into slew-limiting distortion. Running out of bandwidth simply means the output signal will be attenuated relative to mid-band frequencies, but it does not mean it will be distorted. For listeners, attenuation is a lot less irksome than distortion, so we will want to ensure that slew limiting can only happen at frequencies outside the audio range. If we want to take advantage of the *full* bandwidth *at a given signal*

amplitude, then there will be a minimum slew rate requirement that the circuit must meet. But just because a circuit has a high slew rate does not mean it has a wide bandwidth, or *vice versa*; the two are independent of one another.

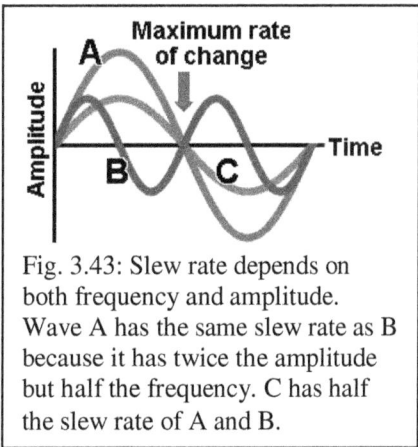

Fig. 3.43: Slew rate depends on both frequency and amplitude. Wave A has the same slew rate as B because it has twice the amplitude but half the frequency. C has half the slew rate of A and B.

To find the slew rate of a sine wave we begin with the general equation:

$$v = V_{pk} \sin(2\pi ft)$$

We want to know how fast the voltage is changing at any given point on the signal, which is found by differentiation:

$$\frac{dv}{dt} = \frac{d\ V_{pk} \sin(2\pi ft)}{dt} = 2\pi f V_{pk} \cos(2\pi ft) \qquad (3.13)$$

However, we are really only interested in the *maximum* rate of change. This occurs at the zero-crossings of the original sine wave, as shown in fig. 3.43. In the above expression, this corresponds to when the *cosine* function reaches its peak value of 1, leaving us with a maximum rate of change, or slew rate (SR), of:

$$SR = 2\pi f V_{pk} \qquad \text{volts per second} \qquad (3.14)$$

For example, at its fastest point (the zero crossing) a $1V_{pk}$ 20kHz sine wave achieves a slew rate of 125664 volts per second. This is a cumbersome number, so it is rather more convenient to work in microseconds by diving by a million, giving 0.125V/μs. However, if the wave is increased to $2V_{pk}$ then it must swing more volts in the same amount of time, so its slew rate must be faster; twice as fast to be precise, or 0.25V/μs. This is represented by A and C in fig. 3.43, which further emphasise how slew rate depends on both frequency *and* amplitude.

If we are designing a gain stage and we know how much output voltage swing is required to drive the *following* stage, how much output current is needed? If the load were purely resistive then it would simply be v/R. But of course, the load is never just resistive –it will include at least some shunt capacitance. The capacitive load will demand a current equal to:

$$i = C\frac{dv}{dt}$$

Assuming we know what C is, we need to know the dv/dt part, which is the rate of change of the voltage across it. Actually we only care about the maximum rate of change, which we already know is the slew rate, so we can substitute equation (3.13) directly:

$$I_{pk} = C2\pi f V_{pk} \qquad (3.15)$$

Notice that bandwidth never appeared in this argument. The driving circuit must therefore be capable of supplying this peak output current if it is to produce a clean 20kHz sinewave across the capacitive load. If it can't, it will simply reach its maximum current limit (the 'current headroom' if you like) and deliver that much into the capacitance until the demand subsides, i.e. we get slew-limiting.

When a constant current is driven into a capacitor, the voltage across it rises linearly. A consequence of this is that a slew-limited sine wave will develop straight sides and begin to resemble a triangle wave – the classic sign of slewing distortion. For example, fig. 3.44 shows a

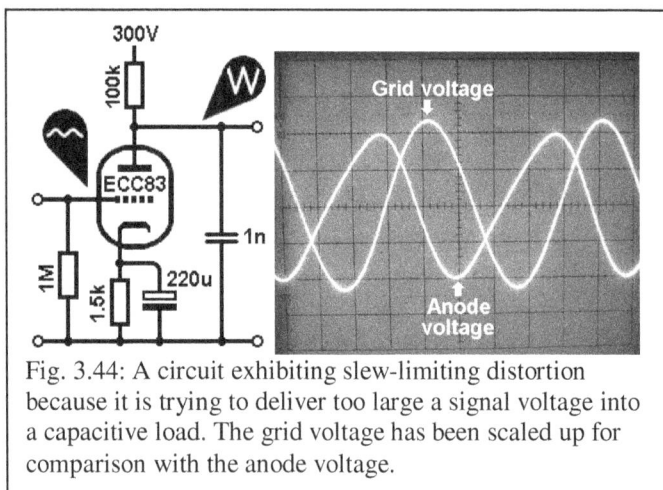

Fig. 3.44: A circuit exhibiting slew-limiting distortion because it is trying to deliver too large a signal voltage into a capacitive load. The grid voltage has been scaled up for comparison with the anode voltage.

circuit with a $1V_{pk}$ 10kHz input voltage, attempting –and failing– to drive a 1nF load (which is unlikely in reality but illustrates the effects very well). The rising edge of the output signal begins to flatten into a straight line, indicating slewing distortion, although the falling edge looks fairly normal because the circuit can sink more current than it can source, i.e. its slew rate is greater for down-going output voltages.

Also notice that the output signal is not 180 degrees out of phase with the input, but is approaching 90 degrees. This is because the capacitive reactance at 10kHz is only 16kΩ, which is much less than the DC load, so it would be fair to say the load now

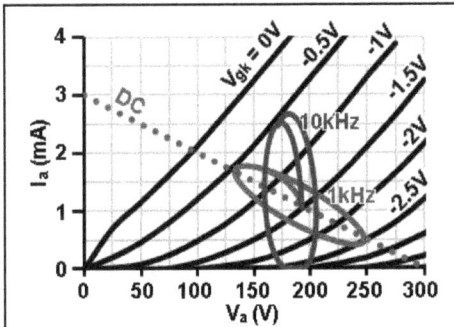

Fig. 3.45: When the load is reactive (in this case a 1nF capacitance) the AC load line opens out into an ellipse, causing distortion to increase and also slew-limiting if the ellipse 'bottoms out' as shown.

appears almost completely capacitive, causing the anode voltage to lag behind the anode current. This phase difference causes the AC load line to open out into an ellipse which grows and shrinks in sympathy with the signal amplitude, and rotates depending on the frequency. The AC load lines for this circuit at $1V_{pk}$ input are shown in fig. 3.45 (they were generated by computer simulation). At 1kHz the capacitive reactance is 160kΩ so the load appears as a combination of resistance (real) and capacitance (imaginary) and the phase shift is only 32°, producing a narrow ellipse.[*] With increasing frequency the ellipse rotates and approaches the vertical as the capacitive part begins to dominate, and in this example it actually bottoms out on the horizontal axis, causing cut-off in the valve. It is this bottoming-out that is the cause of slew-limiting, and it explains why only the rising edge of the signal in fig. 3.44 has a flattened portion. Slew limiting on the falling edge would occur at greater amplitudes when the ellipse becomes so long that its upper end also 'bottoms out' (tops out?) on the 0V grid curve. It is further worth pointing out that because slew limiting is a form of clipping it is not something that global feedback can correct for.[‡]

In the case of an ordinary triode gain stage –or any class-A circuit– the peak output current is at best only equal to the quiescent current, since the operating point can swing symmetrically up and down the load line by an amount equal to the bias point. We must therefore ensure the quiescent current is high enough that there is enough current headroom to drive the expected signal amplitude into the load capacitance.[†]

---

[*] Notice that the operating point traces out the ellipse in an anti-clockwise direction as the signal is applied. If the load were inductive it would trace out a similar ellipse but in a clockwise direction.

[‡] This last point caused something of a stir in the 1970s when it was rebranded as 'transient intermodulation distortion' or TIMD.

[†] In practice the picture may not be quite as bad as it first appears, because *full amplitude* 20kHz signals simply do not exist in real music. Spectral analysis of recorded and acoustic music shows that beyond two or three kilohertz the amplitude of music falls at a second order rate at least. Typically, 20kHz signals will be down by at least 30dB relative to 1kHz, and more like –60dB relative to 100Hz where most musical content lies. Even if we allow for heavy compression or 'experimental' music we still won't get close to the theoretical full-amplitude slew rate during actual listening, as opposed to bench tests.

Nevertheless, before we hit slew limiting, the falling reactance will cause ordinary harmonic distortion to degrade too. Fig. 3.46 shows the measured distortion in the circuit in fig. 3.44. Notice that between 20Hz (dotted) and 200Hz there is no difference in the distortion plots, because the capacitive reactance is so large that it is negligible. But at 1kHz distortion starts to worsen, and beyond this things rapidly decline. Slewing distortion eventually becomes mixed in with the ordinary non-linearity distortion. By 20kHz the distortion is more than a factor of ten worse than at low frequencies.

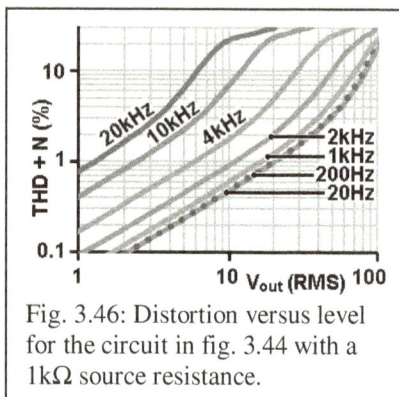

Fig. 3.46: Distortion versus level for the circuit in fig. 3.44 with a 1kΩ source resistance.

In short, reactive loads can be tricky things to drive if you expect low distortion. This can be a trap when driving power triodes, because they require very large input voltage swing to achieve full output. A typical 2A3 output stage might require a drive voltage of $60V_{pk}$ and have a Miller capacitance of 70pF, so the driving stage would in theory need to source and sink a peak current of ±0.5mA at 20kHz, which an ECC83 would have no hope of supplying with low distortion. A low impedance, higher-current driver valve would be required.

3.12: Power Supply Rejection Ratio, PSRR

Until now we have assumed that the HT supply is clean and perfect, whereas in reality it will probably be somewhat noisy. This may include broadband noise, residual ripple from the rectification process, or signal-induced modulation. The last of these is caused by the constantly varying current drawn from the power supply by the various stages of the amplifier. For example, if one stage is amplifying a positive peak it will suck more current from the HT, causing a small voltage drop across the unavoidably finite source impedance of the power supply, i.e. the HT voltage will sag a little. Conversely, when amplifying a negative peak the same stage will reduce its current consumption, so the HT voltage will recover again. The supply voltage will therefore dance along with the program material, and if this signal manages to feed back to previous stages then it can lead to distortion or even oscillation (e.g. motorboating). We are therefore interested in how much noise from the HT will succeed in reaching the output of a given amplifier stage, which is called its power supply rejection ratio, or PSRR. It is defined as:

$$PSRR = \frac{\Delta V_{supply}}{\Delta V_{out}} \quad\quad (3.16)$$

Where Δ means 'a small change in...'.

156

The higher the PSRR figure, the less power supply noise appears at the output of the circuit. Again, this is something that can be determined using equivalent circuits, and fig. 3.47 shows this for a basic gain stage. To find the PSRR the usual audio input must be shorted to ground since the input quantity we are now interested in is the power supply noise, represented by the voltage generator v_{in} (remember, this is a purely AC analysis). We can skip the full treatment because we already know from section 3.8.1 that when looking into the anode we see r_a, and if the cathode is unbypassed then we also see R_k multiplied by $\mu+1$. These two possibilities are depicted in b and c., and it is obvious that we are dealing with a simple potential divider in either case. The gain from input (power supply) to output can therefore be found immediately, using our old friend equation (1.38), but since PSRR is annoyingly defined as the *inverse* of this, we must invert the equation. For the fully bypassed case the PSRR is then:

$$PSRR_{(bypassed)} = \frac{R_a + r_a}{r_a} \qquad (3.17)$$

And for the degenerated case:

$$PSRR_{(degenerated)} = \frac{R_a + r_a + R_k(\mu+1)}{r_a + R_k(\mu+1)} \qquad (3.18)$$

For our example ECC83 gain stage this yields:

$$PSRR_{(byassed)} = \frac{100k + 60k}{60k} = 2.7 \quad (8.6dB)$$

Or:

$$PSRR_{(degenerated)} = \frac{100k + 60k + 1.5k \times (95+1)}{60k + 1.5k \times (95+1)} = 1.5 \quad (3.5dB)$$

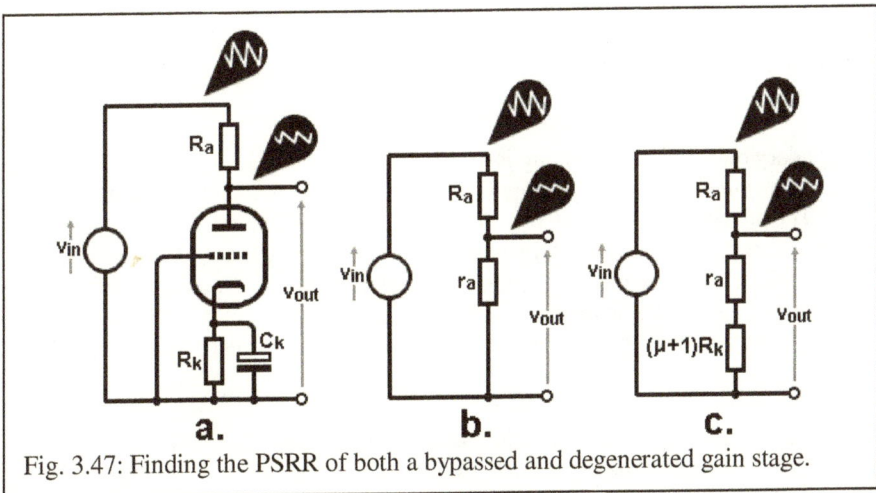

Fig. 3.47: Finding the PSRR of both a bypassed and degenerated gain stage.

157

Neither of these figures is spectacular, but it is clear that degenerating the stage greatly worsens the PSRR, because R_k is multiplied so much. And it should be remembered that even a fully-bypassed stage will appear unbypassed at sufficiently low frequencies, which is another reason to use an oversized cathode bypass capacitor.

3.13: Coupling

The general idea of AC-coupling an audio signal from the output of a gain stage to a load, using a coupling capacitor, was introduced briefly in section 3.6.2.[*] Having now covered some other essential aspects of gain stages – such as input and output impedance, and the effects of loading– some finer details of coupling can now be explored.

The most obvious question to be asked is how large does the coupling capacitance need to be? Fig. 3.48 shows a simplified AC-coupled pair of stages from which is clear that the coupling capacitor creates a CR high-pass filter with a cut-off frequency of:

$$f_{-3dB} = \frac{1}{2\pi C_o (R_o + R_1)} \quad (3.19)$$

Where R_o is the output resistance of the driving stage, implied in fig. 3.48 by the dashed resistor; usually this is so small compared to R_1 that it can be ignored. The effects of several stages coupled together are cumulative, so if we want to avoid attenuation of low audio frequencies then the cut-off frequency must be set well below 20Hz, often sub 1Hz. The capacitor's voltage rating must be sufficient to withstand the full HT voltage, which may appear across it when power is first applied or if V_1 is faulty or removed from its socket.

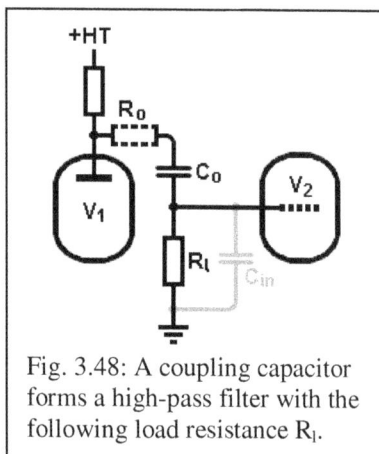

Fig. 3.48: A coupling capacitor forms a high-pass filter with the following load resistance R_1.

A secondary question worth asking is whether there is any *upper* limit on how large C_o can be made, economic considerations aside? The answer is a qualified yes, for various reasons. One reason is that electrolytic capacitors are out of the question. They leak too much, leading to a DC offset at the input of the following stage which would upset its bias (although double coupling could in principle be used –section 7.6). This limits us to film capacitors, which are not normally available in values larger than a few microfarads, especially

---

[*] It is sobering to think that such a commonplace notion as AC-coupling was once novel enough to be patentable: Arnold, H. D. (1922). Vacuum Tube Circuits, US Patent 1403475.

158

with high voltage ratings. Another reason is that if the time constant of C_o and R_l is excessive, e.g. more than a few seconds, it will take a very long time for the coupling capacitor to charge up when the amp is first switched on, and there are limits to the patience of most users. A final and more subtle constraint on the size of the coupling capacitance is blocking distortion.

3.13.1: Blocking Distortion

Blocking distortion is an insidious type of distortion that occurs when a capacitor-coupled valve is driven to the point of grid-current limiting (clipping). Of course, it would be nice if the amplifier was never driven to the point of clipping, but with the best will in the world, it happens. Now, it may be pointed out that valve amplifiers often clip relatively softly, so although a 10W valve poweramp will have less headroom than a 100W solid-state competitor, the occasional clipped transient may well pass unnoticed. The trouble with blocking distortion is that it can make what should be a brief and forgettable clipping event into a drawn out, conspicuous interruption. A good understanding of blocking is therefore essential for the design of high-quality circuits, and too many books gloss over the subject.

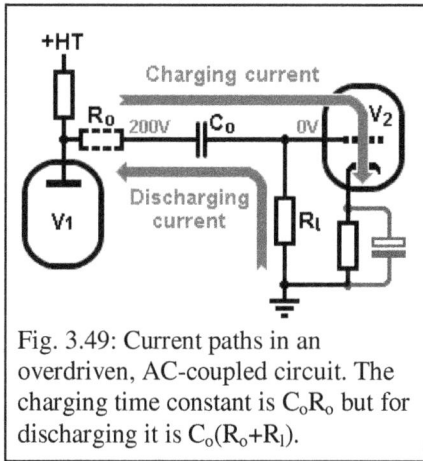

Fig. 3.49: Current paths in an overdriven, AC-coupled circuit. The charging time constant is C_oR_o but for discharging it is $C_o(R_o+R_l)$.

Consider the circuit in fig. 3.49. At quiescence the voltage at the anode of V_1 is 200V and the voltage on the grid of V_2 is 0V, so the coupling capacitor C_1 has 200V across it. Under normal signal conditions, if the anode voltage of V_1 momentarily swings up by 2V, say, a miniscule current will flow via C_o down the grid-leak R_l, causing the voltage across R_l also to rise by 2V. In other words, the audio signal is coupled from one stage to the next as expected. The voltage across C_1 is still 200V since both ends of the capacitor have risen by 2V.

Now, let's suppose that 2V is the maximum peak signal which can be fed to V_2 before it is overdriven (i.e. V_2 is biased to $V_{gk} \approx -2V$). If the anode of V_1 continues to rise by a further 8V, current continues to flow into C_o, but V_2 now draws grid current which prevents the grid voltage from rising any further above 2V. In other words, the left-hand end of C_o rises to 210V but the right-hand end is stuck at 2V. The voltage across C_o has therefore charged up to 208V.

Next, the anode voltage of V_1 swings down to 190V for the other half of the audio waveform. Normally we would expect the voltage across R_l also to fall to −10V, but C_1 is now unable to discharge via the grid of V_2 and instead must source current through R_l, which is large. Now, since the left-hand end of the coupling capacitor has dropped by a total of 20V, and the voltage across the capacitor has not had time to recover, the grid voltage of V_2 must also drop by 20V to −18V, which is likely to

159

drive V_2 deep into cut off. The average grid voltage will remain abnormally negative until C_o has had time to discharge, so during this period V_2 is stuck in cut-off unable to amplify anything; it has been **blocked**.

Fig. 3.50 illustrates this process more clearly with a simple two-stage preamp, modelled on computer. The signal voltages appearing at four labelled nodes are also shown in the figure. The signal applied to the first grid, labelled ①, is a 50Hz 50mV$_{pp}$ sine wave which is interrupted by a loud burst reaching 500mV$_{pp}$. This is amplified by V_1 to produce signal ② at the anode, which ought to be a clean copy. However, the burst is now large enough to overdrive V_2 whose input resistance drops as it draws grid current. This current flows in the coupling capacitor, charging it up with each half cycle, which in turn forces the average grid voltage of V_2 to go more and more

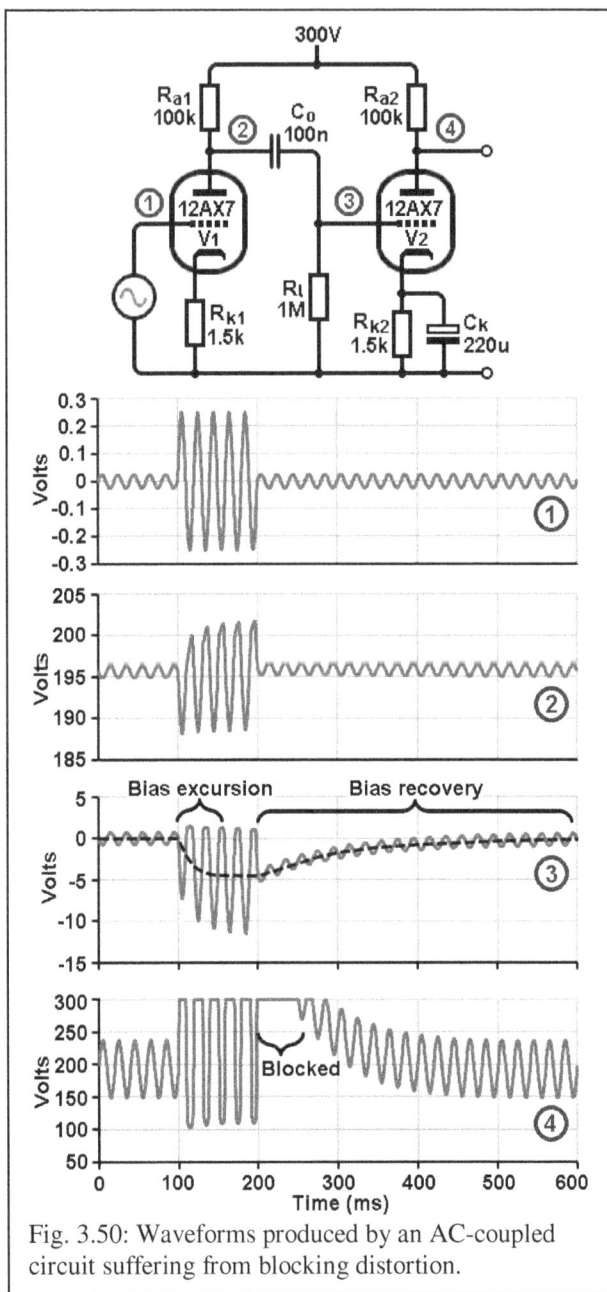

Fig. 3.50: Waveforms produced by an AC-coupled circuit suffering from blocking distortion.

negative, as indicated by the dashed line on signal ③. V_2's cathode bypass capacitor will also charge up somewhat, exacerbating the effect. When the burst ends, the coupling capacitor takes a long time to discharge, so V_2 is held in cut off for many milliseconds during which time it is unable to amplify anything; it has been blocked. This can be seen on the final waveform labelled ④.

Clipping may not be a welcome event, but it will be masked to some extent by the sheer loudness of the passage. By contrast, the blocked 'dead time' before the quiet signal reappears is likely to be more noticeable and subjectively much worse. Here the choice of coupling capacitance comes into play, for the larger it is the longer it takes to charge up. If we only expect overload events to be very transitory and infrequent then a really big coupling capacitor may never get the opportunity to charge up too much before the overload passes. However, if it *does* manage to induce blocking then the blocked period will be long and all the more unpleasant. By contrast, a small coupling capacitance will induce blocking more easily and therefore more often, but it will not last as long so may be less noticeable. Whether a coupling capacitor can be made inordinately large therefore depends on the designer's judgement of subjective discomfort over probability. Since this is not a very satisfying observation, let us examine instead how to minimise rather than manipulate blocking.

3.13.2: Minimising Blocking Distortion

Fundamentally, blocking is caused when the coupling capacitor is able to charge up faster than it can discharge. It can therefore be ameliorated by maximising the charge time and minimising the discharge time, and if the latter can be made equal to or less than the former then blocking will be virtually eliminated.
One direct way to do this might be to make the following grid leak resistance quite small. Unfortunately, this would present a heavy load to the driving stage. A second option might be to increase the charging time by adding a grid stopper to V_2, as shown in fig. 3.51a.[4] Unfortunately, to be effective its resistance would have to be so large that bandwidth would suffer horribly at the hands of Miller capacitance.

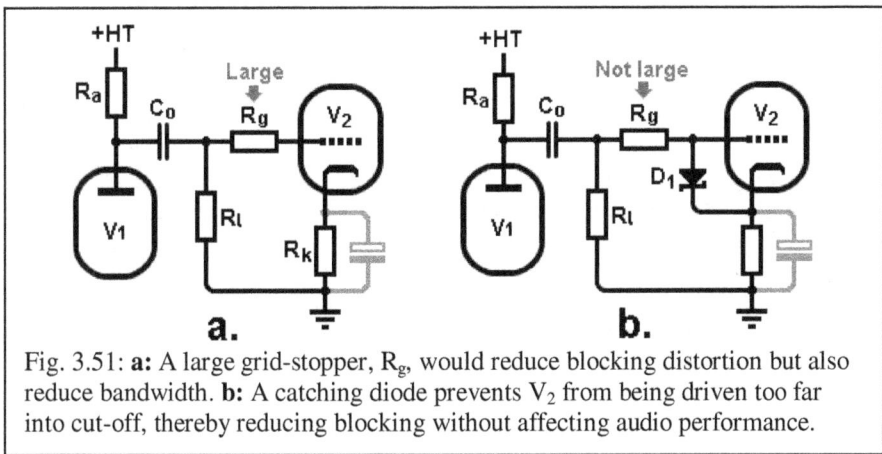

Fig. 3.51: **a:** A large grid-stopper, R_g, would reduce blocking distortion but also reduce bandwidth. **b:** A catching diode prevents V_2 from being driven too far into cut-off, thereby reducing blocking without affecting audio performance.

[4] Schmitt, O. H. (1937). Prevention of Blocking in Resistance Capaciity Coupled Amplifiers, *Review of Scientific Instruments*, 8, March, p90.

Certainly every stage needs a grid stopper to protect against oscillation (omitted earlier from fig. 3.50 for clarity), but it will be a few kilohms at most, which is too small to have any useful effect on blocking.

The most practical way to reduce blocking is to add a **catching diode** between grid and cathode, as shown in fig. 3.51b. Zener diode D_1 is chosen so that it clamps or 'catches' the grid voltage as soon as it tries to drop below the cut-off voltage of the valve. This forcibly prevents it from being driven *really* far into cut off and provides a rapid discharge path for the coupling capacitor. To avoid any reduction of headroom the Zener diode must have a breakdown voltage a little larger than the cut-off voltage (grid base) of the valve, as determined from the load line. For the circuit used earlier the 12AX7 reaches cut-off at about −4V, so a 4.7V Zener is the obvious choice. For the benefit of any sceptics mistrustful of computer

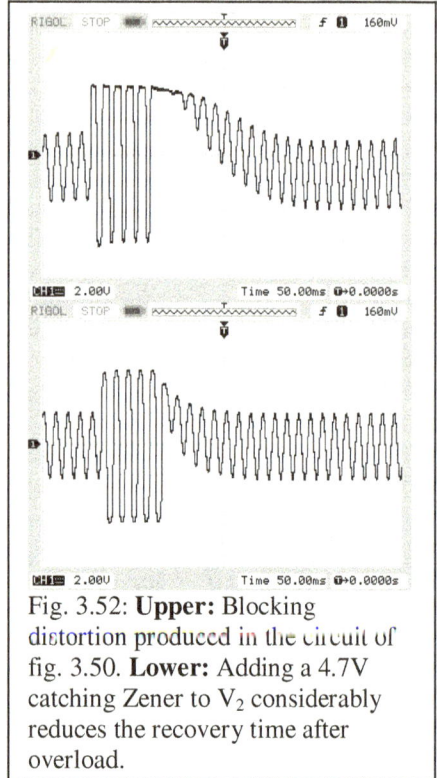

Fig. 3.52: **Upper:** Blocking distortion produced in the circuit of fig. 3.50. **Lower:** Adding a 4.7V catching Zener to V_2 considerably reduces the recovery time after overload.

simulations, the circuit of fig. 3.50 was physically built and tested. Fig. 3.52 shows the oscillograms viewed at node ④ (attenuated twenty times), with the previously stated input signal levels. The agreement with simulation is excellent, and the improvement after adding a BZX79C4V7 Zener diode to V_2 speaks for itself.

3.13.3: DC Coupling

The most common type of coupling used in valve amplifiers is AC coupling, where only the AC signal is passed to the following stage while any DC component is removed. But it is also possible to pass everything down to DC, which is hence called **DC coupling**. This sort of coupling is not always practicable and usually requires a good deal of iterative design adjustment since each stage of the amplifier becomes more interactive with its neighbours. Nevertheless, the two main advantages of DC coupling are that it eliminates blocking distortion and low-frequency attenuation (and therefore also phase shift), all of which can be useful if the circuit is enclosed within a global feedback loop.

The simplest form of DC coupling is a direct connection between the output (usually the anode) of one stage and the input (usually the grid) of the next. This is often referred to simply as **direct coupling**, and a simple example is illustrated in fig. 3.53 with quiescent voltages noted. Notice that the valves are ECC82 / 12AU7. It is easier to achieve DC coupling with low-r_a triodes like these (or pentodes) since they can operate comfortably with less voltage across themselves.

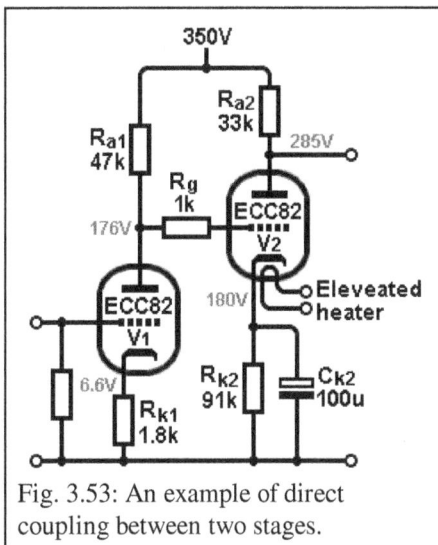

Fig. 3.53: An example of direct coupling between two stages.

It should be obvious that V_1 is an ordinary gain stage, but its anode voltage now directly determines the grid voltage of V_2. For normal biasing the cathode voltage of V_2 must be slightly higher than its grid voltage, so its cathode resistor must be unusually large to achieve this. This means the useful voltage left across V_2 is considerably reduced, and with it the maximum output swing. It is also important to remember that because V_2's cathode voltage is high it becomes necessary to elevate the heater supply to avoid exceeding the $V_{hk(max)}$ rating. In this case the heater supply would probably be elevated to about 90V to place it electrically half way between the two cathodes. The $1k\Omega$ grid stopper is added as good practice to reduce the possibility of parasitic oscillation.

The restriction of output swing, and the fact that C_{k2} must be a large, high-voltage device, are distinct disadvantages when trying to direct-couple two ordinary gain stages. Therefore, DC coupling is more commonly used with cathode followers (chapter 7), which naturally use a large resistance in the cathode circuit and do not use cathode bypass capacitors. We will meet some examples later in this book.

3.13.4: Grid-Cathode Arc Protection

With DC-coupled stages there is an additional concern which has not yet been mentioned. When the power supply is first switched on the cathodes will be cold, so no anode current can flow. The anode voltage of V_1, which is also grid voltage of V_2, will therefore rise to the full HT potential while the cathode of V_2 will still be at zero volts. With such a high voltage between grid and cathode it is possible for an arc to jump between the two, which will seriously damage the cathode surface. The maximum permissible grid-to-cathode voltage is rarely quoted on datasheets, but having destroyed a few valves this way the author is satisfied that ±100V is a practical limit.

163

A traditional solution to this problem was to pre-heat the cathodes before applying the HT, but a much cheaper and more reliable solution is to add a diode or neon lamp between grid and cathode. In fig. 3.54a the grid cannot be driven more than about 0.7V above the cathode because diode D_1 will

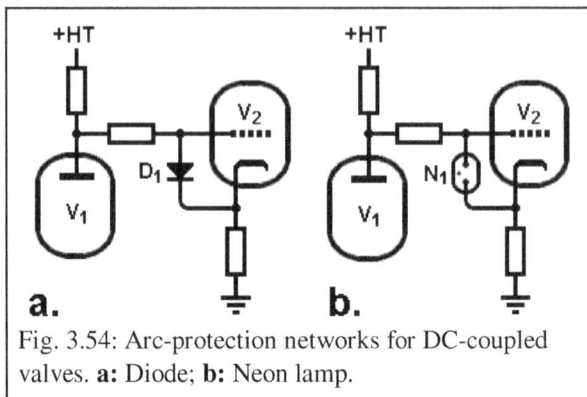

Fig. 3.54: Arc-protection networks for DC-coupled valves. **a**: Diode; **b**: Neon lamp.

become forward biased, safely clamping the grid-cathode voltage. Once the valves have warmed up, the cathode voltage will settle higher than the grid, so D_1 will be reverse biased (off) and play no further part in circuit operation. D_1 can usually be a 1N4148 although in some ambitious circuits the reverse voltage swings may be large enough to demand a high-voltage diode such as a 1N4003 or better. This will unavoidably add a few picofarads of junction capacitance between grid and cathode, but this is usually of no consequence since it is not subject to the Miller effect.

In fig. 3.54b a neon lamp serves the same purpose as the diode previously, but it is likely to have smaller stray capacitance and may appeal more to vacuum purists. Unlike a diode, a neon lamp does not conduct or 'strike' until the voltage across it reaches about 90V, but this is still low enough to protect the valve. Once conducting, the voltage across it immediately drops to about 60V. The neon will of course light up when conducting and it is always nice when a component announces that it is doing its job. Note that a plain neon lamp must be used, not the kind sold as mains power indicators as these have an additional series resistor built in, which is not wanted here.

3.13.5: Level Shifting

The difficulty with direct coupling is trying to marry the naturally high anode voltage of one valve to the grid of the following valve without exceeding any voltage ratings or restricting output swing. One way to dodge this problem is to use a potential divider to lower the DC voltage to a more manageable level, which is called **level shifting**. A practical arrangement is shown in fig. 3.55.

The high anode voltage of V_1 is divided down to 70V by a potential divider (whose resistance is large to avoid loading down V_1). The cathode voltage of V_2 will be biased only a little higher than this, which still leaves a healthy proportion of the supply voltage available for V_2 itself. However, the source resistance of the divider is very large and would therefore cause considerable loss of high frequencies when combined with the Miller capacitance of V_2 –shown dashed in the figure. To avoid this, a **compensation capacitor** C_1 is added in parallel with R_1. A perfectly flat response from DC to perhaps 100kHz can be obtained when the time constant R_1C_1

164

is made equal to R_2C_m. However, in practice it will require a fair amount of tweaking to bring C_1 to exactly the right value to achieve this, not forgetting that C_m will also change with age and when the valve is replaced. Understandably, level shifting is therefore rarely used in audio amplifiers.

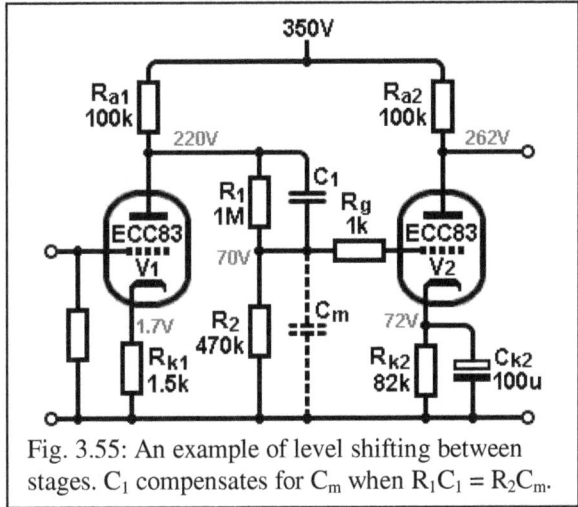

Fig. 3.55: An example of level shifting between stages. C_1 compensates for C_m when $R_1C_1 = R_2C_m$.

Chapter 4: The Small-Signal Pentode

The triode gain stage is the main building block of a preamplifier, but small-signal pentodes (as opposed to the power pentodes found in power-output stages) also appear in some preamplifiers, and this chapter will discuss how they are used. Admittedly they are less common in modern designs than those of the valve-era, and there are various reasons for this. Pentodes are noisier than triodes and suffer more acutely from microphonics, partly because they provide higher gain and partly because they have more physical parts to vibrate. Getting a lot of gain from a single stage was an advantage during the valve era when signal sources were weak and every component was a significant expense, whereas these days the situation is exactly reversed. Another disagreeable feature is that pentodes will sometimes produce stronger odd-harmonic distortion than triodes, although this point is sometimes overemphasised by critics; a favourable choice of operating point can lead to a very triode-like distortion spectrum. However, pentode circuits do tend to be more 'fiddly' to design and optimise; it is easier simply to throw in the towel and use them as triodes instead. Indeed, old-stock radio and TV pentodes constitute a cheap and plentiful source of potential triodes. Nevertheless, this book would not be complete without giving some attention to the design of true pentode stages.

There is a lot more going on inside a pentode than in a triode, which makes the design process for a pentode stage more involved. For competent design it is necessary to have a good grasp of how pentodes actually work –especially the influence of the screen grid– so the following sections will cover the essential theory.

4.1: The Screen Grid

In the 1920s, efforts were made to develop new valves with reduced Miller capacitance, so circuits could be operated at much higher frequencies than was possible with triodes at the time. This was accomplished by inserting a second grid called the **screen grid**, g_2, between the control grid and anode. The resulting device had four working electrodes so it was called the **tetrode** (or 'straight tetrode' to distinguish it from the beam power tetrode which came later).

The screen grid is so called because it shields or 'screens' the control grid and cathode from the anode. The screen voltage is usually held at a constant high voltage so electrons accelerate towards it, but most of them fly between its grid wires after which it is too late for them to slow down, so they crash into the anode, whatever the anode voltage happens to be. The result is that anode current can be sustained even at quite low anode voltages, so tetrodes made more efficient use of the available supply voltage. Furthermore, the anode's varying electric field is hidden behind the fixed electric field of the screen, so C_{ga} is reduced and with it the Miller effect.

The screen grid is the most important electrode in a tetrode or pentode, and understanding its role is fundamental for good circuit design. Many designers, past and present, have failed to appreciate the functioning and fragility of the screen grid and have produced unreliable circuits as a result.

166

4.1.1: Secondary Emission

When electrons hit an object like the anode with enough force they may dislodge one or more other electrons from its surface. This effect is known as **secondary emission** and the dislodged electrons are called **secondary electrons**. Once freed they will be captured by whatever positive electric field happens to be nearby. In a triode they are attracted back to the anode again with no harm done, but if there is another positive electrode close by, such as the screen grid, then they may be captured by its electric field instead, effectively stealing some overall current from the anode. If the screen-grid voltage is higher than the anode voltage then the screen will collect practically all the secondary electrons. Over a limited range it is possible that an increase in anode voltage will *reduce* the anode current, because the screen grid captures more secondary electrons than the anode receives in primary electrons. Over this range the internal anode resistance r_a becomes *negative*. This can be seen in the anode characteristics of the Type 32 straight tetrode shown in fig. 4.1; there is an obvious 'kink' in the curves where the anode voltage is less than the screen voltage.

The screen grid also captures some primary electrons on their way from the cathode, which gives rise to screen current. Notice that in fig. 4.2 the screen current labelled I_{C2} is a mirror image of the 0V grid curve, so the *total* cathode current in the valve is relatively constant at any anode voltage, it simply splits between anode and screen by varying proportions. More on this later.

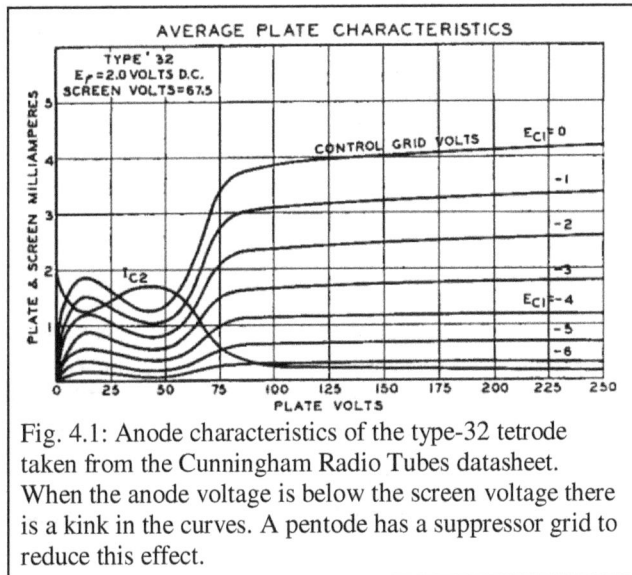

Fig. 4.1: Anode characteristics of the type-32 tetrode taken from the Cunningham Radio Tubes datasheet. When the anode voltage is below the screen voltage there is a kink in the curves. A pentode has a suppressor grid to reduce this effect.

4.1.2: The Suppressor Grid

The kink in the characteristics of the straight tetrode makes it prone to oscillation and unsuited for linear amplification, so valve designers contrived various schemes to suppress it. One solution was to add a third grid between the screen and anode, called the **suppressor grid, g_3**, thereby evolving the tetrode into the pentode. The suppressor grid is wound very coarsely and is made very negative compared to the anode (usually by connecting it to the cathode) so most of the secondary

167

electrons are blocked from reaching the screen grid and are instead repelled back to the anode, as illustrated in fig. 4.2.

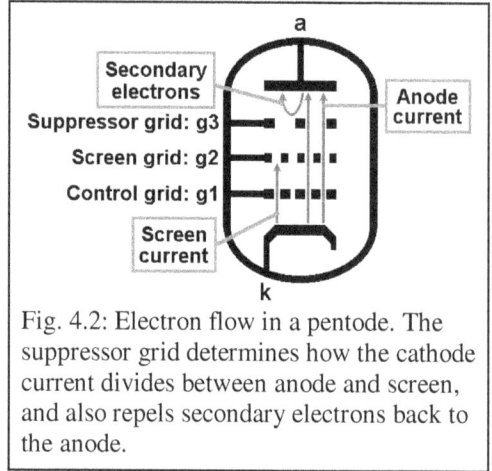

Fig. 4.2: Electron flow in a pentode. The suppressor grid determines how the cathode current divides between anode and screen, and also repels secondary electrons back to the anode.

The suppressor also determines the ratio in which the total cathode current splits between anode and screen. If it is made very negative then more electrons will be diverted to the screen grid, so screen current will increase at the expense of anode current. If the suppressor is made positive then the opposite happens (and it may collect some electrons of its own), and if it is made *really* positive then the characteristic kink will return. This control over the anode/screen current ratio allows the suppressor to be used for modulation, and if there is any impedance in the cathode circuit, such as an unbypassed cathode bias resistor, then connecting the suppressor to ground, rather than to cathode, will dramatically increase the harmonic distortion in the anode current.[1] Obviously this is very undesirable for hi-fi designs, so the suppressor should always be connected directly to the cathode. In many valves this is done internally so we have no choice in the matter. Apart from this the suppressor grid is regarded as benign and plays no other part in the design of the stage.

4.2: The Effect of Screen Voltage

The anode characteristics of a pentode look rather like those of a transistor. The grid curves are nearly horizontal, indicating that r_a and μ are both extremely large and almost impossible to get accurate figures for. Additionally, the anode voltage can swing very low, almost rail-to-rail. However, the characteristics of a transistor are fixed when the device is manufactured so the circuit designer has no control over them. Similarly, the anode characteristics of a triode are also fixed.[*] The characteristics of a pentode, however, are not fixed; they are controlled by the screen voltage.

By manipulating the screen voltage we can 'morph' the anode characteristics from high to low current, and we can even turn the pentode into a triode. This is best understood by looking at the change in the anode characteristics as the screen voltage is varied over a range of values. The oscillograms in fig. 4.3 show the

---

[1] Shepard, W. G., (1953). Suppressor Grid Frequency Doubler. *Electronics.* October, p200. For more explanation of the suppressor's behaviour see: Spangenberg, K. R. (1948). *Vacuum Tubes.* McGraw Hill Book Company Inc., New York.
[*] Unless you count the heater as a control element.

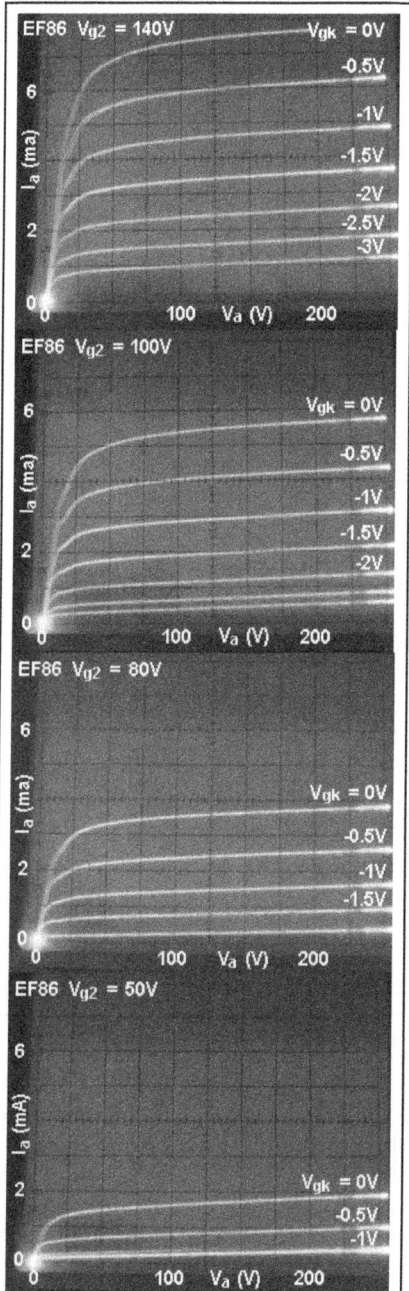

Fig. 4.3: Lowering the screen voltage squashes the grid-curves down, reducing g_m.

measured characteristics of a Mullard EF86. For ease of comparison the axes and grid-curve divisions are the same in each graph.

The top image shows the characteristics when the screen voltage is fixed at 140V, which is relatively high for an EF86. The higher the screen voltage the stronger its electric field, and therefore the more powerfully it will attract electrons, so the greatest anode current is able to flow in this situation. The grid curves are the most widely spaced when the screen voltage is high, indicating the transconductance –and therefore the possible gain– is also greatest, along with input headroom (grid base).

The next image shows how lowering the screen voltage to 100V 'squashes' the grid curves down; the screen's field strength has reduced, so for a given bias voltage the anode current is reduced. The transconductance is also reduced somewhat, so the possible gain is less. Lowering the screen voltage to 80V compresses the curves down even further, continuing the trend. These are typical screen voltages for an EF86 gain stage.

In the bottom image the screen voltage is reduced to 50V. The transconductance is now very low, along with the grid base, since small changes in control-grid voltage now have a greater effect on what little anode current can flow. Reducing the screen voltage to zero would cut-off the anode current completely.

The relationship between screen voltage and anode current is roughly linear. In other words, halving the screen voltage will squash the grid curves down by approximately 50%, while tripling the screen voltage will raise the grid curves by

169

about 300%. It is therefore possible to estimate by eye what the anode characteristics will look like for screen voltages other than those shown on the datasheet. A more precise method is described in section 4.4.

4.3: The Effect of Screen Current

The total current in a pentode is the combination of the anode and screen currents. An important observation is that if the screen and bias (and suppressor) voltages are held constant, then the cathode current will *also* be substantially constant. Changing the anode voltage simply changes the ratio in which the current is shared between anode and screen. This is illustrated in fig. 4.4 which shows the 0V grid curve for an EF86 at a screen voltage of 100V. The screen current is also plotted, and since the total current is nearly constant it must be a mirror-image of the anode current. Remember, this is the screen current we would measure with 0V bias; for other bias voltages we could plot other grid curves and their corresponding screen-current curves, creating two families of curves.

When the anode voltage is about 40V there is a 'knee' in the grid curve, which drops down rapidly towards the origin, while the screen current increases just as rapidly. If the load line passes below this knee then the operating point can be driven into this region of low anode voltage, leading to a sudden increase in screen current. This can be a problem with power pentodes and beam tetrodes because the average screen voltage and current are sometimes pushed to the limit, so any increase in average screen current during full

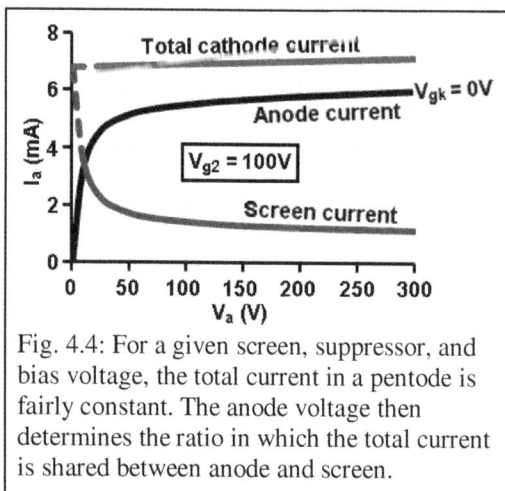

Fig. 4.4: For a given screen, suppressor, and bias voltage, the total current in a pentode is fairly constant. The anode voltage then determines the ratio in which the total current is shared between anode and screen.

drive (or overdrive) may lead to overdissipation of the screen grid. This is probably the main cause of failure in power valves. It is usually mitigated by adding a screen-grid stopper of around 1kΩ to each power valve. However, small-signal pentodes are operated with much lower screen voltages so this is rarely a concern in a preamp, but the effect of adding a screen stopper is worth examining.

Because the screen current varies in sympathy with anode current, if an impedance is placed in series with the screen grid then the screen voltage will be modulated also, i.e. a signal voltage will appear at the screen which opposes that at the control grid. This is called **screen current feedback** or **screen degeneration**, and it reduces the available gain. However, unlike cathode degeneration it does not always improve distortion, as shown later. Fig. 4.5 shows the anode characteristics produced by adding an unusually large 100kΩ screen stopper resistor. The curves become closer

170

together, indicating lower g_m. Also, since the screen current increases at low anode voltages, the effect of degeneration is more pronounced in this region, causing the knee of the curves to soften. However, it must be appreciated that the image on the right only applies to *dynamic* conditions, that is, when the valve is actually amplifying. For biasing or working out DC voltages and currents we would still use the static characteristics in the left image.

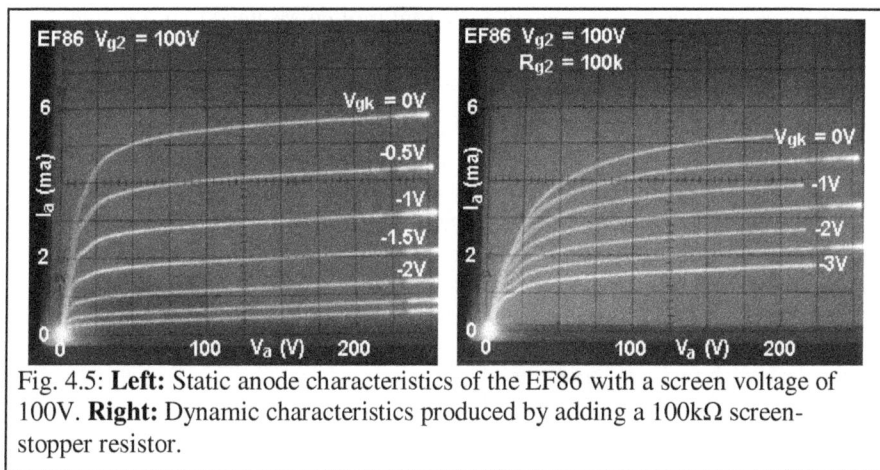

Fig. 4.5: **Left:** Static anode characteristics of the EF86 with a screen voltage of 100V. **Right:** Dynamic characteristics produced by adding a 100kΩ screen-stopper resistor.

4.3.1: The Anode/Screen Current Ratio

With triodes there are three valve constants to work with: μ, r_a and g_m. With pentodes μ and r_a are both very large and difficult to quantify, so they are rarely used in calculations. The useful parameters are instead g_m and the anode/screen current ratio, often denoted by the symbol m. These two things can be thought of as the 'pentode constants'.

The anode/screen current ratio is more-or-less constant for a particular anode voltage, regardless of the screen or bias voltages. For example, the EF86 datasheet states that at an anode voltage of 250V, screen voltage of 140V, and −2.2V bias, the anode current is 3mA and the screen current is 0.6mA, as outlined in table 4.1. The anode/screen

TYPICAL CHARACTERISTICS			
Anode voltage	V_a	250	V
Grid No.3 voltage	V_{g3}	0	V
Grid No.2 voltage	V_{g2}	140	V
Grid No.1 voltage	V_{g1}	−2.2	V
Anode current	I_a	3.0	mA
Grid No.2 current	I_{g2}	0.6	mA
Transconductance	S	2.2	mA/V
Amplification factor	μ_{g2g1}	38	–
Internal resistance	R_i	2.5	MΩ

Table 4.1: Example data for the EF86 taken from the Philips datasheet.

current ratio is therefore: $m = 3 / 0.6 = 5$. It must be emphasised that these are test conditions only and not representative of normal audio usage. Nevertheless, the *m*-ratio still holds at other screen and bias voltages too, so we can be fairly sure that the screen current will always be about one fifth of the anode current, provided the anode voltage is still 250V. At other anode voltages the ratio will be different (falling at lower voltages), but since the grid curves are nearly horizontal it doesn't change *much* at ordinary working voltages, so it is often assumed to be constant over the full range.

4.4: Deriving Grid Curves for any Screen Voltage

The datasheet often provides the anode characteristics corresponding to only one screen voltage, and rarely is it the one we want. However, the datasheet should also provide the mutual characteristics which show how anode current varies with screen and bias voltage, at a fixed anode voltage. With this we can work out the anode characteristics for any screen voltage, at least to a reasonable approximation. Fig. 4.6 shows the mutual characteristics of the EF86.

Suppose we wish to draw curves for a screen voltage of 100V, which is one of the curves plotted on the mutual characteristics (if there is no curve for your desired screen voltage then simply estimate its position using the other curves as a guide). We can take a few points on the mutual characteristic curve and translate them onto the anode characteristics. This is illustrated in fig. 4.6 where points A to D on the mutual characteristic (corresponding to bias voltages from 0V to –3V) have been plotted on a new set of I_a/V_a axes. Note that they must be plotted at the same anode voltage as that indicated on the mutual characteristics, which is 250V in this case.

The grid curves must pass through these points, and if the pentode were ideal then they would be

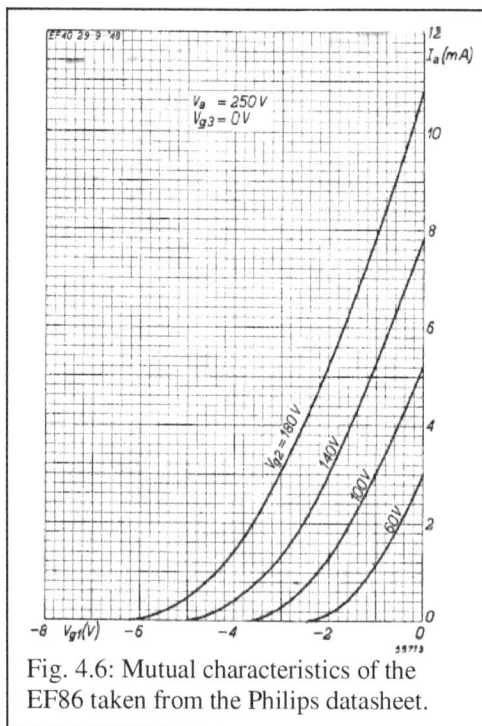

Fig. 4.6: Mutual characteristics of the EF86 taken from the Philips datasheet.

perfectly horizontal and our work would be done. A real pentode is less than ideal, however. Fortunately, examination of other anode characteristics graphs shows that most pentodes have very similarly shaped curves, and that they maintain the same basic form at any practical screen voltage, so it is not difficult to estimate their shape with reasonable accuracy, as shown in the figure. Some datasheets may even give

172

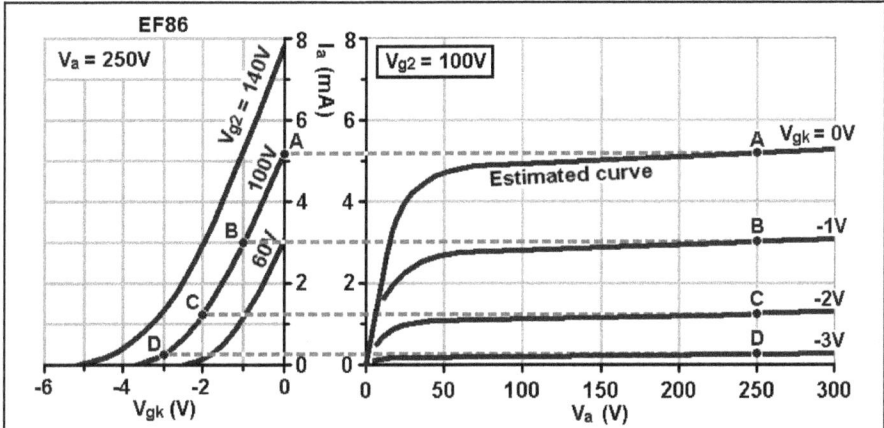

Fig. 4.6: The mutual characteristic (left) can be used to estimate the anode characteristic for a particular screen voltage (right).

two sets of mutual characteristics corresponding to different anode voltages, allowing other points to be plotted, through which the grid curves will pass. The maximum anode dissipation curve could also be added using the method described in section 3.3.5.

4.5: Basic Design Equations

Fig. 4.7 shows the archetypal pentode gain stage. In addition to the usual anode and cathode components there is also a **screen dropping resistor**, R_{g2}, and **screen bypass capacitor**, C_{g2}. R_{g2} provides the necessary voltage drop to bring the screen grid to a suitable working voltage, since the DC screen current flows in this resistor as indicated by the current arrows. C_{g2} serves the dual purpose of smoothing out any power supply noise that would otherwise reach the screen and be amplified, as well as preventing screen degeneration, just as the cathode bypass capacitor prevents cathode degeneration. This component must be rated to withstand the full HT voltage and, strictly, should be connected between screen and cathode as shown, since it is the screen-to-cathode voltage that really matters. Often, however, it is connected between screen and

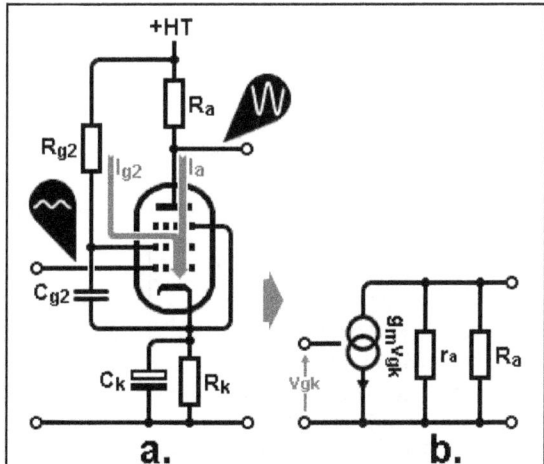

Fig. 4.7: **a**: Archetypal pentode gain stage. **b**: Norton equivalent circuit assuming the screen and cathode are fully bypassed.

173

ground instead, which makes little difference to performance provided the cathode is fully bypassed.

Because μ is very difficult to quantify, whereas g_m is easy to quantify, the Norton equivalent circuit is normally used to model a pentode. Provided the screen and cathode are fully bypassed the circuit appears as in fig. 4.7b. The signal current generated by the valve flows in the parallel combination of r_a and R_a, so the voltage gain of the circuit is:[*]

$$A_{bypassed} = -g_m (r_a \parallel R_a) \tag{4.1}$$

But in a pentode r_a is usually much larger than R_a and so has little effect on the gain. The gain of a fully bypassed pentode can therefore be approximated as:

$$A_{bypassed} \approx -g_m R_a \tag{4.2}$$

The very large internal anode resistance also means the pentode has close to unity PSRR when arranged as a common-cathode gain stage. In other words, any noise on the power supply will appear unattenuated at the output, so a quiet supply is essential.

Provided the screen is bypassed so no signal voltage can appear on it, the control grid is shielded from the anode, so the Miller effect ought to be almost eliminated. In theory, the input capacitance should therefore be determined mainly by the sum of the grid-cathode and grid-screen capacitances. Rather than quote figures for both, the datasheet may simply give the total as 'C_{in}', which for the EF86 is 4 to 5pF depending on which datasheet you read. Unfortunately, this overlooks the valve socket and wiring capacitances. Even with the best efforts it is difficult to get C_{ga} below 0.5pF, so no matter how good the valve may be internally, the Miller effect will still do its

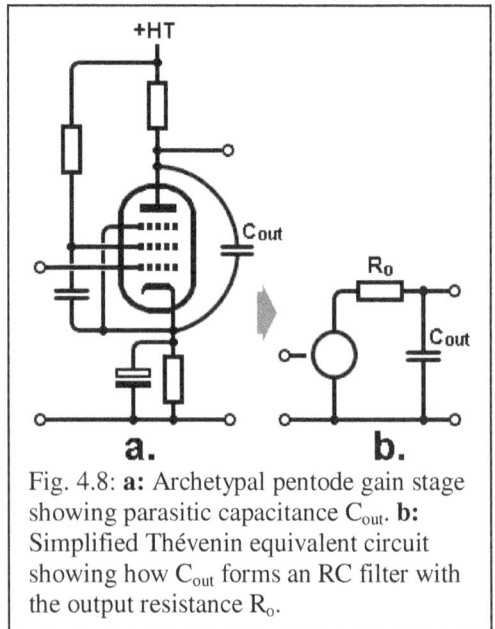

Fig. 4.8: **a:** Archetypal pentode gain stage showing parasitic capacitance C_{out}. **b:** Simplified Thévenin equivalent circuit showing how C_{out} forms an RC filter with the output resistance R_o.

dirty work on the external capacitance. In other words, do not expect to achieve low input capacitance and high gain simultaneously, even with a pentode.

---

[*] The gain of a bypassed triode stage is given by the same equation if we use the Norton equivalent, but unlike the pentode it cannot be simplified any further.

Fig. 4.7b indicates that the output resistance is the parallel combination of r_a and R_a just as it is for a triode but, again, r_a is usually so large that it can be ignored, so we can assume the output resistance is:

$$R_o \approx R_a \qquad\qquad (4.3)$$

If R_k is unbypassed then this becomes even more accurate.

The datasheet will quote a figure called 'C_{out}' which is mainly the capacitance between anode and cathode. This forms an RC filter with the output resistance and so will place an upper limit on the bandwidth. This is illustrated in fig. 4.8 where a simplified Thévenin equivalent is shown to make the situation more obvious. The EF86 datasheet claims C_{out} = 5.5pF, but this seems to be optimistic. Measurement of real circuits indicates a figure a closer to 10pF –too much to be attributed only to strays. Thus although the pentode has reduced Miller capacitance, wide bandwidth may still be out of reach if R_o is large, as it often is in traditional audio pentode circuits.

4.5.1: The Effect of Screen and Cathode Bypassing

In the chapter 3 it was explained how an unbypassed cathode resistor degenerates a triode circuit, and how bypassing it with a capacitor creates a shelf in the frequency response. Exactly the same thing applies to the pentode, except there are two sorts of bypassing to consider: screen and cathode. Having two sources of negative feedback makes the circuit theory rather complex, so here we will just state the facts.

The gain of a pentode stage with unbypassed screen and cathode is approximately:[2]

$$A_{unbypassed} \approx -g_m R_a \cdot \beta \cdot \gamma \qquad (4.4)$$

Where β and γ represent the gain reduction caused by screen and cathode feedback, respectively.

The gain reduction due to an unbypassed screen resistor R_{g2} is:

$$\beta = 1 - \frac{R_{g2}}{R_{g2} + r_{g2}} \qquad (4.5)$$

Where r_{g2} is the internal screen resistance, which is analogous to anode resistance r_a. The internal screen resistance can be found from the vertical separation of the mutual characteristic curve, i.e. for a given change in

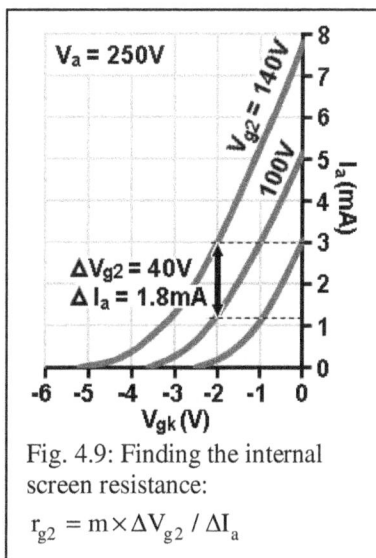

Fig. 4.9: Finding the internal screen resistance:

$$r_{g2} = m \times \Delta V_{g2} / \Delta I_a$$

---

[2] Terman, E. T. *et al.* (1940). Calculation and Design of Resistance Coupled Amplifiers Using Pentode Tubes. *Trans. A.I.E.E.*, **59**. pp879–884.

screen voltage we note the change in anode current at the bias voltage we intend to use, and then multiply by the anode/screen current ratio m.[*] This is illustrated in fig. 4.9, suggesting a value of about: $r_{g2} = 5 \times 40V / 1.8mA = 111k\Omega$ for the EF86.

When a screen bypass capacitor is added, the zero frequency where the gain starts to rise from its lowest level is:

$$f_{lo(screen)} = \frac{1}{2\pi R_{g2}C_{g2}} \qquad (4.6)$$

In general the screen should be fully bypassed, not just to obtain a flat frequency response down to low frequencies but because impedance in series with the screen can degrade linearity, despite it being a form of feedback. There are special cases such as distributed loading and ultra-linear operation where feeding back some signal from the anode to the screen does reduce distortion, but this is more typical of power output stages and is not considered here.

The gain reduction due to an unbypassed cathode resistor R_k is:

$$\gamma = \frac{1}{1 + g_m R_k \beta} \qquad (4.7)$$

If the screen is fully bypassed then β drops out of this equation. When the cathode bypass capacitor is added, the zero frequency where the gain starts to rise from its lowest level is the same as for a triode gain stage:

$$f_{lo(cathode)} = \frac{1}{2\pi R_k C_k} \qquad (4.8)$$

As will shall see shortly, cathode degeneration tends to improve performance, unlike screen degeneration. In fact, a (cathode) degenerated pentode can make a very promising amplifier.

4.6: A Traditional Pentode Gain Stage

Fig. 4.10 shows a traditional pentode circuit taken directly from the 1970 Philips / Mullard datasheet, which is representative of the kind of pentode circuits found in many vintage amplifiers. However, it is not a particularly good choice for a modern amplifier, and it will be instructive to discover why.

The total load on the valve is $r_a\|R_a\|R_l$. The EF86 quotes r_a as 2.5MΩ (which is no doubt an approximation). This brings the total to 168kΩ. The transconductance was estimated to be about 1mA/V, as determined from the slope of the mutual

---

[*] Alternatively, r_{g2} may be found by taking the r_a of the valve when connected as a triode and then multiplying by $m+1$.

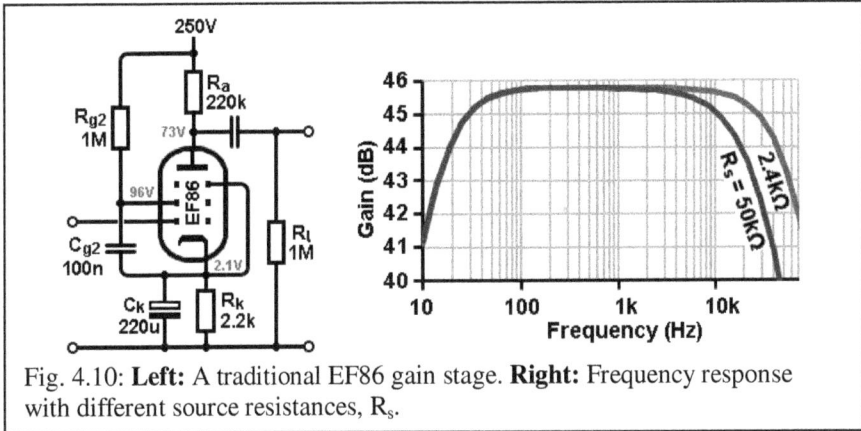

Fig. 4.10: **Left:** A traditional EF86 gain stage. **Right:** Frequency response with different source resistances, R_s.

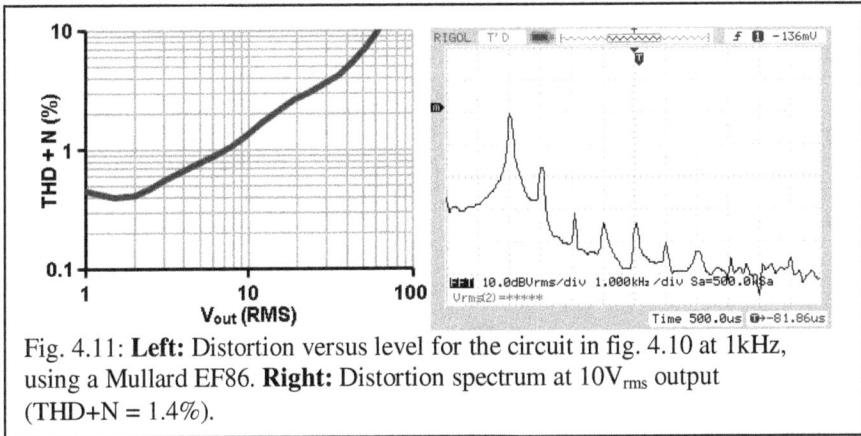

Fig. 4.11: **Left:** Distortion versus level for the circuit in fig. 4.10 at 1kHz, using a Mullard EF86. **Right:** Distortion spectrum at $10V_{rms}$ output (THD+N = 1.4%).

characteristic curve corresponding to the correct screen and bias voltage. This results in a predicted gain of about: $1mA/V \times 168k\Omega = 168$. The test circuit produced a figure of 195, suggesting g_m was closer to 1.16mA/V. Such high gain makes the valve prone to microphonics.

Fig. 4.10 also shows the frequency response of the circuit. The textbooks may boast about the wide bandwidth possible with pentodes, but this particular amplifier didn't get the message. The low-frequency end is limited to about 14Hz owing mainly to C_{g2}, while the high-frequency end is limited to 70kHz with a 2.4kΩ source resistance. Investigation showed to be due to an excessive output load capacitance of around 11pF (the measurement bandwidth was a reliable 130kHz and the circuit was built point-to-point on tag strip which should minimise wiring capacitance, so the author remains somewhat puzzled by this figure). Because the voltage gain is so high, even the slightest stray capacitance between grid and anode leads to significant Miller capacitance –about 110pF in this case. However, this is still much less than a high-gain triode would manage. Incidentally, leaving the internal electrostatic screen

177

unconnected caused the Miller capacitance to increase fractionally, but made no observable difference to C_{out}.

Fig. 4.11 shows distortion versus level, which is of similar order to a mediocre triode (the rise at low levels is the measurement noise floor, not distortion). Fig. 4.11 also shows the distortion spectrum at $10V_{rms}$ output, and it is a decaying series of all harmonics. At the same output level a triode would normally show a much greater dominance of the second harmonic, whereas here everything up to the eighth is clearly visible. Of course, the pentode has a lot more gain than a triode, and so can provide more loop gain if negative feedback is used. Therefore, let us see if a little cathode degeneration can help.

4.6.1: The Effect of Cathode Degeneration

Fig. 4.12 shows the modified circuit from which C_k has been removed. Using equation (4.7) this reduces the effective transconductance by a factor of $1/(1+g_mR_k)$ or $1/(1+1.16\times2.2k) = 0.28$, so to a first approximation the gain ought to be reduced by the same factor, that is, from 195 to $195\times0.28 = 54.6$. The measured figure was 57.5. Fig. 4.12 shows the frequency response which now extends down to about 7Hz. This is not because C_k was a limiting factor before but because the multiplication effect of R_k increases the internal resistance seen by C_{g2}. The upper end of the frequency response has also improved, both because the reduced gain results in less Miller capacitance and because R_k is now in series with any anode-cathode capacitance.

Fig. 4.13 shows that distortion has dropped by a factor of a little over three, which makes sense since the gain has been reduced by a factor of $195/57.5 = 3.4$. The spectrum is essentially the same as previously, however. The total distortion is now about half what most triodes would deliver into the same load, and we have less Miller capacitance too, which is a useful result. Perhaps degenerating the screen will buy even greater improvement?

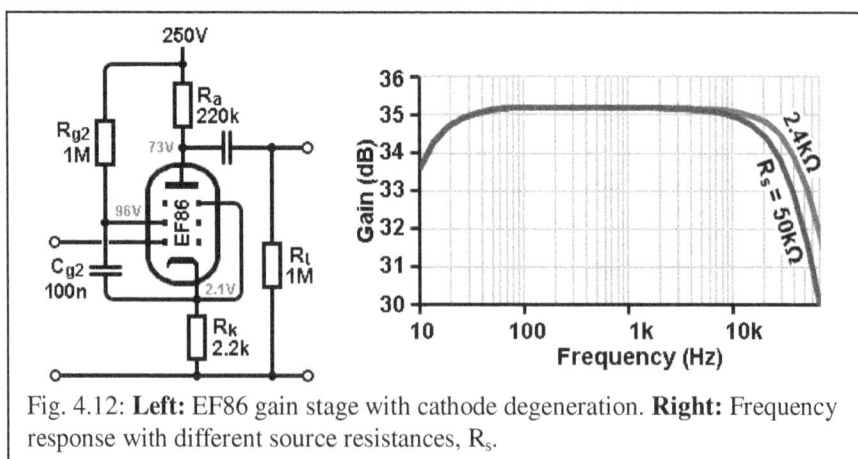

Fig. 4.12: **Left:** EF86 gain stage with cathode degeneration. **Right:** Frequency response with different source resistances, R_s.

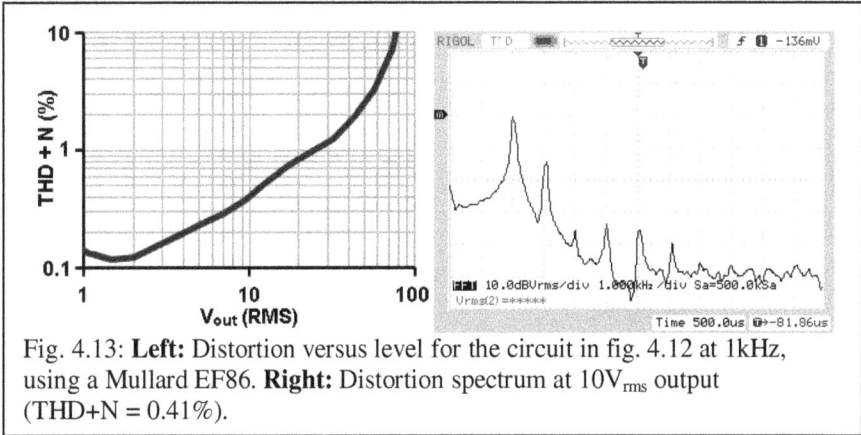

Fig. 4.13: **Left:** Distortion versus level for the circuit in fig. 4.12 at 1kHz, using a Mullard EF86. **Right:** Distortion spectrum at $10V_{rms}$ output (THD+N = 0.41%).

4.6.2: The Effect of Screen Degeneration

Fig. 4.14 shows the circuit with C_{g2} removed, causing the gain to drop to 20.5 (or 29 if C_k was replaced in circuit). However, this does not seem to have bought the performance improvement we might hope for. The frequency response now shows a resonant peak which is damped by increasing the source resistance. Experimentation with a circuit simulator suggests this is caused by certain combinations of stray capacitance between g_2 and the other electrodes, but this is not an arrangement the author has any desire to analyse in detail. The lower end of the frequency response does at least extend down to the limit of the measurement (it would extend down to DC were it not for coupling capacitors).

Fig. 4.15 shows the real surprise: the distortion is markedly worse than before, yet remembering fig. 4.5 which showed the general effect of screen degeneration on the grid curves we would reasonably expect the opposite. Part of the problem is that the bias is too hot, placing the quiescent point very close to the knee of the curves (see

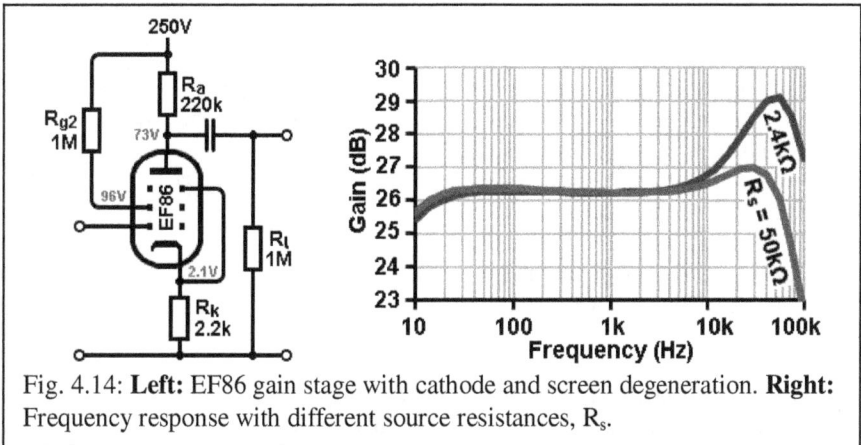

Fig. 4.14: **Left:** EF86 gain stage with cathode and screen degeneration. **Right:** Frequency response with different source resistances, R_s.

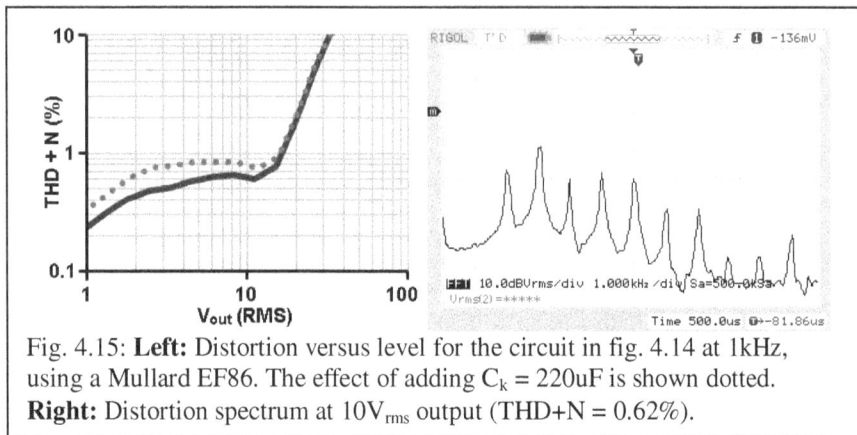

Fig. 4.15: **Left:** Distortion versus level for the circuit in fig. 4.14 at 1kHz, using a Mullard EF86. The effect of adding $C_k = 220uF$ is shown dotted. **Right:** Distortion spectrum at $10V_{rms}$ output (THD+N = 0.62%).

fig. 4.17), which only gets closer with so much screen degeneration. Notice the dramatic rise in distortion at just $14V_{rms}$ output, owing to premature clipping. Fig. 4.15 also shows the distortion spectrum at $10V_{rms}$ out. This is probably the worst the author has ever seen, yet there was no visible clipping (indeed, the spectrum was not much different even at $1V_{rms}$ output).

4.6.3: The Effect of Load on Distortion

To examine the effect of load impedance on distortion we need to keep everything else constant, including the screen voltage. The screen bypass capacitor is a thorn in the side of the pentode since it needs to be quite large to achieve a decent low-frequency response. The best thing to do is to side-step the problem by stabilising or regulating the screen voltage; in this case a couple of 47V Zener diodes were perfect for the task, as shown in fig. 4.16. The screen dropping resistor was reduced to $100k\Omega$ to supply the Zeners with about 1.4mA (the screen consumes a further ~140µA), resulting in a screen voltage of 96V, identical to the traditional circuit tested previously. The Zeners hold the screen voltage constant all the way down to DC, so C_{g2} is now just a noise shunting capacitor added for good measure. Although these components ought to be returned to the cathode rather than to ground, the extra current flowing in R_k would alter the bias voltage, and the author wanted to avoid this complication.

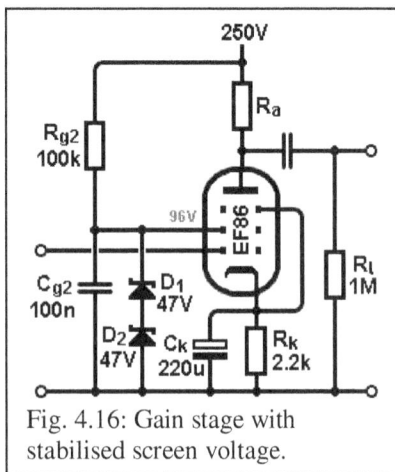

Fig. 4.16: Gain stage with stabilised screen voltage.

Fig. 4.17 shows a selection of load lines to illustrate some possible choices of load, and the resulting bias point when $R_k = 2.2k\Omega$ is also shown. The bias is very close to

2V in all cases since the grid curve is almost horizontal. With the 220kΩ load –as used previously– the bias point is right up near the top of the load line where the grid curves become bunched together. Since they also begin to bunch together at the bottom of the load line it is now easy to see why fig. 4.15 showed so much odd-harmonic distortion; both sides of the waveform were being compressed somewhat, and symmetry means odd harmonics. Fig. 4.18 shows

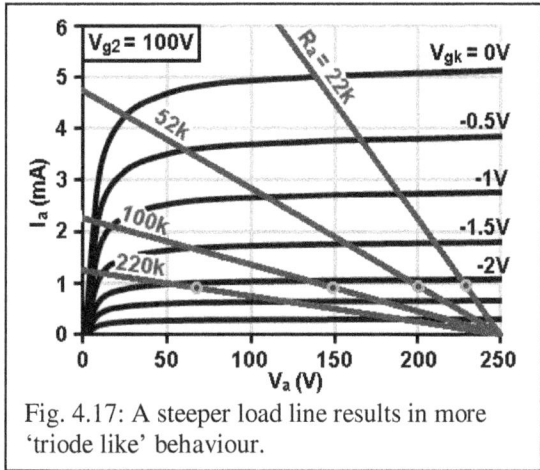

Fig. 4.17: A steeper load line results in more 'triode like' behaviour.

the distortion versus level on a different vertical scale from fig. 4.11 earlier. With the 220kΩ load, clipping occurs at about $40V_{rms}$ output, and from the load line it is clear that this is due to the sudden convergence of the grid curves on the left-hand side of the graph. In other words, the input headroom is barely $300mV_{pk}$ despite the bias being around 2V.

By reducing the load to 100kΩ the gain fell from 195 to 105, but the bandwidth increased to 41kHz with a 50kΩ source resistance. The same 2V bias point is now rather more central on the load line, so there is more equal headroom for positive and negative inputs swings. The operating point can now swing much further before encountering the bunching near the top of the load line, so the distortion spectrum in fig. 4.18 now shows no evidence of third harmonic at $10V_{rms}$ output (compare with fig. 4.11 earlier). The *total* distortion is greater, but much better behaved than before, increasing smoothly up to the point of clipping.

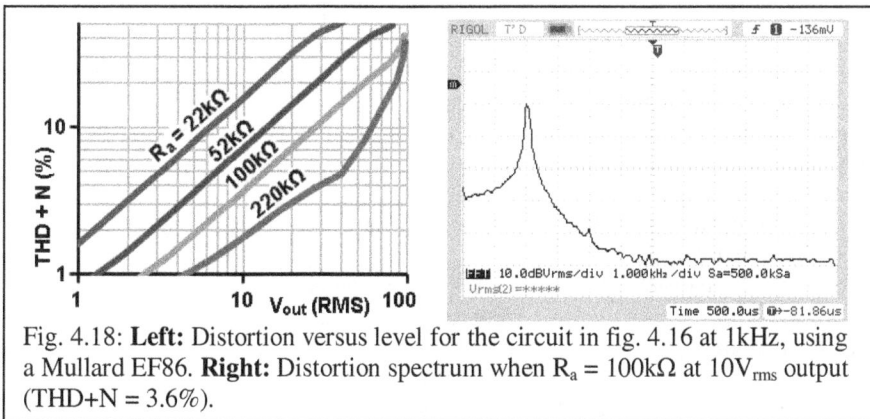

Fig. 4.18: **Left:** Distortion versus level for the circuit in fig. 4.16 at 1kHz, using a Mullard EF86. **Right:** Distortion spectrum when $R_a = 100kΩ$ at $10V_{rms}$ output (THD+N = 3.6%).

With a 52kΩ load the gain dropped to 53. From fig. 4.17 we can see the load line passes right through the knee and the bias point is now towards the cold side of things. Distortion is worsened but was visibly pure second harmonic until clipping.

With a 22kΩ load the gain dropped to 24.3; the bias point is now very cold. There is no bunching of grid curves at the top of the load line because it now passes well above the knee, so what we have could reasonably be described as 'triode like' operation, although the amount of distortion is now an order of magnitude worse than a good triode would produce.

The conclusions we are drawn to are that for good pentode design we do well to stabilise the screen voltage and so avoid the pesky bandwidth limitation of the screen-bypass capacitor, and to set the bias roughly centrally to encourage low-order distortion and plenty of headroom. The popular fear about excessive odd harmonic distortion only applies when the load line passes well below the knee so the grid curves are compressed at *both* ends, and it is easy to avoid this by design, if that sort of thing worries you. A greater load resistance offers lower distortion for a given output but it also means a higher output resistance and therefore greater sensitivity to load capacitance. Thanks to its high open-loop gain, a degenerated pentode will produce less distortion with a given load than most triodes would.

4.7: Designing a Pentode Stage the Easy Way

When designing a pentode gain stage we could go to the trouble of drawing the anode characteristics for a chosen screen voltage, then plot a load line and then find the value of m so that a screen dropping resistor can be found. But this is rather tedious. If we are willing to make some gross assumptions then the design process can be made much quicker. If we are expecting to make many circuit adjustments during the build phase anyway, then a rough starting point may be all we really need.

A good place to start is to choose the HT and anode resistor, either arbitrarily or by deciding the gain we want and then dividing by the g_m, which can be estimated from the datasheet. For example, using the EF86, if we want a gain of 70 and we estimate g_m to be 1.5mA/V from whatever graphs or information are to hand, then we need an anode resistor of: $70 / 1.5\text{mA/V} = 47\text{k}\Omega$. This is also small enough to ensure a

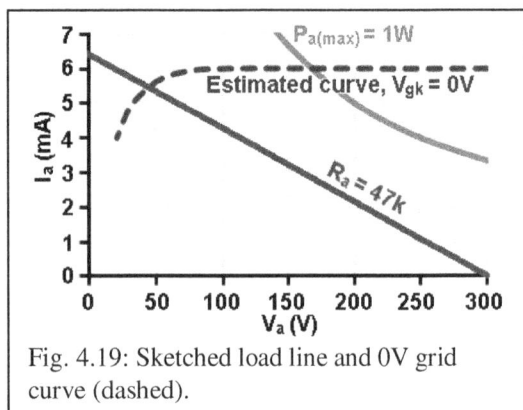

Fig. 4.19: Sketched load line and 0V grid curve (dashed).

decent bandwidth even with a significant load capacitance. We can now draw a load line on a blank graph and sketch in where we would like the 0V grid curve to be, so the load line passes above, through, or below the knee as desired. As we saw earlier, putting it too far below the knee risks some unpleasant distortion spectra, so fig. 4.19 shows an example where it passes right through the knee, using a 300V HT.

Now we know where the 0V grid curve should be, we can find the screen voltage using the mutual characteristics graph.[*] In fig. 4.19 the grid curve levels off at around 6mA anode current, so we find the point corresponding to

$V_{gk} = 0V$, $I_a = 6mA$ on the mutual

characteristics. This is labelled A in fig. 4.20. The characteristic curve can now be sketched in, as shown dotted, and it suggests a screen voltage of just under 120V, say 115V. To set the screen voltage we gratefully abandon the traditional screen dropping resistor and simply fix it with some Zener diodes, or whatever fancy regulator circuit happens to be flavour of the month.

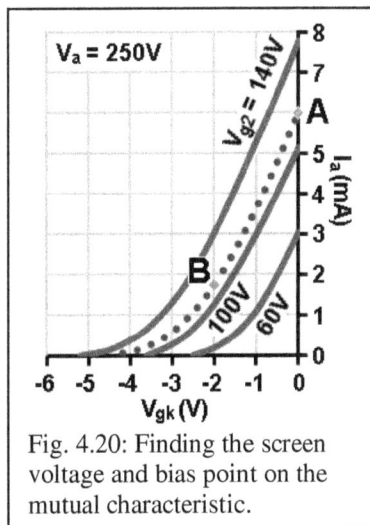

Fig. 4.20: Finding the screen voltage and bias point on the mutual characteristic.

We next choose a suitable bias point, which for maximum headroom and favourable distortion ought to be on the warm side of central. Fig. 4.20 shows that the valve will reach cut-off around $V_{gk} = -4.5V$, so we might choose –2V as the bias point, labelled B. From this we can see the anode current will be about 1.8mA. From section 4.3.1 we know the screen current will be about one fifth of the anode current or 0.36mA, so the total cathode current should be around 2.16mA. The bias could therefore be set with a $2V / 2.16mA = 926\Omega$ cathode resistor, but see shortly.

It is not necessary to know the screen current with great accuracy since the value of R_{g2} is no longer critical. Knowing that the voltage across it is $HT-V_{g2}$, we simply choose a value which will provide more current than necessary for the screen; the rest flows into the Zeners. However, it is worth checking the quiescent screen dissipation before building the circuit, since the screen has only a meagre power rating. In this case our predicted screen voltage and current are 115V and 0.36mA so the estimated dissipation is: $115 \times 0.36 = 41.4mW$. This is well within the 200mW limit quoted on the datasheet.

Rather than use a resistor for biasing, we can alternatively use a suitably chosen LED. The voltage across an LED is fairly constant despite changes in current through it, which eliminates the need for a cathode bypass capacitor and guarantees a flat bandwidth down to DC. It also means the Zener diodes can be returned directly

---

[*] The mutual characteristics are usually given for an anode voltage of 250V which may be completely different from the quiescent anode voltage we end up with. But since we assumed the grid curve was horizontal –and we are doing things the quick way– we shall sweep this discrepancy under the engineering carpet.

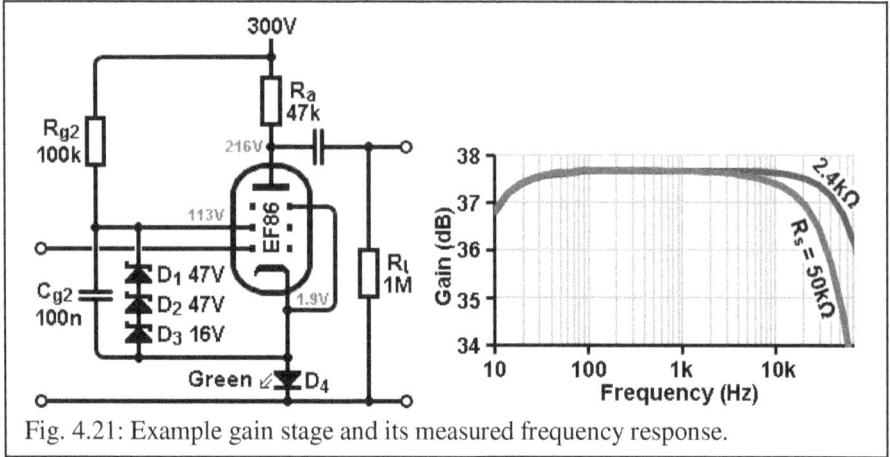

Fig. 4.21: Example gain stage and its measured frequency response.

to the cathode (which is the right and proper thing to do) without the Zener current affecting the cathode bias voltage. LED biasing is covered in more detail in section 6.3.

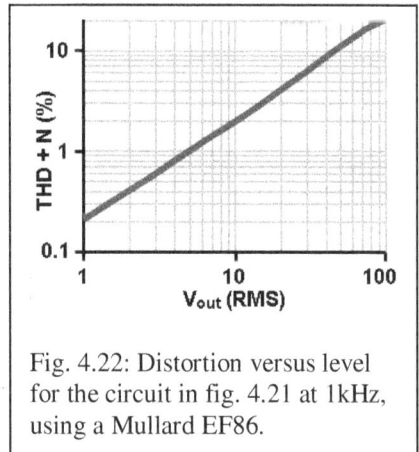

Fig. 4.22: Distortion versus level for the circuit in fig. 4.21 at 1kHz, using a Mullard EF86.

Fig. 4.21 shows the final circuit. A green LED was found to provide 1.9V bias, and a 16V Zener diode was added to the two 47V devices used previously, producing 113V screen voltage. The measured gain was 76.5. Altogether our 'easy' method of pentode circuit design is quite reliable. The frequency response was 52kHz with a 50kΩ source resistance, implying about 61pF of Miller capacitance. This is very good considering how high the gain is (an ECC83 triode could provide a similar amount of gain but with 200pF of Miller capacitance). The low frequency response would extend to DC but for the output coupling capacitor. Fig. 4.22 shows the distortion; it is not bad considering how much gain we have into a 47kΩ load, and was visibly pure second harmonic up until 15V$_{rms}$ output when the third harmonic began to creep in. The equivalent input noise voltage was 2.1μV (20Hz–20kHz), or 2.5μV if C$_{g2}$ was removed (bypassing the LED made no measureable difference); this is at least twice as noisy as most triodes.

The author was also curious to see the effect of screen degeneration on the circuit, so a screen grid stopper R$_{gs}$ was added. A value of 10kΩ reduced the gain to 70 but had little effect on the distortion. However, with a value of 100kΩ, as shown in fig. 4.23, the gain plummeted to 40.2 while the distortion almost doubled. Also, because an amplified signal now appears on the screen, the grid-to-screen capacitance is subject

184

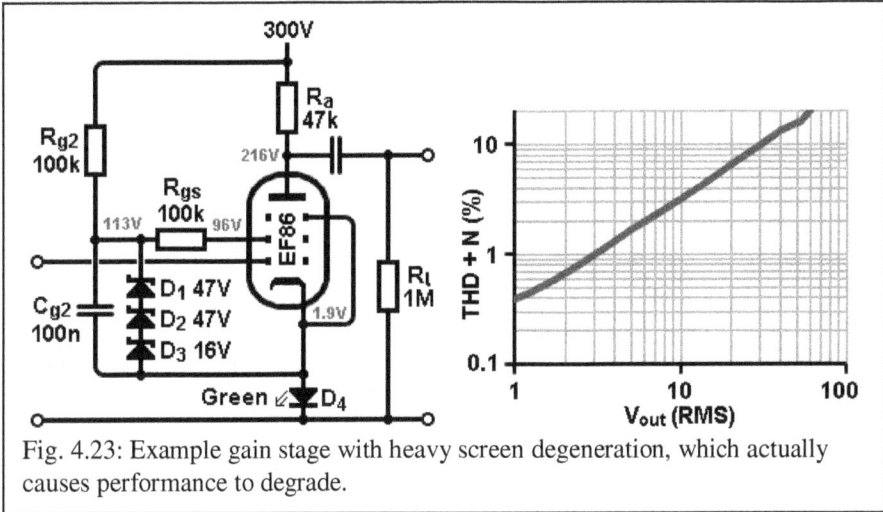

Fig. 4.23: Example gain stage with heavy screen degeneration, which actually causes performance to degrade.

to the Miller effect too, causing the total input capacitance to increase to 64pF despite the gain to the anode being reduced. Heavy screen degeneration, it seems, is something to be avoided.

4.8: Pentodes Connected as Triodes

For reasons already outlined, pentode gain stages are unpopular in modern hi-fi preamps. However, a pentode can also be converted into a perfectly serviceable triode. Triode operation is achieved by connecting the screen-grid to the anode, preferably via a small screen-stopper resistor, as shown in fig. 4.24. Under this condition the screen-grid and anode effectively become one single electrode, and the valve is said to be **triode connected** or **triode strapped**.

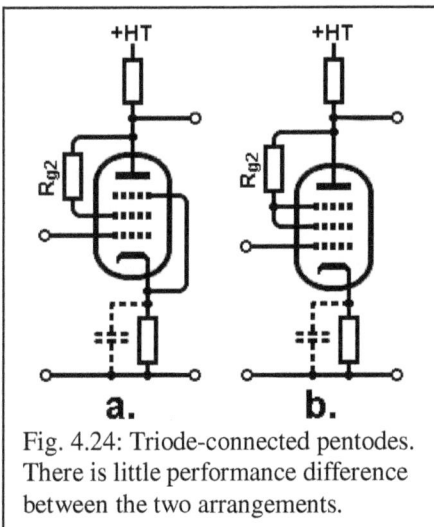

Fig. 4.24: Triode-connected pentodes. There is little performance difference between the two arrangements.

The suppressor grid may be connected to the cathode as normal, or to the screen grid, which slightly increases the perveance. The difference is of little consequence, however, as can be seen from the EF86 anode characteristics in fig. 4.25. The screen stopper serves the dual purposes of damping any parasitics that might cause oscillation, as well as limiting screen current which might otherwise lead to over dissipation. A value of 100Ω to $1k\Omega$ is typical and has negligible effect on performance. In many designs it is omitted altogether, usually without consequence, but any cheap insurance against oscillation is worth having.

185

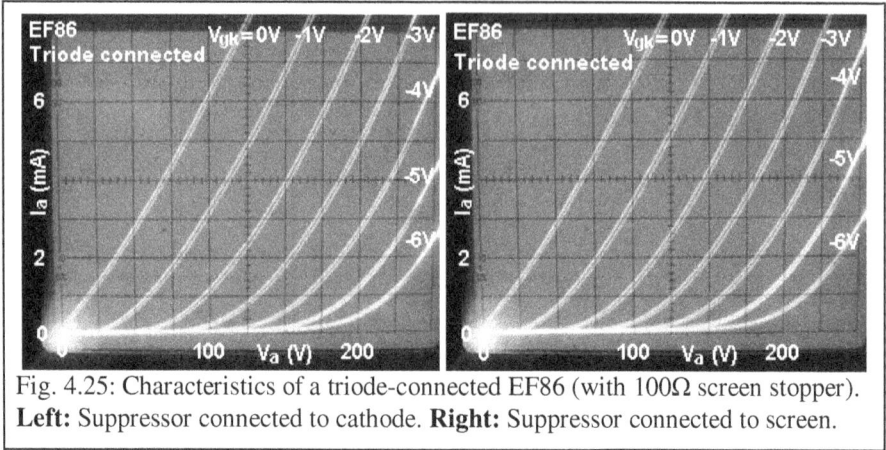

Fig. 4.25: Characteristics of a triode-connected EF86 (with 100Ω screen stopper). **Left:** Suppressor connected to cathode. **Right:** Suppressor connected to screen.

Fig. 4.26 shows an example circuit employing the same Mullard EF86 as the previous circuits, but this time triode connected. The gain was 29 and the distortion was typical of most triodes, reaching 0.91% at $10V_{rms}$ output. It was, needless to say, visually pure second harmonic. Miller capacitance was 100pF which is also typical of a triode with this level of gain.

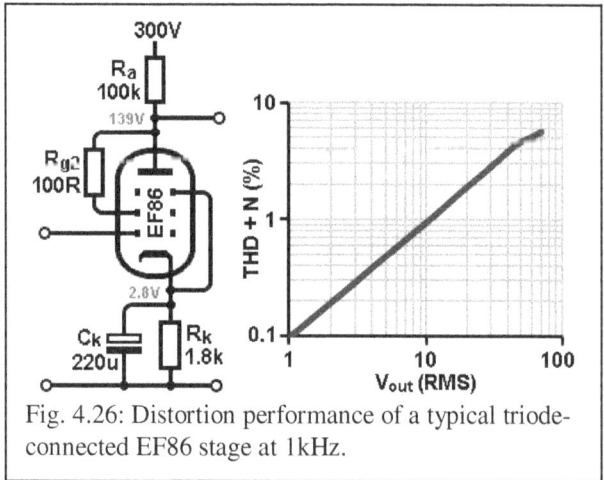

Fig. 4.26: Distortion performance of a typical triode-connected EF86 stage at 1kHz.

4.8.1: Low-Capacitance Mode

A variation on the triode-connected pentode is to ground the anode and use only the screen-grid as the new anode, as shown in fig. 4.27. The original anode then acts as an electrostatic shield around the whole device. Since most pentodes also have a mesh shield, the resulting two-layer shielding may allow the valve to be proudly displayed on top of the chassis without an ugly screening can, even in sensitive applications, though the author has not tried it. The overall characteristics are essentially the same as the ordinary triode-connected characteristics, as can be seen from fig. 4.28 which is indistinguishable from fig. 4.25 –proof of the prepotence of the screen grid. It is of course essential not to exceed the screen dissipation rating with this configuration, which is why the curves in fig. 4.28 have been so severely truncated. The main advantage of this arrangement is that only the grid-to-screen capacitance contributes to the Miller effect, so input capacitance is

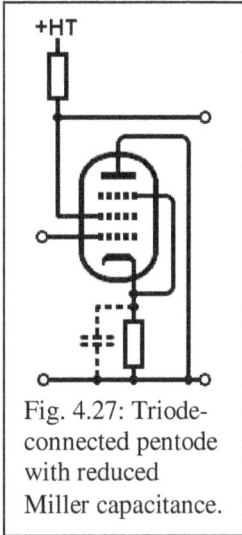
Fig. 4.27: Triode-connected pentode with reduced Miller capacitance.

reduced (though perhaps not as much as might be hoped). Also, depending on the internal construction and arrangement of the valve pins, it may reduce heater hum too.[3]

Fig. 4.29 shows an example circuit. The supply voltage has been reduced in order to keep the screen dissipation within limits, which also makes the distortion marginally worse. The gain was 28.3 but the Miller capacitance dropped to 84pF. Comparing this with the 100pF we had previously this seems to imply the grid-to-screen capacitance is about 2.7pF while C_{ga} is close to 0.5pF including strays.

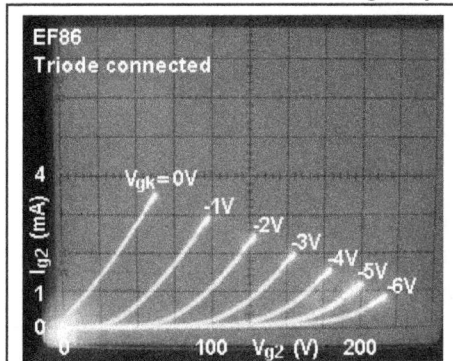
Fig. 4.28: Characteristics of a triode-connected EF86 (anode grounded).

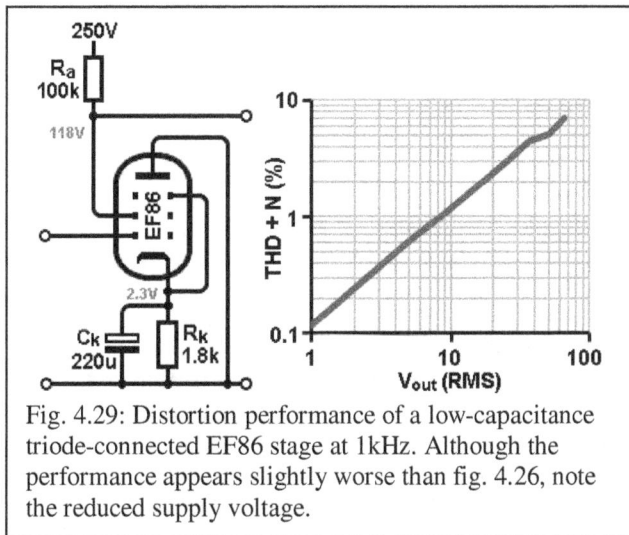
Fig. 4.29: Distortion performance of a low-capacitance triode-connected EF86 stage at 1kHz. Although the performance appears slightly worse than fig. 4.26, note the reduced supply voltage.

[3] Circuit Lab Report Number 2, (1950). Grounded-Plate Type 6AU6 Triode Connection for Preamplifier Use, *Radiotronics*, 142, April, pp45-7.

187

Chapter 5: Noise, Hum and Microphony

Before going any further with audio circuit design it will be beneficial to explore some aspects of noise. Understanding the mechanisms by which this unwanted gremlin creeps in will lead to a set of guiding principles for low-noise design. Whenever low noise is a priority, these principles will steer use quickly towards a good solution. Electrical noise happens to be a special interest of the author's, which is why this chapter is perhaps a little more intensive than would normally be the case in a book of this type. You have been warned.

5.1: Noise

Although we often talk about all sorts of unwanted interference as 'noise', in electronic engineering the word is more specifically reserved for intrinsic circuit noise. This sort of noise is by definition chaotic, whereas other kinds of interference like hum and RF pickup are to some extent deterministic (i.e. predictable) and can be shielded or cancelled out by clever means. Intrinsic noise, however, is generated 'inside' components and is an unavoidable consequence of the physics of materials, and sets the ultimate limit on how quiet a circuit can be.

Since true noise is chaotic we can never predict what it is going to do on an instant-by-instant basis. We can, however, state the *probability* of what it might do next, and from this we can predict its long-term average behaviour, which is usually what we care about most anyway. The sort of noise we are most likely to encounter is called **Gaussian noise** because it

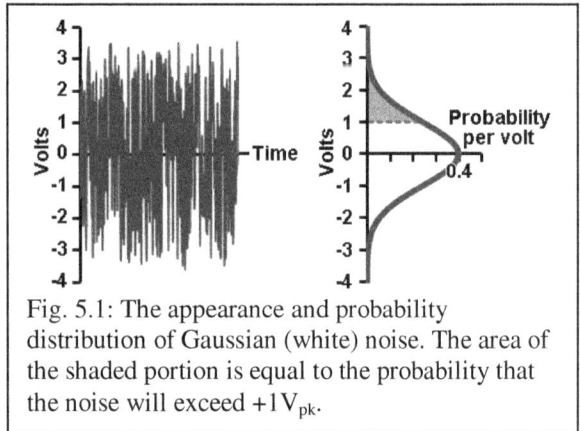

Fig. 5.1: The appearance and probability distribution of Gaussian (white) noise. The area of the shaded portion is equal to the probability that the noise will exceed $+1V_{pk}$.

obeys the classic bell-shaped Gaussian or 'normal' probability curve. Fig. 5.1 shows an example of such noise as it would appear on an oscilloscope, alongside its **probability density function** which has been turned sideways to show its relation to the noise. The area under the curve is equal to the probability that the noise voltage will exceed a particular value at any time, and its total area is 1 by definition, i.e. there is a 100% chance that the noise has *some* value. The curve is symmetrical since the noise spends on average just as much time positive as it does negative, so its average value is zero, which is not very enlightening. We therefore normally work with RMS noise values.[*] A useful rule of thumb is that the peak-to-peak noise, as viewed by eye on an oscilloscope, is typically five to six times the RMS value, e.g. the noise in fig. 5.1 is about $6V_{pp}$ or $1V_{rms}$.

---

[*] Statisticians call the RMS value (of a pure AC signal) the 'standard deviation'.

5.1.1: Johnson or Thermal Noise

If an object is cooled to absolute zero (0K or −273°C) all its atoms are completely frozen still. As the temperature is raised they begin to vibrate and, since atoms contain charged particles, the concentration of charge at any given point fluctuates more and more as temperature increases. Any change in charge between two points amounts to a changing voltage, so within every material there must be an implicit voltage generator producing a noise voltage that increases with temperature. This sort of noise is therefore called **thermal noise**, or **Johnson noise** after its discoverer. This further leads to the perhaps unexpected fact that an isolated resistor, completely unconnected from anything, dissipates some power. This can be thought of as the power delivered to the resistor by the universe at large to keep it at ambient temperature. This power is equal to:

$$P_N = 4kTB \quad \text{watts} \tag{5.1}$$

Where:

k = Boltzmann constant, 1.38×10^{-23} joules/kelvin
T = absolute temperature of the resistor in kelvin.
B = the noise measurement bandwidth in hertz

Notice that the value of the resistance does not appear; all objects at the same temperature dissipate the same noise power. Also, the noise power is the same for any given bandwidth, wherever we choose to place that bandwidth. In other words, the power in the 1Hz-wide band between 1Hz and 2Hz is exactly the same as that between 175Hz and 176Hz. Noise of this type, having the same power per hertz, is referred to as **white noise** since it contains all frequencies with equal power, similar to white light.[*]

Noise *power* is significant for radio engineers, but we are more interested in voltage. But we know that $P = V^2/R$, so by multiplying equation (5.1) by the resistance, R, we get to a figure in mean square volts:

$$v_N^{\,2} = 4kTRB \quad \text{mean square volts} \tag{5.2}$$

The units of mean square volts may seem peculiar at first, but they are really quite literal: measure the noise voltage across a resistor with a true-RMS meter, then square the number on the display; it's a simple as that. The mean square is useful because it is *proportional* to watts, upon which noise theory is based, but it is more directly relevant to us in the land of voltage amplification.

---

[*] One way to measure the frequency response of a circuit is therefore to feed in white noise and monitor the Fourier spectrum of the output to see what frequencies made it through.

Now that we have the mean square volts being generated inside a resistor we could go one step further and take the square root, giving us the root of the mean square volts, better known as the RMS voltage:

$$v_N = \sqrt{4kTRB} \quad \text{volts RMS}$$ (5.3)

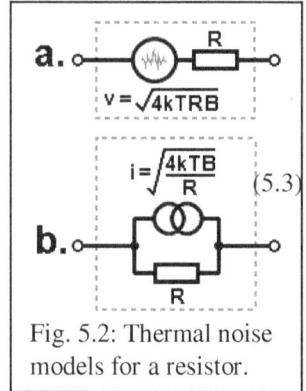

Fig. 5.2: Thermal noise models for a resistor.

Finally we have arrived at the honest RMS voltage that could (in theory at least) be measured across an isolated resistor with an ideal RMS voltmeter. A resistance can therefore be modelled as a noiseless resistor, R, in series with a noise voltage generator equal to $\sqrt{4kTRB}$, as shown in fig. 5.2a. This is a Thévenin equivalent, but it can be transformed into the Norton equivalent consisting

of a noise *current* generator equal to $\sqrt{4kT\dfrac{1}{R}B}$ in *parallel* with the resistance, as

in b. Nevertheless, it is often more convenient to keep things in mean square form when doing noise calculations, i.e. remove the square root sign.

5.1.2: Shot Noise

Charge is carried around a circuit in discrete packets or 'quanta', so rather than being like a smooth fluid it is more like lead shot pouring down a pipe. When a DC current has to cross a 'potential barrier' (i.e. anything with an anode and cathode), the instantaneous amount of charge crossing the anode is constantly fluctuating, even though the average may be quite steady. Hence we can imagine the current to be composed of a steady DC value plus a fluctuating noise current that rides on top of it, which is called shot noise. Shot noise is described by Schottky's theorem:

$$i_{shot} = \sqrt{2qI_{DC}B} \quad \text{amps RMS}$$ (5.4)

Where:
q = charge on an electron, 1.6×10^{-19} coulombs
I_{DC} = mean value of current
B = noise measurement bandwidth

Note that if there is no DC current, there can be no shot noise. Like Johnson noise, shot noise contains equal power per hertz, so it is a type of white noise.

5.1.3: Flicker or Excess Noise

Whenever current has to flow across material boundaries or navigate around microscopic cracks and fissures, strange things happen. It is as if the current is continually avulsing or changing its mind about what path to take, which gives rise to more noise. This sort of noise is due to imperfections in the consistency of the material rather than to its basic physical properties, so it is often called excess noise since it exists 'in excess' of the unavoidable Johnson noise. In valves it is traditionally called **flicker noise**. On an oscilloscope it has more of a 'grassy' appearance rather than the 'furry' appearance of white noise, as illustrated

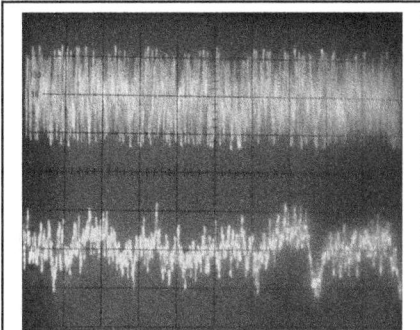

Fig. 5.3: **Upper:** Pure white noise. **Lower:** Pure pink noise. Both traces are 500mV/div and have an RMS value of 200mV in the audio band.

in fig. 5.3. Because flicker/excess noise is due to the consistency of material manufacture almost at the atomic level, it is highly variable and much less easy to predict than Johnson and shot noise. In general, the mean square excess noise riding on a DC current is found to be of the form:

$$i_{excess}^2 = KI_{DC}^a \frac{1}{f^b} \qquad \text{mean square amps (spot frequency)} \qquad (5.5)$$

Where:

K = a constant that depends on the material
I_{DC} = mean value of current
a = a constant that depends on the material and determines the colour of the noise
f = a 1Hz-wide spot frequency band
b = a constant that depends on the material

There is a lot of 'depends' in this list of variables, hence the difficulty in predicting excess noise. However, it is often the case that $a = 2$ and $b = 1$, or near enough, and in such cases the noise power becomes inversely proportional to frequency, in which case it may be called **1/f noise**. 1/f noise has equal power per decade (or per octave, or per any logarithmically proportional bandwidth), and there are more hertz between 100Hz and 1kHz than between 10Hz and 100Hz, so more noise power is crammed in per hertz at low frequencies than at high. Drawing an analogy with light, it is like having an excess of red light mixed in with everything else, so 1/f noise is also called **pink noise**. Unfortunately for us, it often raises its head above white noise somewhere in the audio region, and the point where it does –where the pink noise is equal to the white noise– is called the **noise corner frequency** (e.g. fig. 5.12).

Equation (5.5) only tells us the value of noise current in each and every 1Hz-wide **spot frequency**. To find the *total* excess noise in a given bandwidth (f_{hi}–f_{lo}) we must add all the individually calculated values together, or rather, use integration. If we stick to a = 2 and b = 1, which is good enough for practical design work, then we get:

$$i_{excess}^2 = KI_{DC}^2 \int_{flo}^{fhi} \frac{1}{f}\, df = KI_{DC}^2 \ln\left(\frac{f_{hi}}{f_{lo}}\right) \quad \text{mean square amps} \tag{5.6}$$

We could of course take the square root of this to find the RMS noise current:

$$i_{excess} = \sqrt{KI_{DC}^2 \ln\left(\frac{f_{hi}}{f_{lo}}\right)} \quad \text{amps RMS} \tag{5.7}$$

The awkward variable K can only be determined by actual measurement.

5.1.4: Adding Noise Sources

When two noise sources are entirely independent, such as two separate resistors, they are said to be **uncorrelated**; they each produce their own entirely random voltages and have no control over what the other one is doing. If these sources are mixed together then they will sometimes reinforce each other and at other times cancel each other out. For this reason we cannot directly sum individual noise voltages (or currents) to find the total; instead we must use geometric addition, sometimes called 'RSS addition' meaning the 'root of the sum of the squares'.[*] In other words, if we have several uncorrelated noise voltage sources all combining together, then to find the total RMS noise voltage we must square each individual RMS voltage source, then add them, then take the square root:

$$v_{total} = \sqrt{v_1^2 + v_2^2 + v_3^2 + ...}$$

Exactly the same goes for adding noise currents.

This sort of addition produces some interesting results. For example, if we had two uncorrelated sources each producing $1\mu V_{rms}$ noise, adding them together would give a total noise voltage of:

$$v_{total} = \sqrt{1^2 + 1^2} = 1.4\mu V$$

Thus adding the second noise source increases the total by only 3dB and not 6dB as would have happened with correlated signals. Another important result is that only the noisiest source really matters. If we have, say, a $1\mu V$ source and a $2\mu V$ source then the total is:

$$v_{total} = \sqrt{1^2 + 2^2} = 2.2\mu V$$

This is only 0.8dB greater than the $2\mu V$ source alone. Hence to minimise circuit noise we must concentrate on improving the major noise sources first. Improving or

---

[*] In a previous book I mistakenly called it 'RMS addition' which does not describe the process at all and is frankly meaningless.

even eliminating the smaller ones will have practically no effect on the total and is wasted effort.

5.1.5: Equivalent Input Noise

Noise in circuits can be quantified with the use of equivalent circuits in which the real circuit is modelled as being made up of ideal noiseless components, plus voltage and/or current generators to represent the noise generated inside the real thing (we have already seen this in fig. 5.2). The background noise that a circuit produces all the time is called its **noise floor**. A particularly useful way to

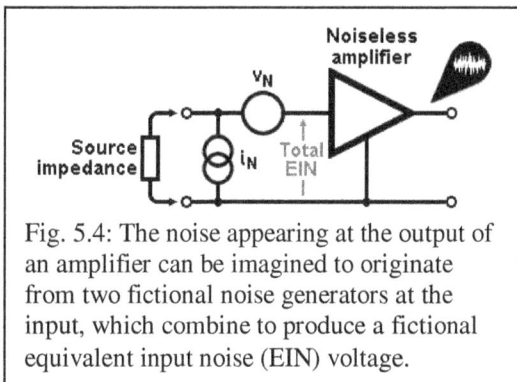

Fig. 5.4: The noise appearing at the output of an amplifier can be imagined to originate from two fictional noise generators at the input, which combine to produce a fictional equivalent input noise (EIN) voltage.

express this is to calculate or measure the total noise voltage appearing at the output of an amplifier, then divide this number by the gain of the circuit to obtain the **equivalent input noise**, or **EIN**. This imagines that the circuit itself is noiseless but has a fictitious noise voltage generator in series with the input, plus a noise current generator in parallel with the input, as shown in fig. 5.4. The input noise current flows in the source impedance and so develops a noise voltage which, together with the input noise voltage generator, forms the total EIN voltage. The EIN is amplified by the circuit gain to produce the noise voltage actually measured at the output. Of course, the output noise is not *genuinely* generated right at the very input –it comes from a plurality of sources within the circuit– but this does not matter from the point of view of quantifying the noisiness of an amplifier. All the noise, wherever it actually comes from, is then said to be **input referred**. Nevertheless, in most cases the majority of the noise really *does* come from two or three critical components close to the input, while the rest are insignificant by comparison, which rather simplifies the problem of low-noise design.

Since we presumably know the how big the audio input signal will be, we can compare it directly with the equivalent input noise to find the **signal-to-noise ratio** or **SNR**. This is normally expressed in decibels according to:

$$\text{SNR} = 20\log_{10}\left(\frac{v_{\text{signal}}}{v_{\text{noise}}}\right) \tag{5.8}$$

An important observation of fig. 5.4 is that the noise contribution due to the *voltage* generator is fixed and always present, but the amount due to the current generator depends on the source impedance and is eliminated if the source impedance is zero. This leads to the general rule of thumb that a high source impedance should be connected to an amplifier with low *current* noise (typically a triode or JFET), but a low source impedance needs an amplifier with low *voltage* noise (typically a bipolar transistor).

In radio electronics it is common to express the EIN voltage as an **equivalent input noise resistance**. This is a fictitious resistor in series with the amplifier input which would generate the same amount of Johnson noise (at room temperature) as the calculated EIN. This resistance is easily found by rearranging equation (5.3):

$$R_{eq} = \frac{v_{EIN}^2}{4kTB} \qquad\qquad (5.9)$$

Where T is normally set to 290 or 300 kelvin. This is makes sense for radio work where power transfer and impedance matching are critical, and noise tends to be white. For audio work it is not a very useful way to express noise since we care more about voltage, and it is rather dubious anyway since our noise tends to be a mixture of white and pink, so it not truly equivalent to Johnson noise. Nevertheless, some authors still use it as a crude means of predicting valve shot noise because the formula is deceptively simple –see section 5.3.2.

5.1.6: Noise Bandwidth

Something that is frequently glossed over in audio textbooks is that the bandwidth we use in noise calculations is *not* the same thing as the −3dB bandwidth commonly used for other purposes. To ensure that all noise power is accounted for we must use the **noise bandwidth**. This is the width of a rectangular or 'brickwall' power gain curve whose total area is the same as that of the actual power gain curve. The power gain of a circuit is equal to the square of its voltage gain, so the noise bandwidth can be formally defined as:

$$B_N = \frac{1}{A_0^2} \int_0^\infty A(f)^2 df \qquad \text{hertz} \qquad\qquad (5.10)$$

Where A(f) is the voltage gain as a function of frequency and A_0 is the mid-band gain. This equation tells us to take the frequency response (voltage gain), square it, then find a rectangle with the same total area. The width of this rectangle is the noise bandwidth B_N, though we will simply use B in this chapter. For a circuit whose frequency response falls off at a first order rate the noise

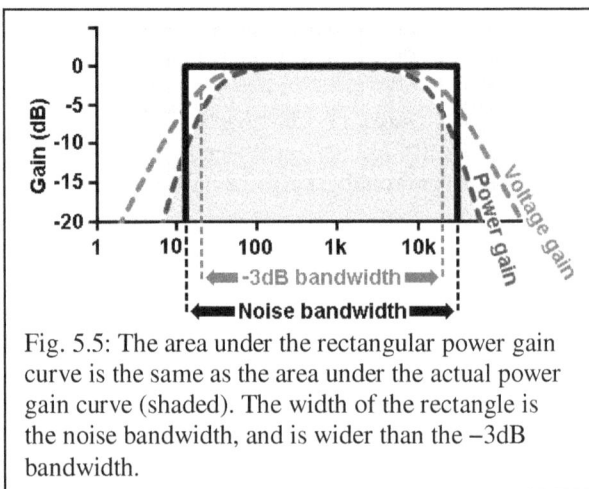

Fig. 5.5: The area under the rectangular power gain curve is the same as the area under the actual power gain curve (shaded). The width of the rectangle is the noise bandwidth, and is wider than the −3dB bandwidth.

Filter order	Noise bandwidth
1st	$1.57 \times f_{-3dB}$
2nd	$1.22 \times f_{-3dB}$
3rd	$1.15 \times f_{-3dB}$
4th	$1.13 \times f_{-3dB}$
5th	$1.11 \times f_{-3dB}$
2nd order Butterworth	$1.11 \times f_{-3dB}$
3rd order Butterworth	$1.05 \times f_{-3dB}$

Table 5.1: Equivalent noise bandwidth of simple filters.

bandwidth is $\pi/2$ or 1.57 times wider than the −3dB bandwidth, as illustrated in fig. 5.5. If the response falls off at a faster rate then the noise bandwidth more closely approaches the −3dB bandwidth, and figures for some simple RC and Butterworth filters are summarised in table 5.1.

Ideally, for audio we would always use a noise bandwidth of 20Hz–20kHz even if the amplifier bandwidth is much greater, since inaudible noise is of no interest to us. Many commercial noise meters employ a 22Hz–22kHz third-order bandpass filter conforming to DIN 45405 (i.e. a noise bandwidth of 21Hz–23kHz) which is close enough for audio work. However, many published noise specifications are decidedly unspecific about the true noise bandwidth used. In some books we are told that noise measurements have been made in a 20Hz–20kHz bandwidth or some such, but the author then fails to say whether this was a true noise bandwidth or simply a −3dB bandwidth. If it was the latter then the noise bandwidth was actually larger, making the measurements appear up to 2dB worse than they should be. Needless to say, this can be a nightmare for anyone trying to compare figures from different authors.

5.1.7: Noise Spectral Density

Unless we are dealing with pure white noise we will probably want to examine the frequency content of noise sooner or later. This can be expressed in the form of **noise spectral density**, in volts squared per hertz, which is essentially the Fourier spectrum of the noise. Fig. 5.6 shows a typical example of the equivalent input noise spectral density of an ECC83/12AX7 (more on this later). The name of the graph makes it sound much grander than it really is. Like any density it tells us how much there is of something in a certain space; in this case how much noise power (alright, *mean square voltage*) there is per hertz

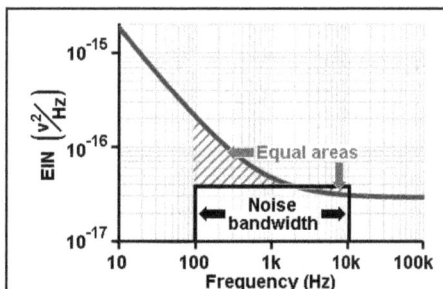

Fig. 5.6: Total noise is equal to the area under a noise density curve between the bandwidth limits, which is the same as the area of the rectangle (the two hatched areas appear unequal because the axes are logarithmic).

of bandwidth. The total equivalent input noise in mean square volts is simply the density multiplied by the available space, i.e. the number of hertz, better known as the noise bandwidth. In other words, the total noise is equal to the area under the curve between the bandwidth limits. This can be found by integration –if we happen to have a formula describing the curve– or by counting squares.[*]

The quickest method is simply to estimate a rectangular area by eye, noting any areas not enclosed by the rectangle must cancel out the bits where the rectangle goes beyond the curve, as shown in fig. 5.6. Remember, however, that the axes are usually logarithmic, so there is actually a lot more area under a curve at high frequencies than it looks like, which is why the hatched areas in fig. 5.6 appear to be different sizes; only when plotted on linear axes would they appear the same. Multiplying the height of the rectangle (the noise density) by its width (the noise bandwidth) gives the total noise in mean-square volts. In this case the total EIN between 100Hz and 10kHz is about:[†]

$$4 \times 10^{-17} \times 9900 = 3.96 \times 10^{-13} \ V^2$$

We could then go on to take the square root of this number to find the RMS volts, which comes out at 629nV. With a little judgement this quick and dirty method gives an immediate and intuitive idea of how the total noise changes with different choices of bandwidth.

An alternative way to present the same information is to take the square root of the whole mean-square density graph, thereby converting mean-square volts into RMS volts right from the start. But this also means the Hz become √Hz, leaving us with 'RMS volts per root hertz', which sound rather peculiar. Nevertheless, finding the total noise follows exactly the same process as previously except that we multiply the noise density by the *square root* of the bandwidth. Fig. 5.7 shows exactly the same information as fig. 5.6 so it ought to give the same answer for total EIN:

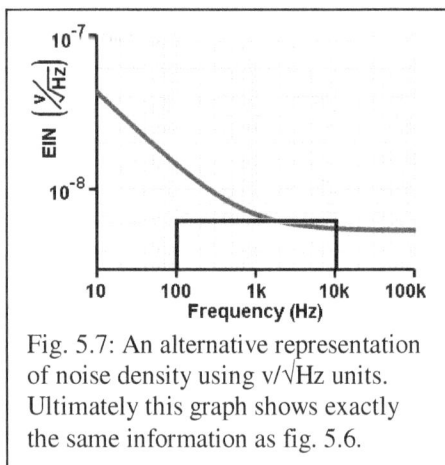

Fig. 5.7: An alternative representation of noise density using v/√Hz units. Ultimately this graph shows exactly the same information as fig. 5.6.

$$6 \times 10^{-9} \times \sqrt{9900} = 597nV$$

The previous figure was 629nV; the 5% difference is due to reading convenient round numbers off the graphs. Despite being rather contrived and further removed from the fundamental maths of noise theory, many data sheets and text books prefer to present noise data in volts per root hertz.

---

[*] For another handy way to determine an oddly-shaped area, look up *Pick's theorem*.
[†] This noise bandwidth was chosen for the example because the smaller hatched area in fig. 5.6 becomes almost invisibly thin when using a 20kHz bandwidth.

5.1.8: Noise Weighting

When measuring audible noise the most unambiguous method to use is a filter with a nominally flat passband with simple, well controlled roll offs, giving a noise bandwidth of 20Hz–20kHz. The measurements are then said to be **unweighted**. But for some noise measurements it may be desirable to consider the *subjective* quality of the noise, and this is done by passing the noise through a **weighting filter** whose frequency response tries to reflect the

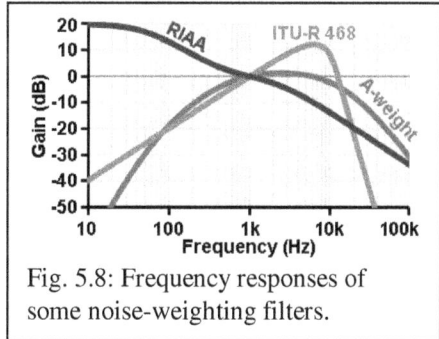

Fig. 5.8: Frequency responses of some noise-weighting filters.

sensitivity of the human ear. The two most common weighting filters are:

- **A-weighting**, commonly used in America;
- **ITU-R 468**, previously called **CCIR-1k** and more commonly used in Europe.

The frequency responses of these filters are shown in fig. 5.8, along with the RIAA equalisation curve (explained shortly). Whether or not such filters reflect subjective noisiness effectively –and whether it is appropriate to use weighted noise measurements at all– is a subject of continued debate. Nevertheless, for better or worse they are often used by hi-fi manufacturers (particularly A-weighting since it makes the figures look better), and some audio test sets –e.g. the Lindos MS series– do not even provide an option for unweighted measurements, as absurd as that sounds.

If we want to find the total weighted noise from a spectral density graph then we need to know the equivalent noise bandwidth of the weighting filter, and the figures are provided in table 5.2.[*] For example, passing broadband noise through an A-

Weighting Filter	Equivalent Noise Bandwidth (Hz)	Gain (dB)
A-weighting	457 – 11575	1.3
ITU-R 468	3577 – 10064	12.2 + 3

Table 5.2: Equivalent noise bandwidths of two popular noise weighting filters.

weighting filter is equivalent to passing it through a 45–11575Hz rectangular noise bandwidth and then adding 1.3dB to the total. For the ITU-R 468 specification we use a 3577–10064Hz noise bandwidth and add 12.2dB to the total but, just to complicate matters, this specification calls for a quasi-peak measurement rather than RMS, so we must add a *further* 3dB (roughly) if we wish to take this into account.

---

[*] Interestingly, this means that when designing a noise meter there is no need to design a noise filter with the actual frequency responses shown in fig. 5.8; any filter shape will do provided it has the correct noise bandwidth.

197

It is also interesting to compare weighted noise with unweighted noise. For example, if white noise is measured with an A-weighting filter the answer will come out 1.26dB quieter than the unweighted (20Hz–20kHz) answer, because some noise power is effectively thrown away in the weighting filter. Unfortunately, audio noise is usually a mixture of white and pink, so the amount of noise power thrown away depends on how much pink noise exists within the noise bandwidth. Fig. 5.9 shows the conversion factor as a function of the noise corner frequency (where white and pink are equal). In practice we are unlikely to know the

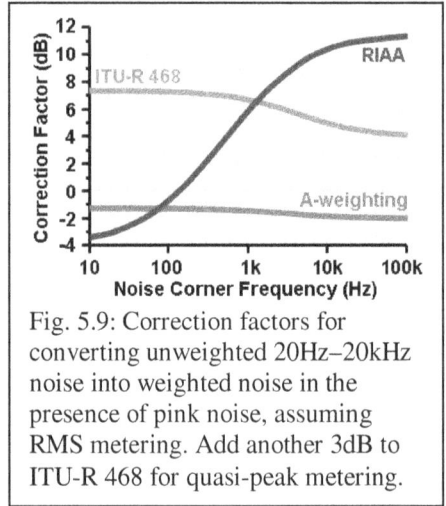

Fig. 5.9: Correction factors for converting unweighted 20Hz–20kHz noise into weighted noise in the presence of pink noise, assuming RMS metering. Add another 3dB to ITU-R 468 for quasi-peak metering.

corner frequency precisely (or at all), but fig. 5.9 at least gives some idea of the scale of the differences that can be expected. For example, if the corner frequency is 1kHz then an A-weighted measurement will be about 1.4dB quieter compared with the unweighted measurement. Conversely, the reading will be about 6.7dB *worse* when using ITU-R 468 weighting, plus another 3dB if quasi-peak metering is used.

A further sort of weighting that may be of interest is the RIAA phono equalisation curve, also shown in fig. 5.9. This is not a noise measurement standard but does affect the noise spectrum measured at the output of a phono amplifier, so it is worth including here. Unfortunately, the shape of the RIAA curve makes it impossible to quote a single equivalent noise bandwidth like those in table 5.2, but the correction factor can be determined nonetheless, and is shown in fig. 5.9. Notably, if the noise is mainly white then RIAA equalisation gives a noise advantage because so much high-frequency noise is discarded. But if significant pink noise is present then we get a noise *dis*advantage because the low frequency boosting overtakes the high frequency loss.

5.2: Noise in Resistors

There are two types of noise that need to be considered in resistors: Johnson noise and excess noise (wirewound resistors do not have the latter). Pure capacitors and inductors do not generate any noise at all, and though practical ones will have some parasitic resistance that does produce noise, it is small enough to be insignificant for our purposes.

As explained in 5.1.1, the Johnson (thermal) noise can be modelled as a noise voltage generator in series with the resistor, or as a noise current generator in parallel with it. The noise voltage is given by:

$$v_{johnson} = \sqrt{4kTRB} \tag{5.11}$$

198

And the noise current is given by:

$$i_{johnson} = \sqrt{4kT\frac{1}{R}B}$$

(5.12)

Where:
k = Boltzmann constant, 1.38×10^{-23} J/K
T = absolute temperature in kelvin
R = resistance in ohms
B = bandwidth in hertz

Excess noise was covered in section 2.1.5 and can be modelled in the same way. It tends to be worse in low-wattage, high resistance devices and is usually quoted on the data sheet as a 'noise index', NI, in RMS microvolts per DC volts applied to the device, per decade of bandwidth, i.e. µV/V/decade. Table 5.3 summarises the figures for various resistor types.

Resistor type	Excess noise (µV/V/decade)
Wirewound	~0
Metal film	0.01 – 0.3
Metal oxide	0.1 – 1.0
Carbon film	0.01 – 0.4
Carbon composition	0.1 – >1.0

Table 5.3: Resistor excess noise indices.

The number of decades in a given bandwidth is equal to $\log_{10}(f_{hi}/f_{lo})$ and the noise in each decade must be added geometrically. There are three decades in the audio bandwidth so the total audible excess noise voltage generated in a resistor is:

$$v_{excess} = NI \times V_{dc} \times \sqrt{3}$$

(5.13)

For example, a film resistor with NI = 0.1µV/V, and with 100V drop across it, generates a Thévenin excess-noise voltage of: $0.1 \times 100 \times \sqrt{3} = 17$µV.

5.3: Noise in Triodes

Noise in triodes comes from three sources:
- Grid current shot noise;
- Anode current shot noise;
- Anode current flicker noise.

Other text books may also include 'induced grid noise' but this is only significant at megahertz frequencies. Fig. 5.10 shows a basic noise model for a triode incorporating the three noise current sources. Note that r_a must be included in the model because it affects circuit gain, but since it is not a physical resistor it does not generate any noise of its own. Directional current arrows are not needed since we are dealing with uncorrelated AC signals.

As explained earlier, shot noise is due to the quantised or 'granular' nature of current. Shot noise current in the grid is given by Schottky's theorem:

$$i_{g(shot)} = \sqrt{2qI_g B} \quad \text{amps RMS} \tag{5.14}$$

Where I_g is the average grid current, typically 1 to 100nA under normal bias conditions. This current flows in the source impedance and will generate a noise voltage which is amplified like any other voltage at the grid.

However, Schottky's theorem predicts too much shot noise on the anode current. This is because the space charge introduces a degree of correlation between the electrons flowing in the valve, and it therefore has a reducing or 'smoothing' effect on the shot (and flicker) noise current. This is taken into account by modifying Schottky's theorem to include a **space charge smoothing factor, Γ^2**:

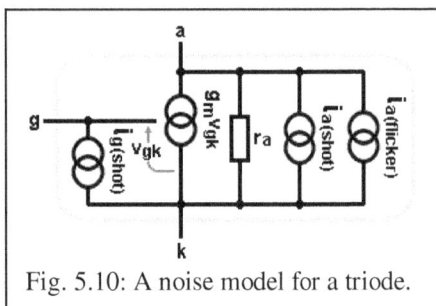

Fig. 5.10: A noise model for a triode.

$$i_{a(shot)} = \sqrt{\Gamma^2 2qI_a B}$$

For an ideal triode Γ^2 is found to be roughly equal to:[1]

$$\Gamma^2 = \frac{2k(0.644T_k)g_m}{qI_a}$$

Substituting this into the previous equation and simplifying produces:

$$i_{a(shot)} = \sqrt{4k(0.644T_k)g_m B}$$

However, even this is not enough, because real triodes have somewhat less g_m than ideal theory predicts, so we must introduce yet another correction factor, σ (sigma), to take this into account. This leaves us with the shot current in the anode being equal to:

$$i_{a(shot)} = \sqrt{4k(0.644T_k)\frac{g_m}{\sigma} B} \quad \text{amps RMS} \tag{5.15}$$

Where
k = Boltzmann constant, 1.38×10^{-23} J/K
T_k = cathode temperature, normally 1000K
g_m = valve transconductance in siemens or A/V
σ = empirical constant which normally varies between 0.6 and 1.
B = bandwidth in hertz.

---

[1] Robinson, F. N. H. (1962). *Noise in Electrical Circuits*. Oxford University Press.

Flicker (1/f) noise in valves is thought to be due to random fluctuations in the work function of the cathode, causing corresponding changes in the number of emitted electrons. It varies considerably between valve samples and with aging (increasing seriously if interface resistance is present), and is much less predictable than shot noise, which is annoying because it is often the dominant noise source. Nevertheless, to a good approximation the RMS flicker noise spectral density in the anode is:

$$i_{a(flicker)}^2 = K \frac{I_a^2}{f} \qquad \text{amps squared per hertz (spot frequency)} \qquad (5.16)$$

The total RMS flicker noise current over a bandwidth is then found by integration to be:

$$i_{a(flicker)} = \sqrt{K I_a^2 \ln\left(\frac{f_{hi}}{f_{lo}}\right)} \qquad \text{amps RMS} \qquad (5.17)$$

Where:
K = an empirical constant that typically varies between about 10^{-14} for a very quiet sample to 10^{-12} for a very poor one.
I_a = average anode current in amps
f_{hi} = upper bandwidth limit in hertz
f_{lo} = lower bandwidth limit in hertz

Some values for the empirical constants σ and K measured by the author[2] are shown in table 5.4.

Valve type	σ	K
ECC81/12AT7	0.7	2.1×10^{-13}
ECC82/12AU7	1.0	6.2×10^{-14}
ECC83/12AX7	1.0	1.9×10^{-14}
ECC88/6DJ8	0.6	6.2×10^{-14}
6J52P/6Ж52П (triode connected)	0.8	3.0×10^{-14}

Table 5.4: Average noise constants for some triodes.

5.3.1: EIN Spectral Density

As explained in the previous section, a triode contains two noise current sources in the anode: shot (white) and flicker (pink). Although these noise currents are actually generated inside the valve we can imagine them to be due to two fictitious noise voltage sources at the input, which are multiplied by the transconductance to become the currents we already know them to be. In other words, we can translate the noise model in fig. 5.10 into the input-referred noise

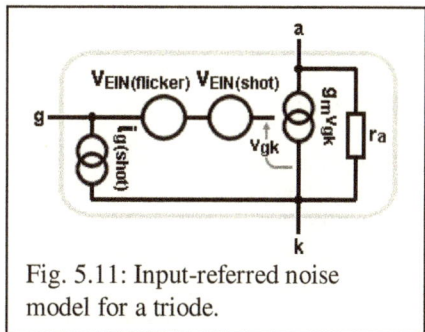

Fig. 5.11: Input-referred noise model for a triode.

[2] Blencowe, M. (2013). Noise in Triodes with Particular Reference to Phono Preamplifiers, *JAES*, 61 (11), November, pp911-16.

model in fig. 5.11. These equivalent input noise voltages are simply the anode current noise densities divided by $g_m{}^2$ (squared because we are dealing with mean-square quantities):

$$v_{EIN(shot)}{}^2 = \frac{4k(0.644T_k)}{\sigma g_m} \quad \text{volts squared per hertz} \qquad (5.18)$$

$$v_{EIN(flicker)}{}^2 = \frac{K}{g_m{}^2}\frac{I_a{}^2}{f} \quad \text{volts squared per hertz} \qquad (5.19)$$

In fig. 5.12 these two equations are plotted for an average ECC83/12AX7 using values from table 5.4 and letting $T_k = 1000K$, $I_a = 1mA$, $g_m = 1.3mA/V$ (ignoring grid current noise for the time being). The point where the two types of noise are equal, i.e. the point where the two dashed lines cross, is the noise corner frequency. The total is the sum of the two noise sources, which can be added directly together since

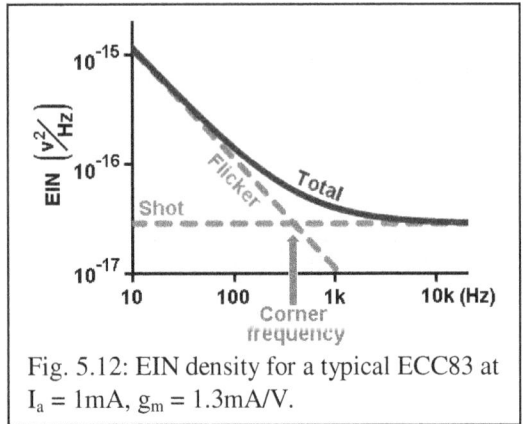

Fig. 5.12: EIN density for a typical ECC83 at $I_a = 1mA$, $g_m = 1.3mA/V$.

we're dealing with mean-square quantities. The basic shape of this plot is typical of all amplifying devices; only the relative levels of white and pink noise change, and with them the corner frequency. Fig. 5.13 compares the EIN densities of the triodes listed in table 5.4, all operating at the same anode current of 2mA.

Things get really interesting if we study equations (5.18) and (5.19) more closely. Notice that the shot noise power is inversely

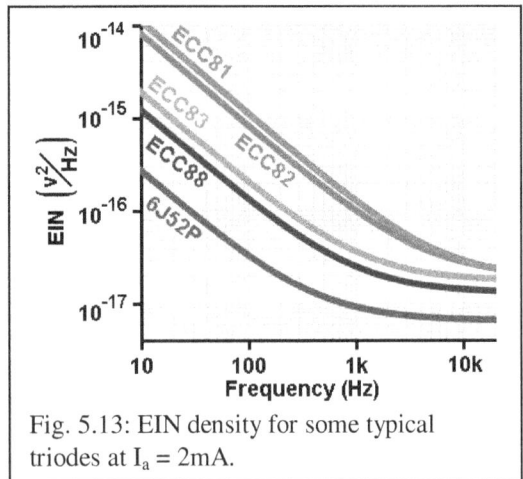

Fig. 5.13: EIN density for some typical triodes at $I_a = 2mA$.

proportional to g_m, whereas the flicker noise power is proportional to $I_a{}^2/g_m{}^2$. But g_m is approximately proportional to $\sqrt{I_a}$, so these two things are at odds; increasing anode current will *reduce* shot noise but *increase* flicker noise. The whole EIN spectral density does a kind of see-saw as anode current increases, as shown in fig. 5.14. Logically, therefore, the total noise in the audio band must reach a minimum at some intermediate, optimum value of anode current.

202

5.3.2: Total EIN

The spectral density graphs and noise constants discussed previously are conveniently neat and well defined, but it must be remembered that they are actually the statistical result of averaging many measurements of different valve samples, no two of which are ever quite the same. For

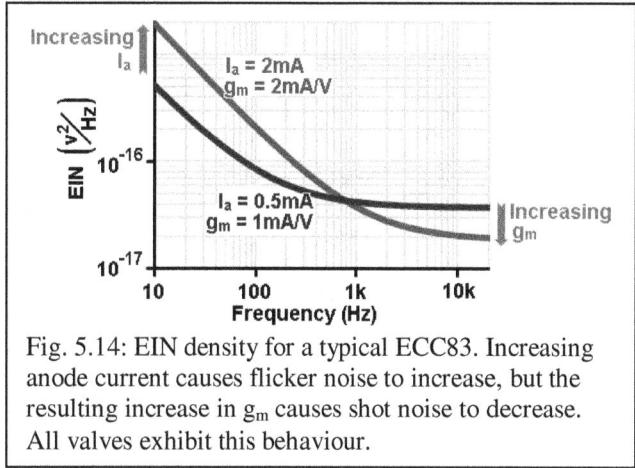

Fig. 5.14: EIN density for a typical ECC83. Increasing anode current causes flicker noise to increase, but the resulting increase in g_m causes shot noise to decrease. All valves exhibit this behaviour.

example, fig. 5.15 shows the total measured EIN voltage of a collection of ECC82/12AU7s of various brands and ages. But while there is a good ±6dB spread in the data, there is one unifying feature: they all show the same 'hockey stick' response whose minimum, crucially, nearly always occurs at the same value of anode current. This is enormously useful to us as circuit designers because it means that although we can't predict in advance the

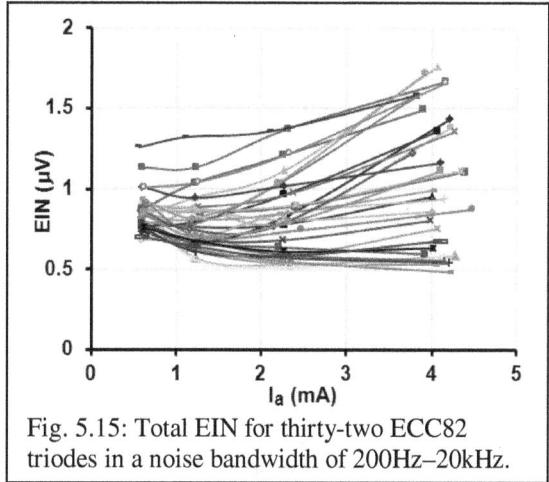

Fig. 5.15: Total EIN for thirty-two ECC82 triodes in a noise bandwidth of 200Hz–20kHz.

absolute value of the noise very accurately, we *can* still optimise the SNR with confidence. Beyond this, further improvement can only come from actually measuring and cherry-picking lucky low-noise samples.

The total EIN is easily found by combining equations (5.15) and (5.17) and dividing by g_m.

$$EIN_v = \frac{\sqrt{4k(0.644T_k)\left(\frac{g_m}{\sigma}\right)B + KI_a^{\,2}\ln\left(\frac{f_{hi}}{f_{lo}}\right)}}{g_m} \quad \text{volts RMS} \qquad (5.20)$$

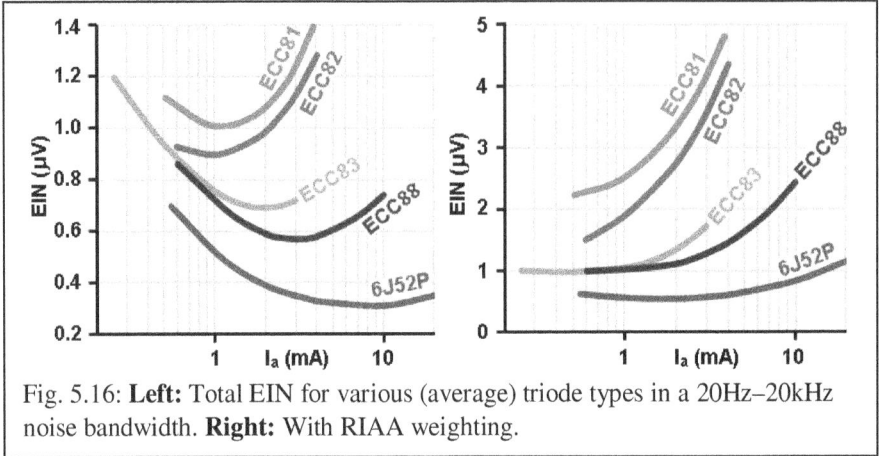

Fig. 5.16: **Left:** Total EIN for various (average) triode types in a 20Hz–20kHz noise bandwidth. **Right:** With RIAA weighting.

If the noise constants are unknown then a good approximation is $\sigma = 0.8$, $K = 10^{-13}$. If we also take the cathode temperature to be 1000K then the previous equation simplifies to a formula for the EIN voltage of any typical triode in a 20Hz–20kHz bandwidth:

$$EIN_v = \frac{\sqrt{8.9 \times 10^{-17} g_m + 6.9 \times 10^{-13} I_a^2}}{g_m} \qquad \text{volts RMS} \qquad (5.21)$$

Equation (5.20) is plotted in fig. 5.16 for a 20Hz–20kHz bandwidth using values from table 5.4, measured values of g_m, and letting $T_k = 1000K$. The ECC83 is quieter than might be expected because it exhibits comparatively low flicker noise, or at least the author's fourteen measured samples do. High-g_m valves tend to have smaller EIN and a broader noise minimum, so the optimum operating current is less critical. On the other hand, they often suffer more acutely from microphonics. Fig. 5.16 also shows the result with RIAA weighting. The optimum anode current is dramatically reduced since the RIAA curve enhances the effects of flicker noise (by contrast, with ITU-R 468 or A-weighting the optimum current is *raised* slightly, though not enough to be worth reproducing the graph).

In section 5.1.5 it was mentioned that an alternative way to express the EIN is as an equivalent resistance that would generate an equal amount of Johnson noise. The equivalent input noise resistance of a triode could therefore be found by substituting equation (5.20) into (5.9), though it would not be a true equivalency since it involves both white and pink noise. But *radio* engineers do not worry about flicker noise, so they can happily discard it from the proceedings and concentrate only on shot noise, resulting in:

$$R_{eq} = \frac{4k(0.644T_k)^{g_m/\sigma} B}{4kTBg_m^2} = \frac{0.644T_k}{T\sigma g_m} \qquad (5.22)$$

204

Where T is the temperature of the fictional resistance, often standardised at 300K. If we also assume $T_k = 1000K$ then the above boils down to:

$$R_{eq(shot)} = \frac{2.15}{\sigma g_m} \qquad (5.23)$$

Depending on the value of σ you are happy to use, R_{eq} might range anywhere from $2.2/g_m$ to $3.6/g_m$. Such misleadingly simple formulas are often found in radio textbooks and have been borrowed by some naïve audio designers. Applying the same technique (somewhat dubiously) to audible flicker noise leaves us with:

$$R_{eq(flicker)} = 2.1 \times 10^{16} K \frac{I_a^2}{g_m^2} \qquad (5.24)$$

Depending on the value of K this will vary from about $200 I_a^2/g_m^2$ to $20000 I_a^2/g_m^2$.

An interesting variation in noise voltage is observed when the cathode temperature of the valve is varied. Fig. 5.17 shows the total EIN of each triode in a Brimar ECC88/6DJ8 at heater voltages of 5V (low), 6.3V (normal) and 7V (high). It is clear from the graph that at anode currents below about 1mA the EIN actually improves when the heater voltage is low, whereas at higher currents the situation is reversed (this figure also gives some idea of the degree of matching between triodes in the same bottle). Fig. 5.18 shows the same effect in one triode of a Tungsram ECC82 (the other triode is omitted as it was very similar and only made the graph look confused). The author has measured many triodes of

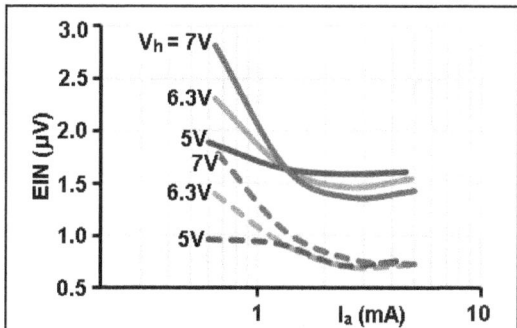

Fig. 5.17: The effect of heater voltage on the EIN of a Brimar ECC88. Solid lines are triode A, dashed lines are triode B. Noise bandwidth 200Hz–20kHz.

different types and all showed this EIN see-saw effect to some extent, with the crossover point usually occurring close to I_a = 1mA. Fig. 5.19 shows the result of the same experiment on a Mullard ECC81/12AT7 that had a known interface-resistance problem. Clearly, interface resistance can dramatically increases noise (mainly flicker) but the curves still show signs of trying to cross over, just off the left-hand side of the graph.

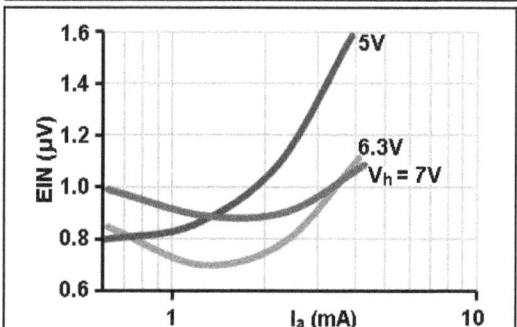

Fig. 5.18: The effect of heater voltage on the EIN of a Tungsram ECC82. Noise bandwidth 200Hz–20kHz.

205

A rough and ready rule of thumb to have in one's head is that most triodes have an EIN of about one microvolt RMS in the audio band. But let's be generous and suppose we have a device that has a more respectable 0.6μV of voltage noise, but a ghastly 100nA of grid current. From equation (5.14) the noise current in the grid will be 25pA in the audio band and will develop a noise voltage across the source resistance. This noise voltage will add 3dB to the total EIN when it is equal to the noise voltage we already have, which occurs when the source resistance is 0.6μV/25pA = 24kΩ. With higher source resistances the grid current noise will dominate the total, while for lower resistances the voltage noise will dominate. It is therefore safe to assume that grid current noise can be ignored completely if the source resistance is not more than 10kΩ say, and this is a very conservative estimate considering the assumptions used to derive it.[*]

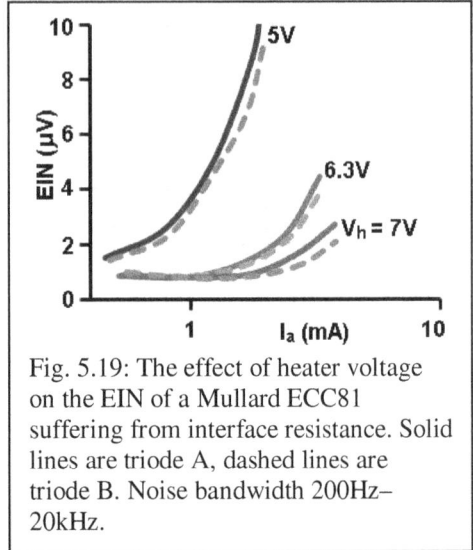

Fig. 5.19: The effect of heater voltage on the EIN of a Mullard ECC81 suffering from interface resistance. Solid lines are triode A, dashed lines are triode B. Noise bandwidth 200Hz–20kHz.

5.4: Noise in Pentodes

Pentodes produce all the same types of noise as triodes, plus one more called partition noise, as illustrated in fig. 5.20. If the screen is unbypassed then the noise current in the screen will also develop a noise voltage which will be amplified, making the circuit noisier still –another reason to bypass the screen grid, then.

The RMS flicker noise current in the anode is the same as for a triode so equation (5.17) applies. The shot noise current in the *cathode* is also the same as for a triode, so equation (5.15) applies, provided we remember to use the triode-connected g_m. Of this total cathode shot current only $m/(m+1)$ times as much flows in the anode, where m is the anode/screen current ratio; the

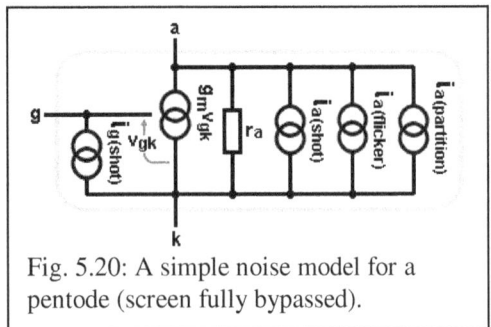

Fig. 5.20: A simple noise model for a pentode (screen fully bypassed).

---

[*] Readers might point out that a 10kΩ resistance itself generates 1.8μV noise which would swamp the valve anyway. However, the source resistance could be reactive, in which case it generates no noise of its own.

rest is diverted to the screen grid.

Partition noise is caused by the tiny fluctuations in the way the current divides or partitions between anode and screen grid. The RMS partition noise current in the anode (or screen) is given by:

$$i_{a(partition)} = \sqrt{2q\frac{I_a \times I_{g2}}{I_a + I_{g2}}B} = \sqrt{2qI_{g2}\frac{m}{m+1}B} \qquad \text{amps RMS} \qquad (5.25)$$

Where:
q = electron charge, 1.6×10^{-19} coulombs.
I_a = average anode current in amps
I_{g2} = average screen current in amps
m = anode/screen current ratio
B = noise bandwidth in hertz

However, if the pentode is connected as a triode then partition noise is eliminated and we are left with the same noise as any standard triode. Partition noise will tend to make a pentode 2 to 6dB noisier than when it is triode connected, although this does rather depend on operating current. At higher currents, flicker noise may dominate all other noise sources, in which case the pentode may appear no quieter even when triode connected.

Some authors have suggested methods for eliminating partition noise while retaining the high gain and other features of the pentode. For example, the partition noise current in the screen is equal to, but out of phase, with that in the anode. Therefore, by inserting a small impedance in series with the screen grid an out of phase noise voltage will develop at the screen, which could be fed back to the control grid. When this signal is then amplified (and inverted again) it would cancel the partition noise voltage already at the anode. This technique has been suggested for use at radio frequencies[3] but whether it is practical at audio frequencies is questionable. Blöbaum[4,5] has suggested eliminating partition noise by allowing the screen current to recombine with the anode current using the scheme in fig. 5.21. This circuit is in fact a cascode, where the screen grid forms the anode of the lower 'intrinsic' triode.

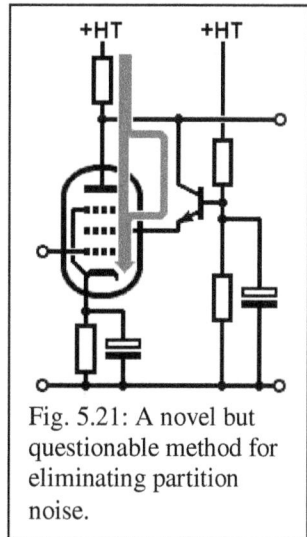

Fig. 5.21: A novel but questionable method for eliminating partition noise.

---

[3] Bonham, L. L. (1949). Reduction of Partition Noise in a Pentode Amplifier. Unpublished B.S. Thesis, University of Maryland.
[4] Blöhbaum, F. A New Low-Noise Circuit Approach for Pentodes, *Linear Audio*, 0, pp46-65.
[5] Blöhbaum, F. (2009). Rauscharmer Pentodenverstarker, Patent DE102008017678.

However, most current still flows via the original anode, which cleverly reduces dissipation in the transistor. The EIN of the whole circuit is no better than ordinary triode connection, but linearity, PSRR and output resistance are considerably poorer.[6]

5.5: Noise in Bipolar Transistors

In spite of our affection for all things thermionic, it has to be said that silicon is often quieter. But since this is not primarily a book on solid state technology we will keep treatment of transistor noise to the essential minimum. A simple noise model for a bipolar junction transistor (BJT), based on the well-known π–model

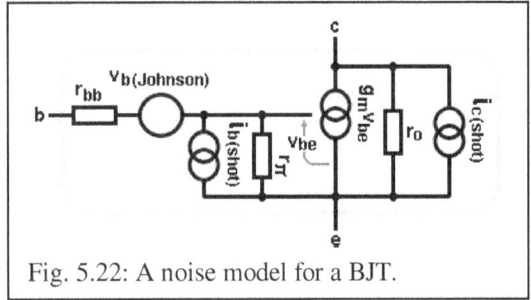

Fig. 5.22: A noise model for a BJT.

(which can be found in any modern textbook) is shown in fig. 5.22. The principle sources of BJT noise are:

- Base current shot noise;
- Collector current shot noise;
- Johnson noise in the **base spreading resistance**, r_{bb} (this is the physical resistance of the silicon in the base region and can be though of as a sort of built-in stopper resistor).

Shot noise in the collector is given by:

$$i_{c(shot)} = \sqrt{2qI_cB} \quad \text{amps RMS} \tag{5.26}$$

Shot noise in the base is:

$$i_{b(shot)} = \sqrt{2qI_bB} \quad \text{amps RMS} \tag{5.27}$$

And the Johnson noise of r_{bb} is of course:

$$v_{b(Johnson)} = \sqrt{4kTr_{bb}B} \quad \text{volts RMS} \tag{5.28}$$

Excess noise also exists, but with so many devices on the market it is difficult to make a useful statement about its significance compared with the readily predictable white-noise sources. In general it seems to be ignored, or is small enough that it is not worth trying to put a value on it. Fig. 5.23 shows the EIN voltage density for a ficticious BJT, and the dotted line shows the effect of adding an entirely speculative but plausible amount of 1/f noise. Usually it is the noise of r_{bb} that limits the

---

[6] Blencowe, M. (2015). Partition Noise and the 'BestPentode' Revisited, *Linear Audio*, 10, pp111-24.

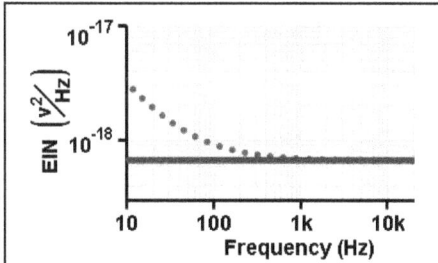

Fig. 5.23: EIN density for a BJT at $I_c = 1mA$, $g_m = 40mA/V$, $r_{bb} = 40\Omega$.

performance, and in this case it sets the total EIN voltage to about $0.12\mu V$ in the audio band. Sadly and inexplicably, r_{bb} is seldom quoted on data sheets. It is likely to be smaller in high-current devices since the base region must be thick, so a low noise transistor is therefore one designed for high current (minimising r_{bb}) and high β (minimising base current).

Although BJTs have extremely small input noise voltage, we have no choice but to operate them with significant base current, and therefore high input noise current. Usually this can only be neglected if the source resistance is less than 500Ω or so. BJTs are therefore ideal when the smallest EIN voltage is needed and the source impedance is small enough for current noise not to swamp it. For higher source impedances we switch to JFETs. The rules of thumb for low-noise BJT circuit design are to maximise g_m by maximising the collector current, minimise base current by using a high-β device, and use multiple devices in parallel.[7] Annoyingly, the best low-noise transistors such as the 2SB737, 2SC3329, and 2SA1316, with $r_{bb} \approx 2\Omega$, all seem to be out of production. The author's favourite –and readily available– devices are the BC337/327 ($r_{bb} \approx 30\Omega$). The ZTX653/753 are probably even better.

5.6: Noise in Field Effect Transistors

Fig. 5.24 shows a simple noise model for a field effect transistor (FET), and it shares much in common with that of a triode. The principle noise sources are:

- Gate current shot noise;
- Drain current shot noise (or Johnson noise of the channel resistance, depending on how you look at it);
- Drain current flicker noise.

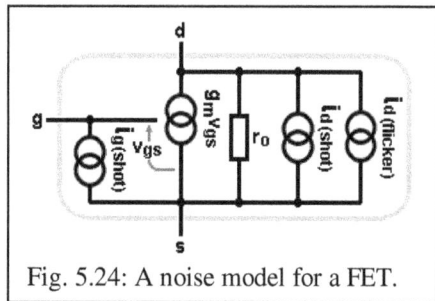

Fig. 5.24: A noise model for a FET.

Gate current shot noise is given by:

$$i_{G(shot)} = \sqrt{2qI_G B} \qquad \text{amps RMS} \tag{5.29}$$

Drain current shot noise is given by:

---

[7] For further advice on low-noise solid-state design, see: Self, D. (2015). *Small Signal Audio Design*. 2nd ed., Elsevier, Oxford.

$$i_{D(shot)} = \sqrt{4kT\gamma g_m B} \quad \text{amps RMS} \tag{5.30}$$

(Where γ is an empirical constant between about 0.7 and 1.)

Drain current flicker noise is given by:

$$i_{D(flicker)} = \sqrt{KI_D \ln\left(\frac{f_{hi}}{f_{lo}}\right)} \quad \text{amps RMS} \tag{5.31}$$

Where K is an empirical constant around 10^{-16}.

Fig. 5.25 shows the EIN spectral density of a J111, which is quite typical of modern JFETs (compare with fig. 5.12). Since the gate current of a JFET is usually less than 1nA, current noise is lower than that of a triode and can be ignored for source impedances less than 100kΩ. All in all, a JFET comfortably achieves an EIN that is less than what the best triode struggles to achieve, and without the problem of microphonics too. MOSFETs have even lower gate current than JFETs but they are on the whole noisier in other respects, so are of little use as low-noise amplifiers.

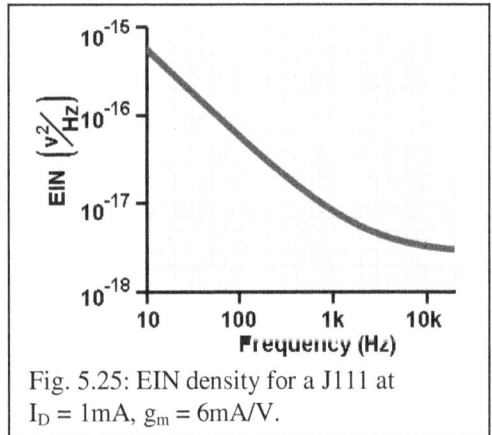

Fig. 5.25: EIN density for a J111 at $I_D = 1mA$, $g_m = 6mA/V$.

5.7: Noise in Diodes

For an ordinary forward-biased diode such as a rectifier or LED, the noise current riding on the anode current is given by Schottky's theorem. The same applies to true Zener diodes even though they are operated in reverse bias, but note that true Zener diodes are those with breakdown voltages less than about 7V. So-called Zener diodes with breakdown voltages higher than 7V are actually **avalanche diodes** and generate a much higher levels of white noise, plus another sort called multistate noise which is mainly pink. Diodes also exhibit excess noise, though it is often negligible in the audio band.

How much noise *voltage* manifests at the output of a circuit depends on the internal resistance of the diode. Infrared LEDs tend to have the smallest internal resistance at a given anode current and cheap red (not ultra-bright) LEDs come a close second. Both are popularly used as low-noise, low-voltage references. Progressing along the colour spectrum the forward voltage increases and with it the internal resistance. Zener diodes occupy the range between about 2.7V and 7V and have slightly higher internal resistance, while beyond them lie avalanche diodes with higher resistance still, so they are by far the noisiest. Audiophiles therefore tend to sneer at avalanche diodes, but it has to be said that voltage reference noise is often a moot point because

210

in practice the diode will be bypassed by a large decoupling capacitor which shunts (i.e. attenuates) the noise quite effectively.

Gas reference diodes / glow discharge tubes are the valve version of modern avalanche diodes. They are usually regarded as being noisy, but data presented by Jones[8] appear to show that they generate mainly shot noise, making them just as quiet as LEDs and Zeners. Unfortunately, they have much higher internal resistance, so in a given circuit the output noise *voltage* will be high, hence their poor reputation. They also become accustomed to steady operating conditions; substituting one into a new circuit or altering its quiescent current may lead to excessive noise until it has had several hours to get used to the new enviroment.

5.8: Noise Calculations

In order to appreciate the guiding principles of low-noise design it is worth doing some example calculations. This should give the reader an intuitive grasp of what the major noise sources are in a typical circuit.

5.8.1: Triode Gain Stage

Fig. 5.26a shows a gain stage using an ECC88, which is a popular choice for a low-noise amplifier. Fig. 5.26b shows an output-referred equivalent circuit ready for noise analysis. There are four noise currents to consider:

- Noise generated by the source resistance R_s, which in practice might be a grid stopper plus any other resistance associated with the signal source. This noise begins as a noise voltage at the grid which is amplified by the transconductance of the valve to form a noise current in the anode circuit.
- Valve shot noise current.
- Valve flicker noise current.
- Noise current generated in the anode resistor.

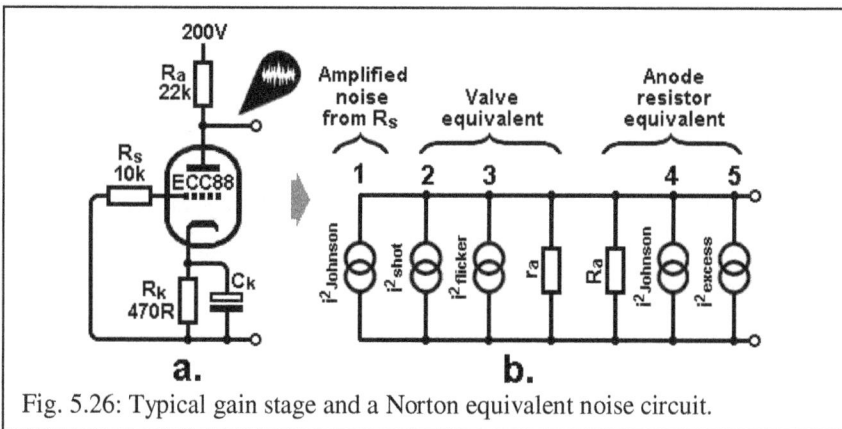

Fig. 5.26: Typical gain stage and a Norton equivalent noise circuit.

[8] Jones, M. (2012). *Valve Amplifiers*, 4th ed., Elsevier, Oxford.

The valve's internal anode resistance r_a is not a physical resistor so it generates no noise of its own. It does affect voltage gain, however, which is why it must be included in the model. The cathode resistor is fully bypassed which shunts any noise it produces. There is a resistor in series with the grid, e.g. a grid stopper, and although in reality there would also be a grid leak resistor, here the input is shorted to ground to

Boltzmann constant (k)	1.38×10^{-23} J/K
Ambient temperature (T)	300K
Cathode temperature (T_k)	1000K
g_m	6mA/V
r_a	5kΩ
σ	0.6
K	6.2×10^{-14}
Upper bandwidth limit (f_{hi})	20kHz
Lower bandwidth limit (f_{lo})	20Hz
DC anode current (I_a)	5mA
Voltage gain	23.6
Excess noise index (NI)	0.1μV/V

Table 5.5: Figures relating to fig. 5.26.

represent the effect of a near-zero source impedance plugged into this amplifier.

For convenience we shall assume R_s generates only Johnson noise, but that R_a produces Johnson noise and excess noise, with a noise index of 0.1 (typical of film resistors). The noise bandwidth will be 20Hz to 20kHz unweighted. Other important figures are given in table 5.5. Since all these noise sources are uncorrelated it is straightforward to find the total output noise once we have found the individual noise currents, as follows:

1. The RMS noise *voltage* generated by R_s is given by equation (5.3):

    $$V_{johnson} = \sqrt{4kTRB} = \sqrt{4 \times 1.38 \times 10^{-23} \times 300 \times 10000 \times 19980}$$

    $= 1.8 \times 10^{-6}$ or 1.8μV. This voltage appears at the grid and is multiplied by g_m to form a noise current of 10.9nA in the output circuit.

2. The RMS shot noise flowing in the anode is given by equation (5.15):

    $$i_{shot} = \sqrt{4k(0.644T_k) \frac{g_m}{\sigma} B} = \sqrt{4 \times 1.38 \times 10^{-23} \times 0.644 \times 1000 \times \frac{0.006}{0.6} \times 19980}$$

    $= 2.7 \times 10^{-9}$ or 2.7nA.

3. The RMS flicker noise flowing in the anode is given by equation (5.17):

    $$i_{flicker} = \sqrt{KI_a^2 \ln\left(\frac{f_{hi}}{f_{lo}}\right)} = \sqrt{6.2 \times 10^{-14} \times 0.005^2 \times \ln\left(\frac{20000}{20}\right)}$$

    $= 3.9 \times 10^{-9}$ or 3.27nA.

4. The RMS Johnson noise current generated by R_a is given by equation (5.12):

$$i_{johnson} = \sqrt{4kT\frac{1}{R}B} = \sqrt{4\times1.38\times10^{-23}\times300\times\frac{1}{22000}\times19980}$$

$$= 0.12\times10^{-9}A \text{ or } 0.12nA.$$

5. The RMS excess noise current generated by R_a is given by equation (5.13) divided by the resistance. There is $110V_{dc}$ dropped across R_a, so:

$$i_{excess} = \frac{NI\times V_{dc}\times\sqrt{3}}{R} = \frac{0.1\times110\times\sqrt{3}}{22k}$$

$= 0.87nA$. Note that this is a good deal greater than the Johnson noise.

Remembering to use RSS addition, the total noise current flowing in the anode circuit is therefore:

$$i_{total} = \sqrt{10.9^2 + 2.7^2 + 3.27^2 + 0.12^2 + 0.87^2} = 11.7nA$$

This current flows in the parallel combination of R_a and r_a to produce a total output noise voltage of:

$v = iR = 11.7nA\times(5k\Omega \| 22k\Omega) = 47.6\mu V$

Alternatively, this can be expressed as an equivalent input noise (EIN) voltage by dividing by the voltage gain of the stage, giving: $47.6 / 24.4 = 1.95\mu V$. Comparing this with the answer in step 1 we find that nearly all the noise comes from R_s! Most of the rest is valve noise; the contribution of the anode resistor is utterly negligible, which is always the case for a sensibly-designed gain stage. If the input audio signal was 50mV, say, then the signal-to-noise ratio would be:

$$SNR = 20\log_{10}\left(\frac{V_{signal}}{V_{noise}}\right) = 20\times\log_{10}\left(\frac{50\times10^{-3}}{1.95\times10^{-6}}\right) = 88.2dB$$

Of course, this assumes the audio signal was noise-free to begin with.

Now, if R_s could be reduced to zero ohms then the EIN would be only $0.72\mu V$, which is basically the irreducible noise of the valve and corresponds to an equivalent input noise resistance of $1.6k\Omega$. In other words, if R_s is larger than $1.6k\Omega$ it will dominate the SNR, but if it is less than $1.6k\Omega$ then the valve dominates. It is also worth pointing out that if R_k was left unbypassed then it would degenerate the stage gain and noise from R_k would itself add to the total, thereby worsening the EIN. In short, designing for minimum noise means avoiding degeneration, so it is hardly worth doing the calculations for this arrangement. If you really want to know, leaving the 470Ω cathode resistor unpassed would increase the EIN to $0.77\mu V$ when $R_s = 0\Omega$; a 0.6dB degradation.

5.8.2: Parallel Valves

When valves are connected directly in parallel we effectively create a new valve with twice the transconductance and half the r_a (and therefore the same μ) as before. However, the individual noise contributions of the two devices remain independent of each other, i.e. they are uncorrelated, so they add by RSS addition. In other words, when we put devices in parallel, the gain increases faster than the noise, so the SNR improves.

Fig. 5.27 shows a dual triode whose sections have been connected in parallel, together with an equivalent circuit from the point of view of noise. Even though the grids and cathodes are connected together, both triodes still have their own independent EIN figures, which are assumed to be equal in magnitude (but uncorrelated) since they are each operating at the same anode current. These EINs are amplified by the transconductance of their respective devices and sum together in the output circuit, and from section 5.1.4 it will be remembered that when two equal uncorrelated noise sources are added, the total increases by 3dB. However, if r_{a1} and r_{a2} are both infinite (pentodes and transistors are a good approximation to this) then the *voltage gain* will be double (6dB) what it was for a single device alone. Thus the SNR would improve by $6 - 3 = 3$dB compared with a single device. Indeed, this is a general rule for low-noise design: the SNR improves by up to 3dB every time we double the number of parallel devices. With practical devices the improvement is likely to be somewhat less since the two devices will never be exactly matched, and r_{a1} and r_{a2} will not be infinite, particularly with triodes. Nevertheless, a 2dB improvement in SNR is usually achievable even with triodes.

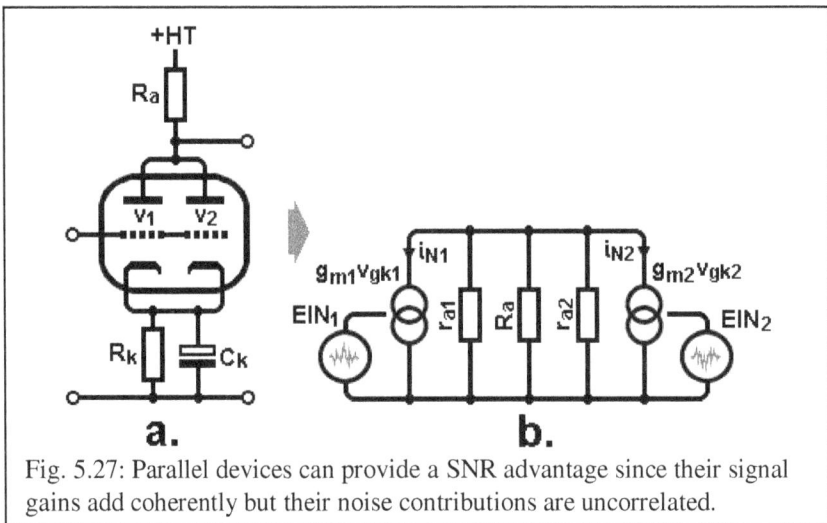

Fig. 5.27: Parallel devices can provide a SNR advantage since their signal gains add coherently but their noise contributions are uncorrelated.

5.8.3: Principles of Low-Noise Design

When low noise is a priority, the following points should be kept in mind:

- Resistances in series with the signal directly add Johnson noise so they should be made as small as possible;
- Resistances that shunt the signal should be made as large as possible to minimise attenuation;
- Attenuating the signal degrades the SNR. A −6dB potential divider has worst effect since it has the highest Thévenin source resistance, i.e. the most Johnson noise;
- The first amplification stage nearly always dominates the SNR of the whole amplifier;
- Cathode degeneration degrades noise performance;
- Valve shot and flicker noise are limiting factors only when the resistance in series with the input grid is less than about 10kΩ;
- Grid current noise is significant only when the source impedance is more than about 10kΩ;
- Noise in the anode resistor can usually be ignored unless it is a carbon composition type;
- Reducing the bandwidth will reduce the noise floor but does not necessarily improve the SNR since audio frequencies are attenuated along with the noise;
- Pentodes are noisier than triodes;
- Paralleling devices improves the SNR by up to 3dB.

5.9: Hum

Hum (and buzz) is frequently caused by interference from electromagnetic fields at mains frequency, and may originate inside or outside the amplifier. Outside sources include the electric fields produced by fluorescent lights, televisions, computers and so forth, and the magnetic fields produced by transformers and high-current carrying cables.

Inside the chassis, hum comes mainly from the electromagnetic fields originating in the power transformer, mains wiring, and AC heaters. It may also be caused by residual ripple on the power supply. Single-ended power amplifiers are the most susceptible to this since the current burden on the power supply is high and the PSRR is usually poor, whereas balanced circuits naturally tend to cancel hum thanks to their common-mode rejection ratio.

5.9.1: Electric Fields

The electric field strength –also called the **electrostatic field intensity**– around a conductor is related to the *voltage* on the conductor, and for simple parallel plates or wires it is simply:

$$E = \frac{V}{d}$$ (5.32)

Where:
E = electric field strength, in volts per metre
V = potential difference between the conductors, in volts
d = distance between conductors, in metres

If the electric field is varying then it will induce a noise *current* into any nearby circuit. The more rapidly the electric field changes, the greater the induced current. This is really just a physical way of describing how voltage signals couple from one wire to another through the stray capacitance between them, and since stray capacitance forms a CR filter with the resistances in each circuit, high frequencies will be coupled more easily than low frequencies.

5.9.2: Magnetic Fields

A wire carrying current generates a magnetic field around itself according to Ampere's law:

$$B = \frac{\mu I}{2\pi r}$$ (5.33)

Where:
B = magnetic flux density in tesla, or webers per square metre
μ = permeability of the material around the conductor, which for a simple wire can be taken to be the same as air: $4\pi \times 10^{-7}$ H/m
I = current in the wire, in amps
r = radial distance from the wire, in metres

If there is any fluctuation in the current then there will be a corresponding fluctuation in the magnetic field which, in accordance with Faraday's law, will induce a fluctuating EMF into any wire loops which happen to be within the magnetic field at the time. The larger the loop area, the greater the induced voltage. Wires carrying AC mains and AC heater wires, are the main offenders here since they carry the largest AC currents in the amp. Transformers also generate strong magnetic fields, but steps can be taken to minimise their effect on other circuitry.

5.9.3: Transformers and Hum

Transformers (and to some extent smoothing chokes) generate noisy magnetic fields around themselves where the magnetic flux leaks from the core. This occurs most strongly where the flux has to go around corners or across the grain of the core material. Toroidal transformers have no corners or grain and so leak less than EI transformers, and are preferred where possible. However, they do leak around the lead-out wires where the winding geometry departs from the ideal, so this area should be orientated away from sensitive circuitry. The magnetic field strength around a transformer is approximately proportional to the inverse of the cube of the distance from its core, so the cheapest weapon against transformer hum is physical separation.

Transformers will also couple signals between one another, notably between the power and output transformers in a power amplifier. This cross coupling can again be minimised with physical separation, but also by optimising the orientation of the transformers. In theory, the axis of one transformer core should be kept orthogonal to the other to minimise coupling but, in practice, an apparently random orientation sometimes provides the best performance. The optimum arrangement can be found at an early stage of construction by applying mains voltage to the power transformer and connecting headphones to the secondary of the other transformer. No other connections are required, but obviously great care must be taken when doing this since one transformer is live. The second transformer can then be moved around the chassis while listening for the changes in hum, and it is easy to find the relative orientation that results in the least coupling.

Transformers are usually mounted to the chassis with non-ferrous bolts and an insulating gasket or fibre washers, to prevent the core flux from flowing in the chassis which might otherwise lead to hum creeping into the audio ground. Note that the transformer core is still *electrically* grounded to the chassis via the metal bolts; it should not be left floating.

5.9.4: Electric Shielding

Shielding against electric fields is very easy; any bit of metal will do, it just has to be connected to ground or to some other low-impedance potential. The chassis provides the ultimate shield against external electric noise, but shielding may still be required internally, either to protect sensitive wiring against noise coming from other areas or to enclose noisy conductors and prevent their electric fields from permeating the air. In particular, the wires leading from the input connector to the first valve will sometimes be shielded to discourage parasitic feedback.

Noisy wiring should run close to the chassis wall or, better still, be pushed into the corners to maximise the capacitance to earth. The earthed chassis will then have a similar effect to a Faraday cage and will tend to draw the electric field towards itself. This often avoids the need to shield wires completely (especially heater wires).

Shielding-cans may be used for small valves, and if the valve socket has a central spigot then this can be grounded too at some convenient point, as it will offer some shielding between opposing pins. Some valves like the EF86 and ECC88 include an internal shield or screening cage which can also be grounded.

Large power supply smoothing capacitors can sometimes be judiciously placed to provide some screening between different parts of a circuit since the metal can is normally connected to the negative terminal, which is often grounded. Likewise, other grounded metal objects like smoothing chokes or reinforcing brackets can be used to advantage. In extreme cases the chassis may be partitioned into sections with bulkheads to create isolated chambers that are shielded from one another.

5.9.5: Magnetic Shielding

Shielding against magnetic fields is much more difficult than shielding electric fields, for various reasons. At low frequencies the metal used for shielding must have high magnetic permeability; the usual choice is mu-metal, which has one hundred times higher permeability than steel. Be aware, however, that this permeability degrades when the metal is machined, so it must be annealed once the final shape has been formed.

Furthermore, magnetic shielding is most effective when it forms form a completely unbroken band or box, since the idea is to encourage any interfering flux to flow in the shield rather than in the thing being screened. Any gaps or seams will prevent the flux from flowing in a complete loop, so magnetic screens are often stamped out, rather than cut and folded. This is an expensive process. Explicit magnetic shielding is therefore rarely used except on phono/microphone signal transformers. A mild-steel shroud may have a slight magnetic advantage over a stainless steel or aluminium one, but it is hardly noteworthy unless the steel is unfeasibly thick. Instead, the most practical techniques for minimising the effects of magnetic fields are to adopt a tight audio layout and good lead dress to minimise loop area; to orientate transformers favourably; and to keep transformers and high-current carrying wires well away from sensitive circuitry.

5.9.6: Lead Dress

Lead dress refers to the physical arrangement of conductors in a circuit, and it is important with regards to hum. Electromagnetic fields decay with distance from the source, so physical separation is our friend. If wires that may interfere with one another must cross then they should do so in a perpendicular fashion, to minimise the area of wire 'seen' by each other. Allowing a noisy wire to run in parallel with a sensitive signal wire is a cardinal sin.

Signal wires leading to valve pins should be kept as short as possible. This obviously requires mounting the valves physically close to the circuit area they serve. Although it may be aesthetically pleasing to arrange the valves in a neat line, fewer problems will be encountered if they are positioned in a way that favours the shortest lead runs, i.e. form should follow function.

Pairs of noisy AC wires (e.g., mains and heater feeds) should be neatly twisted so the opposing magnetic fields around each wire are forced to occupy the same space, causing them to cancel each other out. Note that a loose twist is useless, only a tight twist will do! But neither should the twists be *too* tight as this may lead to fatigue and internal breakage; the old rule of three twists per inch is usually ideal. Twisting is easily done by anchoring the wires at one end and holding the other end in the chuck of a drill, keeping reasonable tension on the wire while twisting. Stretching it gently before releasing will discourage it from twirling back on itself. It is better to use stranded wire for this as solid-core wire will quickly develop metal fatigue and break internally.

Heater supplies normally daisy-chain from one valve socket to the next, supplying many heaters in parallel, so current flowing in the *supply* end of the chain is greater than the current flowing near the last valve in the chain. Power valves require the most current and are the least sensitive to heater hum, so they should be at the beginning of the chain. The heater chain should then progress logically through the amplifier with the input valve last in the chain where the current –and therefore field strength– is smallest. Heavier gauge wire may be needed at the supply end of the chain where more current is being carried, but a common error is to use wire which is *too* heavy, making it difficult to manipulate. Ordinary 16×0.2mm (~20 AWG) equipment wire has a resistance of about 0.04Ω per metre and is sufficient for up to about ten amps.

When heater wires approach a valve socket the twisting must be kept tight right up to the socket since the other valve pins are in close proximity here, and we must suppress the magnetic fields as much as possible. Allowing the twists to become loose near the socket will spoil a lot of hard work! It is also important not to create a loop of heater wiring around a valve socket since any other wires or valve pins inside the loop will then be subject to increased EM interference. The heater wiring should approach from one side of the socket and, if it must cross it, should jump directly across the socket

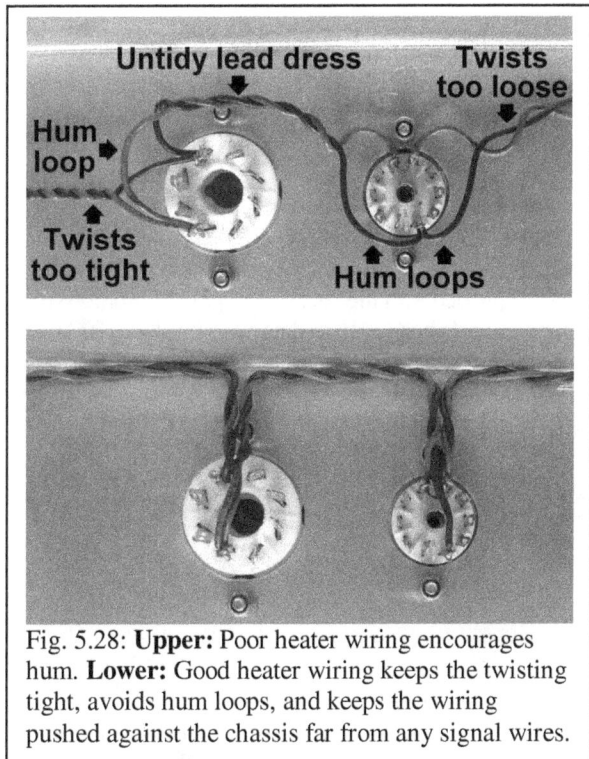

Fig. 5.28: **Upper:** Poor heater wiring encourages hum. **Lower:** Good heater wiring keeps the twisting tight, avoids hum loops, and keeps the wiring pushed against the chassis far from any signal wires.

219

and straight back. It helps to orientate valve sockets so the heater feeds can approach from the same side, usually from a trunk feed pushed into the corner of the chassis.

An example of good heater wiring is shown in the lower image in fig. 5.28. This takes care and patience, and it will usually be obscured by other wiring once the amp is complete, so it is worth spending the time on getting it right from the beginning. The upper image shows some typical mistakes. Electrical methods for dealing with intrinsic heater hum are dealt with in section 11.8.

5.10: Microphony

Microphony refers to the effect where components act like microphones and produce spurious signals due to mechanical shock and vibration. Valves are the main offenders since they contain fine wires and airy structures. As a general rule, higher-g_m valves exhibit the worst microphonic problems. Pentodes suffer more strongly from microphonics than triodes,[*] both because they have extra grids and because they usually develop more gain, so vibrations are amplified more. In new designs it may be advisable to operate small signal pentodes at 'reasonable' gain levels (not more than 100, say), rather than the very high-gain levels found in vintage circuits and data sheets. Ceramic capacitors can also be quite microphonic compared with other passive components.

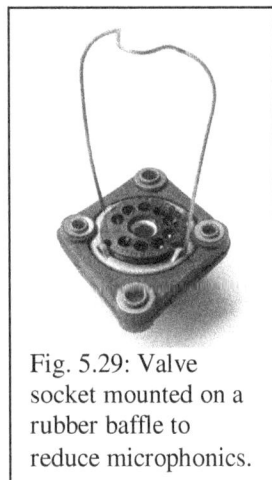

Fig. 5.29: Valve socket mounted on a rubber baffle to reduce microphonics.

Anything rigid will couple vibrations more readily than compressible materials. Rubber feet for the chassis are usual, and stranded wire is preferred over solid-core for connections to valve sockets (this also reduces the stress on the valve pins when they are pushed into the socket). Slipping automotive O-rings over preamp valves is a popular and cheap way to dampen vibrations, at least in the upper audio range. In extreme cases the valve sockets themselves may be mounted on rubber grommets or a sprung sub-chassis. Beyond this there is not much we can do to tackle microphonic problems except cherry-pick valves that happen to be more robust.

---

[*] Although, the author has a 12AT7 that is so microphonic it will readily pick up a speaking voice on the opposite side of the room. It would probably make a good seismometer!

220

Chapter 6: Advancements in Amplification

Chapter 3 introduced the basic gain stage in a traditional way, which is essential for getting a feel for the various factors that influence circuit performance. This chapter will expand on the basics by introducing some of the alternative techniques that modern electronic devices have made possible. It will also serve as an introduction to the most common triodes used in modern audio (plus some less common ones) and give the reader a broad appreciation for the sort of levels of distortion that are possible with certain valves and by certain means. This should provide a useful point of reference when designing new circuits or choosing what valves to use to attain a desired performance level.

6.1: Current-Source Loading

In chapter 3 it was explained how increasing the load resistance tends to lower distortion. Of course, there is a limit to how far this can be taken with ordinary resistors. For example, if we tried to use a $1M\Omega$ anode resistor and we also wanted 1mA of quiescent current then we'd need in excess of $1mA \times 1M\Omega = 1kV$ of supply voltage, which isn't particularly inviting. In the interests of practicality what we need is something which behaves like a high-valued resistor as far as AC signals are concerned, but which doesn't need so much DC drop across it. Such a circuit is usually called a **current source** or a **current sink**. Whether or not it is a 'source' or 'sink' is a matter of convention, depending on whether the amplifier –or whatever it happens to be– notionally receives positive current *from* the *source*, or feeds it *into* the *sink*, as illustrated in fig. 6.1.

A source or sink whose internal impedance is many times greater than the rest of the circuit –here represented by a generic impedance Z_1– can be considered to be a *constant*-current source/sink, or **CCS**. The impedance of the CCS then dominates the total impedance, so the current through the network will barely change even if the impedance of Z_1 varies by a large percentage –like a triode caught in the act of amplifying a signal. The CCS simply varies the voltage across itself to whatever is necessary to maintain the same current through itself. Such a circuit can be used as the anode load for a valve, thereby presenting a near-infinite impedance to AC signals. This has certain advantages over a simple load resistor, including:

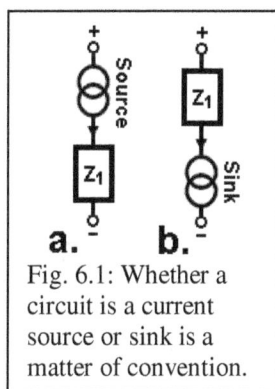

Fig. 6.1: Whether a circuit is a current source or sink is a matter of convention.

- Maximum gain, which becomes equal to the μ of the valve;
- Low distortion, limited only by the linearity of μ;
- Excellent PSRR;
- Possible noise advantage when used with low-r_a triodes.

However, the disadvantages compared with simple resistor loading include:
- More components (and more expensive components);

- Possible need to hand-select components and adjust on test;
- Reduced headroom for a given HT;
- Possible impolite clipping effects;
- Output resistance equal to r_a and therefore at the mercy of valve variation;
- Gain totally dependent on μ and therefore at the mercy of valve variation;
- Greater sensitivity to load capacitance, e.g. narrower bandwidth;
- Second-harmonic less dominant compared to the rest of the spectrum;
- Distortion not always constant with frequency or directly proportional to level;
- Poorer signal-to-noise ratio when used with high-r_a triodes.

6.1.1: Triode Current Source

There are many ways to build a current source, so let's begin with something familiar: a degenerated triode as shown in fig. 6.2. To avoid confusing novice readers it is worth stating explicitly that we're not looking at a gain stage here, but a circuit which can serve as an electronic load *for* a gain stage. The nice thing about this circuit is that it has only two terminals; the current that goes in through the top must come out at the bottom, so it can be used freely as either a current source or a sink.

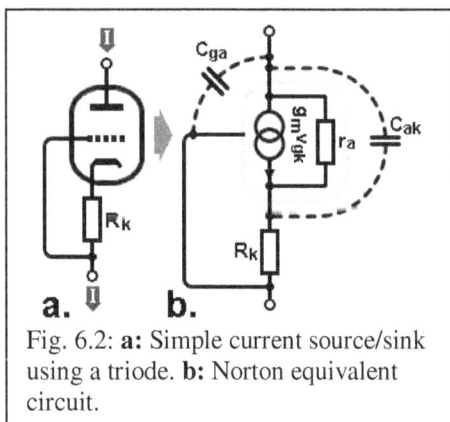

Fig. 6.2: **a:** Simple current source/sink using a triode. **b:** Norton equivalent circuit.

The question is, what resistance does it present to AC signals?

We already know that when looking into the anode of a degenerated triode we see a resistance consisting of r_a, plus R_k multiplied by $\mu+1$:

$$r' = r_a + R_k (\mu + 1) \tag{6.1}$$

Or if you prefer to work in terms of g_m rather than μ, substituting $\mu = g_m r_a$ into the above gives:

$$r' = r_a + R_k (g_m r_a + 1) \tag{6.2}$$

Taking a plausible example using an ECC83 with r_a = 65kΩ, μ = 100 and R_k = 1.5kΩ, the equation above predicts 216kΩ. This can also be determined from a cathode load line (section 3.3.3) as shown in fig. 6.3. The slope of the line is equal to the internal resistance of the source.

222

To put theory to the test, the author used an Audio Precision System 1 to measure the internal impedance using the arrangement in fig. 6.4. The same setup was used for other current sources described shortly. The current source forms a potential divider with resistor R, so by injecting a signal at one end of the current source and measuring the voltage across R, the attenuation –and therefore the internal impedance– is readily calculated.

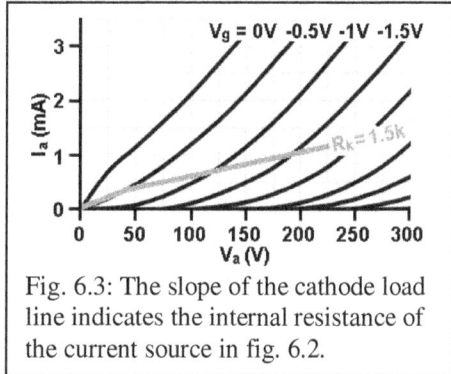

Fig. 6.3: The slope of the cathode load line indicates the internal resistance of the current source in fig. 6.2.

Impedances up to and above 100MΩ were measurable with this method.

Fig. 6.5 shows that the measured impedance of the 12AX7LPS current source was very close to the predicted value, and constant with frequency. However, as indicated in fig. 6.2b, stray capacitances bypass the current source and will inevitably reduce its impedance at high frequencies, though in this case the strays are too small to have any effect within the audio band, which is nice. Of greater concern is the fact that this current source requires a large DC

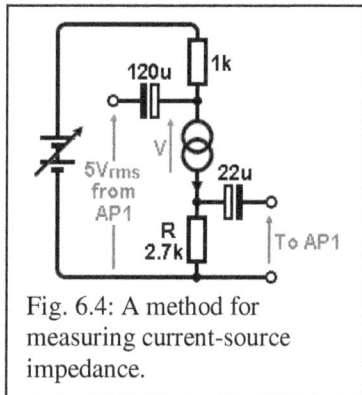

Fig. 6.4: A method for measuring current-source impedance.

drop across the triode just to work properly. For example, setting the quiescent current to 1mA requires a 200V drop across the valve, as indicated by the cathode load line. In short, the overall performance is hardly any better than an ordinary 200kΩ resistor. We need a more efficient current source. A pentode could provide a much higher internal impedance with less voltage drop (though still relatively high), but it would have the added complication of a screen-grid supply. This is really a job for transistors.

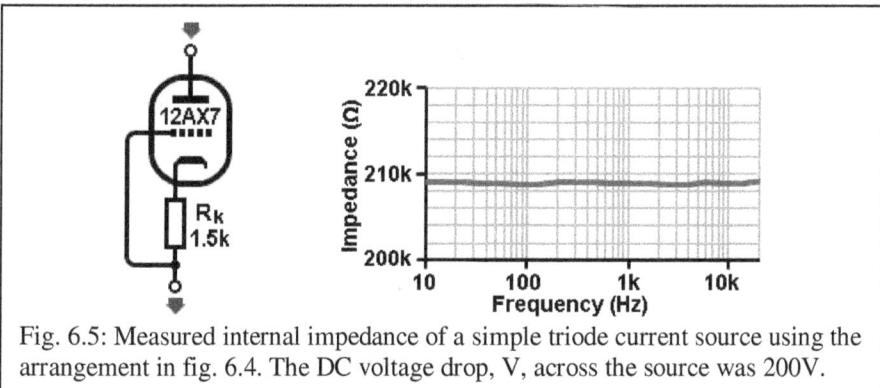

Fig. 6.5: Measured internal impedance of a simple triode current source using the arrangement in fig. 6.4. The DC voltage drop, V, across the source was 200V.

223

6.1.2: Simple BJT Current Source

Fig. 6.6a shows a textbook current-sink using a bipolar junction transistor (BJT). Unlike triodes, BJTs are enhancement devices, meaning they are normally 'off' until enough positive voltage is applied between base and emitter to turn them on. Here this is achieved using the forward drop across an LED. A cheap red LED has a typical forward voltage of about 1.7V, and a BJT needs about 0.7V between base and emitter to switch it on, which leaves about 1V across the emitter resistor R_e. The quiescent (DC) current can therefore be set using Ohm's law: $R_e = 1V/I_e$, where I_e is the emitter current. If the transistor has fairly high current gain –greater than 100, say– then the collector 'sink' current I_c will be practically equal to I_e. An incidental advantage of using a red LED as the voltage reference is that it has practically the same tempo as the transistor's base-emitter voltage (about $-2mV/°C$), so the programmed current should drift very little with temperature. Admittedly, valve circuits seldom require such precision, but all the same it is nice when it happens.

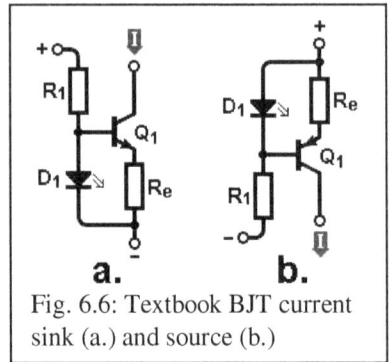

Fig. 6.6: Textbook BJT current sink (a.) and source (b.)

The obvious disadvantage of the circuit in fig. 6.6a is that a third terminal is needed to feed current to the LED, which is why it is really only suitable as a current *sink*. This is not a great handicap though, since the whole circuit can be transformed into a current *source* by using a PNP transistor as in fig. 6.6b. In either case the transistor must have a V_{ce} rating greater than the maximum voltage which will appear across it during use, and a power rating sufficient to manage its average dissipation.

The internal resistance looking into the collector is approximately:

$$r' \approx g_m r_o (R_e \parallel R_\pi) \tag{6.3}$$

Where:
g_m = the device transconductance;
r_o = the internal collector resistance (analogous to internal anode resistance);
R_π = the internal base-emitter resistance.

Unfortunately, you will not find any of these terms on the transistor datasheet; they must be calculated. One of the nice things about BJTs is that it is very easy to estimate their transconductance:

$$g_m = \frac{I_c}{V_T} \approx \frac{I_c}{0.025} \tag{6.4}$$

Where I_c is the collector current in amps.[*]

---

[*] V_T is called the **thermal voltage** and is equal to kT/q volts, where k is the Boltzmann constant, T is the absolute temperature, and q is the electron charge.

224

The internal collector resistance is equal to:

$$r_0 = \frac{V_{ce} + V_A}{I_c}$$

(6.5)

Where V_{ce} is the collector-emitter voltage, I_C is the collector current, and V_A is the **Early voltage** which depends on transistor type, often falling between 50V and 200V. Unfortunately the datasheets won't tell you the Early voltage either (yes, the development of transistor engineering was as confused and unhelpful as possible), so it is common just to estimate it as 100V and forget about V_{ce} altogether.

The internal base-emitter resistance is equal to:

$$r_\pi = \frac{\beta}{g_m} = \frac{V_T \beta}{I_c} = \frac{0.025 \times \beta}{I_c}$$

(6.6)

Where β is the current gain of the transistor, usually called 'h$_{FE}$' on the datasheet.

Substituting all these things into equation (6.3) produces:

$$r' \approx 4000 \frac{\beta R_e}{\beta + 40 R_e I_c}$$

(6.7)

Thus if the circuit in fig. 6.6a uses a transistor with $\beta = 50$, a red LED, and a 470Ω resistor to set the current to 2.1mA, equation (6.7) predicts a total internal resistance of just over 1MΩ. This is a very crude approximation, but at least it suggests we are making progress.

Fig. 6.7 shows the measured performance of a simple current-source using an MPSA92 transistor –a popular 300V-rated device– using the setup from fig. 6.4. The circuit was first tested without C_1 fitted. As the graph shows, the internal impedance is relatively constant but does not reach the calculated figure of 1MΩ. This is mainly because the internal dynamic resistance of the LED forms a potential divider with R_1, so the voltage across R_e and the base-emitter junction is not held as constant as it

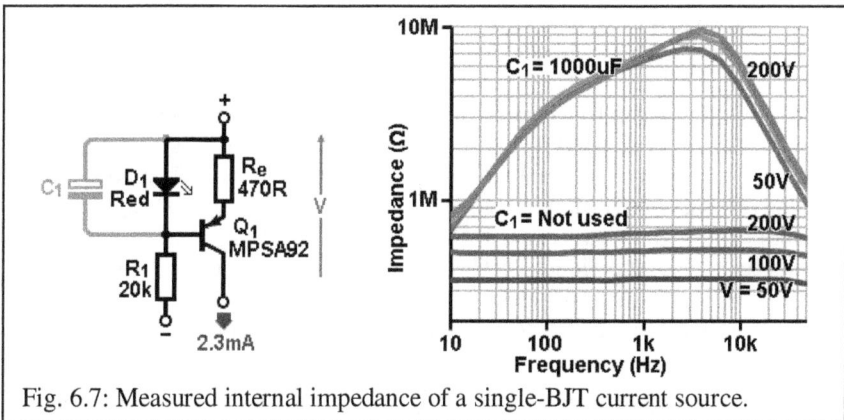

Fig. 6.7: Measured internal impedance of a single-BJT current source.

225

ideally should be. It is also clear that the internal resistance of the source increases with the voltage dropped across it. This is partly because the transistor's r_o increases with V_{ce}, and partly because the current through R_1 and the LED increases too, thereby reducing the LED's dynamic resistance.

Bypassing the dynamic resistance of the LED with a large capacitor exposes the true potential of the current source; its internal resistance reaches almost $10M\Omega$ at 4kHz. Above this it falls at a first order rate due to stray shunt capacitance,[*] while below 100Hz it is limited by C_1 failing to bypass the LED effectively. Why there is a break in slope between 100Hz and 4kHz is not clear. The fact that the internal resistance is not constant with frequency is troubling, because if the circuit were used as a load for a valve stage, distortion in that stage would then also vary with frequency. With low-impedance valves like the ECC88/6DJ8 the effect may be negligible since a load of $1M\Omega$ appears just as 'infinite' as one of $10M\Omega$, but to a high-impedance ECC83/12AX7 this simplification may not apply. Time for a better current source.

6.1.3: Cascode BJT Current Source

In pursuit of a better current source we may note that most of the apparent resistance comes from the multiplication of R_e, so what happens if we replace R_e with *another* current source? Fig. 6.8 shows such an arrangement where a second transistor is now stacked on the first to create a **cascode** current source and sink. Thanks to the double-multiplication the internal resistance of this circuit will be extremely high, and fiendishly difficult to calculate. A further advantage of this arrangement is that the collector-emitter voltage of Q_1 is always small (equal to the drop across D_2), so it can

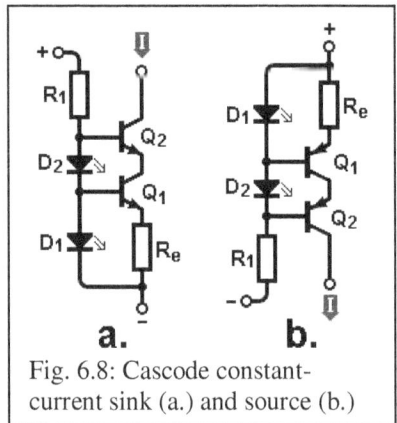

Fig. 6.8: Cascode constant-current sink (a.) and source (b.)

be a low-voltage device. Low-voltage transistors tend to have superior β, r_o and junction capacitances to high-voltage devices. Q_2 can be a high-voltage transistor that does the hard work of withstanding the voltages in a valve circuit.

Fig. 6.9 shows the circuit tested by the author. With the LEDs unbypassed the internal resistance is only slightly higher than the single-transistor source tested previously, but it is more constant with frequency, which is certainly a good thing. To see the real superiority of the cascode we have to bypass the LEDs again (interestingly, a single bypass capacitor produced a slightly superior result to bypassing both LEDs individually). The internal resistance then shoots up to over $30M\Omega$ at 1kHz, and with a sufficiently large bypass capacitance it is beautifully flat until shunt capacitance starts to degrade matters at high frequencies. It does,

[*] If heatsinking is required for a CCS then it is preferable to use an isolated sink (rather than using the chassis, say) in order to minimise capacitance to ground.

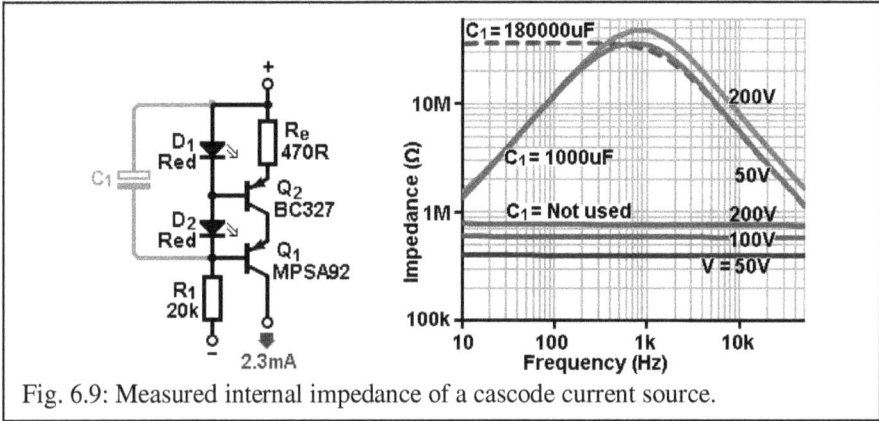

Fig. 6.9: Measured internal impedance of a cascode current source.

however, take a rather large bypass capacitor to achieve this flatness (yes, that's 0.18 *farads*!). Still, with a few thousand microfarads the impedance can be kept above 3MΩ from 20Hz to almost 20kHz, which is quite respectable. On the other hand, we might settle for the unbypassed version simply because the impedance is politely flat (within ±1.3% to be exact) across the whole audio band, even though it is not neck-craningly high.

It is clear that the performance of a BJT current source depends very much on the dynamic resistance of its voltage reference. Perhaps it is worth replacing the simple LEDs with something a little more sophisticated? The ultimate in voltage references is the **band-gap** reference. The science of this device is beyond the scope of this book, but suffice it to say that it can be treated like an ultra-precision Zener diode. The LM385 is such a device, and provides a 1.24V reference with extremely low dynamic resistance. Fig. 6.10 shows this part substituted for the LEDs in the cascode current source; since this leaves only about 0.6V across R_e it was reduced to 270Ω to achieve approximately the same idle current as previously. Reducing R_e in turn caused the maximum internal impedance to fall to about 14MΩ, but it is clear from the figure that the impedance is now reassuringly flat down to low frequencies (presumably down to DC), without needing a bypass capacitor. Again, the upper end

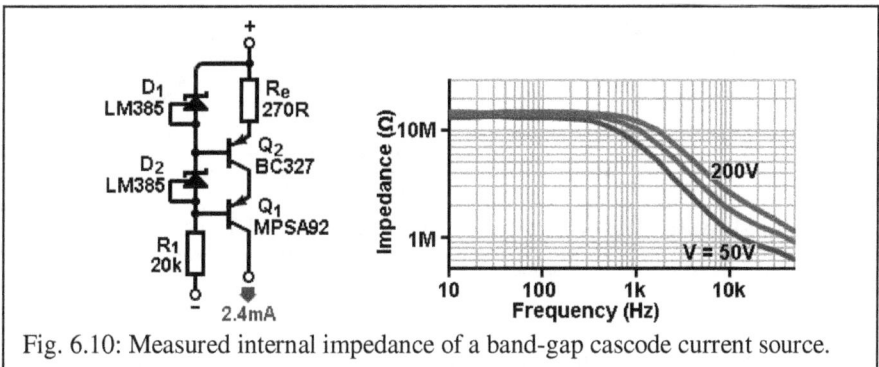

Fig. 6.10: Measured internal impedance of a band-gap cascode current source.

227

is limited by stray capacitance but improves as the DC drop across the source increases (this is because the capacitance across a silicon junction falls as the reverse voltage across it increases). The falling response is a little wobbly because the dynamic resistance of the LM385 increases with frequency, but this is a minor criticism.

6.1.4: FET Current Sources

A disadvantage of the previous BJT current sources is that they have three terminals; it would be nice to have a transistor equivalent of the triode source from fig. 6.2, which can be used as both a source and sink. To achieve this we need a depletion-mode transistor, i.e. one that requires a negative voltage to bias it, just like a valve. JFETs meet this requirement but are only available with low voltage ratings. MOSFETs are readily available with high voltage ratings but most of them are enhancement devices. However, notable exceptions are the DN2535 (350V), DN2540 (400V), DN3545 (450V), and LND150 (500V) depletion MOSFETs, which have become very

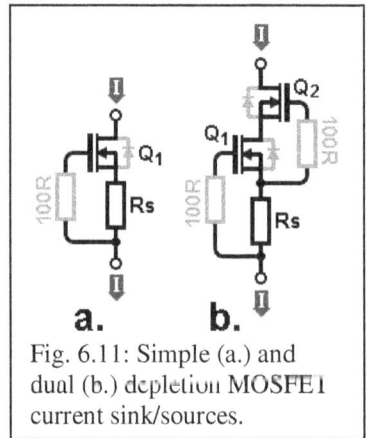

Fig. 6.11: Simple (a.) and dual (b.) depletion MOSFET current sink/sources.

fashionable for valve circuits. Fig. 6.11 shows examples of single and dual (quasi-cascode) circuits using these devices.[1] The 100Ω resistors are simply gate stoppers added to discourage oscillation, while the source resistor R_s sets the DC current just like R_e did previously. Unfortunately, the gate-source voltage of FETs varies considerably between devices, so R_s will nearly always need to be adjusted by hand to achieve a given current, unlike when using BJTs.

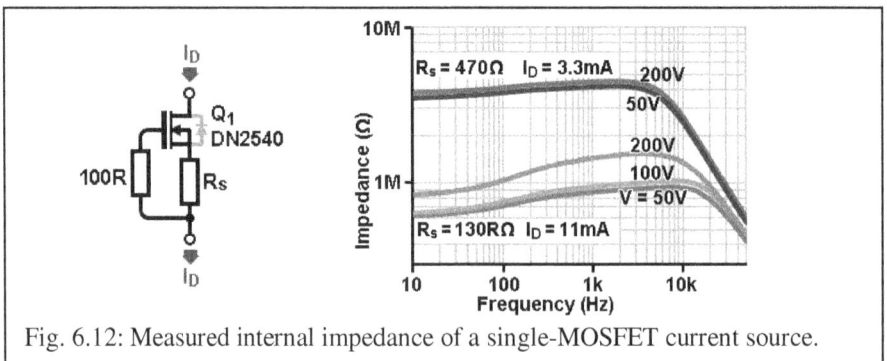

Fig. 6.12: Measured internal impedance of a single-MOSFET current source.

[1] Jung, W. (2009). High Performance Current Regulators Revisited (letter), *Audio Xpress*, April, pp40-2.

Fig. 6.12 shows the measured impedance of the single-MOSFET arrangement. At low current the performance is rather better than the single BJT current source from fig 6.7, although at higher current there is a fair amount of variation with frequency. The results for the dual-MOSFET arrangement are shown in fig. 6.13. It is worth mentioning that this dual version suffers from a complication in that the voltage across Q_1 is equal to the V_{GS} of Q_2, and so varies between samples. An unlucky combination of devices can result in a circuit that refuses to operate at the desired current no matter how much R_s is fiddled with. Be prepared, therefore, to spend time sorting through MOSFETs to find one with a large enough V_{GS} to allow the lower device to function, e.g. larger than $-1.5V$. Case in point, in the author's test circuit it was impossible to set the idle current to ~2.4mA to match previous tests, which is why R_s was set arbitrarily at 470Ω, yielding a current of 3.3mA. Reducing it to 130Ω increased this to 11mA. The DC voltage across the source proved to have negligible effect on its impedance, so only the 100V traces are shown. The impedance is now really quite impressive, reaching over 120MΩ at low current. More importantly, it is still 9MΩ at 20kHz in either case, which is enough to qualify the circuit as a true constant-current source over the whole audio range in almost any application.

Since depletion MOSFETs are not common devices it might be desirable to use as few as possible in a design. The lower transistor Q_1 does not, in fact, need to be a high-voltage device, and can be replaced with an ordinary JFET. Such an arrangement tends to suit applications under 10mA, and the internal impedance is

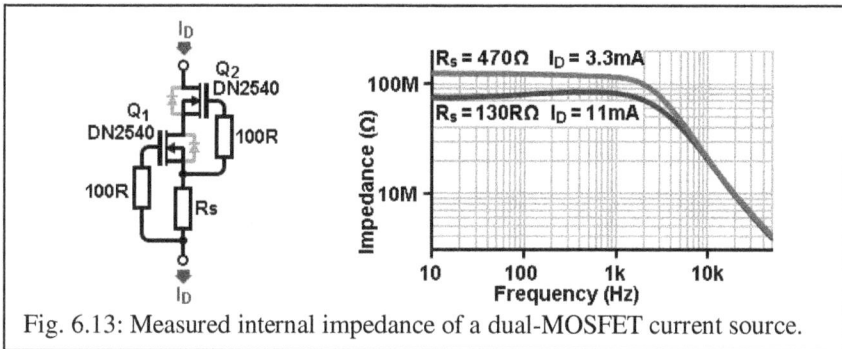

Fig. 6.13: Measured internal impedance of a dual-MOSFET current source.

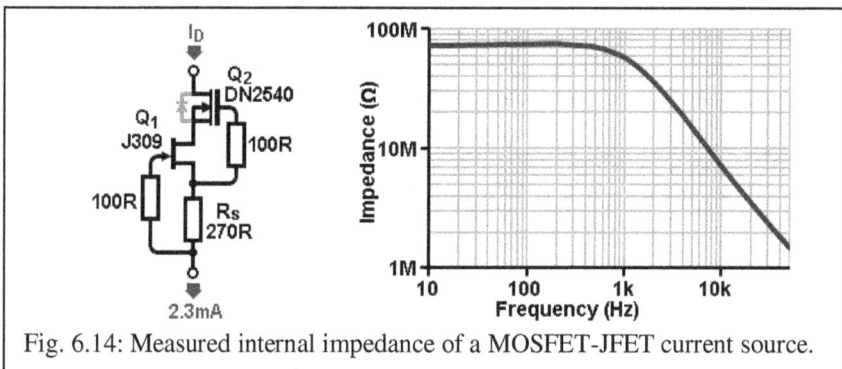

Fig. 6.14: Measured internal impedance of a MOSFET-JFET current source.

229

likely to be slightly poorer than for dual MOSFETs. The author used a randomly selected J309, but almost any JFET should work. A source resistor of 270Ω yielded a constant current of 2.3mA and, as fig. 6.14 shows, the internal impedance exceeded 3MΩ at 20kHz. Again, the voltage across the circuit had negligible effect on this.

6.2: Valve Distortion Measurements

Having identified a few useful CCS circuits we are now in a position to take another look at the effects of loading on triode distortion. Most of the audio measurements presented in this book were made using an Audio Precision System One (AP1) audio analyser. The analyser has a distortion-residual output which allowed the harmonic spectrum to be monitored using the FFT function of a Rigol DS1102E digital oscilloscope. Some measurements have already been seen in previous chapters, but to avoid unnecessary distraction the exact test method was not discussed at the time; it is described here instead. The following methods were used for other distortion measurements presented in this book, too.

The AP1 has an input impedance of 100kΩ||350pF plus 150pF for the connecting cable, which is much too heavy to attach directly to the output of a valve stage as it would lead to misleading results. However, if the valve has a resistive load then the solution is simple: take the output from a tapping point on the anode resistor. Fig. 6.15 shows the arrangement used by the author, where the anode resistor is split into two parts R_a and R_3. Since the input impedance of the AP1 is effectively in parallel with R_3, which can be made quite small in comparison, the AP1 will have negligible effect on the loading of the valve itself. Output-level measurements were then scaled to correspond to the actual anode voltage before being presented in

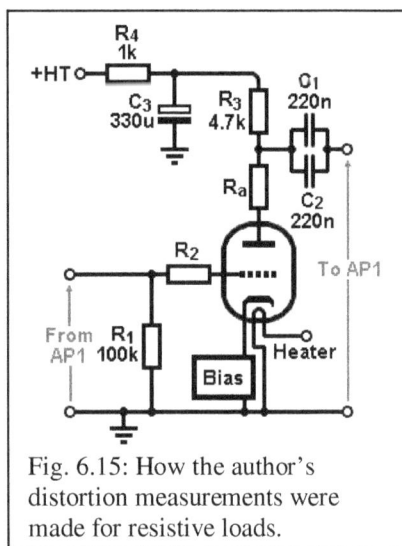

Fig. 6.15: How the author's distortion measurements were made for resistive loads.

this book. R_3 was 4.7kΩ for most tests, but was reduced to 2kΩ for wide-band measurements such as frequency response. R_2 is a grid stopper and was set at 10kΩ for most tests. This is representative of real-world source impedances and should expose any serious grid current effects. All resistors were metal film, and the output coupling capacitors were polypropylene (not that this makes the slightest difference, as the author later confirmed). The HT was supplied from a simple but adequate adjustable supply –described in section 11.7.5– and the heater from a separate, regulated bench power supply. The whole circuit was built into an earthed metal box to exclude ambient hum.

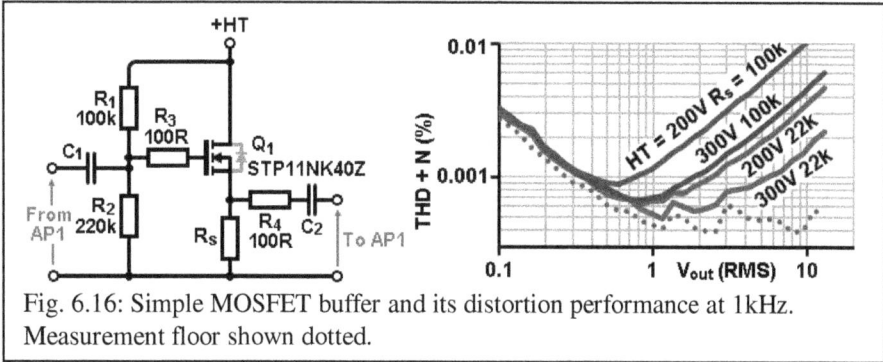

Fig. 6.16: Simple MOSFET buffer and its distortion performance at 1kHz. Measurement floor shown dotted.

The previous arrangement is not suitable when the valve has a CCS load, since the current in R_3 would remain constant and therefore no signal voltage would appear across it to measure. For such tests there is no alternative but to buffer the whole valve stage from the AP1, and for simplicity a MOSFET source follower was used. Of course, it is essential that the buffer itself produces much less distortion than the valve stage, so it was first tested in isolation, as shown in fig. 6.16. The functioning of this circuit will be obvious to some readers and will not

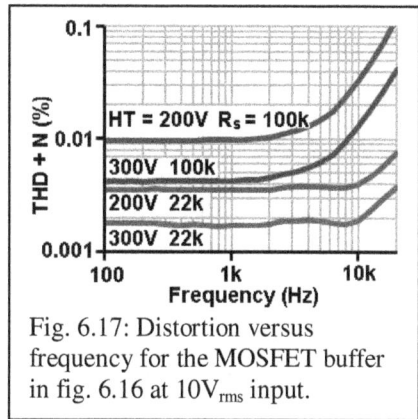

Fig. 6.17: Distortion versus frequency for the MOSFET buffer in fig. 6.16 at $10V_{rms}$ input.

be explained here (see chapter 7 instead), since the purpose is only to show that its distortion is low enough not to interfere with measurements of valve distortion.

Fig. 6.16 indicates that distortion improves as the quiescent current in the MOSFET increases, so the author was careful to ensure the drain current was always >5mA during tests, by adjusting R_s. Note that where the traces converge below $1V_{rms}$ is simply the limit of measurement, not actual distortion in the buffer. Fig. 6.17 shows how distortion genuinely does worsen at high frequencies as the MOSFET struggles to drive the input

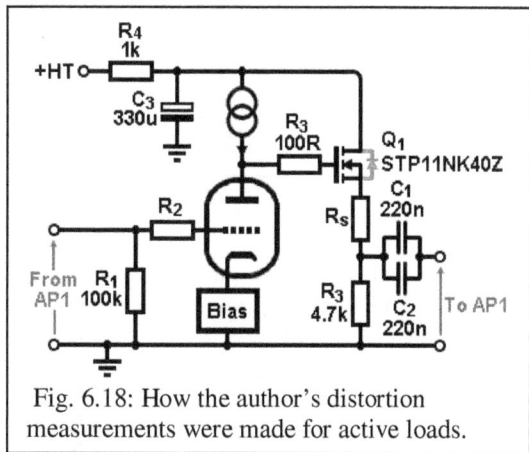

Fig. 6.18: How the author's distortion measurements were made for active loads.

capacitance of the analyser. Nevertheless, it is still low enough not to interfere with valve distortion measurements, provided the drain current is >5mA. Fig. 6.18 shows

231

how the buffer was actually mated to CCS-loaded valves during testing. The input capacitance of the buffer was measured at ~29pF which agrees with the datasheet value for gate-drain capacitance. The output was taken from a tapping point on R_s simply to protect the AP1 from large signal voltages, and results were then scaled to correspond to the actual anode voltage of the valve.

The following subsections present a library of measurements for different valves, made using the methods outlined above. The distortion measurement bandwidth was 80kHz in all cases. These data are by no means exhaustive and should not necessarily be taken as recommendations for certain operating conditions; they are meant only to give the reader an intuitive feel for how triode distortion can be expected to vary under some plausible conditions.

6.2.1: ECC81 / 12AT7

The ECC81 / 12AT7 is a high-μ device ($\approx$60) originally advertised for general RF applications, but now often used for audio purposes too. It has much lower r_a ($\approx$15kΩ) than its high-μ brother, the ECC83, so it a common choice when wider bandwidth is needed with relatively high gain. However, its noise, grid-current, and general distortion are all much worse than the ECC83.

The ECC81 was tested with a MOSFET CCS as shown in fig. 6.19. Adjusting R_s to 500Ω resulted in an anode current of 3.2mA. The load line is also shown and is of course horizontal at 3.2mA. However, it cannot remain horizontal all

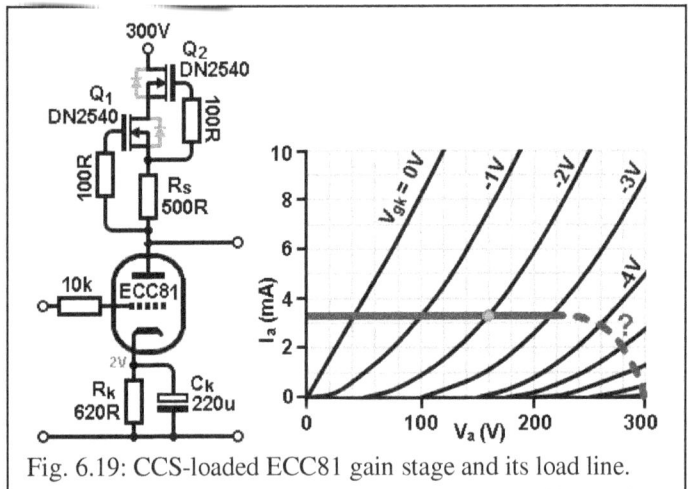

Fig. 6.19: CCS-loaded ECC81 gain stage and its load line.

the way up to the 300V HT since there comes a point where the anode voltage swings so high that there is not enough voltage left across the CCS for it to function properly. Exactly what happens in this region depends on the peculiarities of the CCS, and the premature clipping it introduces is one of the drawbacks of using a CCS. Nevertheless, where the load line is reliably horizontal the gain becomes equal to the μ of the triode, and therefore the linearity of the stage is in turn determined by the linearity of μ.

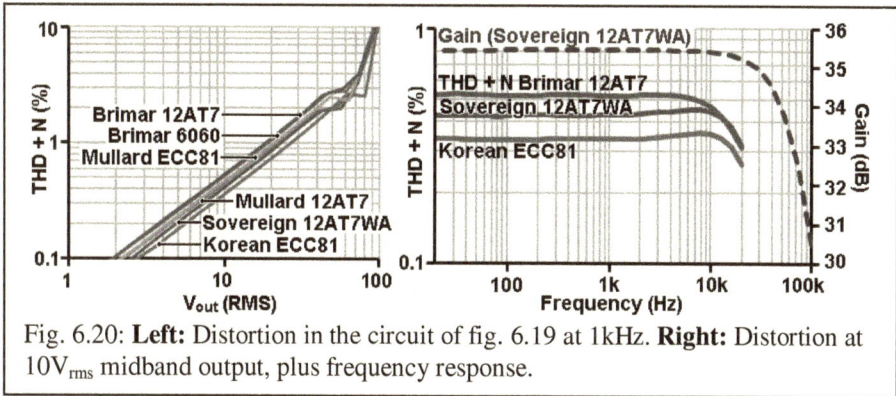

Fig. 6.20: **Left:** Distortion in the circuit of fig. 6.19 at 1kHz. **Right:** Distortion at $10V_{rms}$ midband output, plus frequency response.

Six valves were checked on a curve tracer for good health and reasonable matching before being tested in the circuit of fig. 6.19 (only one triode per bottle was tested). All produced voltage gains between 50 and 60 with Miller capacitance between 150pF and 180pF respectively (not forgetting that this includes the effect of the plastic McMurdo valve socket).

Distortion results are given in fig. 6.20, and the spread is so tight that some of the traces lie on top of one another. Note that clipping is reached at only $50V_{rms}$ output and is very abrupt –a consequence of the CCS running out of voltage. Fig. 6.20 also shows distortion versus frequency for three representative valves from the initial batch. The input signal was adjusted to give an output of $10V_{rms}$ at 250Hz and was then swept between 20Hz and 20kHz. Above a few kilohertz the distortion starts to creep up due to the falling impedance of the CCS and the 29pF load capacitance presented by the MOSFET buffer, but above 10kHz the falling frequency response begins to filter out distortion products, so the THD falls again. The third harmonic was about 32dB below the second in every case; higher harmonics could not be resolved.

In recent years there has been a trend towards operating valves at ever higher anode currents in an attempt to reduce distortion. However, this is not nearly as effective as is sometimes

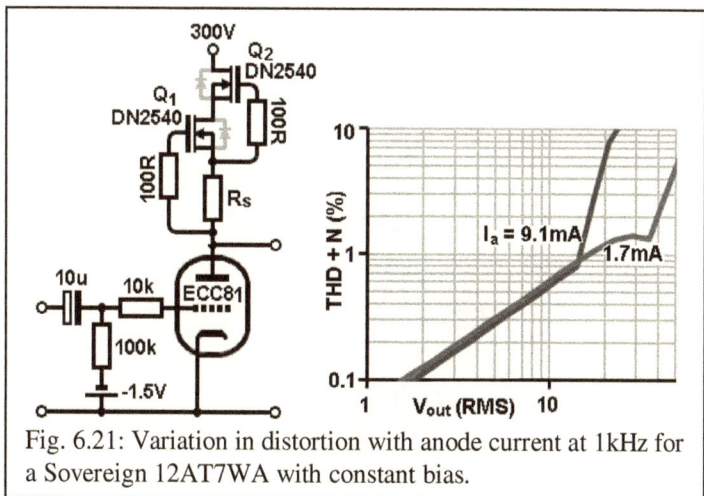

Fig. 6.21: Variation in distortion with anode current at 1kHz for a Sovereign 12AT7WA with constant bias.

233

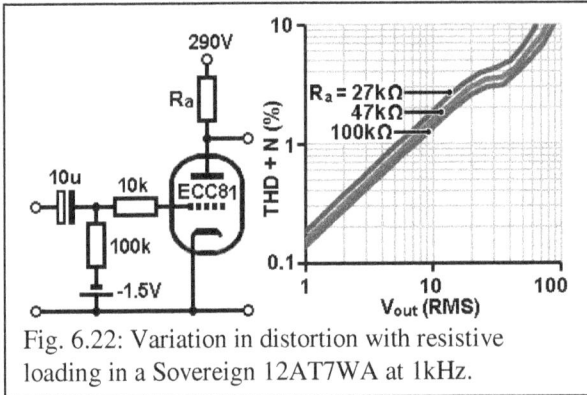

Fig. 6.22: Variation in distortion with resistive loading in a Sovereign 12AT7WA at 1kHz.

supposed. Fig. 6.21 shows how distortion varied in one typical 12AT7 specimen as the anode current was varied from 1.7mA (R_s = 1kΩ) to 9.1mA (R_s = 160Ω) with the bias voltage held constant with a battery. The voltage gain varied from 57 to 70 respectively, but the distortion barely improved, and this is characteristic of all ordinary triodes the author has encountered. Not only does an excessive idle current not improve distortion very much, it lowers the clipping threshold since V_{ak} must be larger, puts greater strain on the power supply, and shortens the life of the valve. The only useful trend seems to be the increase in voltage gain and some reduction of r_a (output resistance).

For comparison, fig. 6.22 shows some distortion results with ordinary resistive loading, varying from 27kΩ to 100kΩ. The distortion is at least doubled compared with CCS loading, but it was constant across the whole bandwidth. At 10V_{rms} output the third harmonic was around 24dB below the second in each case, illustrating that CCS loading reduces distortion mainly by suppressing the second harmonic. Notice also how clipping (due to grid-current limiting) occurs relatively soon at about 20V_{rms} –a consequence of this valve type's high grid current.

6.2.2: ECC82 / 12AU7

The ECC82 is a low-μ device ($\approx$18) marketed for both audio and radio use. It has lower r_a ($\approx$10kΩ) and grid current than the ECC81, and is sometimes described as the noval version of the 6SN7. For

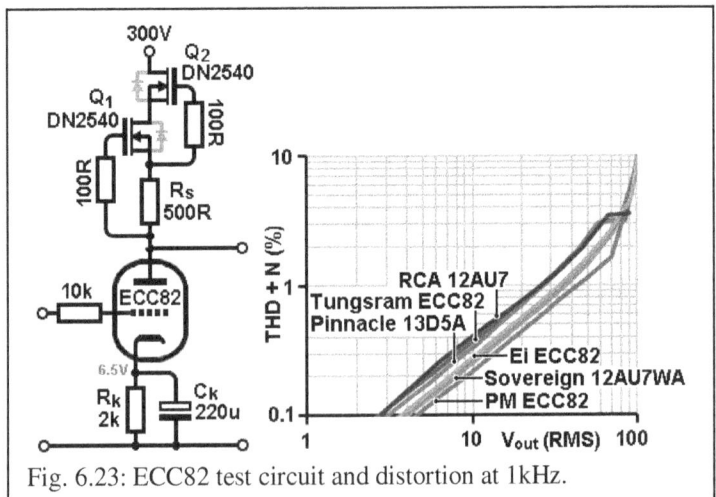

Fig. 6.23: ECC82 test circuit and distortion at 1kHz.

234

some reason, however, the ECC82 seems to have acquired a reputation for poor linearity that is does not deserve. Jones[2] has published some very uninspiring distortion measurements which are perhaps due to an unfortunate batch of valves

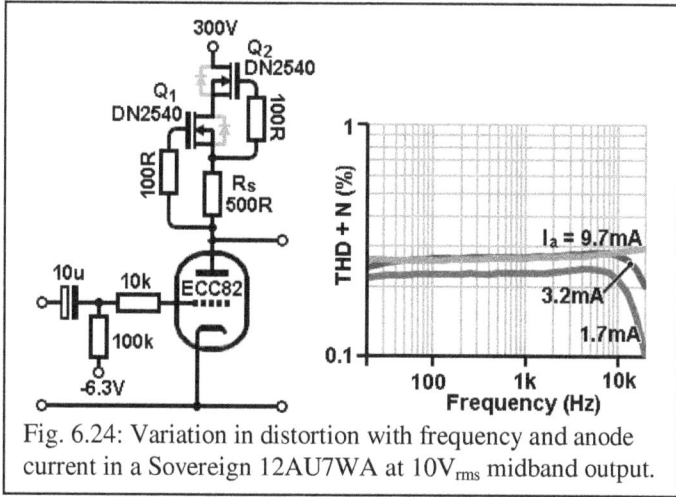

Fig. 6.24: Variation in distortion with frequency and anode current in a Sovereign 12AU7WA at $10V_{rms}$ midband output.

because, as will be seen, ECC82s are actually quite respectable (or at least average) performers.

The ECC82 was tested using the same 3.2mA CCS as previously. As shown in fig. 6.23, distortion in six samples is on average better than for an ECC81 (but about twice as bad as the 6SN7s tested later). Gain varied between 14 and 17 with Miller capacitance between 50pF and 60pF, respectively.

Fig. 6.24 shows the variation in distortion with frequency and with changes in CCS current, for one typical specimen, with the bias voltage held constant at $-6.3V$. Higher current reduces r_a and so widens the frequency response but, unusually, in this valve the distortion is actually

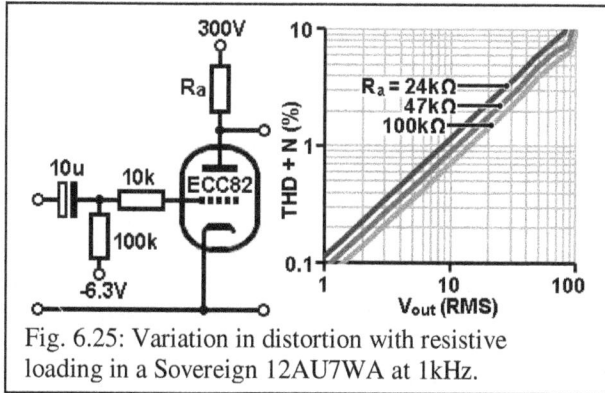

Fig. 6.25: Variation in distortion with resistive loading in a Sovereign 12AU7WA at 1kHz.

slightly better at low current (all six samples showed this behaviour to some extent).

For comparison the distortion was also measured with resistive loading, as shown in fig. 6.25. Similarly unusual was that the third harmonic was roughly 26dB below the second for both CCS and resistive loading.

---

[2] Jones, M. (2012). *Valve Amplifiers* (4th ed.), Elsevier, Oxford. p205.

235

6.2.3: ECC83 / 12AX7

The ECC83 / 12AX7 has already been introduced in chapter 3. It offers the highest μ of all the common triodes (≈100), but its high r_a (≈70kΩ) means its performance quickly degrades when faced with even slightly heavy load impedances.

The ECC83 was tested with a 0.9mA MOSFET/JFET CCS as shown in fig. 6.26. Six healthy valves were tested and all produced voltage gains between 90 and 105 with Miller capacitance close to 270pF. Distortion results are given in fig. 6.27, and comparing them with results from chapter 3 they are an order of magnitude lower than what was possible with resistive loading. The ECC83 is a high impedance valve and more sensitive to the falling impedance of the load above 1kHz, causing distortion to rise at high frequencies.

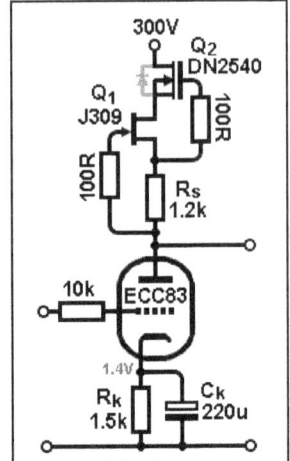

Fig. 6.26: Circuit used to test the ECC83.

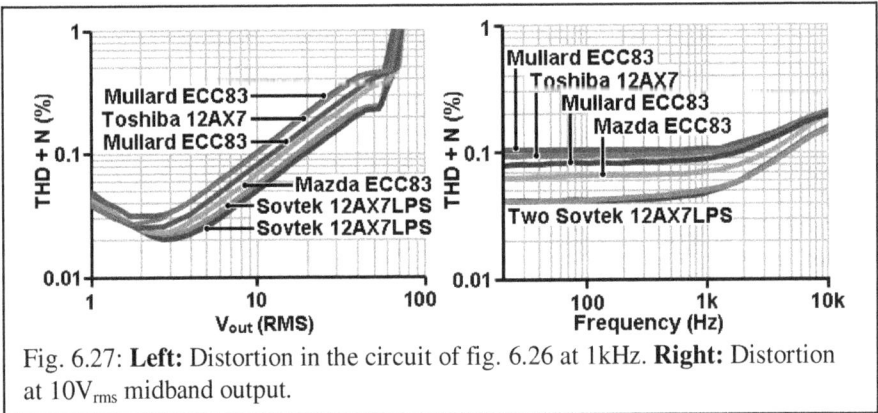

Fig. 6.27: **Left:** Distortion in the circuit of fig. 6.26 at 1kHz. **Right:** Distortion at $10V_{rms}$ midband output.

At $10V_{rms}$ output the third harmonic was around 26dB below the second. Rather than repeat the results for resistive loading from chapter 3, fig. 6.28 instead shows the variation in distortion for a circuit in which the HT voltage is varied instead. Clearly, the higher the voltage, the better.

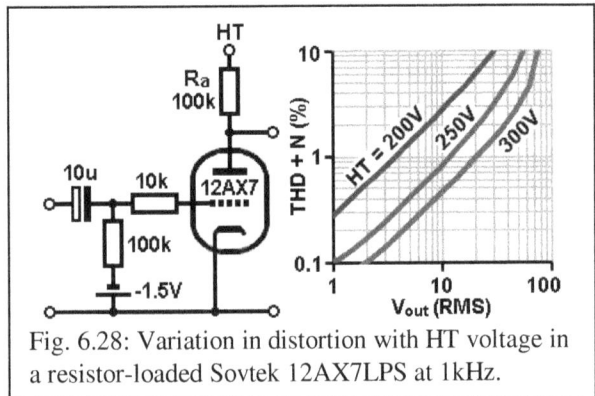

Fig. 6.28: Variation in distortion with HT voltage in a resistor-loaded Sovtek 12AX7LPS at 1kHz.

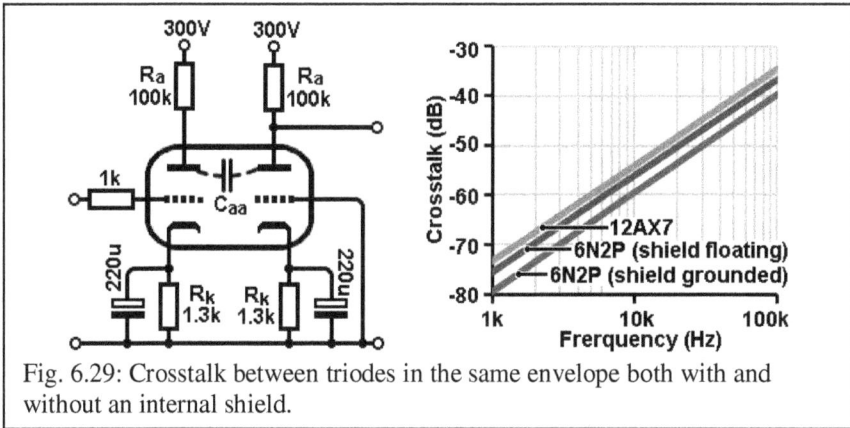

Fig. 6.29: Crosstalk between triodes in the same envelope both with and without an internal shield.

An alternative to the ECC83 is the Russian 6H2Π / 6N2P, which is practically identical except for having a much smaller C_{ga} ($\approx$0.7pF) and an internal shield connected to pin 9 (i.e. a 9AJ pinout). This shield sits between the two triodes in the bottle and should be grounded to reduce the capacitance between the anodes, C_{aa}, thereby reducing the stray coupling between them. Fig. 6.29 shows the effectiveness of the shield in reducing crosstalk in an example circuit. Compared with a 12AX7LPS which has no shield, the improvement is almost 6dB. The case for the 6N2P with the shield left floating is also shown.[*] Ideally, crosstalk would continue to fall as frequency falls. However, if the triodes happen to share the same power supply connection and are not well decoupled from one another then crosstalk will begin to rise again at low frequencies, though this is rarely cause for concern.

6.2.4: ECC84 / 6CW7

The ECC84 / 6CW7 is a low r_a ($\approx$5kΩ), medium-μ ($\approx$23) valve specially designed for use as a cascode RF amplifier. It is rarely used in audio design, but since the author had a couple on hand it seemed only fair to try them out. Like most high-g_m triodes it is intended to operate at relatively high current but low voltage; the maximum permissible quiescent anode voltage is quoted as 180V. The CCS test circuit was therefore set for 9.7mA anode current with a 200V HT.

The author had two ECC84s and two PCC84s available (all Mullard brand), the latter being a 7V-heater version of the valve. However, all were tested at a heater voltage of 6.3V initially. The gain was close to 21 in each case, with Miller capacitance around 90pF for one of the triodes in the envelope but only 55pF for the other (this is a result of the special design of the valve; refer to the datasheet).

---

[*] The shield should never be left floating in practice as it can collect enough charge to cause some very weird and unpleasant noise and distortion effects.

As indicated in fig. 6.30, distortion was the worst so far encountered for CCS loading, as well as showing some deviation from a straight line. The PCC84s were also retested at their rated heater voltage of 7V. This had minimal effect on gain (it fell by less than

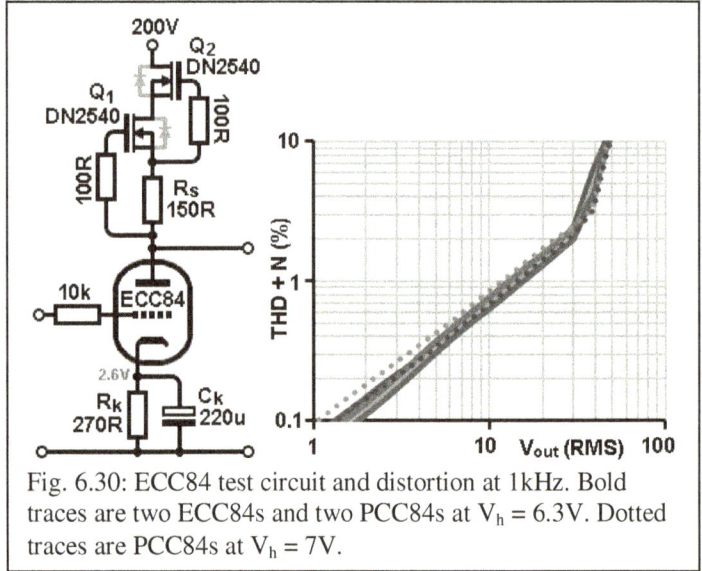

Fig. 6.30: ECC84 test circuit and distortion at 1kHz. Bold traces are two ECC84s and two PCC84s at $V_h = 6.3$V. Dotted traces are PCC84s at $V_h = 7$V.

1%), but the distortion got very slightly worse! In either case the distortion was at least ruler flat over the audio frequency range and not worth showing. Unlike the valve types discussed previously, harmonics higher than the third were just discernible, and an example spectrum is presented in fig. 6.31 (remember, the fundamental is nulled out by the analyser).

This valve is not totally useless, however; fig. 6.32 shows some distortion results for resistive loading which are actually slightly better than the ECC81 in fig. 6.22.

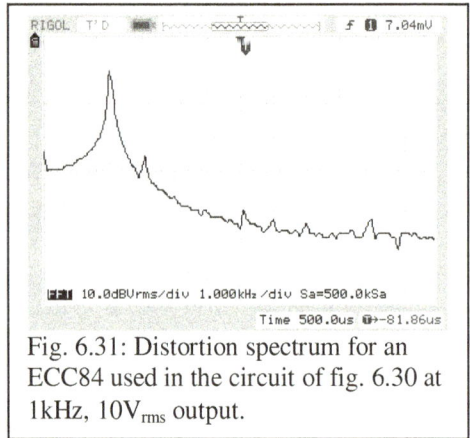

Fig. 6.31: Distortion spectrum for an ECC84 used in the circuit of fig. 6.30 at 1kHz, 10V$_{rms}$ output.

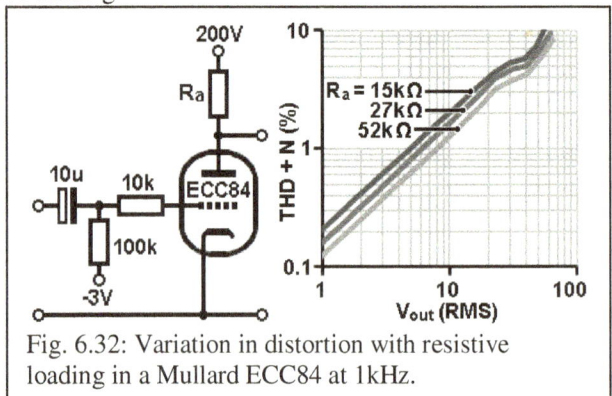

Fig. 6.32: Variation in distortion with resistive loading in a Mullard ECC84 at 1kHz.

6.2.5: ECC88 / 6DJ8

The ECC88 / 6DJ8 is another low-r_a ($\approx 4k\Omega$), medium-μ (≈ 30) valve specially designed for cascode RF use, but which has become enormously popular for hi-fi purposes. Like the ECC84 its maximum permissible quiescent anode voltage is very low at 130V.

Fig. 6.33: ECC88 test circuit and distortion at 1kHz.

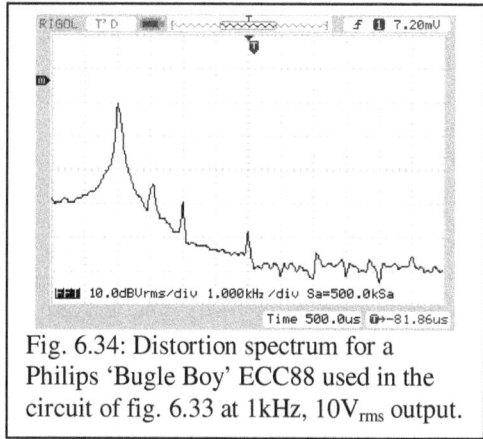

The same 9.7mA CCS as previous was used to test the ECC88, again with a 200V HT. Measured gain for four triodes was between 29 and 34 with Miller capacitance between 95pF and 110pF respectively. Fig. 6.33 shows that distortion is much better than the ECC84, and about the same as the ECC82 (perhaps to the dismay of some fans), but there are noticeable deviations from direct proportionality with level. As with the ECC84, some higher harmonics were readily apparent, and the distortion spectrum for a representative specimen is shown in fig. 6.34. The distortion was constant with frequency across the whole audio band, as usual.

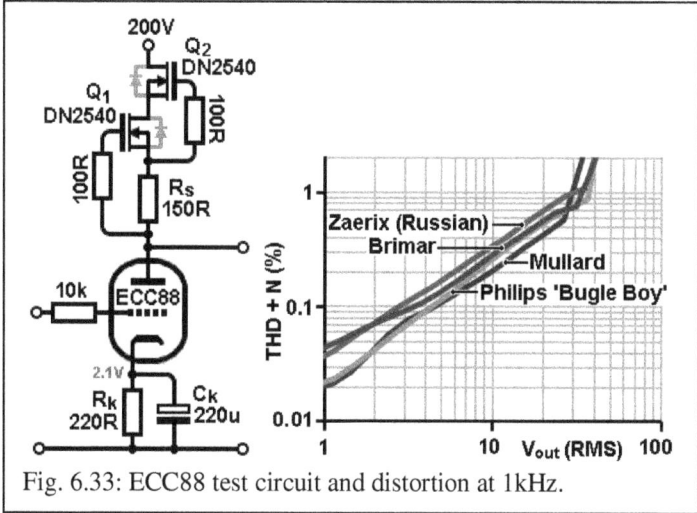

Fig. 6.34: Distortion spectrum for a Philips 'Bugle Boy' ECC88 used in the circuit of fig. 6.33 at 1kHz, $10V_{rms}$ output.

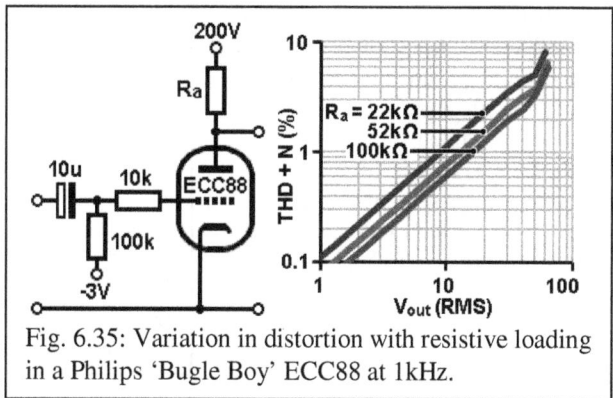

Fig. 6.35: Variation in distortion with resistive loading in a Philips 'Bugle Boy' ECC88 at 1kHz.

239

Fig. 6.35 shows some distortion results for resistive loading, which are also very similar to the ECC82. Under these conditions the third harmonic was barely visible at just over 30dB below the second, with all others lost in the noise floor.

6.2.6: 6Ж52П / 6J52P

Towards the end of the valve era a number of very-high-g_m frame-grid pentodes were developed to compete with transistors, and when operated as triodes they boast some of the lowest noise and best linearity of all valves. One such device is the Russian 6Ж52П / 6J52P ($\mu \approx 80$, r_a

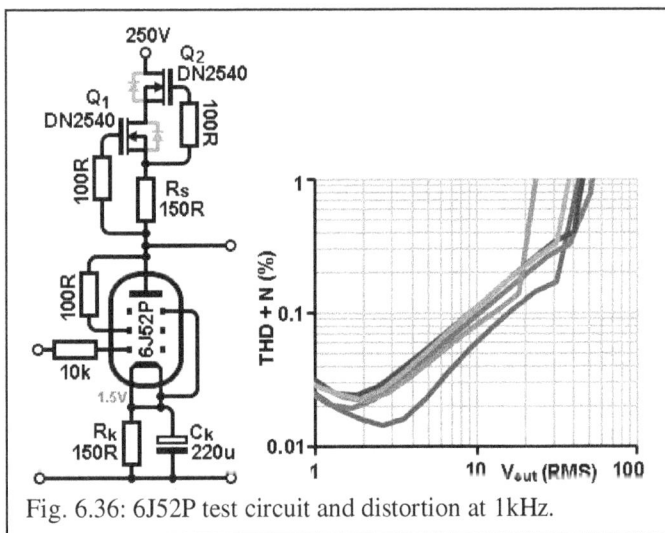

Fig. 6.36: 6J52P test circuit and distortion at 1kHz.

$\approx 3k\Omega$). It is reportedly a cheap alternative to the European D3a which has recently become quite popular for phono preamps. However, these high-g_m valves rely on such tiny electrode spacings that they tend to be highly microphonic, have high grid current, and large interelectrode capacitances. Also, their construction is so critical that they often suffer a wider spread in parameters than 'ordinary' valves, so finding a matched pair may be tricky (and you only get one per bottle), so for these reasons their use in phono preamps may be counterproductive.

The author tested five specimens, connected as triodes (section 4.8). The valve has two cathode connections for low-inductance RF purposes, and the author initially made the mistake of joining the suppressor grid to one cathode pin while using the other cathode pin to connect to R_k/C_k. This resulted in excessive distortion at low output levels, which was cured by connecting both cathode pins together as in fig. 6.36.

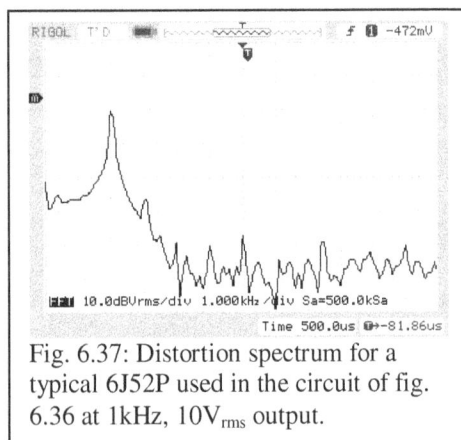

Fig. 6.37: Distortion spectrum for a typical 6J52P used in the circuit of fig. 6.36 at 1kHz, $10V_{rms}$ output.

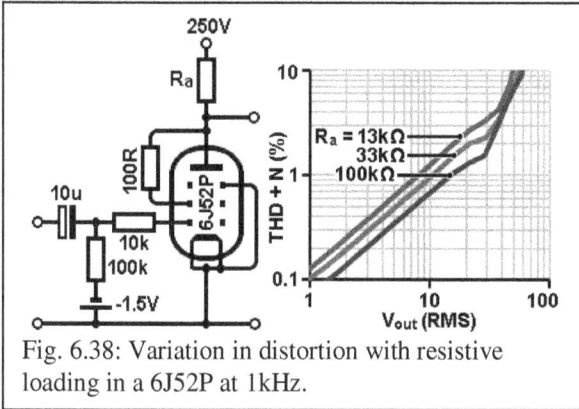

Fig. 6.38: Variation in distortion with resistive loading in a 6J52P at 1kHz.

All the valves tested were exceedingly sensitive to vibration and interference –the lid of the metal enclosure had to be firmly on when making measurements, unlike with previous the valve types.

Gain at 9.7mA anode current was between 70 and 84, with Miller capacitance at a crippling 310pF to 360pF respectively. To its credit, the distortion was very low –similar to the ECC83– and the harmonic spectrum was respectable, as shown in fig. 6.37. However, the high grid-current causes clipping at barely $0.6V_{pk}$ input (or $0.3V_{pk}$ for one very unfortunate specimen) despite a bias of −1.5V. Fig. 6.38 shows distortion for some resistive loads using one of the average-performing samples. Into 100kΩ the distortion is similar to an ECC83, but for a heavy load like 13kΩ the low-impedance 6J52P copes far better.

6.2.7: 6SN7GT

The 6SN7GT is the big octal brother of the ECC82, having similar μ (≈20) and r_a (≈10kΩ). However, its larger dimensions permit almost twice the anode dissipation (5W), but also incur greater interelectrode capacitances. It is much loved by audiophiles for it boasts low distortion, but we pay for this performance with a hefty 600mA of heater current at 6.3V (or 300mA for the 12.6V version –the 12SN7).

Five 6SN7GTs and a two Russian equivalents – 6H8C / 6N8S– were tested with the same 9.7mA CCS as previously. Gain for the seven valves was between 20 and 21 with Miller capacitance

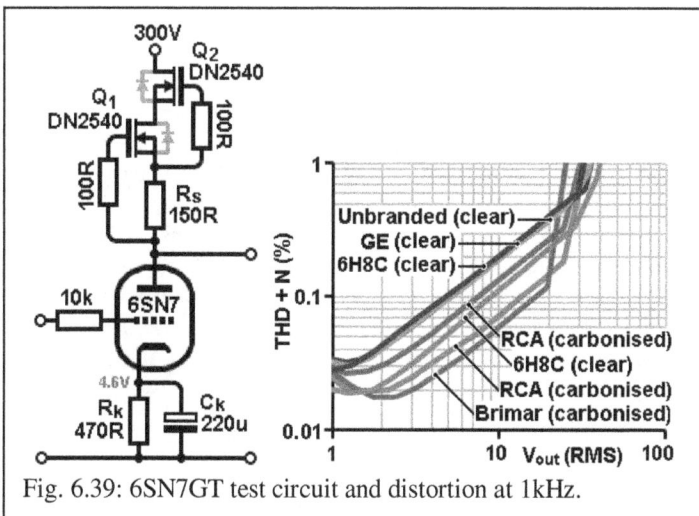

Fig. 6.39: 6SN7GT test circuit and distortion at 1kHz.

241

around 120pF – twice that of the ECC82. As fig. 6.39 shows, the distortion is indeed low, though there is a fairly wide spread (three traces lie on top of one another in the figure). Jones[3] has reported that versions of the valve with carbonised envelopes

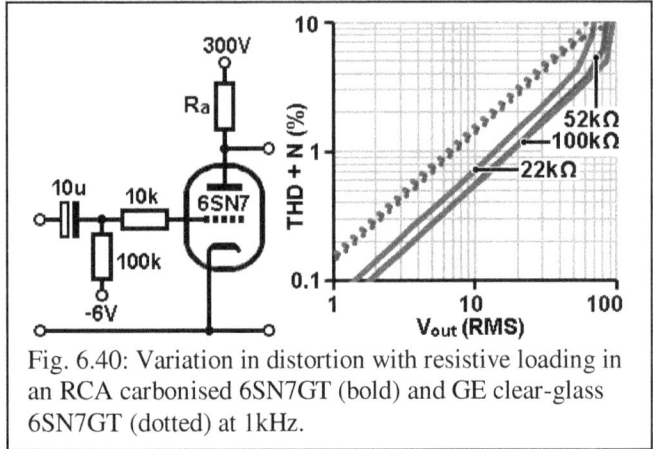

Fig. 6.40: Variation in distortion with resistive loading in an RCA carbonised 6SN7GT (bold) and GE clear-glass 6SN7GT (dotted) at 1kHz.

consistently produce less distortion than clear-glass samples, and fig. 6.39 shows a similar trend. Nevertheless, even the clear-glass valves are better than the ECC82. What's more, the distortion was visually pure second harmonic, the third harmonic being lost somewhere in the noise floor. Clipping happens rather soon in fig. 6.39, but this is due to the limitations of the CCS and the available supply voltage rather than the valve.

Fig. 6.40 shows some results for resistive loading of two selected devices, which are an order of magnitude worse than for CCS loading. The clear-glass version is somewhat disappointing (three traces lie on top of one another), being worse than the ECC82 tested earlier, though this could be an unlucky sample. The carbonised version, however, is very good indeed, performing much better than the ECC83 into heavy loads and roughly the same for high resistance loads. Interestingly, the 6SN7GT seems to show much less variation in distortion with load resistance than most other valve types.

6.2.8: Summary

The foregoing measurements show that there is an enormous difference in total distortion levels when a triode is operated with a CCS load as opposed to ordinary resistor loading. Since CCS loading pushes distortion down to the irreducible minimum, dependent only on variations in μ, the differences between valve types are most apparent under these conditions, and the harmonic spectrum is the most colourful. High-g_m valves are more likely to show deviation from direct proportionality with level. With resistor loading the performance differences between valves types shrinks dramatically and may in fact be swamped by individual sample variation and choice of operating conditions. In other words, with resistor loading it may not be worth selecting a valve type based on distortion expectations; most of them perform quite similarly.

---

[3] Jones, M. (2012). *Valve Amplifiers* (4th ed.), Elsevier, Oxford. p203-7.

Increasing anode current and/or biasing hotter will usually reduce distortion, but often by such a trivially small amount that is this too is more likely to be swamped by sample variation. Maximising the load impedance and supply voltage are usually the most effective ways to improve distortion.

The natural spread in performance leads to some inevitable crossover between valve types, but based on the foregoing measurements a very general list, from most linear to worst, appears to be:

- ECC83 / 12AX7
- 6SN7GT (carbonised)
- 6Ж52П / 6J52P
- 6SN7GT (clear glass)
- ECC88 / 6DJ8
- ECC82 /12AU7
- ECC81 / 12AT7
- ECC84 / 6CW7

Note that this list assumes the valves are operated under 'typical conditions', e.g. with a load impedance greater than the valve's own internal resistance. With heavy loading the ranking will be different, particularly with regard to the ECC83.

6.3: Diode Biasing

Cathode bias using a resistor has the useful effect of being self-adjusting, helping to maintain a reasonable operating point over the life of the valve. However, to prevent degeneration it must be bypassed by a large –and therefore electrolytic– capacitor, which brings with it a certain amount of baggage. Discerning builders will no doubt worry about capacitor distortion, though this is not a genuine problem if enough capacitance is used (in fact, the author has never been able to detect any changes in THD caused by cathode bypass capacitors even when they are deliberately too small). Nevertheless, without *infinite* capacitance, degeneration will eventually kick in at low frequencies and so introduce phase shift which may be unwanted if the stage is part of a global feedback loop. Also, the bypass capacitor will make the stage somewhat more prone to blocking distortion because

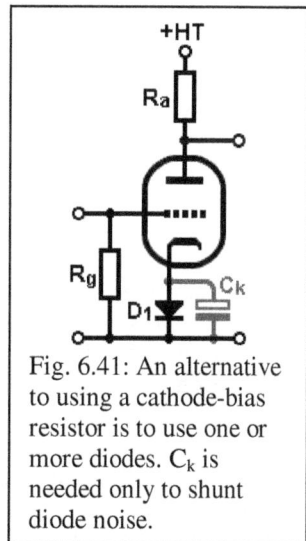

Fig. 6.41: An alternative to using a cathode-bias resistor is to use one or more diodes. C_k is needed only to shunt diode noise.

during overdrive the bypass capacitor will charge up somewhat, but may take much longer to discharge back to normal after the overload has passed, leaving the valve temporarily cold-biased. Moreover, electrolytics have poor tolerance and dry out over time. How serious these problems are depends on exact circuit details and usage, but a simple way to side-step them altogether is to use diode biasing. Here the cathode resistor is replaced by a solid-state diode; fig. 6.41 shows the basic principle.

243

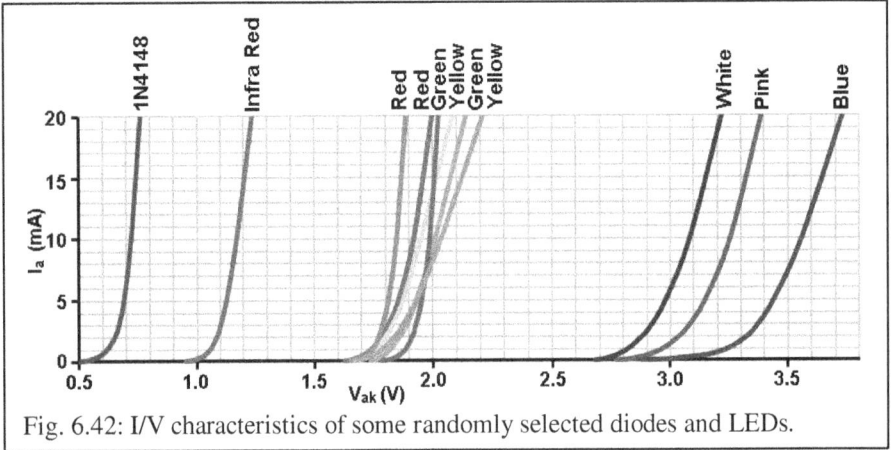

Fig. 6.42: I/V characteristics of some randomly selected diodes and LEDs.

The forward voltage across a solid-state diode is more-or-less constant over a wide current range. A signal diode like the 1N4148 will provide a bias of around 0.7V, while LEDs can be used to obtain higher voltages, and there is nothing to stop us stacking several diodes in series. Zener diodes could be used too, though they are often avoided on account of their greater excess noise, multi-state noise, and higher dynamic resistance.

The advantage of diode bias is that because the diode naturally maintains a near-constant forward voltage, whatever current happens to be flowing in the valve, no signal voltage can develop across it, so there is no degeneration and no need for a cathode bypass capacitor. The valve can therefore achieve maximum gain and a flat frequency response all the way down to DC. Of course, LEDs also have the aesthetic appeal of lighting up and indicating the valve is actually working. The disadvantage is that we lose the auto-stabilising property of ordinary cathode bias, but this is often a painless sacrifice. However, it is important to note that the noise current generated in the diode will be of similar order to that generated in the valve, and will therefore worsen the signal-to-noise ratio to some degree. In very low-noise applications the diode must therefore be bypassed with a capacitor after all (100µF will do), even though this has no effect on the gain or phase of the circuit.

There is nothing new about diode biasing,[4,5] but it has become something of a hallmark of modern design thanks to the wide range of diodes and LEDs now available. Fig. 6.42 shows the I/V characteristics of some devices measured by the author. The astute reader will of course spot that the forward voltage is not quite as constant as might be hoped, especially below 5mA where we often operate preamp

---

[4] Telefonaktiebolaget L. M. Ericcsson, (1942). Improvements in or Relating to Thermionic Valve Circuits, *GB549484*.
[5] Walker, A. H. B. (1952). The Application of the Metal Rectifier in Television Receivers. *Proceedings of the IEE*, **99**(19), pp560-71.

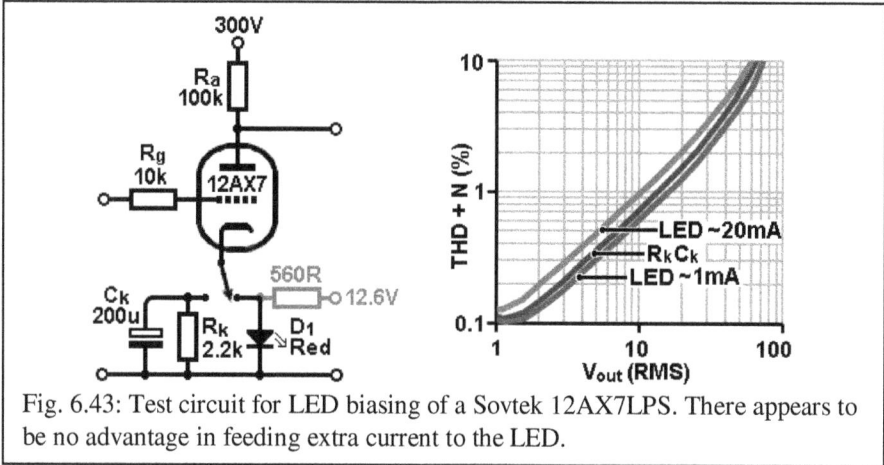

Fig. 6.43: Test circuit for LED biasing of a Sovtek 12AX7LPS. There appears to be no advantage in feeding extra current to the LED.

valves. In other words, even solid-state diodes have some internal dynamic resistance, albeit small, as indicated by the gradient of the curves. To a rough approximation the dynamic resistance of a silicon diode or LED is inversely proportional to the current through it, so doubling the current will roughly halve its resistance. This is not a problem if the valve is CCS loaded since the current through both the valve and LED will be constant, but with a resistor load the LED resistance will contribute some degeneration and, worse still, here the resistance is not constant and therefore neither is the degeneration. It therefore seems prudent to investigate how much effect this may or may not have on valve distortion before we get too carried away with diode biasing.

To test whether LED biasing might introduce extra distortion the author constructed the circuit shown in fig. 6.43. Initially the 12AX7LPS was biased with a single, cheap red LED, yielding a bias voltage of 1.47V at 0.86mA, and producing the lower distortion trace in the figure. Under these conditions the LED operates very low on its I/V curve where its dynamic resistance varies most significantly. Next, to reduce the LED's dynamic resistance, an extra 19mA of current was fed into it from the $12.6V_{dc}$ heater supply (shown faint in the figure). However, this also caused the bias to increase to 1.79V (i.e. colder), which in turn caused distortion to degrade. Since this change in bias makes the comparison an unfair one, the LED was finally substituted with a conventional bias resistor and bypass capacitor, specifically chosen to produce an intermediate bias voltage (1.61V). As shown in the figure, the distortion is likewise intermediate between the two previous traces and there were no conspicuous changes in the harmonic spectrum, which suggests that the LED's dynamic resistance does not have a significant effect on performance. This is not entirely surprising when you realise that, at ~1mA, the dynamic resistance of the LED is about 70Ω, which is an order of magnitude smaller than the internal cathode resistance of the 12AX7.

245

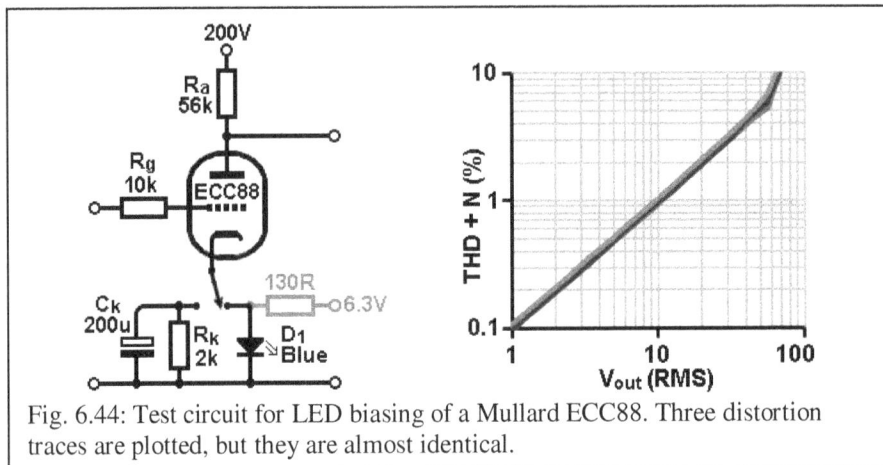

Fig. 6.44: Test circuit for LED biasing of a Mullard ECC88. Three distortion traces are plotted, but they are almost identical.

Therefore, in a further attempt to expose any degradation which could be blamed on the LED the author performed the same general test using an ECC88 –which has a much smaller internal cathode resistance– and a blue LED –which has a higher dynamic resistance. Fig. 6.44 shows the test circuit.

With the LED alone the bias was 3.44V at 1.77mA. At this level of current the LED has a dynamic resistance of about 100Ω while the internal cathode impedance of the valve was estimated to be about 280Ω –a much smaller ratio than with the 12AX7. With an extra 20mA of feed-current the bias increased to 3.9V at 1.6mA and the dynamic resistance fell to <20Ω. A 2kΩ cathode resistor produced 3.5V bias at 1.7mA. However, despite these deliberately contrived conditions the distortion trend was the same as in the previous example; the LED alone produced the best trace and the 'boosted' LED the worst. But as fig. 6.44 demonstrates, the differences between the three are extremely small, almost merging into one solid trace. Based on these data the author is satisfied that the dynamic resistance of a bias LED has no deleterious effect on distortion, and that extra 'boosting' current is not required, indeed, it is best avoided. As is so often the case, the valve produces more than enough distortion by itself to swamp any secondary phenomena.

246

Chapter 7: The Cathode Follower

Almost the whole of analog audio design could be said to revolve around three basic ingredients: amplifiers, attenuators, and buffers. Amplifiers are of course concerned with making signals bigger; attenuators (which include filters) make them smaller. Buffers come somewhere in between, often literally, because they isolate one circuit section from another. A buffer is normally intended to have unity gain and act only as an impartial middle-man between two otherwise incompatible circuits. Since the output of the buffer is notionally an identical copy of its input, i.e. the output 'follows' the input, buffer circuits are generally called 'followers'. In the world of valves the quintessential buffer is the *cathode follower*, while in the world of transistors there is the BJT *emitter follower* and the FET *source follower*, but all of them share the same basic characteristics. Naturally, in this book we will concentrate attention on the valve kind, but the fundamental principles and concepts are exactly the same for all.

Credit for the first cathode follower appears to belong to one Anthony Winther for a radio-circuit patent applied for in 1925[1]. However, circuit theory was in its infancy at the time, and a more explicit understanding of the cathode follower as we know it today did not appear until 1936 when it was patented by the inimitable Alan Blumlein[2] (applied for in 1934). The full name 'cathode follower' first appears in subsequent patents obtained simultaneously by Blumlein and Eric White, who worked together.[3,4]

7.1: Design Equations

Fig. 7.1 shows a cathode follower in essence, without any explicit biasing scheme or coupling capacitors to cloud matters. The input is applied between grid and ground, and the output is taken between cathode and ground. A positive input signal will cause the valve to increase its conduction so more current flows down through the load resistor R_k. The voltage across R_k therefore rises in sympathy with the grid, so the circuit is non-inverting. But the rising output voltage counteracts the rising input voltage, i.e. we have cathode-current degenerative feedback. If fact, this is a special case where *all* of the output voltage is now working to cancel

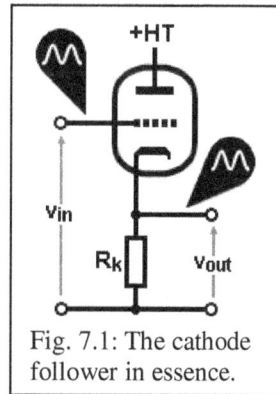

Fig. 7.1: The cathode follower in essence.

[1] Winther, A. (1929) Radio Frequency Amplification Circuits. US Patent 1700393.
[2] Blumlein, A. D. (1936). Improvements in and Relating to Thermionic Valve Circuits. British Patent 448421.
[3] Blumlein, A. D. (1936). Improvements in or Relating to Electric Circuits for Reducing the Effective Shunt Capacity Introduced by Circuit Elements Such as, for Example, Electric Batteries. British Patent 462530.
[4] White, E. L. C. (1937). Improvements in and Relating to Thermionic Valve Apparatus. British Patent 462536.

out most of the applied input voltage, so only a tiny portion actually appears between grid and cathode to be amplified. In other words, the circuit has 100% negative feedback, trading all of the gain we would normally get from the valve in exchange for various improvements in performance, such as lower distortion, wider bandwidth, higher input impedance, and lower output impedance. These are all the characteristics we want from a good buffer.

With an ordinary gain stage it is easy to appreciate the circuit using load lines, but this can be less immediately intuitive with the cathode follower. We will therefore develop some general design equations first, and look at load lines later. It is worth going through the derivations in full, not just as an exercise in algebra but because it will lead to an expert understanding of why cathode followers have the properties they do. Nevertheless, we will keep the treatment as simple as possible.

7.1.1: Gain

Fig. 7.2a shows a simplified cathode follower ready for analysis; biasing has been ignored since we're only interested in small AC signals at this juncture. A smoothing capacitor C_s is also shown to remind the reader how the HT is effectively shorted to ground as far as AC is concerned. In fig. 7.2b the valve has been replaced with its Norton equivalent, while fig. 7.2c is simply a tidying up (note that the VCCS has been 'folded over' relative to the previous diagram). This makes it obvious that the valve's internal anode resistance r_a, and the cathode load resistor R_k, are directly in parallel as far as AC is concerned. Any external load will also appear in parallel, so we can make the maths simpler by treating the whole lot as one combined load impedance, Z, as shown in the final diagram of fig. 7.2d. Finding the gain of the circuit is now very easy.

From Ohm's law the output voltage is the current generated by the valve, $g_m v_{gk}$, multiplied by the impedance Z:

$$v_{out} = g_m v_{gk} Z$$

It is also clear from the figure that $v_{in} = v_{out} + v_{gk}$ which rearranges to:

$$v_{gk} = v_{in} - v_{out}$$

Notice that what the valve actually amplifies (v_{gk}) is the difference between the

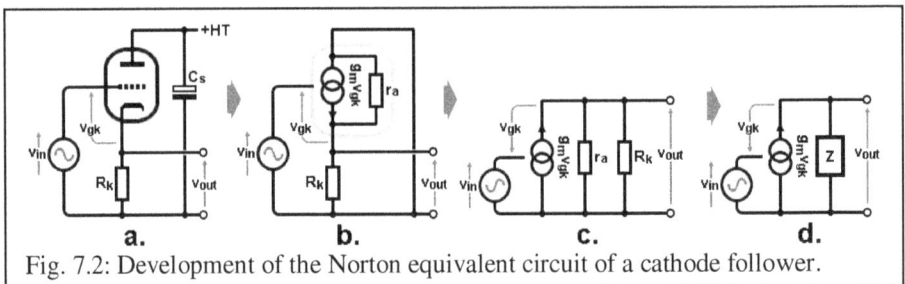

Fig. 7.2: Development of the Norton equivalent circuit of a cathode follower.

output and input, i.e. the error voltage; this is classic feedback theory. Substituting this into the previous equation gives:

$$v_{out} = g_m(v_{in} - v_{out})Z = g_m v_{in}Z - g_m v_{out}Z$$

The above equation can immediately be solved to find the gain, v_{out}/v_{in}:

$$A = \frac{g_m Z}{1 + g_m Z} \tag{7.1}$$

Remember, Z is actually the parallel combination of r_a, R_k, and any external load.

Since $g_m = \mu/r_a$ the gain can also be written as:[*]

$$A = \frac{\mu Z}{r_a + \mu Z} \tag{7.2}$$

If $\mu Z \gg r_a$ (which is normally the case) then this simplifies to a handy approximation:

$$A \approx 1 \tag{7.3}$$

Whichever equation we examine, we find that the circuit strives to produce a close copy of the input signal, and the higher the μ, g_m, or Z, the closer it gets. A practical cathode follower might achieve a gain of about 0.9 (−0.9dB) so there is always a small loss, but the other properties of the circuit make this well worth putting up with.

7.1.2: Input Impedance

One of the essential properties of a buffer is that it must not load down the preceding stage, i.e. it must have a very large input impedance. Fig. 7.3a shows a cathode follower with some extra components connected between grid and cathode: the interelectrode capacitance C_{gk}, and a resistor representing grid-leak resistance. This circuit has then been transformed into the Norton equivalent in fig. 7.3b and, to simplify the maths, the various components have been lumped together into two generic impedances Z_1 and Z_2.

The input impedance is the input voltage divided by the input current. From the figure it is clear that $i_{in} + g_m v_{gk} = i_{out}$, and from Ohm's law we can see that $i_{out} = v_k/Z_2$ and $v_{gk} = i_{in}Z_1$. Substituting these in gives:

$i_{in} + g_m i_{in}Z_1 = v_k/Z_2$

---

[*] The same result can be obtained using the universal feedback equation by setting the feedback fraction to B = 1 and the open-loop gain to $A_o = \mu Z/(r_a+Z)$.

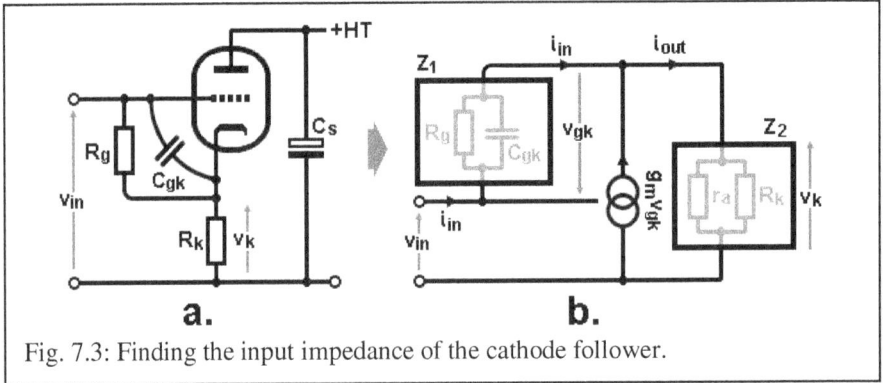

Fig. 7.3: Finding the input impedance of the cathode follower.

We need to eliminate v_k, and looking at the figure we see that:

$$v_k = v_{in} - v_{gk} = v_{in} - i_{in}Z_1 .$$

Substituting this in produces:
$$i_{in} + g_m i_{in}Z_1 = (v_{in} - i_{in}Z_1)/Z_2$$

The above expression can now be solved for v_{in}/i_{in}, which is the input impedance:

$$Z_{in} = v_{in} / i_{in} = g_m Z_1 Z_2 + Z_1 + Z_2 \qquad (7.4)$$

If this looks strangely familiar it is because this equation appeared in section 3.10 where it was used to explain why a valve can oscillate.

This equation can be used to find the input impedance with any combination of components in place of Z_1 and Z_2. At low audio frequencies we can ignore the interelectrode capacitances and concentrate on the input resistance only. We therefore replace Z_1 with the grid-leak resistor R_g, and Z_2 with the parallel combination of r_a and R_k (and mentally include any external load too):

$$R_{in} = g_m R_g (r_a \| R_k) + R_g + (r_a \| R_k) = R_g [g_m (r_a \| R_k) + 1] + (r_a \| R_k)$$

But astute readers will spot that $g_m(r_a \| R_k)$ is the open-loop gain A_o, that is, the gain of the circuit if it were reconfigured as an ordinary gain stage. This can also be written as $1/(1-A)$ where A is the cathode follower gain, so the input resistance can be boiled down to:

$$R_{in} = R_g / (1 - A) + (r_a \| R_k) + R_g \qquad (7.5)$$

The important result is that the input resistance is not merely equal to the grid-leak as it is with an ordinary gain stage. R_g in fact appears multiplied by A_o+1 or, in other words, it is *bootstrapped*. Physically this happens because the valve produces a close copy of the input signal at its cathode, that is, at the bottom of the grid-leak resistor. Hence, with almost the same signal voltage at both ends of the resistor there can be very little signal current in it, so its value appears artificially magnified as far as the source is concerned. In a typical cathode-follower a grid-leak of a few hundred kilohms can behave like tens of megohms. But before we become drunk with power we ought to look at the input capacitance too.

Let us shift our point of view to higher frequencies where the interelectrode capacitances might not be insignificant anymore. Returning to fig. 7.3a we will suppose the grid-leak is either too large to be significant at these frequencies or it is not used at all (e.g. the valve is DC coupled). We will also ignore any anode-cathode capacitance because $r_a \| R_k$ is in parallel with it and is assumed to dominate the total impedance. Entering the impedances into equation (7.4) gives:

$$Z_{in} = g_m \frac{1}{j\omega C_{gk}}(r_a \| R_k) + \frac{1}{j\omega C_{gk}} + (r_a \| R_k) = \frac{1}{j\omega C_{gk}}\left[g_m(r_a \| R_k) + 1\right] + (r_a \| R_k)$$

The above shows the same general result as previously: the impedance connected between grid and cathode is bootstrapped by a factor A_o+1 or $1/(1-A)$. In this case it is the reactance of C_{gk} which appears multiplied, which is the same as saying the capacitance appears *smaller*, so it is a sort of inverse Miller effect. On the other hand, the anode is effectively grounded, so grid-anode capacitance C_{ga} enjoys no special multiplication, it simply loads the signal source. The total input capacitance is therefore:

$$C_{in} = C_{ga} + C_{gk}(1 - A) \tag{7.6}$$

This amounts to just a few picofarads usually, so the cathode follower lives up to the expectation of high input impedance even at high frequencies.

7.1.3: Output Impedance

To find the output impedance we must ground the input, apply a fictitious voltage to the output, and work out how much current then flows into the output. Fig. 7.4 shows the transformation from the basic circuit into its Norton equivalent. A voltage v_1 is applied to the output and a current i_1 flows into it, and it is clear from the diagram that i_1 splits into three components:

$$i_1 = i_2 + i_3 + (-g_m v_k)$$

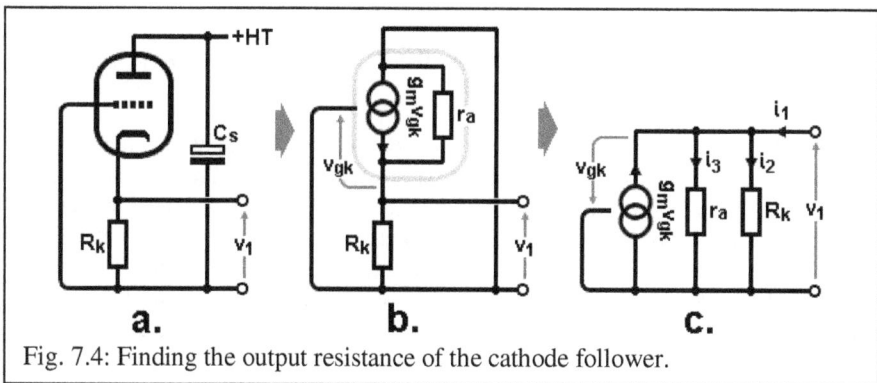

Fig. 7.4: Finding the output resistance of the cathode follower.

The last term is negative because it is pointing in the opposite direction.

Using Ohm's law it is easy to see that $i_2 = v_1/R_k$ and $i_3 = v_1/r_a$. From the voltage arrows we also see that $v_1 = -v_{gk}$, so whatever voltage we apply to the circuit, the current generator takes the negative of it and multiplies it by g_m, i.e. it generates a current of $g_m \times (-v_1)$. The negative sign in this term cancels out the negative sign in the previous equation. Substituting in:

$$i_1 = \frac{v_1}{R_k} + \frac{v_1}{r_a} + g_m v_1$$

Solving the above for v_1/i_1 (which is the impedance seen looking into the output) by dividing both sides by v_1 and inverting, we finally obtain:

$$R_o = \frac{1}{\dfrac{1}{R_k} + \dfrac{1}{r_a} + g_m} \qquad (7.7)$$

Notice that this is the formula for three resistances in parallel, except one of them is actually a conductance, g_m. Indeed, the parallel combination of r_a and g_m has a special significance: it is the output impedance of the cathode itself. When looking into the anode we see the internal anode resistance r_a, but looking into the cathode we see the internal **cathode resistance**, usually given the symbol r_k. This happens to be equal to $r_a/(\mu+1)$. The output impedance of the whole cathode follower circuit is therefore the parallel combination of r_k and R_k, which is exactly what equation (7.7) is telling us. In most cases R_k and r_a are fairly large (kilohms) so they actually have little effect on the total, so the above simplifies to the handy approximation:

$$R_o \approx \frac{1}{g_m} \qquad (7.8)$$

Even a mundane valve with a transconductance of 2mA/V will therefore achieve an output resistance of <500Ω, which for a valve circuit is commendably low and exactly the sort of thing we want from a buffer. Cathode followers are therefore the essential driving circuit for difficult loads such as long cables, where the cable capacitance would otherwise cause loss of high frequencies with a higher driving resistance.

However, when textbooks boast about the low output impedance of the cathode follower it is easy to be duped into thinking it is a kind of magic bullet; that even a puny valve will happily drive punishingly low load impedances like headphones or loudspeakers. This sort of thinking will lead to great disappointment in practice. It is essential to appreciate the distinction between output impedance and drive capability. Overestimation of the cathode follower's talents (i.e. bad circuit design) is surely the main cause of its undeserved poor reputation among some audiophiles.

252

The drive capability –i.e. maximum output power, clipping and slew-rate limits– of the cathode follower are *identical* to those of an ordinary gain stage built using the same components. The output resistance might be <500Ω and therefore capable of achieving a >100kHz bandwidth into a 3nF load capacitance, but it will still run into slew-rate limiting at the same amplitude-frequency combination as it would when arranged as a regular gain stage driving the same 3nF. Remember, slew rate depends only on the peak current that the valve can swing and has nothing to do with bandwidth. Similarly, an output resistance of 500Ω does not mean the valve will deliver hundreds of volts of signal into a load of a few hundred ohms. It will deliver only the same amount of swing that a regular gain stage would when driving 500Ω. The difference is that before the maximum output limit is reached, distortion in the cathode follower will be much lower and the bandwidth much wider than for the ordinary gain stage. But once the clipping or slewing limit is reached, the negative feedback loop collapses and distortion increases sharply back towards what it would be if the circuit was configured as an ordinary gain stage. So while a cathode follower *can* deliver a signal into a load with less distortion than an equivalent gain stage, its maximum output drive limit is no different (unless perhaps your definition of this limit is a particular distortion figure).

In fact, one of the best ways to design a good cathode follower circuit is to start by designing an ordinary gain stage, which we already understand intimately. We ignore its gain, however, because we automatically know it will be close to unity when we eventually convert into a cathode follower. Choose the valve, draw a load line or otherwise select all the components so the circuit could deliver the required *output* voltage into the intended load without clipping or slew limiting (but perhaps only with 'medium fidelity'); then simply turn the circuit on its head to convert it to a cathode follower. The bias, clipping level, and slew rate, all remain the same, but the bandwidth and distortion will be improved by the feedback factor; medium fidelity is converted into high fidelity, as it were. The gain will be close to unity, and PSRR and immunity to external hum will also improve. This is an example of designing a circuit to be as good as possible before feedback is added, so that adding feedback only makes it better.

7.1.4: PSRR

To find the PSRR of the cathode follower we will skip the full analysis and cheat. Remember from section 3.8.1 that when looking into the anode of an unbypassed gain stage we see the internal anode resistance r_a, plus the cathode resistor R_k, plus a sort of 'phantom'

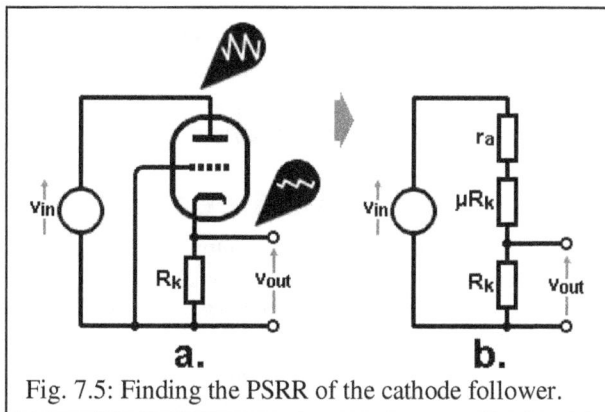

Fig. 7.5: Finding the PSRR of the cathode follower.

253

version of R_k which is multiplied by μ. Any ripple on the power supply therefore encounters the potential divider shown in fig. 7.5b, so using the general equation for a potential divider:

$$\frac{V_{out}}{V_{in}} = \frac{R_k}{r_a + R_k + \mu R_k}$$

But PSRR is the inverse of this, so we must flip the previous equation over:

$$PSRR = \frac{r_a + R_k(\mu+1)}{R_k} \tag{7.9}$$

From the above it is clear that the cathode follower has relatively good PSRR, at least when compared to a simple gain stage. Physically this improvement occurs because the grid effectively shields the cathode from voltage changes at the anode. Furthermore, if $\mu R_k \gg r_a$ then the above can be simplified to a very memorable formula:

$$PSRR \approx \mu \tag{7.10}$$

7.2: Noise

It is worth pointing out something about the noise performance of the cathode follower. Shot and flicker noise are generated as currents directly inside the valve and are unchanged whether it is arranged as a gain stage, cathode follower, or whatever. These currents flow into the load, generating an output noise voltage which is subject to the same degeneration as wanted signals. In other words, the shot and flicker equivalent input noise voltages of the cathode follower are exactly the same as for a gain stage built with the same components. For a triode this is typically in the region of $1\mu V_{rms}$ for unweighted audio-band noise. However, since the signal voltage gain, A, is always less than unity, the audio signal will be slightly attenuated as it passes through the circuit, so the resulting SNR will be 1/A times worse than if the same valve were used as a gain stage. A cathode follower is therefore not the best choice for the input stage of a very low-noise circuit.

Any noise voltage created by grid-current flowing in the source impedance simply adds to the signal voltage quite ordinarily, so the contribution of grid current noise is exactly the same as for an ordinary gain stage. This is of minimal interest since grid current noise is usually negligible anyway.

7.3: Oscillation and Stoppers

For any practical cathode follower circuit the input impedance works out to be a capacitance in parallel with a resistance. The resistance is positive if the load is resistive or inductive, but it can become negative if the load is capacitive. Negative input resistance is a nightmare because it can lead to oscillation (note that this result is the opposite of an ordinary gain stage, which is stable when driving a capacitive load but may oscillate when driving an inductive one).

254

Since a cathode follower barely has a gain of unity it might be thought that it would be very stable and well behaved. However, looking at it from another point of view, it is a stage with a lot of negative feedback (the maximum possible!) and whenever we use feedback, the stability margin suffers. It seems likely that undiagnosed oscillation in the cathode follower is at least partly responsible for the bad press it has received from some quarters. Certainly it will wreak revenge on those who are lackadaisical with their designs, just like any feedback amplifier, but it is easy enough to tame when you understand the root of the problem.

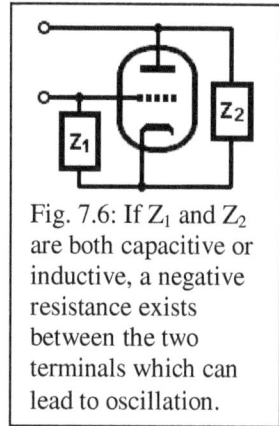

Fig. 7.6: If Z_1 and Z_2 are both capacitive or inductive, a negative resistance exists between the two terminals which can lead to oscillation.

From equation (7.4) the impedance between anode and grid of a valve, as illustrated in fig. 7.6, is:

$$Z_{in} = g_m Z_1 Z_2 + Z_1 + Z_2$$

With the ordinary gain stage an awkward situation arose when Z_1 and Z_2 were both inductive as this created a negative resistance and an inductance, which would oscillate when combined with the unavoidable anode-to-grid capacitance. However, we didn't have to worry about the opposite situation –Z_1 and Z_2 both capacitive– because it is very unlikely that we might accidentally build a circuit with some hidden *inductance* between anode and grid.

With the cathode follower it is the other way around. The anode is connected directly to the HT which, as far as AC is concerned, is the same as ground, so anything connected between grid and ground is effectively connected between grid and anode. This includes grid lead inductance, which in the megahertz range can be effectively shorted to ground by stray capacitance. If Z_1 and Z_2 are both capacitive then we have our potential oscillator, and it so happens that these capacitances are already built into the valve in the form of C_{gk} and C_{ak}, with the added inconvenience that cathode-heater capacitance and plain old load capacitance are also effectively in parallel with C_{ak}. However, it is very unlikely that we will get any grid inductance shorted directly to *cathode*, so we don't have to worry about the opposite condition, when Z_1 and Z_2 are both inductive. This leads to a general engineering maxim: an ordinary gain stage is more likely to oscillate when

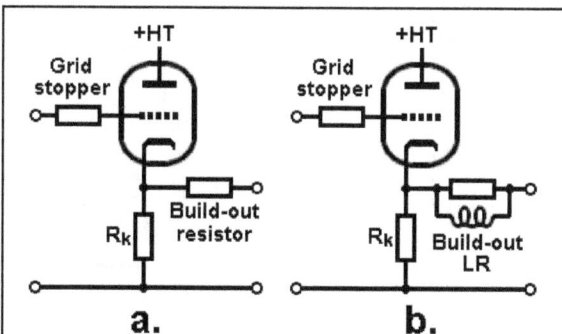

Fig. 7.7: Precautions against oscillation include a grid stopper, build-out resistor, or build-out LR combination.

driving an *inductive* load, whereas a cathode follower is more likely to oscillate when driving a *capacitive* load.

The solutions for stabilising a cathode follower are the same as for an ordinary gain stage: add a grid stopper to cancel out any negative input resistance, and add a build-out resistor to isolate the stage from load capacitance (though this doesn't help with C_{ak} and C_{hk}). The grid stopper should again be around $2/g_m$ or larger, and should be mounted very close to the valve base for maximum effect. The build-out resistor is normally around 100Ω to $1k\Omega$, and in the case of a self-biased cathode follower the bias resistor itself can serve as a build-out resistor if it is left unbypassed (section 3.10). A build-out resistor will of course add to the output resistance of the stage, but usually this is of no consequence because we only need the output resistance to be low, not vanishingly low. However, in cases where it *is* essential to achieve the lowest possible output resistance –such as output-transformerless amplifiers– the standard solution is to add an inductor in parallel with the resistor, as shown in fig. 7.7. A value of a few tens of microhenries is typical, which effectively shorts the resistor at audio frequencies but still provides stabilisation at very high frequencies where the inductive reactance is significant. Such a circuit must be thoroughly checked for any sign of oscillation into the intended load, however.

7.4: Anode Characteristics

When considering load lines for the cathode-follower we cannot read values such as gain directly from the usual anode characteristics, because all its information is given relative to the cathode. Since the cathode is now following the grid, anything measured relative to it will vary similarly, so trying to pin down certain information could result in a major headache.

The anode voltage, however, is fixed at the HT voltage, so it would make more sense to measure all voltages relative to the anode. We could then draw a new set of grid curves for V_{ga}, rather than V_{gk}.[5] In actual practice it is not necessary to draw these curves when designing a cathode follower, but they are presented here to give a better understanding of the behaviour of the circuit. It is easier to demonstrate the process on a low-µ valve so we will use the ECC82 / 12AU7 as an example.

Suppose we choose a point on the original 0V grid curve where the anode-to-cathode voltage is 50V. Then suppose we rearrange the valve as a cathode follower, and set it up so V_{gk} and V_{ak} are still 0V and 50V respectively; fig. 7.8 shows the voltage relationships. If we reverse the probes on our voltmeter then the *grid-to-*

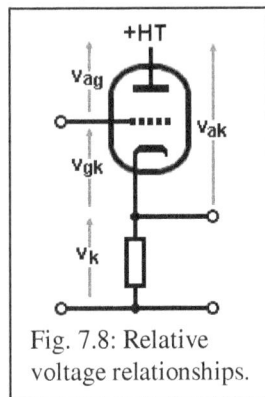

Fig. 7.8: Relative voltage relationships.

[5] 'Cathode Ray' (1955). Cathode Followers – with Particular Reference to Grid Bias Arrangements. *Wireless World*, June. pp292–296.

anode voltage V_{ga} must be −50V. Now suppose we make the grid 2V negative of the cathode. V_{ga} would then be −52V, so in order to keep it at −50V (the curve we are trying to draw) we must reduce V_{ak} by 2V to 48V. Thus if we make the grid more negative we must always reduce V_{ak} by the same amount in order to keep V_{ga} at a constant value. Formalising this process we will call V_{ak} for the cathode follower $V_{ak(cf)}$, to distinguish it from V_{ak} on the original graph.

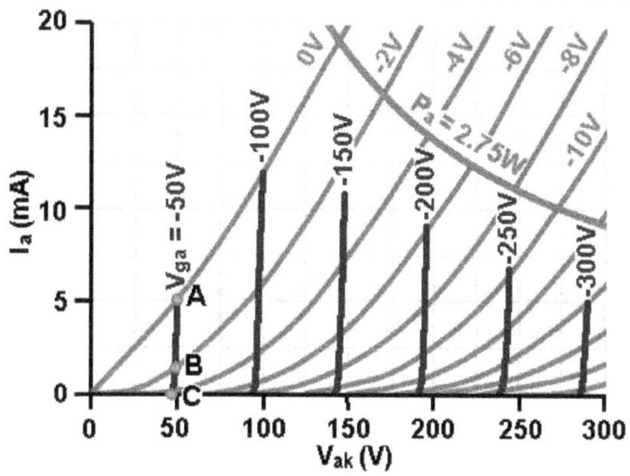

Fig. 7.9: Anode characteristics for cathode follower operation, drawn on the ordinary ECC82 / 12AU7 anode characteristics.

On the original curves, if V_{ak} = 50V and V_{gk} = 0V, then:
$V_{ak(cf)}$ = 50 − 0 = 50V
This point is then plotted at V_{ak} = 50V, V_{gk} = 0V, labelled A in fig. 7.9.

If V_{ak} = 50V and V_{gk} = −2V, then:
$V_{ak(cf)}$ = 50 − 2 = 48V
This point is plotted at V_{ak} = 48V, V_{gk} = −2V, labelled B.

If V_{ak} = 50V and V_{gk} = −4V, then:
$V_{ak(cf)}$ = 50 − 4 = 46V
This point is plotted at V_{ak} = 46V, V_{gk} = −4V, labelled C.

Joining these points we obtain a new V_{ga} = −50V grid curve for the cathode follower. Other curves can be plotted in exactly the same fashion. The most obvious result is that the new curves are very straight and evenly spaced, even down at the very bottom of the graph, implying very low harmonic distortion. Another point to note is that the slope of these curves represents the cathode impedance r_k, just as the slope of the ordinary grid curves represents the anode resistance r_a. From their steepness it is obvious that r_k is very small.

7.4.1: Using the New Anode Characteristics

Load lines can be drawn on the new anode characteristics in the usual way. As an example, suppose the cathode follower is to be used as a simple line output driver that may be called upon to drive up to $10V_{rms}$ into a load impedance of 10kΩ‖500pF, which might represent the input of a solid-state amplifier plus a couple of metres of cable capacitance. With such a heavy capacitive load we need to consider bandwidth *and* slew-rate requirements. To ensure >100kHz bandwidth the output resistance must not be more than:

$$R = \frac{1}{2\pi fC} = \frac{1}{2\pi \times 100000 \times 500 \times 10^{-12}} = 3183\Omega$$

This will be quite easy to achieve, and a lower value is further desirable as it improves immunity to interference being picked up on the cable.

The required peak voltage is $\sqrt{2}\times10V = 14.1V_{pk}$, so using equation (3.15) the peak output slewing current at 20kHz would be:

$$I_{pk} = 2\pi fCV_{pk}$$

$$I_{pk} = 2\pi \times 20000 \times 500 \times 10^{-12} \times 14.1 = 0.88 \times 10^{-3} A = 0.9mA$$

The quiescent current in the valve must therefore be at least equal to this, and more would be better. We will suppose that for some unspecified reason the HT is limited to 250V, which is perhaps a little low for an ECC82, but still useable.

Fig. 7.10 shows a tentative DC load line for an 18kΩ load resistor. A warm bias point at $I_a = 6mA$ has been selected, but this ultimately appears fairly central once the AC load line is drawn through it. The estimated transconductance at the bias point is 1.9mA/V, so the output resistance should be less than $1/0.0019 = 526\Omega$. These component values and bias point were deliberately chosen for illustrative purposes –so the load lines

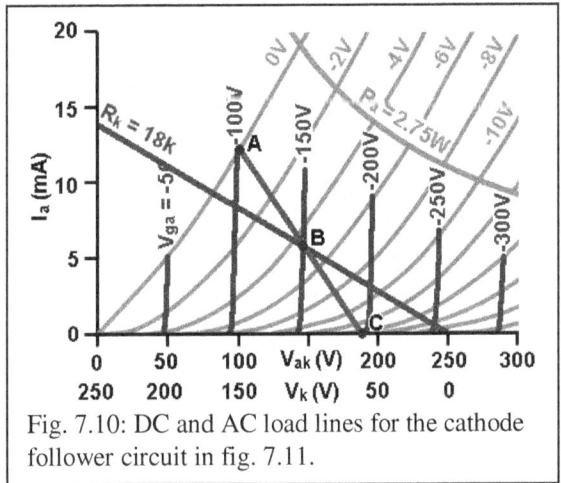

Fig. 7.10: DC and AC load lines for the cathode follower circuit in fig. 7.11.

neatly intersect with the new grid curves– but they would be very reasonable choices in reality too. Fig. 7.10 also demonstrates visually that in order to drive a 10kΩ load we must start with a valve which is capable of driving 10kΩ, regardless of the fact that it is in cathode-follower form.

The horizontal axis normally shows the anode-to-cathode voltage, but we are now also interested in the cathode-to-*ground* voltage, V_k, so a second voltage scale can be added. If the full HT appeared across the valve (anode to cathode) then the cathode would be at zero volts, whereas if the valve were a short circuit the cathode voltage would be equal to the HT. The new voltage scale for V_k must therefore run from the HT down to 0V, as shown.

With the load lines drawn on the new anode characteristics graph it is easy to appreciate how the cathode follower works. Remembering to read the correct voltage scale, the AC load line indicates that the cathode can swing at most from about 150V (point A) down to 62V at cut-off (point C), which is a total of $88V_{pp}$. This more than accommodates the $10V_{rms}$ ($28.2V_{pp}$) requirement. Looking at the original grid curves, shown faint, this would correspond to an input swing of about 0V to −14V, implying an *open*-loop gain of about 88/14 = 6.3 (although for small signals it is closer to 10).

Looking at the new grid curves in bold, the input voltage swings from $V_{ga} = -100V$ to about $V_{ga} = -200V$, a total of $100V_{pp}$, which implies a *closed*-loop voltage gain of 88 / 100 = 0.88. Notice how the input voltage can swing over a huge range without overdriving the valve; only if the grid is driven above 150V does the cathode fail to 'catch up' and we enter grid-current territory.

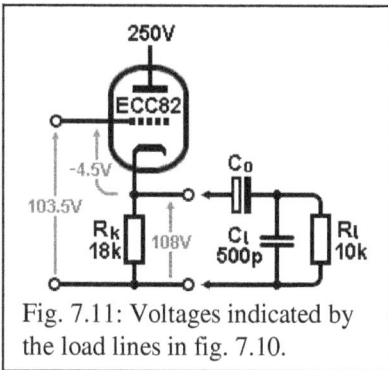

Fig. 7.11: Voltages indicated by the load lines in fig. 7.10.

From the original grid curves it is clear that the bias point corresponds to a grid-to-cathode voltage of about −4.5V. However, by reading off the new horizontal axis we see that this also corresponds to a cathode voltage of 108V. The only way to reconcile these two figures is to bias the grid voltage up to 103.5V, so it is 4.5V below the cathode. Fig. 7.11 shows these voltage relationships. The ECC82 / 12AU7 datasheet quotes a $V_{hk(max)}$ rating of 180V, so in theory no heater elevation will be required here (although this is a suspiciously bold claim so we might apply some elevation for good measure anyway, say 50V). All that remains is to apply the grid bias voltage, and there are several ways to do this.

7.5: Cathode Bias

Since the grid needs to be at a slightly lower potential than the cathode, one way to bias it is to 'tap off' the grid voltage from a point on the cathode resistor, as shown in fig. 7.12a. An input coupling capacitor C_1 is needed to block the grid voltage from upsetting any preceding circuit. A pot is shown for illustrative purposes but in practice a bias resistor, R_b, would be used, as shown in b. In fact, this is exactly the same sort of self-biasing used with an ordinary gain stage except everything is now stacked on top of the load resistor, rather than being underneath it. R_b is therefore found in exactly the same way as for a normal gain stage, e.g. by drawing a cathode load line or simply by calculating its value from the bias point. R_g is the grid-leak resistor for which the usual design considerations apply; 1MΩ is a common choice.

Sometimes a bypass capacitor C_b is added in parallel with R_b to maximise the bootstrapping of R_g (C_b can be a low-voltage part). However, for modern hi-fi this is deprecated because extreme values of input resistance are seldom required, and nobody likes

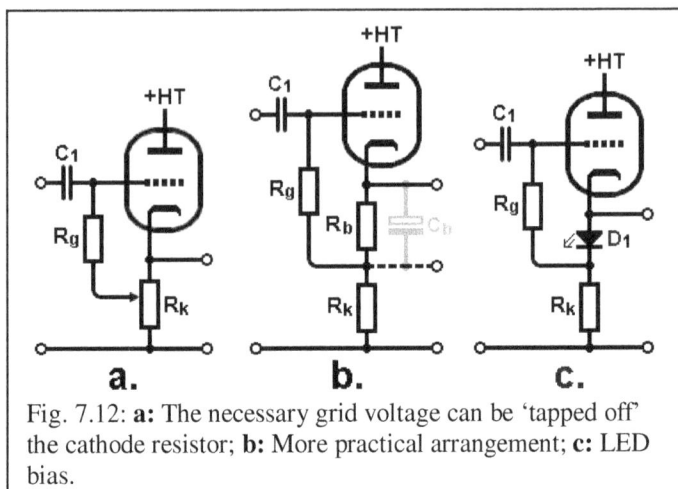

Fig. 7.12: **a:** The necessary grid voltage can be 'tapped off' the cathode resistor; **b:** More practical arrangement; **c:** LED bias.

electrolytic capacitors much. Also, by omitting the bypass capacitor, R_b can do double duty as a build-out resistor if the output is taken from the junction of R_b and R_k as shown dashed in the figure. The alternative to a bypass capacitor is to replace R_b with an LED, as in fig. 7.12c.

Referring to fig. 7.10 the bias was found to be around $-4.5V$ and the quiescent anode current was 6mA. If a bias resistor is used –rather than diode bias– then it would need to be $4.5V / 6mA = 750\Omega$. This means the *total* cathode load resistance would be increased by the addition of this resistor, which will alter the load line somewhat, so we might choose to reduce R_k to $17.25k\Omega$ so the total resistance is restored to $18k\Omega$, but in practice the difference is negligible.

Fig. 7.13 shows the example circuit with component values, including an obligatory grid stopper. The circuit was built using a Sovereign 12AU7WA and the measured voltages are indicated in the figure. The anode current was 6.4mA and the bias was $-4.8V$, very close to the expected values, thanks to the self-adjusting nature of the circuit. However, something which often catches out beginners is trying to measure the grid voltage, shown in brackets in the figure. The grid is a very high-impedance node; attaching an ordinary voltmeter between the grid and ground will pull the grid voltage down and produce a reading which is misleadingly low (about 60V when the author tried it using a digital multi-meter). The bias can instead be checked by measuring the voltage directly across R_b.

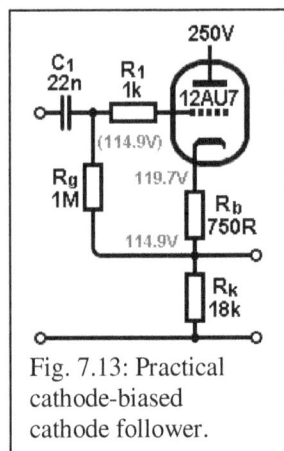

Fig. 7.13: Practical cathode-biased cathode follower.

260

Although cathode-biasing is commonly used in traditional circuits, it suffers from some non-obvious problems. Firstly, the input resistance, although potentially very large, is dependent on valve characteristics and loading, rather than being nicely constant under all conditions. Secondly and more subtly, R_g provides a positive feedback path from cathode to grid which subtracts from the otherwise 100% negative feedback existing inside the valve itself. The total amount of negative feedback is therefore not unity but depends on the source impedance of the previous stage; the greater the source impedance, the smaller the feedback fraction.[6] This means all the characteristics of the stage –in particular the output impedance– are dependent on the source impedance, which ideally must be small. This is highly unwelcome behaviour considering the whole point of a buffer is to isolate a *high* impedance source from the load, so they *don't* affect one another. The cathode-biased cathode follower should therefore be avoided; these problems can all be solved quite easily by using fixed-bias instead.

7.6: Fixed Bias

The simplest way to raise the quiescent grid voltage to the value required for biasing is to use a potential divider from the HT, as shown in fig. 7.14a. This eliminates any external feedback path between the cathode and grid, which was the problem with the cathode-biased circuit. The input and output impedances, bandwidth and distortion, will no longer vary with source impedance (except at very high frequencies where stray capacitance becomes significant, but this is not a concern within the audio band).

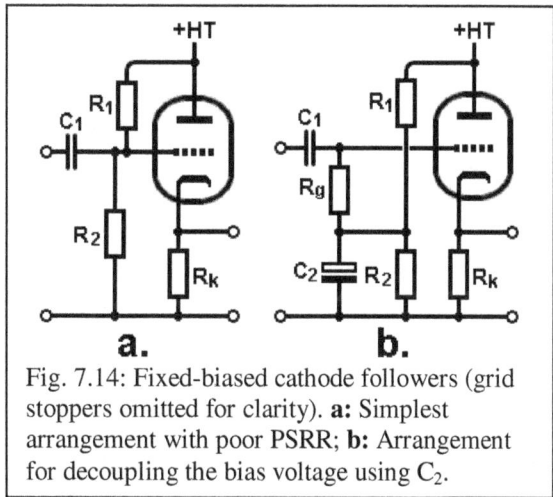

Fig. 7.14: Fixed-biased cathode followers (grid stoppers omitted for clarity). **a:** Simplest arrangement with poor PSRR; **b:** Arrangement for decoupling the bias voltage using C_2.

In most applications we will want to keep the input impedance large, so high-value resistors must be used in the divider (this will also minimise current drain from the power supply). But at first sight it might be thought that the maximum data-sheet value for grid-leak resistance would be a severe limitation in this regard, but that is not the case. A little-known advantage of the fixed-bias cathode follower is that the grid leak resistance can be much larger than normal, since cathode-current feedback

---

[6] 'Cathode Ray' (1955). Cathode Followers, *Wireless World*, June. pp292-6.
Short, G. W. (1961). The Bootstrap Follower, *Wireless World*, January. pp21-5.

stabilises the circuit against grid current.[7] To appreciate why, consider the cathode biased version from the previous section. If a grid current flowed it would develop a voltage i_gR_g almost directly between grid and cathode, which would therefore be amplified by the *open-loop* gain, A_o. In other words, the DC stability is exactly the same as for an ordinary gain stage, so the usual limitation on grid-leak resistance applies. With fixed-bias, however, any voltage i_gR_g developed across the grid leak is only amplified by the *closed-loop* gain, A. The total grid-leak resistance can therefore be A_o/A times larger than the normal datasheet value. For most circuits this means at least $10M\Omega$ can easily be tolerated.

Returning to the design example, a grid voltage of 103.5V was required (fig. 7.10). If R_1 is chosen arbitrarily to be $10M\Omega$ then R_2 must be:

$$R_1 = R_2 \frac{V_g}{HT - V_g} = 10M\Omega \times \frac{103.5}{250 - 103.5} = 7.06M\Omega$$

The nearest standard in the E24 range is $7.5M\Omega$ and yields a total input resistance of $R_1 \| R_2 = 4.3M\Omega$ with a grid voltage of 107V, which is close enough. Being off by 3.5V does not mean the bias voltage V_{gk} will be similarly incorrect, since the cathode naturally follows the grid and will compensate for the error.

A drawback of the simple circuit in fig. 7.14a is that any noise on the HT will be passed to the grid (albeit attenuated by the potential divider formed by R_1 and R_2 in parallel with the source impedance of the preceding stage) so the PSRR is degraded. This can be overcome with the arrangement shown in fig. 7.14b. Here the same potential divider provides the DC bias voltage, but it is heavily filtered by an additional capacitor C_2. This noise-free voltage is then applied to the valve via R_g, so this arrangement is sometimes ambiguously called **noiseless biasing**. To achieve good decoupling down 1Hz requires a capacitor larger than:

$$C = \frac{1}{2\pi R}$$

Where:
$R = R_1 \| R_2 \| R_g$

R_g appears in the expression above because we assume it is effectively shorted to ground by a small source impedance. The capacitor only has to be rated to withstand the bias voltage, plus some safety margin to allow for variations in the HT. In this case a 150V-rated capacitor would suffice. Fig. 7.15 shows the example circuit tested by the author, with measured voltages. The input resistance is equal to R_g which is $10M\Omega$, while R_1 and R_2 have been reduced in value to keep the total DC leak resistance within reasonable limits. A grid-stopper, build-out resistor, and arc-protection diode have also been added out of good engineering practice.

---

[7] Hemmingway, T. K. (1962). Bootstrap Follower Characteristics, *Wireless World*, July. pp322-4.

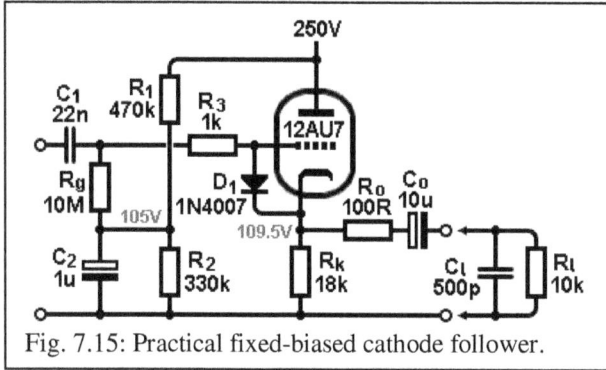

Fig. 7.15: Practical fixed-biased cathode follower.

Note that the output coupling capacitor must be relatively large if we expect to drive heavy loads. However, when driving cables it is still important to use a large coupling capacitor even if the load impedance at the far end of the cable is *not* heavy. For example, a 1μF

capacitor might seem like plenty if the load impedance is 100kΩ, but its reactance at 50Hz is over 3kΩ which may make the cable sensitive to picking up mains hum, despite the low output impedance of the cathode follower itself. A larger value is therefore preferable when driving signals into the outside world. Unfortunately, a larger value implies an electrolytic capacitor, and electrolytics leak DC current, resulting in a DC offset across any high-impedance load. If this is a concern then you may have to use a 1μF plastic capacitor after all. An alternative is to use two electrolytic capacitors is series with a pull-down resistor in between. This double-coupling will reliably eliminate any DC offset but may take up rather a lot of space (but see section 10.6.4 for a sneaky way around this problem). For the purposes of this chapter, however, DC offset is not a problem since the audio analyser has its own input blocking capacitor.

The measured small-signal gain of the circuit in fig. 7.15 was 0.91 with no external load, or 0.90 into a 10kΩ‖500pF load. Fig. 7.16 shows the distortion performance for both cases up to 10V$_{rms}$ output (roughly the limit of the signal generator within the AP1). Thanks to the high negative feedback factor inherent within the cathode follower, the difference in performance is quite small despite the enormous difference in loading conditions. In either case the total is about an order of magnitude better than when arranged as an ordinary gain stage, and visually it was

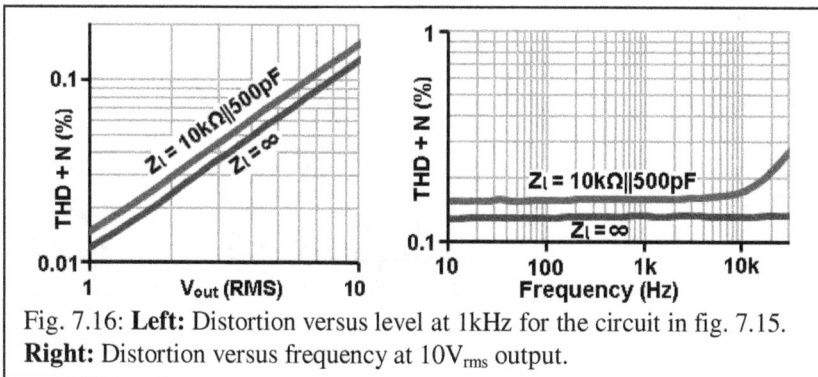

Fig. 7.16: **Left:** Distortion versus level at 1kHz for the circuit in fig. 7.15. **Right:** Distortion versus frequency at 10V$_{rms}$ output.

pure second harmonic. Only above 10kHz does the capacitive load begin to degrade performance. The output impedance was measured by loading the circuit with a 1kΩ resistor and noting the drop in output amplitude of a small signal. The drop at 1kHz was from 500mV$_{rms}$ to 285mV$_{rms}$ implying an output impedance of:

$$Z_o = 1000\Omega \times \left(\frac{500mV}{285mV} - 1\right) = 754\Omega \text{ including the}$$

build-out resistor; higher than predicted but still good enough for a line driver. The bandwidth was flat to just over 400kHz into 500pF. The PSRR (input shorted) was constant all the way up to 100kHz at 22.2 (26.9dB), which as expected is roughly equal to the valve's μ.

7.7: DC Coupling

The cathode follower is a prime candidate for direct coupling to the preceding stage, since its cathode naturally needs to be at a high voltage. This of course has the advantage of eliminating a coupling capacitor, thereby removing any low frequency attenuation, phase shift, and blocking distortion. Fig. 7.17 shows the basic arrangement, and it is such a neat and ubiquitous marriage that it is often regarded as a single compound-pair. Filling in the component values is fairly straightforward by examining the load lines of each stage more-or-less simultaneously. The important thing to remember is that the biasing of V$_1$ now affects the biasing of V$_2$, since the

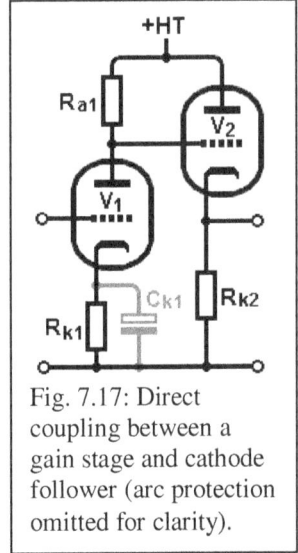

Fig. 7.17: Direct coupling between a gain stage and cathode follower (arc protection omitted for clarity).

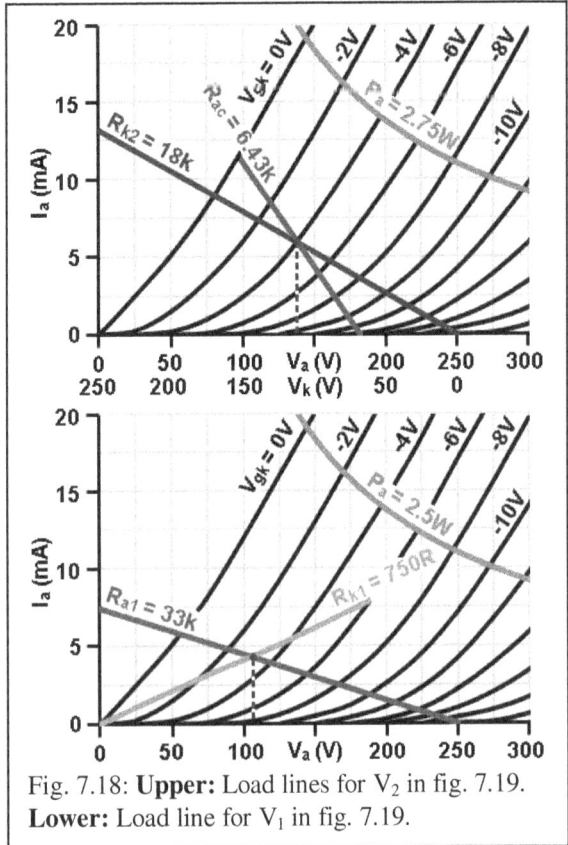

Fig. 7.18: **Upper:** Load lines for V$_2$ in fig. 7.19. **Lower:** Load line for V$_1$ in fig. 7.19.

anode voltage of the former is also the grid voltage of the latter.

Taking the previous example of an ECC82 cathode follower with an 18kΩ cathode resistor, fig. 7.18 reproduces the DC load line but dispenses with the cathode-follower grid curves as they are not actually needed. Again, an external load of 10kΩ is assumed, and a bias point of −4V has been selected, resulting in the AC load line shown in the figure. This indicates an anode-to-cathode voltage of 137V, which is a cathode-to-ground voltage of $250-137 = 113V$. The grid needs to be 4V below this, or 109V.

For the sake of simplicity the second triode in the same envelope will be used for V_1, although this is not mandatory. The anode voltage of V_1 decides the grid voltage of V_2 since they are directly connected, so we must aim for an anode voltage of 109V. This is relatively low, so V_1 will need a fairly large anode resistor if a reasonable bias point is going to be found. The lower image in fig. 7.18 shows that it could be achieved with a 33kΩ anode load and a 750Ω cathode bias resistor, giving a bias voltage of about −3V for V_1. If for some reason this was inappropriate for the rest of the amplifier design then further juggling of values would be necessary, indeed, it is inevitable when DC coupling is used. R_{k1} can of course be bypassed if desired. This is probably not the best place to use LED

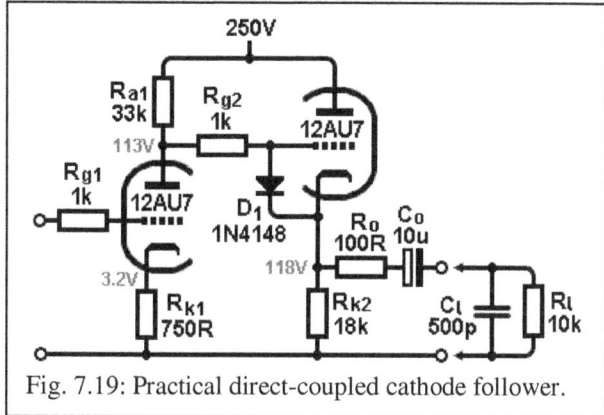

Fig. 7.19: Practical direct-coupled cathode follower.

biasing, however, since it would rigidly fix the bias voltage of V_1. As the valves age this would have the knock-on effect of biasing V_2 hotter over time, so the circuit would prematurely lose headroom. The self-adjustment provided by a cathode resistor is particularly desirable to stabilise DC-coupled stages.

Fig. 7.19 shows the completed circuit tested by the author, with the usual accoutrements of grid stoppers, build-out resistor, and arc-protection. The measured voltages were slightly higher than predicted, but not too bad. The gain of V_1 was 20dB and the total gain of

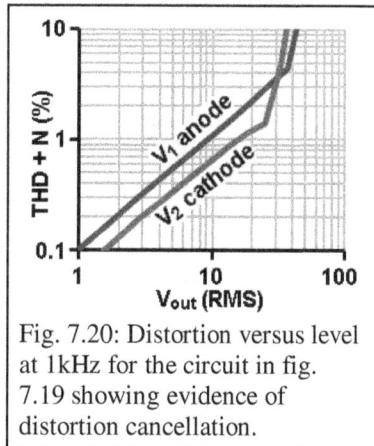

Fig. 7.20: Distortion versus level at 1kHz for the circuit in fig. 7.19 showing evidence of distortion cancellation.

the whole circuit was 19.4dB. Fig. 7.20 shows distortion versus level, as measured at the anode of V_1 and at the cathode of V_2.

7.8: Distortion Cancellation

It may come as a surprise to some readers to discover that the distortion performance measured at the output of fig. 7.19 is better than when measured at the anode of V_1, i.e. passing the signal through two imperfect stages is somehow better than one alone. This is an example of distortion cancellation. If a signal is passed through a non-linear transfer characteristic then its waveform will be altered and harmonics are introduced. But if the distorted waveform is inverted and then passed again through an identical transfer characteristic, it will be bent back into its original shape and emerge completely undistorted. In other words, if the harmonics generated in the second stage are of equal magnitude but opposite phase to those generated in the first stage, they will cancel one another out. It even works for intermodulation distortion too.

This effect occurs with a pair of cascaded triodes, although the cancellation is never perfect. In this case V_1 amplifies and inverts the signal and adds some unavoidable distortion in the usual way; the signal at the anode will appear slightly compressed on the up-going half cycle and expanded on the down-going half. V_2 is more linear than V_1, but it also has to handle the much larger amplified signal; it compresses on the down-going half and expands on the up-going half. This tends to bend the signal back into shape a little, i.e. it introduces some distortion cancellation, as proven by fig. 7.20.

The natural variance between valves makes the exact degree of cancellation unpredictable, so it is rarely exploited to any precise degree during the design process, but is taken as a happy accident. For example, fig. 7.21 shows how

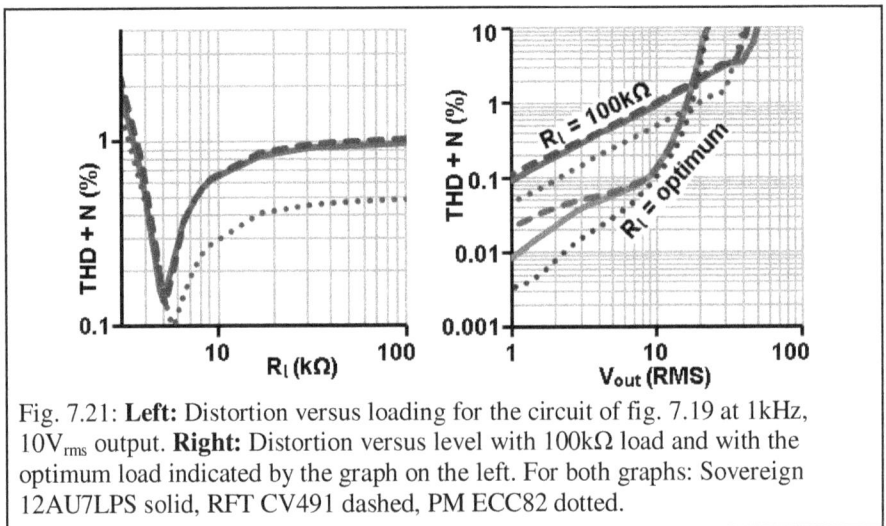

Fig. 7.21: **Left:** Distortion versus loading for the circuit of fig. 7.19 at 1kHz, 10V_{rms} output. **Right:** Distortion versus level with 100kΩ load and with the optimum load indicated by the graph on the left. For both graphs: Sovereign 12AU7LPS solid, RFT CV491 dashed, PM ECC82 dotted.

distortion in the circuit varies with external load resistance at $10V_{rms}$ output, with three different valve samples. In each case a deep distortion null is achieved with a certain load, where the non-linearity of the cathode follower is degraded to the point where it more closely complements the non-linearity of the preceding stage. However, this 'optimum' load varies with different valves *and with signal level*. For example, with the 12AU7LPS the optimum load at $10V_{rms}$ was $5.2k\Omega$, but at $5V_{rms}$ it was $4.7k\Omega$. Ageing is sure to cause the null to drift too.

Fig. 7.21 also shows how wildly the total distortion varies when significant cancelation is in effect. The available output swing is of course reduced when the cathode follower is more heavily loaded, and the transition into clipping becomes softer, so it is hard to put an exact figure on the clipping limit. Fig. 7.22 shows the harmonic spectrum for the 12AU7LPS sample, again at $10V_{rms}$ output. When operating in the distortion

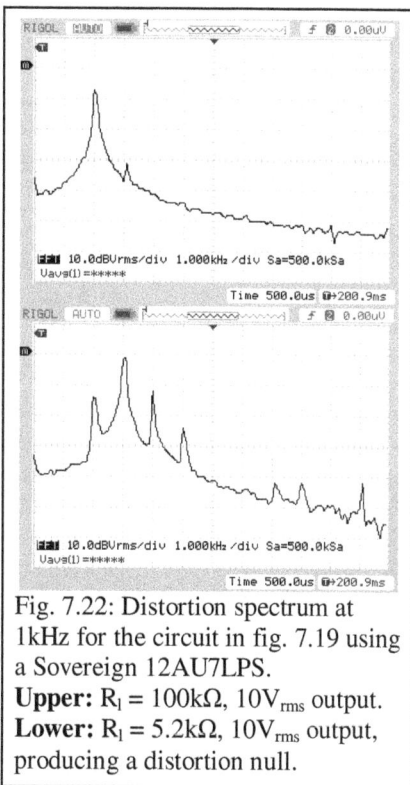

Fig. 7.22: Distortion spectrum at 1kHz for the circuit in fig. 7.19 using a Sovereign 12AU7LPS.
Upper: $R_l = 100k\Omega$, $10V_{rms}$ output.
Lower: $R_l = 5.2k\Omega$, $10V_{rms}$ output, producing a distortion null.

null the second harmonic is suppressed (which allows the distortion analyser to pick up on the higher harmonics that previously were below the noise floor, making the spectrum look rather unpleasant). These are all further reasons why distortion cancellation is not usually exploited to its fullest and most component-sensitive extent.

7.9: The Bootstrapped Pair

In section 7.1.2 it was shown that an impedance connected between the grid and cathode of a cathode follower will be bootstrapped, making it appear much larger. This effect can be exploited to make the anode load resistor of the preceding stage appear much larger too, and so increase the stage gain and reduce distortion. This technique is commonplace in transistor amplifiers but, surprisingly, has enjoyed only sporadic interest from the valve world.[8,9,10,11,12]

---

[8] Jeffrey, E. (1947). Push-Pull Phase-Splitter, *Wireless World*, August, pp274-7.

[9] Daniels, H. L. (1949). Stabilized High-Gain Amplifier, US Patent 2489272.

[10] Young, J. F. (1962). Bootstrap D.C. Amplifier. *Wireless World*, November, pp553-6.

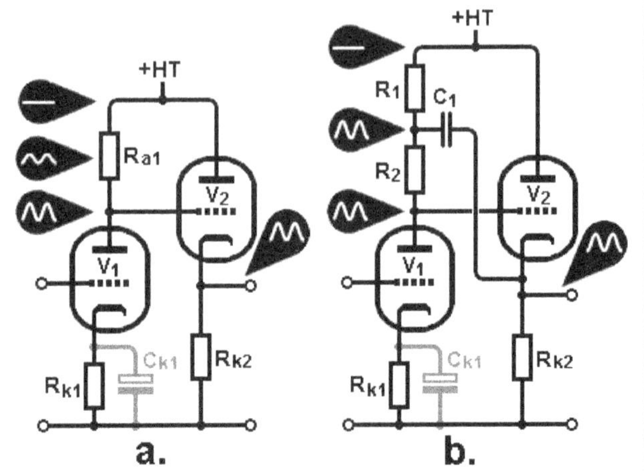

This basic idea is illustrated in fig. 7.23. The first image shows a standard DC-coupled circuit with relative signal waveforms indicated. The signal is largest at the anode of V_1 and zero at the power supply, so the signal at the point half-way along the length of the anode resistor R_1 must be half the amplitude of the signal at the anode. The circuit in fig. 7.23b shows the bootstrap transformation. R_1 has

Fig. 7.23: **a:** Ordinary DC-coupled cathode follower arrangement. **b:** C_1 allows V_2 to drive the mid-point of V_1's anode load. This bootstraps R_2, making it behave like a constant-current source.

been split into two parts and the output of the cathode follower is coupled to the junction via C_1. If C_1 is large and the cathode follower is perfect and has unity gain, then the signal at the anode of V_1 is fed to the cathode follower and is immediately buffered and passed back to the junction of R_1-R_2, so exactly the same signal voltage appears at both ends of R_2. With no difference in AC voltage across this resistor, there can be no AC current in it, so it would appear to have infinite resistance. Another way of looking at it is to say that if C_1 is large then as far as AC signals are concerned R_2 appears to be connected directly between grid and cathode of V_2, hence it is bootstrapped and behaves like a constant-current source load for V_1. The gain of V_1 will then be equal to its μ,

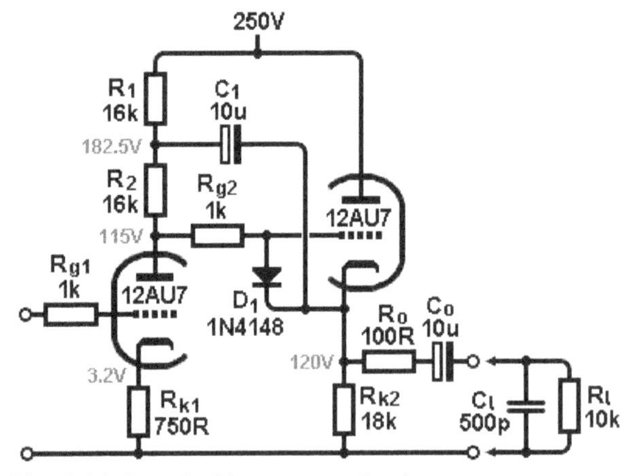

Fig. 7.24: Practical bootstrapped pair.

[11] Bennett, R. (2006). An Improved Split-Load Phase Inverter, Audio Xpress, July, pp25-7.
[12] Blencowe, M. (2012). *Designing Valve Preamps for Guitar and Bass*. Lulu, pp129-32.

268

and the distortion should be very low, as explored at length in chapter 6. R_1, however, is not bootstrapped and simply loads the cathode follower.

In practice the cathode follower will have slightly less than unity gain so R_2 will only be bootstrapped to a value of $R_2/(1-A)$, where A is the cathode follower gain. But this is still potentially a very large value. Coupling capacitor C_1 is of course required to block DC and should be rated for the full HT voltage. Its reactance must be small compared with R_1 down to sub-audio frequencies if a flat frequency response is expected. Using the circuit from the previous section, R_{a1} might be split into two $16k\Omega$ resistors, so C_1 must be greater than:

$$C_1 = \frac{1}{2\pi R_1} = \frac{1}{2\pi \times 16000} = 9.9 \times 10^{-6}\,F \approx 10\mu F$$

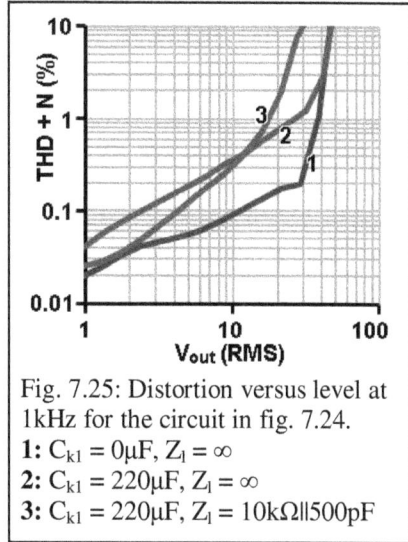

Fig. 7.25: Distortion versus level at 1kHz for the circuit in fig. 7.24.
1: $C_{k1} = 0\mu F$, $Z_l = \infty$
2: $C_{k1} = 220\mu F$, $Z_l = \infty$
3: $C_{k1} = 220\mu F$, $Z_l = 10k\Omega \| 500pF$

Fig. 7.24 shows the shows the circuit tested. Initially V_1 was left without a cathode bypass capacitor and the total gain of the circuit was found to be 14.6. As shown in fig. 7.25, the distortion (measured without an additional AC load on the cathode follower) was very low but not directly proportional to level, and the harmonic spectrum was fairly colourful, as shown in fig. 7.26. These are typical characteristics of CCS loading. Adding a $220\mu F$ bypass capacitor caused the gain to increase to 15.7 but the distortion (mainly second and third harmonic this time) roughly tripled. This demonstrates that the load impedance on V_1 is so large that degeneration has very little effect on the voltage gain, but it does still improve linearity (it also increases the output resistance of V_1, but this is of little consequence since the input capacitance of V_2 is just a few picofarads). The cathode follower was finally loaded with the external $10k\Omega \| 500pF$ impedance. The gain

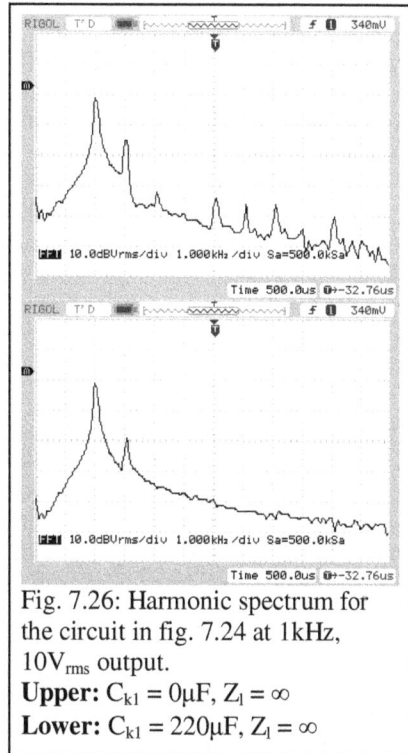

Fig. 7.26: Harmonic spectrum for the circuit in fig. 7.24 at 1kHz, $10V_{rms}$ output.
Upper: $C_{k1} = 0\mu F$, $Z_l = \infty$
Lower: $C_{k1} = 220\mu F$, $Z_l = \infty$

269

dropped to 14.9 and at low levels the distortion improved somewhat –no doubt the result of changes in harmonic cancellation– but the headroom reduced considerably as V_2 struggled to drive the heavy combination of R_1 and the external load. This is another situation where the cathode follower does not provide perfect buffering between V_1 and the external load, so it should be avoided when the cathode follower is expected to drive reactive or unknown impedances, e.g. line outputs. It should also be appreciated that bootstrapping is a form of controlled positive feedback which subtracts from the cathode follower's usual 100% negative feedback fraction. Its general performance is therefore degraded, and in this case the output resistance (including build out resistor) increased to 1105Ω.

7.10: The Constant-Current-Draw Amplifier

An interesting use of the compound-pair formed by a gain stage and cathode follower is to create a circuit which draws a constant current from the power supply. Indeed, this is one of the very earliest applications for the cathode follower, patented by Eric White[13,14](creator of the White cathode-follower discussed in section 7.14.2). Fig 7.27 shows the arrangement in essence. Here the cathode follower serves as a buffer but also as a sort of counter-balance for V_1. When V_1 increases its conduction its anode voltage falls, and this is passed to the grid of V_2 which in turn will *decrease* its conduction. If V_2 has unity gain then the signal voltage

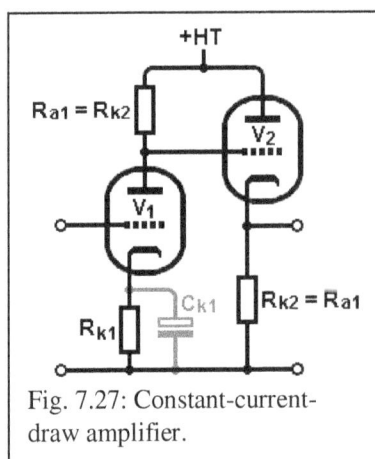

Fig. 7.27: Constant-current-draw amplifier.

across R_{a1} will be exactly reproduced across R_{k2}, so if these resistors are identical then the signal current through them will be exactly equal but opposite in phase. These two signals currents therefore balance each other out, so the total current drain from the power-supply remains perfectly constant. This eliminates any modulation of the power supply voltage, reducing the burden of filtering and decoupling from any power supply smoothing capacitors or regulators (chapter 11).

In reality V_2 will have slightly less than unity gain, so R_{k2} should be made equal to $A{\times}R_{a1}$ (where A is the cathode follower gain) to obtain exactly equal but opposite current swings in each valve. However, in most cases V_2 will come so close to unity gain that it is not worth troubling over this detail. Like the bootstrapped pair this arrangement is not an ideal choice when the cathode follower has to drive a heavy or unknown load, since it relies on both valves always driving equal loads. It is more

---

[13] White, E. L. C. (1937). Improvements in and Relating to Thermionic Valve Apparatus. British Patent 462536.
[14] Cocking, W. T. (1954). Constant-H.T.-Current Amplifier, *Wireless Engineer*, pp29-31.

suitable when we need low output resistance but not great drive capability, e.g. when driving volume or tone controls, or RIAA equalisation networks.

Very often V_1 is biased so that its anode voltage is half the supply voltage, leaving half across R_{a1}. This therefore allows half the supply voltage across V_2 and half across R_{k2} also

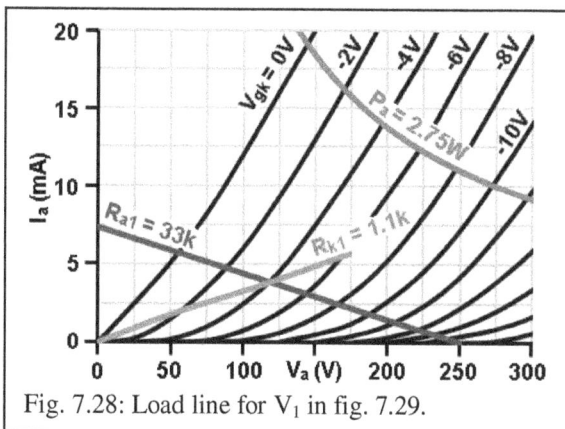

Fig. 7.28: Load line for V_1 in fig. 7.29.

(ignoring the small bias voltage needed for V_2), so both valves will end up biased to roughly the same anode current. This is not an actual requirement of the circuit –it is only the AC currents we are trying to balance out, not the DC ones– but it does make the design process particularly easy, and the all-round symmetry has pleasing aesthetic appeal. Nature seems to approve of symmetry wherever if occurs, so we might as well apply it here.

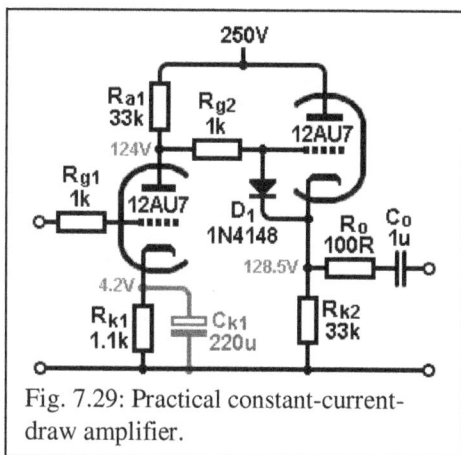

Fig. 7.29: Practical constant-current-draw amplifier.

Modifying the previous example circuit, R_{k2} was increased to 33kΩ to match R_{a1}, and the external load impedance was removed. Fig. 7.28 shows the load line for V_1 indicating that a 1.1kΩ bias resistor should result in a quiescent anode voltage close to half the supply voltage. The bias of V_2 will then sort itself out to something very similar. Fig. 7.29 shows the circuit with measured voltages.

Fig. 7.30 shows the measured distortion at the anode of V_1 (dashed) and cathode of V_2 (solid), but they are virtually identical. The cathode follower is now driving such an easy load that its linearity is too good to produce any significant harmonic cancellation. The overall gain with R_{k1} bypassed was 12, falling to 8 when degenerated. The distortion likewise fell by a factor of about 1.5, as fig. 7.30 demonstrates. The figures are quite mediocre either way, but no doubt they would improve if a higher supply voltage were available.

271

By way of some light relief, this seemed like a convenient point to try out some other valve types in the same configuration. The various types chosen were all 9-pin dual triodes since they could be tested with minimal rewiring. In each case R_{k1} was adjusted to obtain an anode voltage of about half the supply voltage (the HT was reduced to 200V when testing the less hardy valve types). Each valve was operated at its rated heater voltage, which was elevated to one fifth of the HT, and V_1 was left degenerated. Fig. 7.31 shows the resulting distortion performance at the cathode of V_2. Table 7.1 lists the various operating conditions, and the clear winners in this case are the ECC88 / 6DJ8 and the Russian 6H6П (similar to the ECC99 made by JJ). Note, however, that the ECC88 achieves this with a lower HT and higher gain, which could be traded for even more feedback. It also requires 365mA heater current compared with 750mA for the 6H6P (which is really a small power valve). It is also worth pointing out that the ECC88 performs much better than the ECC82 from fig. 7.30 even though measurements from chapter 6 suggest they normally perform quite similarly; the difference here is that the ECC88 enjoys more cathode degeneration. The PCC84 is similarly improved. The poorly performing PCC189 and PCC805 are variable-g_m triodes, included here for a bit of fun.

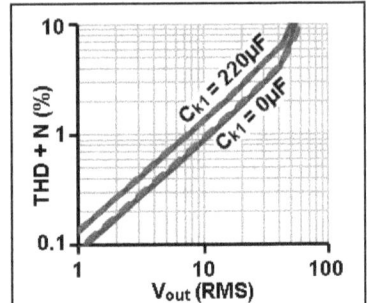

Fig. 7.30: Distortion versus level at 1kHz for the circuit in fig. 7.29. **Dashed:** Anode of V_1; **Bold:** Cathode of V_2.

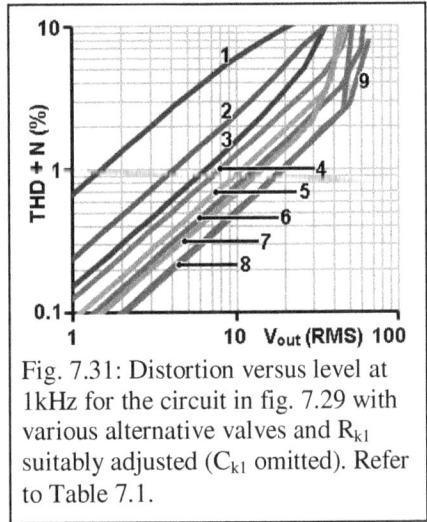

Fig. 7.31: Distortion versus level at 1kHz for the circuit in fig. 7.29 with various alternative valves and R_{k1} suitably adjusted (C_{k1} omitted). Refer to Table 7.1.

Trace	Type	HT	R_{k1}	V_{gk1}	Gain
1	Pinnacle PCC189	200V	1.2kΩ	3.8V	6.7
2	Mullard PCC805	250V	1.2kΩ	4.6V	6.8
3	Sovereign 12AT7WA	250V	390Ω	1.4V	25.5
4	Mullard ECC85	250V	390Ω	1.5V	23.8
5	Mazda ECC804	200V	1.5kΩ	4.4V	7.1
6	Mullard PCC84	250V	1.2kΩ	4.6V	8.8
7	6H3П/6N3P	250V	560Ω	2.1V	15.1
8	Mullard ECC88	200V	910Ω	3.0V	12.8
9	6H6П/6N6P	250V	1.5kΩ	5.8V	7.1

Table 7.1: Figures relating to the distortion measurements in fig. 7.31.

7.11: Transistor Followers

In the interests of economy it is often convenient to use a transistor follower instead of a valve cathode follower. Fig. 7.32 shows a typical arrangement using a MOSFET. This works in fundamentally the same way as a cathode follower but with the obvious advantages of smaller package footprint, no heater supply or heater elevation being required, and almost rail-to-rail headroom. This in turn means that the quiescent anode voltage of the previous stage can be chosen freely without having to account for transistor biasing. Ordinary enhancement MOSFETs are ideal for the task, and almost any type will do, provided the drain-source voltage rating exceeds the HT. The author has used the 400V-rated STP11NK40Z in many situations, mainly because it is available in an all-plastic TO-220FP package. This is convenient since the (metal) tab of most TO-220 devices is connected internally to the drain, which invites accidental shorting.

A disadvantage of MOSFETs is that their drain-gate capacitance, also called the reverse transfer capacitance C_{rss}, is often rather large –commonly around 40pF– which is much worse than for a valve cathode follower (the gate-source capacitance, also called the input capacitance, is also large but is bootstrapped down to a small value). However, this is still quite small compared to the Miller capacitances we are commonly used to dealing with, so it is rarely a barrier to achieving decent bandwidth. Nevertheless, low capacitance MOSFETs such as the 2N60 and STP4NK60 are available for design emergencies.

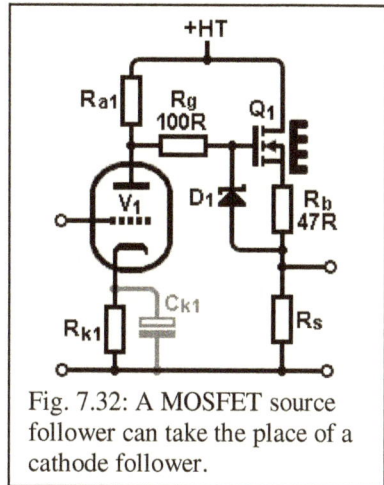

Fig. 7.32: A MOSFET source follower can take the place of a cathode follower.

Provided the MOSFET is operated at more than a couple of milliamps drain current, its transconductance will be high and its linearity far better than a valve cathode follower, as shown already in chapter 6. In short, a MOSFET can serve as an entirely transparent buffer (and it is therefore not likely to provide any distortion cancellation). It can, of course, be arranged as a bootstrapped pair or constant-current-draw amplifier, too.

Most MOSFETS are enhancement devices, meaning they are biased with the gate slightly more positive than the source, quite the opposite of valves. For a DC-coupled source follower the bias will sort itself out, and will usually settle with the source about 4 to 5V below the gate, i.e. below the anode voltage of V_1. This is so small that it can usually be ignored, meaning the quiescent current in the MOSFET can be set using the approximation: $I_D = V_{a1}/R_s$, where V_{a1} is the anode voltage of V_1. Alternatively, R_s can be replaced with a CCS. The power dissipation in Q_1 can in turn be determined from $P = IV = I_D(HT-V_{a1})$. A device in a TO-220 package can

273

usually dissipate up to 1W by itself, while a clip-on heatsink will increase this to 2 or 3W, which is enough for most applications.

MOSFETs can only withstand about 20V between gate and source before the gate insulation –a thin layer of quartz– breaks down. Some MOSFETs (including the STP11NK40Z) have built-in Zener diodes to protect this junction against static discharges, but it cannot hurt to add a more robust Zener diode externally. A 10V to 16V 400mW device should be sufficient for any MOSFET type. What's more, the external Zener diode D_1 and build-out resistor R_b can do triple duty by protecting the gate-source junction from overvoltage, discouraging oscillation, *and* providing a current-limiting function. The last point is easily overlooked, but when power is first applied the output may be pulled down to a low voltage as heavy charging currents flow into coupling capacitors. The power dissipation in Q_1 may therefore be surprisingly large for the first few milliseconds. By arranging D_1 and R_b as in fig. 7.32 the output current will be limited to $(V_z-V_{GS})/R_b$, or little more than 100mA when using a 10V Zener and 47Ω resistor. R_g is a gate-stopper resistor of around 100Ω to 1kΩ and is essential to discourage oscillation; it must be soldered very close to the MOSFET.

7.12: Active Loading

Cathode followers benefit from current-source loading in the same way as ordinary gain stages. A current-source presents a high-impedance to the valve while still allowing it to operate at a healthy current, thereby minimising open-loop distortion and maximising open-loop gain. This in turn means that as a cathode follower the loop-gain will be higher, so the circuit will achieve lower closed-loop distortion and gain closer to unity.

Since the ECC88 performed so well in section 7.10 it may be worth seeing how well it performs as a cathode follower alone. Fig. 7.33 shows two example circuits tested by the author. The first has a 22kΩ resistive load and 5.5mA anode current, chosen

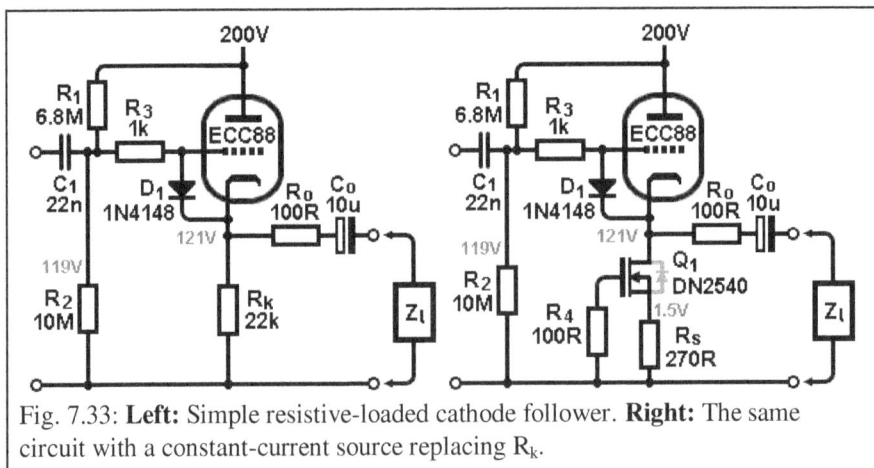

Fig. 7.33: **Left:** Simple resistive-loaded cathode follower. **Right:** The same circuit with a constant-current source replacing R_k.

so that the bias point appears roughly central when driving a 10kΩ AC load. For simplicity, ordinary fixed bias was used, and the heater supply was again elevated to one fifth of the HT. For the second circuit R_k was replaced by a single depletion-MOSFET constant-current source (CCS), with the source resistor R_s adjusted on test to give the same 5.5mA current as in the first example.

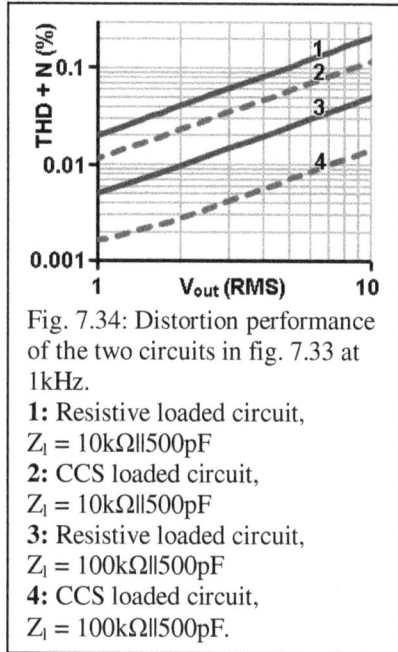

Fig. 7.34: Distortion performance of the two circuits in fig. 7.33 at 1kHz.
1: Resistive loaded circuit, $Z_l = 10kΩ\|500pF$
2: CCS loaded circuit, $Z_l = 10kΩ\|500pF$
3: Resistive loaded circuit, $Z_l = 100kΩ\|500pF$
4: CCS loaded circuit, $Z_l = 100kΩ\|500pF$.

Fig. 7.34 shows the distortion performance of the two circuits. The results for the simple circuit are shown solid and correspond to AC loads of $Z_l = 10kΩ\|500pF$ and $100kΩ\|500pF$ (the latter is simply the input impedance of the distortion analyser). Into the high-impedance load the circuit achieves less than half the distortion of the 12AU7 circuits shown earlier (pure second harmonic), and in both cases it was constant up to 20kHz, whereafter it just started to rise, owing to the capacitive load. The results for the CCS-loaded circuit are shown dashed, and the additional improvement is startling. Into a high-impedance load this cathode follower now produces an order of magnitude less distortion than the 12AU7 examples. The gain was about 0.96 into the high-impedance load, falling to 0.94 with the low-impedance load, and the small-signal output resistance was 230Ω including build-out resistor.

7.13: Output Voltage Limiting

One of the most common reasons for using a cathode follower is to drive line-level signals into the (possibly) low and (probably) unspecified load impedances presented by cables and the equipment they're connected to. If that equipment happens to be solid-state then this presents a worrying problem, because valve circuits can deliver high voltages, whereas solid-state circuits invariably operate from low voltages, use low-voltage components, and expect to receive low voltage inputs. If we are lucky, the receiving device will have some input over-voltage protection, but this fairly uncommon. It is therefore not a good idea to go shoving hundred-volt transients down audio cables, yet a cathode follower can and will do just that.

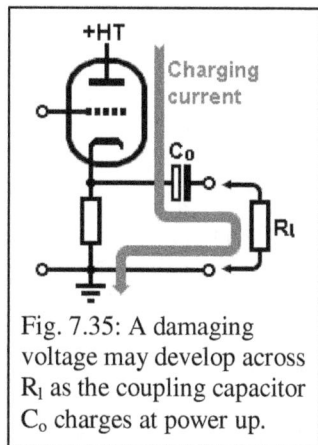

Fig. 7.35: A damaging voltage may develop across R_l as the coupling capacitor C_o charges at power up.

275

High-voltage transients can be created as the coupling capacitors charge at start up. This is particularly bad if the valve is allowed to pre-heat before the HT is switched on, since the largest possible charging current will then flow down the interconnect and develop a high voltage across the unfortunate device (or person!) at the other end. Fig. 7.35 shows the charging path from which it is clear that the voltage across R_l could potentially peak at $HT \times R_l/(r_a + R_l)$. But even with the natural soft-start provided by heater warm up, transients are easily created by intermittent contacts, hot-swapping of cables, or simply by turning the signal level up too high. Failing to provide some form of protection for external equipment against these high voltages is negligent design.

The simplest and most reliable way to avoid this problem is to add a pair of back-to-back Zener diodes across the output, which will clamp the peak output voltage at any desired level. Most solid-state circuits can handle input voltages of $15V_{pk}$ without damage, even if this is outside their range of intended use. Very few systems expect signal levels higher than this to be sent down cables, so a pair of 15V or 16V Zeners is ideal for domestic applications, although 22V may be preferred for studio environments where 26dBu signals are occasionally encountered. However, diodes do not conduct with razor-sharp abruptness (although avalanche diodes are better than true Zeners in this regard), so readers will no doubt want to know if they introduce any significant distortion below the clipping threshold.

Fig. 7.36 shows a test circuit in which R_1 represents the source resistance of a valve stage, and a pair of ordinary 16V 400mW Zener diodes (actually avalanche diodes) provide voltage limiting. Fig. 7.37 shows the distortion performance. The plot on the left shows distortion versus *input* level, and the horizontal axis uses a linear scale to reveal more clearly that the onset of clipping is not precisely $16V_{pk}$, but is quite sharp nonetheless. When R_1 is up to $1k\Omega$ the trace is purely noise floor, which is reassuring. However, the

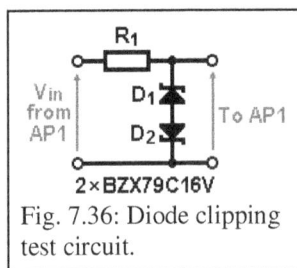

Fig. 7.36: Diode clipping test circuit.

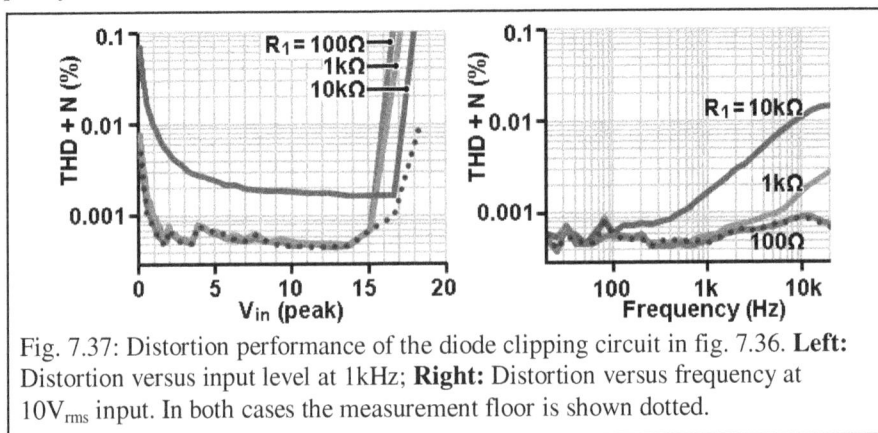

Fig. 7.37: Distortion performance of the diode clipping circuit in fig. 7.36. **Left:** Distortion versus input level at 1kHz; **Right:** Distortion versus frequency at $10V_{rms}$ input. In both cases the measurement floor is shown dotted.

plot on the right shows that when R_1 is large, distortion degrades at higher frequencies –even though the frequency response was flat across the audio band– presumably due to nonlinear junction capacitance. Fortunately, if R_1 is much smaller than $10k\Omega$ then the distortion remains well below the level of valve distortion, which points towards Zener limiting as being a harmless addition to a cathode-follower line driver.

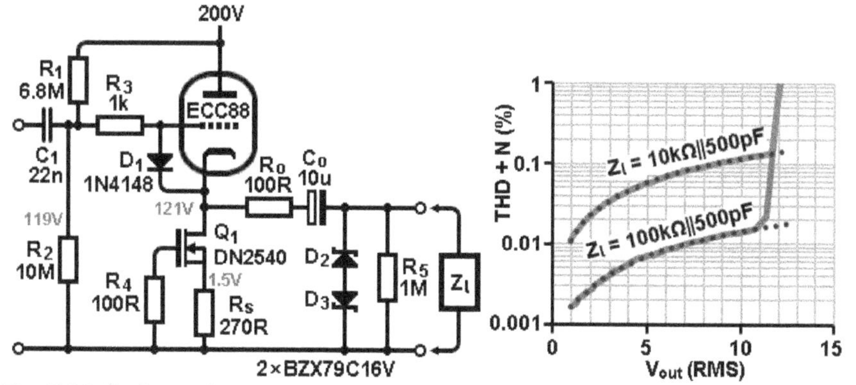

Fig. 7.38: **Left:** Buffer or line driver with output voltage limiting.
Right: Distortion performance at 1kHz. Performance without clipping diodes is shown dotted.

Fig. 7.38 shows the CCS-loaded cathode follower from the previous section, with output voltage protection added. Since this is an inherently low-distortion circuit it stands the best chance of revealing any shortcoming of the clipping circuit. R_5 is a pull-down resistor that provides a complete charging path for C_o, which reduces pops when hot-swapping cables. Incidentally, although the Zeners can only withstand $0.4W / 16V = 25mA$ of continuous current, they can handle many times this level for a few milliseconds. The instantaneous charging current will also be limited by the valve itself to perhaps a few tens of milliamps, so the clipping diodes should have no trouble handling ordinary start-up conditions. As an added bonus they also speed up the charging time of the output coupling capacitor. It is perhaps worth pointing out that if this circuit were used as a stand-alone buffer (rather than as the final stage in a larger circuit) then a pull-down resistor would be added to the input too, but Zeners probably wouldn't be needed here since the charging current for C_1 is limited to a miniscule value by R_1.

Fig. 7.38 shows distortion versus output level; note that the horizontal axis is linear to show the clipping threshold clearly, and it is in RMS rather than peak volts. The results with the clipping diodes removed are shown dotted, proving that they cause no degradation at all below $11V_{rms}$ or $15.5V_{pk}$. Distortion was also constant across the audio band. Based on these results the author is confident that there is no excuse not to provide protective output clipping in a valve preamp.

7.14: Compound Cathode Followers

In chapter 6 the possibility of using a triode as a current-source load was briefly investigated and promptly dismissed as being inadequate, at least for the point being made at the time. However, there are situations where active loading using valves is hard to resist, because it creates a rather harmonious topology, especially when using dual-triode bottles. Fig. 7.39 shows a simple example which should not require much explanation since it is functionally the same as fig. 7.33, except the MOSFET has been replaced by a triode to create a two-valve buffer circuit, or compound cathode follower. It should, however, be pointed out that the ECC88 datasheet quotes $V_{hk(max)}$ of only 50V for one of the triodes in the envelope (the one connected to pins 6/7/8) but 130V for the other (pins 1/2/3). The former should be used for the lower position in the circuit, so in this example the heater supply does not need to be elevated. But if it is, the elevation voltage should not exceed 50V.

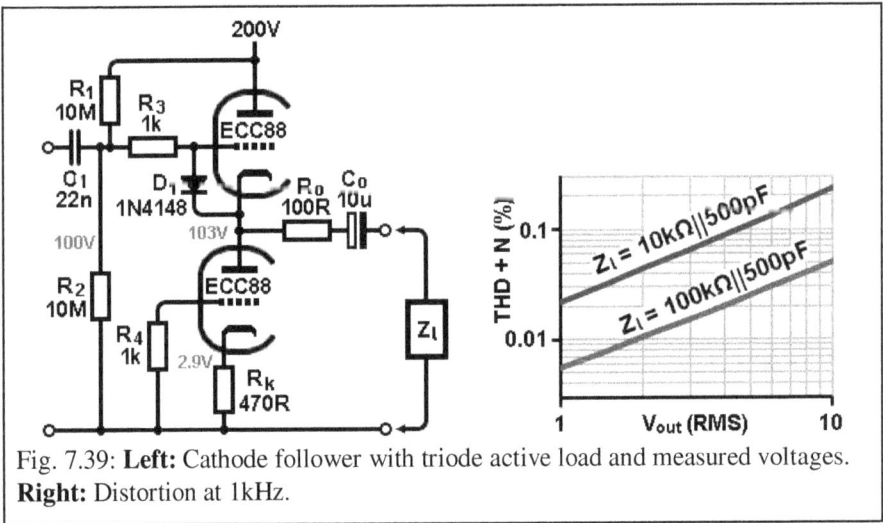

Fig. 7.39: **Left:** Cathode follower with triode active load and measured voltages. **Right:** Distortion at 1kHz.

There are also two further observations worth making. In order to exploit the symmetry of the circuit, the upper grid has been biased to exactly half the supply voltage. The cathode will settle just a few volts above this, so we know in advance that approximately half the supply voltage will be dropped across each triode. Assuming both triodes are the same, the bias of the upper triode will therefore have no choice but to track the

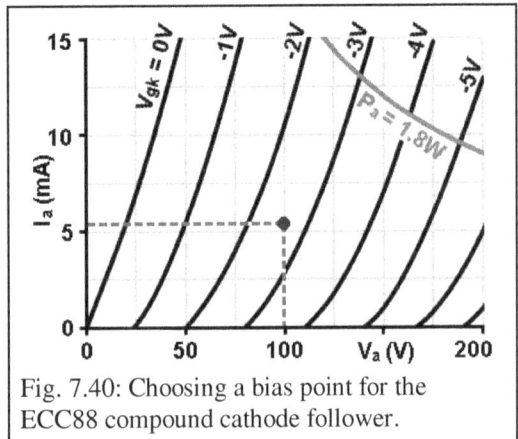

Fig. 7.40: Choosing a bias point for the ECC88 compound cathode follower.

bias of the lower triode. Choosing a value for R_k is therefore quite easy: simply look at the anode characteristics to find the bias voltage at the desired anode current when V_{ak} = HT/2, as illustrated in fig. 7.40. Unlike with a MOSFET-CCS this is not likely to require any adjustment on test.

The second observation is that the internal resistance of this triode current source is not large, in this case being about:

$$r' = r_a + R_k(\mu+1) = 4300 + 470 \times (30+1) = 18870\Omega$$

Performance is therefore not likely to be much different from the circuit of fig. 7.33a which used a simple 22kΩ load resistor. For example, fig. 7.39 shows the distortion performance, which is indeed not much different from fig. 7.34. The output impedance was 240Ω including R_o.

As it stands, the lower valve serves little purpose other than to use up a spare triode. However, this is only the first step on an instructive journey. The next step is to notice that the lower triode is really a gain stage (albeit one that is rather heavily loaded by the cathode impedance of the upper triode) and its grid is begging to have some signal or other fed into it.

7.14.1: Feed-forward Noise Cancellation

One non-obvious and quite devious use for the lower triode in fig. 7.39 is to feed it a sampling of the power supply noise. It will amplify and invert this signal so that it subtracts from the noise already leaking through the upper valve to the output (which is not inverted). In other words, we deliberately 'feed-forward' a fraction of noise so that it partially or wholly cancels any noise that we already had, thereby improving the overall PSRR of the circuit, as illustrated in fig. 7.41. Such examples of feed-forward noise cancellation have always been rare in amplifiers,[15] but in recent years the technique – under the title of 'Audio Aikido'– has been justly championed by Broskie.[16]

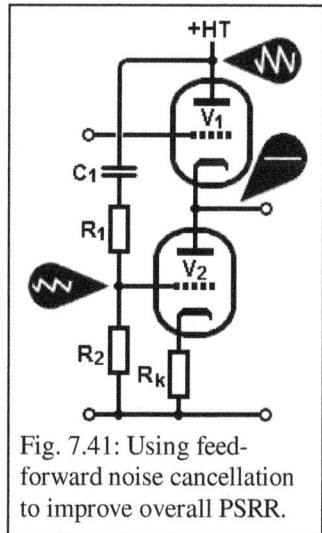

Fig. 7.41: Using feed-forward noise cancellation to improve overall PSRR.

Referring to fig. 7.41, let subscript '1' refer to V_1 and subscript '2' to V_2. Since V_2 is degenerated its effective transconductance is roughly

---

[15] Britton, K. G. (1948). Reducing Heater Hum, *Wireless World*, October, p360.

[16] Broskie, J. http://www.tubecad.com. The name 'aikido' is a reference to the Japanese martial art which attempts to redirect an opponent's strength against himself.

$g_{m2}/(1+g_{m2}R_k)$. This transconductance operates into the output impedance of the upper valve, which is roughly $1/g_{m1}$. The voltage gain of the lower triode is therefore approximately:

$$A_2 \approx -\frac{g_{m2}}{1+g_{m2}R_k} \cdot \frac{1}{g_{m1}} \tag{7.11}$$

If both valves are the same then this simplifies to $-1/(1+g_mR_k)$. Therefore, however much power-supply noise leaks through V_1 to the output, roughly $1+g_mR_k$ times as much needs to be fed forward to V_2 to cancel it out. R_1 and R_2 can then be selected to provide this fraction. C_1 blocks the DC supply voltage and must be large enough to pass all the power supply noise without significant phase shift. Continuing with the circuit from the previous section, the μ of the upper valve is about 30 which is roughly equal to the PSRR before the feed-forward cancellation is added. In other words, roughly 1/30 times the power supply noise already leaks through the upper valve to the output. The g_m of the lower valve is estimated to be 7mA/V, so we must feed-forward a fraction of power-supply noise equal to:

$$\frac{1}{30}(1+0.007\times470) = 0.143$$

Fig. 7.42 shows the circuit tested. A trimpot was used for the lower resistor in the potential divider and was initially set to 166kΩ to provide the calculated attenuation of 0.143. The PSRR was measured by inserting a 1kΩ resistor in series with the power supply and feeding in a signal to the anode of the upper valve while monitoring the output. The usual audio input was grounded, as shown dashed in the figure. As indicated, the PSRR without any cancellation (i.e. that of fig. 7.39 earlier) was 29.7dB (×30.5), which agrees well with the expectation that it should be

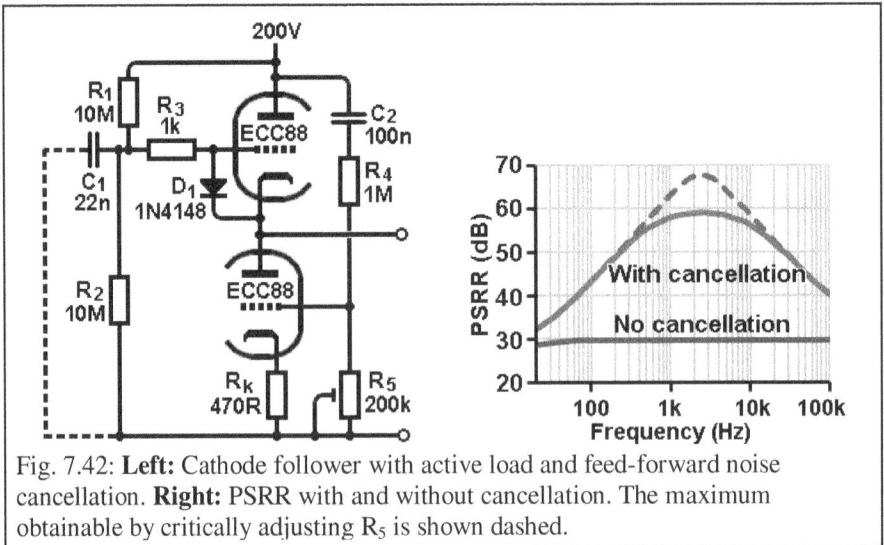

Fig. 7.42: **Left:** Cathode follower with active load and feed-forward noise cancellation. **Right:** PSRR with and without cancellation. The maximum obtainable by critically adjusting R_5 is shown dashed.

approximately equal to µ. When the calculated amount of feed-forward cancellation was added, the PSRR increased to an impressive 59dB at mid frequencies. This could be increased a little further to 68dB by adjusting R_5 to 175kΩ –a division ratio of 0.149, so the approximate calculations are fairly reliable. However, at high frequencies the PSRR falls at a first-order rate due to the input capacitance of the lower valve (this could in theory be corrected by adding a very small compensation capacitance across R_4), while at low frequencies it falls because C_1 does not fully decouple the upper grid, allowing more noise to reach the output. Still, 43dB at 100Hz is not to be sniffed at.

However, to test this circuit in isolation is rather contrived. In reality it would be part of a larger system, and the preceding amplifier stage –whatever it may be– will have its own PSRR shortcomings and so will pass on some of its own power supply noise to this circuit, meaning more would have to be fed-forward to the lower triode if we were seeking the best overall PSRR. Depending on the circuit complexity and natural component variation, it may be fiendishly difficult to calculate the optimum value. It can always be found by adjustment on test, but this is rather inconvenient as it means making the power supply deliberately noisy in order to measure or listen to the output hum while making the adjustment. This is another case where we might prefer to forget about aiming for the 'optimum' and instead apply the simplified formulae and leave the rest to fate. This is sure to give *some* PSRR improvement, even if it is not as large as it could be. On the other hand, we might forget about noise cancellation altogether and instead make a further step in the development of the circuit by turning it into a White cathode follower.

7.14.2: The White Cathode Follower

The White cathode follower is shown in fig. 7.43 and is so-named after Eric White who applied for the patent in 1940.[17] The important difference between this circuit and the previous one is that the lower valve is not just a current-sink but actually helps to deliver more current into the load. This is because V_1 has a (small) anode resistor R_a, so as well as acting as a cathode follower it also produces a small inverted signal at the anode which is promptly fed to the lower valve via coupling capacitor C_1. V_2 is therefore driven in antiphase to V_1, so when V_1 conducts more, V_2 conducts less, and the difference in the two anti-phase currents will flow in the external load. It is thus a push-pull amplifier, but one which makes very efficient use of parts by providing its own phase inversion and

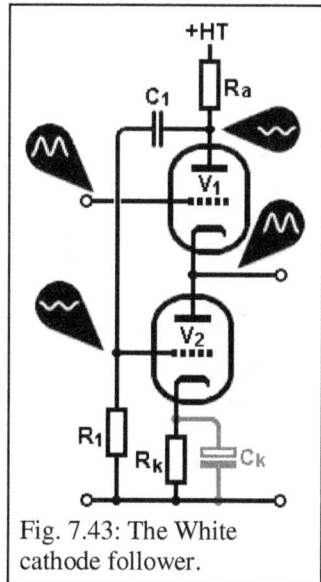

Fig. 7.43: The White cathode follower.

---

[17] White, E. L. C. (1944). Improvements in and Relating to Thermionic Valve Amplifier Circuit Arrangements. British Patent 564250.

allowing both valves to share the same quiescent current (some readers will appreciate the kinship with the SRPP discussed in chapter 8[*]). Since we now have two valves actively driving the load, the output power is increased and the output impedance greatly reduced. The circuit becomes capable of driving unusually heavy loads and forms the basis of many output-transformerless headphone amplifiers. Here preamp design merges with (mini) power-amp design.

If the load impedance is heavy, e.g. headphones, then we will normally be interested in maximising the output *power*, and like any push-pull amplifier this occurs when it is properly balanced. If R_a is too large then the signal developed at the anode of V_1 will become large enough to overdrive V_2 long before V_1 itself runs out of headroom, while if it is too small then V_1 will clip before V_2 has delivered its full quota. Hence there must be an optimum value for R_a which exploits both valves to their full extent, thereby yielding maximum headroom and output power. In theory, this can be determined from an equivalent circuit such as the one shown in fig. 7.44. We are interested in the (AC) currents i_1 and i_2 delivered into the load Z_l, by V_1 and V_2 respectively, so using KVL we can write:

$$v_o = v_2 + (-\mu_2 v_{gk2}) + v_k$$

From KCL it is clear that $i_o = i_1 - i_2$, and by applying Ohm's law:

$$(i_1 - i_2)Z_l = i_2 r_{a2} - \mu_2 v_{gk2} + i_2 R_k$$

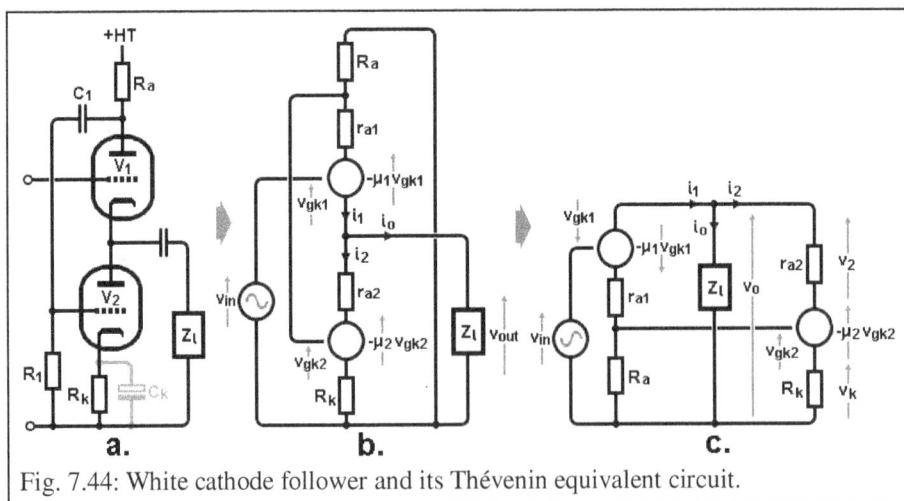

Fig. 7.44: White cathode follower and its Thévenin equivalent circuit.

[*]In fact, this circuit *is* an SRPP, but in cathode-follower form. Their equivalent circuits are the same except for the connection of the input voltage and the polarity of the output voltage.

But $v_{gk2} = -i_1 R_a - i_2 R_k$ since R_a 'samples' the current in V_1. Substituting this in:

$$i_1 Z_1 - i_2 Z_1 = i_2 r_{a2} - \mu_2 (-i_1 R_a - i_2 R_k) + i_2 R_k = i_2 r_{a2} + i_1 \mu_2 R_a + i_2 \mu_2 R_k + i_2 R_k$$

By collecting all the i_1 terms on one side and all the i_2 terms on the other we have:

$$-i_1 (\mu_2 R_a - Z_1) = i_2 (Z_1 + r_{a2} + \mu_2 R_k + R_k)$$

Push-pull balance occurs when both valves deliver equal but opposite currents into the load. By setting i_2 equal to $-i_1$ they drop out of the equation above, allowing it to be solved for the critical value of R_a:

$$R_{a\,(optimum)} = \frac{2Z_1 + r_{a2} + R_k}{\mu_2} + R_k \tag{7.12}$$

Since $r_a/\mu = 1/g_m$ this can also be written as:

$$R_{a\,(optimum)} = \frac{1}{g_{m2}} \left(1 + \frac{2Z_1 + R_k}{r_{a2}} \right) + R_k \tag{7.13}$$

Notice that the characteristics of V_1 do not come into it, so the two valves do not have to be matched or even of the same type. Unfortunately, all this theory applies to small signals, yet the whole point of finding the 'optimum R_a' was to achieve *maximum* output, i.e. the largest signals. For large signals, valve non-linearity spoils this theory, and it is usually the case that maximum output obtains when R_a is slightly larger than the value calculated above.

If the load impedance is large then maximum *power* is not the issue; what we want is minimum distortion. An interesting result of the foregoing theory is that if $R_a = Z_1/\mu_2$ then the current in V_2 becomes unvarying, i.e. it behaves like a true CCS. We might expect this to be the optimum condition when driving high-impedance loads, but in practice a larger value again seems to work best in practice. As a simple example, the circuit from the previous section was reconfigured as a White cathode follower driving a 10kΩ load, as shown in fig. 7.45. Distortion versus level is shown for a selection of values of R_a. A value of 330Ω should result in constant current in the lower triode, and the measured performance is indeed the same as that of the MOSFET-CCS loaded circuit from fig. 7.33. The theoretical $R_{a(optimum)}$ should be about 1.3kΩ, but the improvement actually continues all the way up to 4kΩ (visually it was mainly the second harmonic which was progressively reduced). However, 8kΩ goes too far, as the DC drop across it causes the available headroom to shrink. The output impedance (including R_o) was 204Ω when $R_a = 330$Ω, or 137Ω when $R_a = 4$kΩ. The gain ranged from 0.94 to 0.96.

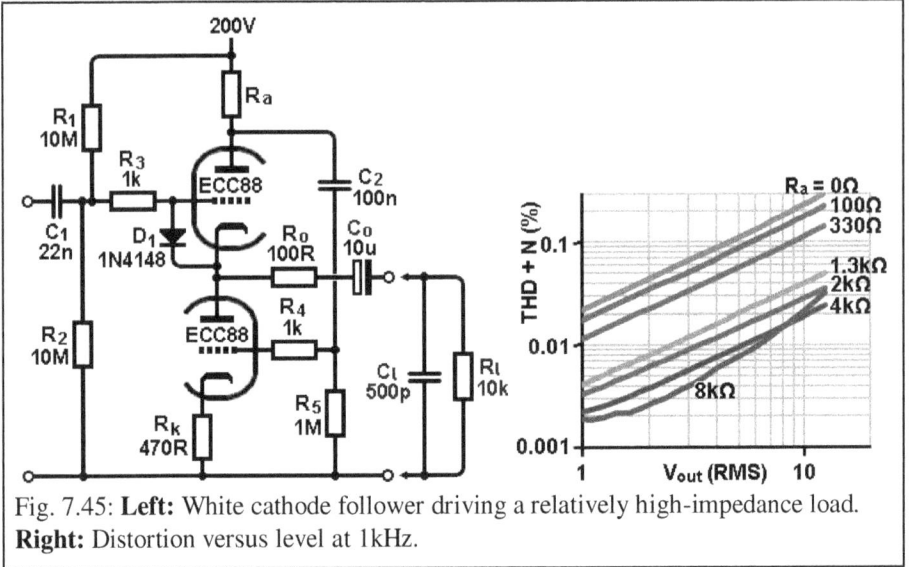

Fig. 7.45: **Left:** White cathode follower driving a relatively high-impedance load. **Right:** Distortion versus level at 1kHz.

Sparing the reader the full treatment, the formula for the output resistance of the White cathode follower is:

$$R_o = \frac{(r_{a1} + R_a)(R_k + r_{a2})}{(\mu_1 + 1)(r_{a2} + R_k) + (\mu_1 + 1)\mu_2 R_a + r_{a1} + R_a} \qquad (7.14)$$

When driving a heavy load, R_k will sometimes be bypassed or replaced with diode bias in order to minimise the output impedance. If that is the case then the gain is given by:

$$A = \frac{\mu_1 Z(1 + g_{m2} R_a)}{(\mu_1 + 1)(g_{m2} Z R_a + Z) + r_{a1} + R_a} \approx 1 \qquad (7.15)$$

The PSRR is:

$$PSRR = \frac{(\mu_1 + 1)(g_{m2} Z R_a + Z) + r_{a1} + R_a}{Z(1 - g_{m2} r_{a1})} \qquad (7.16)$$

Where in each case Z is equal to r_{a2} in parallel with the load impedance. If R_k is not bypassed then the gain and PSRR both become a tiny bit worse, but the formulae become far too long and cumbersome to be worth reproducing here.

7.14.3: A Transformerless Headphone Driver

The White cathode follower's ability to drive relatively heavy loads with fairly low distortion makes it suitable as an output-transformerless headphone driver. However, the vast majority of modern headphones are *very* low impedance (16Ω to 32Ω) better suited to battery-powered equipment which delivers low voltage but relatively high current. It is therefore worth investing in some high-impedance (>50Ω) headphones if you are to get the best out of a valve headphone driver. It is still possible to find phones up to 300Ω, and Beyerdynamic and AKG® even make a few 600Ω models.

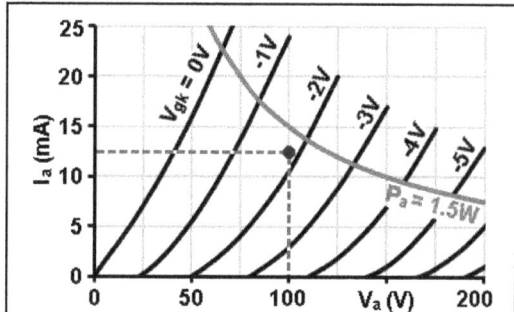

Fig. 7.46: Choosing a bias point for the ECC88 headphone driver in fig. 7.47.

Despite the wide range in impedances, most modern headphones have similar sensitivities of around 100dB$_{SPL}$ at 1mW. Average listing level is usually around 0.2mW to 0.5mW, though loud passages may reach a hundred times more, or 50mW. As a rough rule of thumb we might therefore aim for a headphone amplifier to be capable of delivering 50mW to 100mW maximum continuous power. Some may argue for even more headroom, but beyond a couple of hundred milliwatts there is the risk of dangerously-high sound pressure levels in the ear, and some headphones may even burn out.

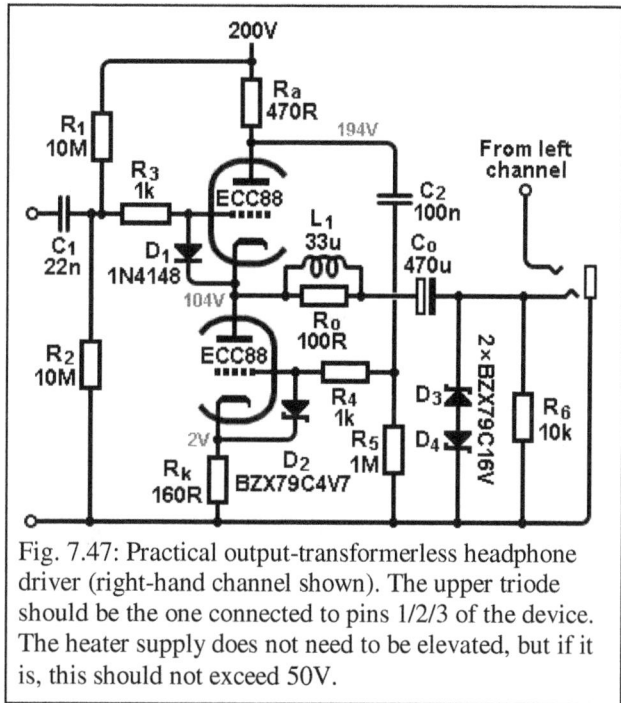

Fig. 7.47: Practical output-transformerless headphone driver (right-hand channel shown). The upper triode should be the one connected to pins 1/2/3 of the device. The heater supply does not need to be elevated, but if it is, this should not exceed 50V.

Let us adapt the White cathode follower circuit shown previously. When driving a very low-impedance load the AC load line for each valve will be almost vertical. In order to maximise the output power the bias point needs to central on this line, or if this is beyond the acceptable dissipation limit or too close to the region of grid current, then as hot as we are willing to go. In this case the quiescent current will be set to about 12.5mA as indicated in fig. 7.46. This will result in just under

285

$100V \times 0.0125A = 1.25W$ dissipation in each triode, which is within the datasheet limit of 1.5W for both triodes simultaneously.

Because both triodes contribute to the output, the peak current into the load can theoretically reach twice the quiescent current, or in this case $25mA_{pk}$ which is $17.7mA_{rms}$. Using $P = I^2R$ the circuit should therefore muster 94mW into 300Ω phones, which is good, but only 10mW into 32Ω. Since it may have to drive a range of load impedances we also have no choice but to optimise it for something roughly in the middle of the expected range, say 150Ω. The optimum value for R_a predicted by equation (7.12) is then about 270Ω, but adjustment on test indicated that 470Ω was a better compromise. The complete circuit is shown in fig. 7.47.

The White cathode follower is very sensitive to blocking distortion, because if the lower valve blocks, the upper one is forced to cut-off too, resulting in a rather unfortunate latch-up condition. A 4.7V Zener diode, D_2, was therefore added to improve overload recovery, as described in section 3.13.2. This made the clipping characteristic *considerably* more polite. It should not be omitted.

There is no 'correct' source impedance for driving headphones, but it should certainly be less than 100Ω. Many commercial amplifiers hover around 10Ω to 50Ω. With headphones as with loudspeakers, the lower the better. The usual build-out resistor is therefore bypassed with an inductor L_1 whose value is not critical – anything from $22\mu H$ to $100\mu H$ will do. The resulting output impedance was found to be 38Ω at 1kHz which is quite respectable.

Distortion versus output power for different load impedances is shown in fig. 7.48. The circuit just about manages the expected figures, depending on where you draw the limit. Distortion was manly second harmonic with increasing third for heavier loading, and was constant with frequency, so no surprises. The voltage gain ranged from 0.96 for high-impedance loading down to 0.3 for a 16Ω load. If an extra stage of gain is required then it could readily be DC-coupled to the White cathode follower, and global feedback could also be applied to reduce distortion and output impedance further.

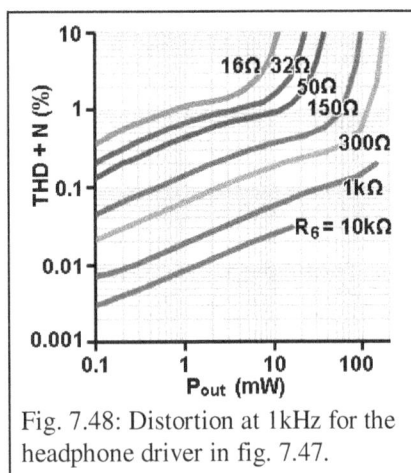

Fig. 7.48: Distortion at 1kHz for the headphone driver in fig. 7.47.

The author also tried bypassing R_k with a $470\mu F$ capacitor and adjusting R_a to a new optimum of 180Ω, which reduced the output impedance to 33Ω. The distortion was slightly improved for loads around 100 to 500Ω (distortion cancellation was evident), but for both higher and lower impedances it was not much different. However, the clipping characteristics and overload recovery became really quite

286

ugly, so this does not seem like a modification worth pursuing. Another popular modification to circuits like this is to add a plastic capacitor in parallel with the (usually electrolytic) output capacitor C_o, placed there in the naïve hope that high frequencies will navigate around the electrolytic and so avoid any defects it may have. This doesn't actually happen, of course, but it does add some redundant 'high-end' ornamental value

Chapter 8: Compound Amplifiers

There is a collection of circuits which use two or more amplifying devices in such close, inter-dependent configurations that they are best treated as individual gain stages or amplifier building blocks. Such stages may be grouped together under the title of 'compound amplifiers'. We have already met some in chapter 7, and this chapter will present a few more which crop up time and again in modern hi-fi design, for better or worse.

8.1: Totem-Pole Gain Stages

A family of circuits that has enjoyed a particular renaissance among valve enthusiasts are the 'totem-pole' amplifiers, so-called because they stack multiple devices on top of one another. But despite their new-found fashion they include some of the oldest circuits, dating back to the earliest days of electronics when designers had very few components to work with. The evolution of these circuits began with the idea of using one valve as an active load for another valve.[1] This idea was initially disregarded in chapter 6 on the grounds that a triode does not make a good *constant*-current source, and a pentode makes an inconvenient one. However, if we forgo the pursuit of a rigid constant-current source and take a more flexible attitude we may discover more subtle reasons to use a triode as a simple active load.

8.1.1: The Half-µ Stage

In chapter 6 it was explained that by loading a triode with a constant-current source its load line becomes horizontal and its gain becomes equal to µ (unless there is an external AC load aswell, of course). Since µ is also the most linear of the valve 'constants' this tends to minimise distortion. However, there is another possible approach for improving linearity, which is to devise a load line that is curved in the opposite fashion to the amplifying valve. Fig. 8.1 shows

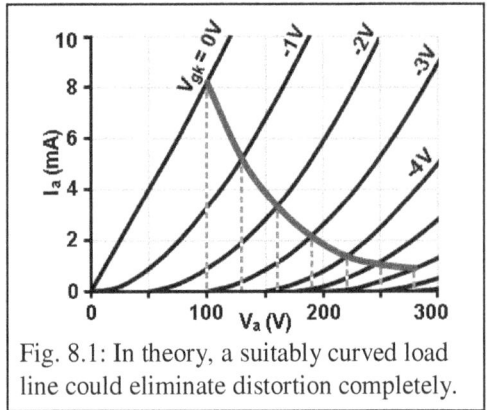

Fig. 8.1: In theory, a suitably curved load line could eliminate distortion completely.

how this might work; note that the intersections of the grid curves with the (cleverly chosen) load line are evenly spaced, indicating perfectly constant gain. This is another example of distortion cancellation –two opposing nonlinearities cancelling out to produce a linear transfer function. The practical difficulty, of course, lies in devising a load with the right curvature.

---

[1] Read, H. S. (1922). Vacuum-Tube Repeater Circuits. US Patent 1403566.

One way to make a load whose resistance decreases as the current through it increases is to use another triode, such as the totem-pole circuit in fig. 8.2a. Let subscript-1 indicate V_1 and subscript-2 indicate V_2. The lower valve is an ordinary common-cathode amplifier, so ignoring any external AC load its gain will be:

$$A = -\mu_1 \frac{R_a}{r_{a1} + R_a + R_{k1}(\mu_1 + 1)}$$

But in this case the anode load resistance R_a is actually formed by the active load created by V_2, which has an internal resistance of:

$$r' = r_{a2} + R_{k2}(\mu_2 + 1)$$

Remember, this is never very large, so it does not qualify as a *constant-current source*, it is just a non-linear resistance, as represented in fig. 8.2b. Substituting one into the other, the gain of this totem-pole circuit becomes:

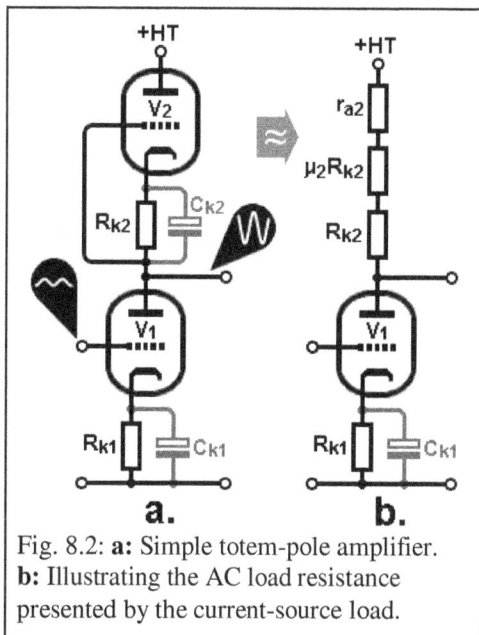

Fig. 8.2: **a:** Simple totem-pole amplifier. **b:** Illustrating the AC load resistance presented by the current-source load.

$$A = -\mu_1 \frac{r_{a2} + R_{k2}(\mu_2 + 1)}{r_{a1} + r_{a2} + R_{k2}(\mu_2 + 1) + R_{k1}(\mu_1 + 1)} \tag{8.1}$$

But something special happens if we make $R_{k1} = R_{k2} = R_k$ and use matched triodes so that $r_{a1} = r_{a2} = r_a$ and $\mu_1 = \mu_2 = \mu$:

$$A = -\mu \frac{r_a + R_k(\mu + 1)}{2[r_a + R_k(\mu + 1)]} = -\frac{\mu}{2} \tag{8.2}$$

Thus everything except μ drops out of the equation, so the linearity ought to be as good as the linearity of μ –just as it is with a true CCS load. This is not quite the same thing as fig. 8.1 where the load *exactly* mirrors the non-linearity of the amplifier valve, but it is still an improvement over a simple resistor load, at least in theory. This circuit has been around for a long time[2] and really deserves a name of its own so the author suggests 'half-μ stage' for obvious reasons.[3]

[2] Meissner, E. R. (1933). A High Gain AC-DC Amplifier, *Electronics*, July, p195
[3] Blencowe, M. (2010). The Optimised SRPP Amp, *Audio Xpress*, May, pp13-9.

When both cathode bypass capacitors are omitted the output resistance of the half-μ stage is:

$$R_{o(degenerated)} = \frac{r_a + R_k(\mu + 1)}{2} \tag{8.3}$$

From the point of view of the power supply the upper and lower halves of the circuit form a 1:1 potential divider, so half the power-supply noise will appear at the output, i.e. the PSRR is thoroughly unimpressive:

$$PSRR = 2 \tag{8.4}$$

However, note that the gain and PSRR remain constant provided the top and bottom halves of the circuit are always matched. This also applies for aging and variations in heater voltage, and this long-term stability is one of the reasons for the circuit's original invention. Bypassing the cathode resistors (equally) or replacing them with LEDs merely reduces the output resistance to $r_a/2$ without affecting the gain *or introducing phase shift*. Since V_1 and V_2 are biased the same, the junction of the two will always settle at ½HT (except for minor mismatching between the triodes). The bias point can therefore be chosen from the anode characteristics graph by noting the bias voltage at the desired anode current when $V_a = $ ½HT, just as we did with the compound cathode followers in chapter 7.

Fig. 8.3 shows a test circuit using a 12AT7 / ECC81, together with its distortion performance (obligatory grid leak not shown). R_2 is an ordinary grid stopper, as is R_1 except that it is larger than it would be in a real amplifier because here it also simulates a typical source resistance. The heater supply was elevated to ¼HT so that it lay half-way between the two cathodes.

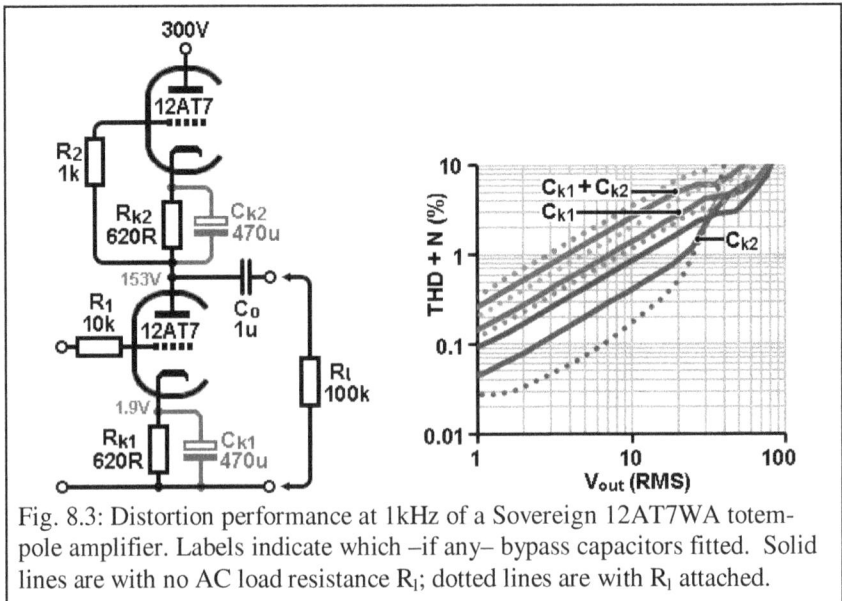

Fig. 8.3: Distortion performance at 1kHz of a Sovereign 12AT7WA totem-pole amplifier. Labels indicate which –if any– bypass capacitors fitted. Solid lines are with no AC load resistance R_1; dotted lines are with R_1 attached.

When driving a near-infinite load, and with both bias resistors left unbypassed, the gain was 25 and the output impedance was 24kΩ. Adding both bypass capacitors caused the gain to increase to 25.3 (a negligible change, as expected) and the output impedance to fall to 8.1kΩ. Sadly, fig. 8.3 shows that distortion tripled. What's more, the gain is less than would normally be achieved with ordinary resistor loading, yet the distortion for a given output level is no better (and quite possibly worse). The ideal theory is clearly rather naïve. Another fact which counts against the half-μ is that it includes *two* valves that inject equal noise currents into the output, making it theoretically 3dB noisier than the simple resistor-loaded alternative. In this case the EIN of the degenerated half-μ was 3.8μV (20Hz–20kHz, with the grid grounded to eliminate R_1), but replacing V_2 with a 47kΩ resistor yielded the same overall gain and quiescent current with an EIN of 2.4μV –some 4dB quieter, indicating the upper triode was actually noisier than the lower one. Remembering that the half-μ also has a pathetic PSRR of 2, it is obviously not a suitable choice for the input stage of a low-noise preamplifier.

However, before we abandon all hope it is worth seeing what happens if we depart from strict symmetry by fitting only one of the cathode bypass capacitors. Fitting only C_{k1} produced a gain of 38.2 since the lower half of the circuit now presents a smaller impedance than the upper half, so it is no longer a true half-μ stage. The output impedance was 12.2kΩ and the distortion was intermediate between the degenerated and bypassed half-μ variants, as shown in the figure. What came as more of a surprise was that by fitting only C_{k2} the gain fell to a mere 12.4 (output impedance 11.9kΩ) but the distortion *also* fell considerably, despite the fact that the lower valve now has a very low-impedance load to drive, equal only to r_{a2}. Evidently the curvature of r_{a2} provides significant distortion cancellation, although this was not obvious from the harmonic spectrum which was almost pure second harmonic in every case. On the other hand, clipping sets in sooner since V_1 is driving a steep load line.

A 100kΩ AC load resistance (R_1) was then attached and the measurements were repeated. The results are shown by the dotted lines in the fig. 8.3. In the first three cases the distortion was slightly worsened, as one might expect. But with only C_{k2} fitted the distortion dropped *again* (several other ECC81s were tried and all showed similar behaviour). Apparently, the blend of linear and non-linear loads makes for an even better match to the transfer characteristic of V_1. Presumably, therefore, there is a particular value of load resistance which gives the very best match, i.e. the most distortion cancellation, and indeed this proves to be the case. Fig. 8.4 shows distortion versus loading at $10V_{rms}$ output for the four cases. It is clear that distortion cancellation is significant only when C_{k2} alone is fitted. The greatest distortion null was achieved with a 150kΩ load, resulting in 0.05% THD at $10V_{rms}$. Fig. 8.4 also shows the harmonic spectrum at this point, and it is now easy to see that the second harmonic has been considerably suppressed. At this point the overall gain was 11.5.

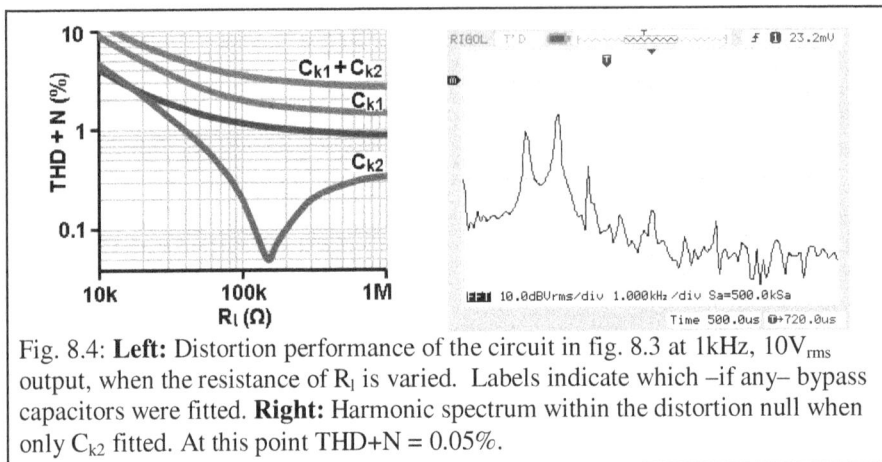

Fig. 8.4: **Left:** Distortion performance of the circuit in fig. 8.3 at 1kHz, $10V_{rms}$ output, when the resistance of R_l is varied. Labels indicate which –if any– bypass capacitors were fitted. **Right:** Harmonic spectrum within the distortion null when only C_{k2} fitted. At this point THD+N = 0.05%.

Unfortunately, distortion cancellation is an unpredictable phenomenon, depending innately on valve type, operating conditions, and sample variation. For example, fig. 8.5 shows the results for an ECC88 totem-pole amplifier. Here the distortion with only C_{k2} fitted is *worse* than the other three cases when the load impedance is large, although a distortion null does exist with a 9kΩ load. It should also not be forgotten that the degree of distortion cancellation in any circuit varies with signal level too. Incidentally, the unloaded gain of this circuit was 15.4 for the degenerated half-μ case, 15.6 for the bypassed half-μ, 24.6 with only C_{k1} fitted, and 6.7 with only C_{k1} fitted. The output impedance was 5.9kΩ, 1.8kΩ, 2.8kΩ, and 2.9kΩ respectively. Note that the coupling capacitor C_o is a fairly large, electrolytic type, simply to facilitate full-bandwidth testing with unrealistically heavy loads.

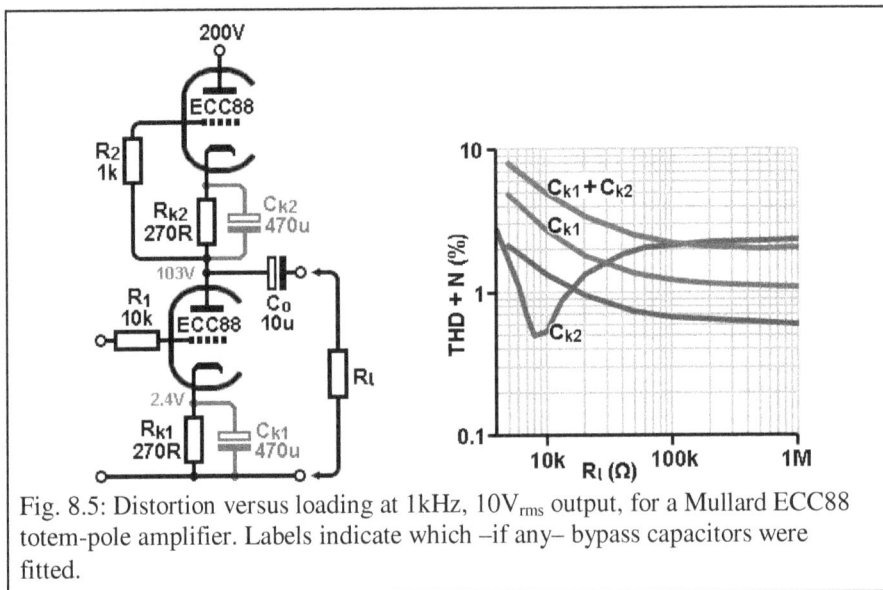

Fig. 8.5: Distortion versus loading at 1kHz, $10V_{rms}$ output, for a Mullard ECC88 totem-pole amplifier. Labels indicate which –if any– bypass capacitors were fitted.

292

8.1.2: The SRPP

The **shunt-regulated push-pull** (SRPP) is a totem-pole amplifier that looks superficially like the half-μ stage, except the output is taken from the upper end of R_{k2} as shown in fig. 8.6. Nevertheless, this small but crucial change results in some distinct improvements in performance over the half-μ, and the circuit has become very fashionable in modern valve hi-fi (wearisomely so). Part of its appeal must be its minimalism, as it requires only a couple of resistors and two triodes –usually in the same bottle– and almost no thought or understanding on the part of the circuit designer.

The SRPP is intimately related to the White cathode follower in exactly the same way that an ordinary gain stage is related to an ordinary cathode follower. Like the White cathode follower, the circuit stops being push-pull if the AC load impedance is large, in which case it is

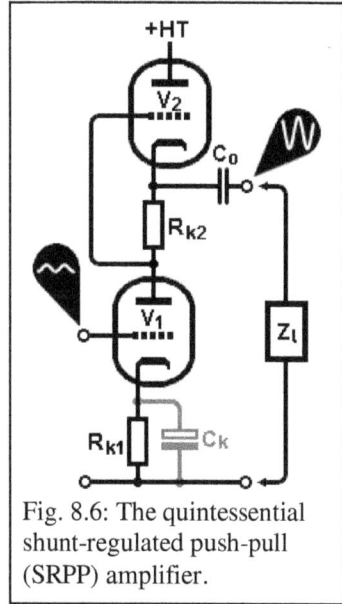

Fig. 8.6: The quintessential shunt-regulated push-pull (SRPP) amplifier.

somewhat perverse to call it an SRPP at all, so the author prefers 'quasi-SRPP' to distinguish such cases. The earliest reference to the true SRPP circuit appears to be a patent applied for in 1939 by Newsome Clough of the Marconi company,[4] although a quasi-SRPP was also published by Horton in 1933.[5] The circuit was patented again (this time with an extra resistor) in 1940 by Maurice Artzt working for RCA.[6] By the 1950s the circuit was acknowledged in textbooks as a member of the 'shunt regulated' amplifier family, but the exact name 'shunt-regulated push-pull' seems to have popped-up first in relation to transistor circuits exploiting similar principles.[7] This name was re-borrowed for the valve circuit in the 1970s[8] and has firmly stuck.

---

[4] Clough, N. H. (1940). Improvement in or Relating to Modulator Arrangements. British Patent 526418.

[5] Horton, J. W. (1933). The Use of a Vacuum Tube as a Plate-Feed Impedance, *J. Franklin Institute.* **216** (6), December , pp749-62.

[6] Artzt, M. (1943). Balanced Direct and Alternating Current Amplifiers. US Patent 2310342. Artzt's modification was to add another resistor equal to $R_{k1}+R_{k2}$ in series with the anode of V_2. This makes the total AC resistance in each 'half' of the circuit equal, which ought to encourage more symmetrical operation and therefore lower distortion (the same principle can be applied to the White cathode follower). In practice the extra resistor is so small compared with the valves' internal anode resistance that it makes negligible difference.

[7] Chba, M. & Hirai, A. (1969). A Pulsed ESR Apparatus, *Japanese Journal of Applied Physics*, 8 (12), December.

[8] Hiraga, J. (1977). Le Préamplificateur SRPP, *l'Audiophile*, **2**, December.

The SRPP can be push-pull because R_{k2} not only provides bias for V_2 but also serves as a current-sampling resistor. If V_1 increases its conduction, current will be pulled out of the load and down through R_{k2}. This resistor samples the current and so develops a voltage across it that appears directly between the cathode and grid of V_2, thereby driving V_2 simultaneously but with the opposite phase to V_1. Conversely, if V_1 conducts less current then the voltage across R_{k2} will fall, thereby causing V_2 to conduct more current, which is diverted into the load. It is in essence a phase inverter and class-A push-pull amplifier combined, just like the White cathode follower. Unlike the White cathode follower, however, V_2 cannot suffer from blocking distortion since it is directly coupled to V_1.

The current source formed by V_2 presents the lower valve with an apparent load resistance equal to:

$$r' = r_{a2} + R_{k2}(\mu_2 + 1)$$

This can also be expressed in Norton form by substituting $g_m r_a = \mu$:

$$r' = r_{a2} + R_{k2}(g_{m2}r_{a2} + 1)$$

This is a more convenient expression when dealing with the SRPP because any external load, Z_l, appears directly in parallel with r_{a2} as far as AC signals are concerned. We can then mentally substitute $r_{a2}\|Z_l$ wherever we find r_{a2} in the (unloaded) design equations, which makes it easy to visualise the effect of loading the circuit.

Life is too short to go through the full derivations, so we will just state the facts. The gain of the SRPP is:

$$A = -g_{m1}r_{a1}r_{a2} \frac{g_{m2}r_{a2} + 1}{r_{a1} + r_{a2} + R_{k1}(g_{m1}r_{a1} + 1) + R_{k2}(g_{m2}r_{a2} + 1)} \qquad (8.5)$$

If the top and bottom halves of the circuit are matched, as is normally the case, then this simplifies to just under $-\mu/2$. In other words, the gain of the quasi-SRPP (i.e. unloaded) is almost the same as that of the half-μ stage. This makes sense since all we have done is to move the output 'along' the upper current-source impedance by an amount equal to R_{k2}, which is quite small. If R_{k1} is bypassed then it disappears from the formula above and the gain increases.

The big difference between the half-μ and the SRPP is its output resistance:

$$R_o = \frac{r_{a2}(r_{a1} + R_{k2}) + r_{a2}R_{k1}(g_{m1}r_{a1} + 1)}{r_{a1} + r_{a2} + R_{k1}(g_{m1}r_{a1} + 1) + R_{k2}(g_{m2}r_{a2} + 1)} \qquad (8.6)$$

If the top and bottom halves are matched then this simplifies to approximately $r_a/2$, which is the same as the output resistance of the *fully bypassed* half-μ stage. In other words, we can have our cake and eat it too: a degenerated quasi-SRPP gives

294

practically the same gain (and distortion) as a degenerated half-μ, but the SRPP's output impedance is much lower. Thus we can extend the bandwidth and improve immunity to hum pickup at no extra cost. If R_{k1} is bypassed then again it disappears from the equation above and the output resistance is made lower still.

So far the SRPP has given only improvements over the half-μ, so we should expect to pay the price somewhere and, indeed, the PSRR is slightly worse. But then, it was already so poor with the half-μ that this is not much of a penalty. For reference the formula is:

$$PSRR = \frac{r_{a1} + r_{a2} + R_{k1}(g_{m1}r_{a1}+1) + R_{k2}(g_{m2}r_{a2}+1)}{r_{a1} + R_{k2} + R_{k1}(g_{m1}r_{a1}+1)} \qquad (8.7)$$

If a quasi-SRPP is one which drives such a high-impedance load that it is not, in fact, push-pull, then what load impedance pushes it into genuine push-pull 'true SRPP' territory? This can be answered in exactly the same manner as for the White cathode follower and, again, there must be a critical condition at which point both valves contribute equally to the output. Fig. 8.7 shows the Thévenin equivalent circuit. We are interested in the (AC) currents i_1 and i_2 delivered into the load, Z_l, by V_1 and V_2 respectively, so using KVL we can write:

$$v_o = -v_2 - (-\mu_2 v_{gk2})$$

From KCL it is clear that $i_o = i_2 - i_1$, and by applying Ohm's law:

$$(i_2 - i_1)Z_l = -i_2 r_{a2} + \mu_2 v_{gk2}$$

But $v_{gk2} = -i_1 R_{k2}$ since R_{k2} 'samples' the current in V_1. Substituting this in:

$$(i_2 - i_1)Z_l = -i_2 r_{a2} + \mu_2(-i_1 R_{k2}) = -i_2 r_{a2} - i_1\mu_2 R_{k2}$$

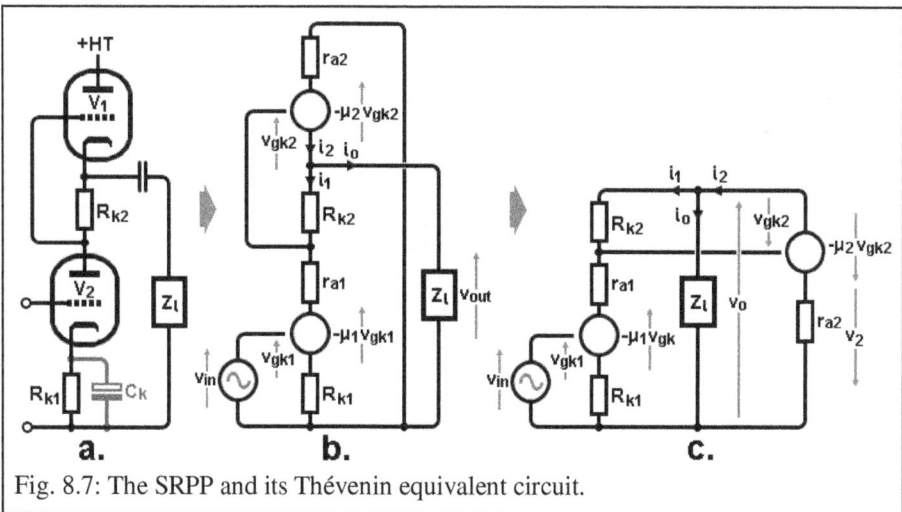

Fig. 8.7: The SRPP and its Thévenin equivalent circuit.

By collecting all the i_1 terms on one side and all the i_2 terms on the other we have:

$$i_1(\mu_2 R_{k2} - Z_1) = -i_2(Z_1 + r_{a2})$$

Push-pull balance occurs when both valves deliver equal but opposite currents into the load. By setting $-i_2$ equal to i_1 they drop out of the equation above, allowing it to be solved for the critical value of R_{k2}:

$$R_{k2(optimum)} = \frac{2Z_1 + r_{a2}}{\mu_2} \qquad (8.8)$$

This is exactly the same result as for the White cathode follower, which should remove any residual doubt the reader might have as to their brotherhood. Again, the characteristics of the 'master valve' V_1 do not come into it, so the two valves do not need to be the same. As with the White cathode follower the circuit stops being push-pull and becomes a quasi-SRPP when $Z_1 \ge \mu_2 R_{k2}$.

With the White-cathode follower we were free to choose the current-sampling resistor to suit the load impedance. However, with the SRPP this resistor also provides bias for V_2, which rather limits our choices. It is therefore more useful to *start* with a suitable value for R_{k2} and then make the load impedance suit the circuit. Solving equation (8.8) for the optimum load impedance:

$$Z_{1(optimum)} = \frac{\mu_2 R_{k2} - r_{a2}}{2} \qquad (8.9)$$

In practice the true optimum will be a little larger than this, owing to non-linearity on large signals. For the degenerated SRPP this optimum load also tends to minimise distortion in the circuit, as demonstrated shortly, but first let us compare an SRPP with a half-μ or ordinary totem-pole stage.

Fig 8.8 shows the ECC81 / 12AT7 circuit from the previous section converted into an SRPP. The degenerated gain was 24.6 –virtually the same as the half-μ– and from the figure it is easy to see that the distortion is the same too (compare with fig. 8.3). However, the output impedance was 8.1kΩ which is the same as the *bypassed* half-μ. With C_{k1} fitted the gain was 37.7 and output impedance 4.1kΩ. This is roughly the same gain as the similar totem-pole stage, and the distortion (dashed) is also the same, but the output impedance is now three times lower. Thus by converting the stage from a half-μ into a quasi-SRPP we have effectively gained something for nothing.

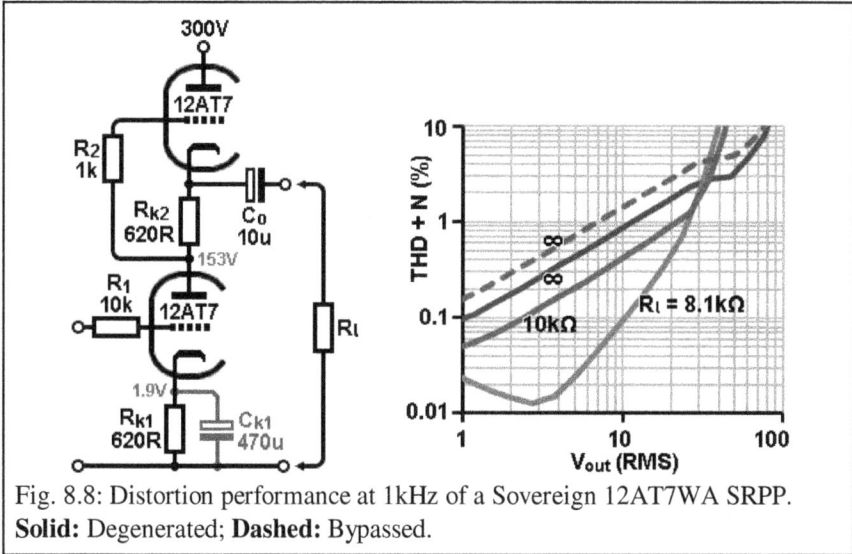

Fig. 8.8: Distortion performance at 1kHz of a Sovereign 12AT7WA SRPP.
Solid: Degenerated; **Dashed:** Bypassed.

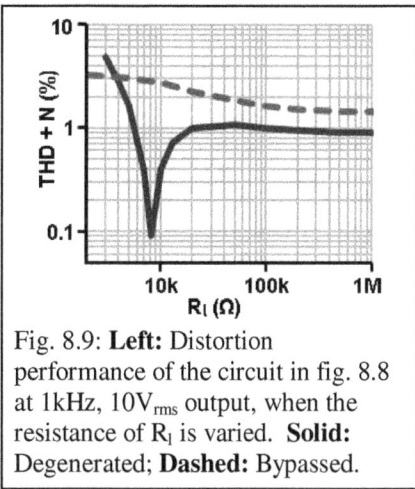

Fig. 8.9: **Left:** Distortion performance of the circuit in fig. 8.8 at 1kHz, 10V_rms output, when the resistance of R_l is varied. **Solid:** Degenerated; **Dashed:** Bypassed.

Fig. 8.8 also shows the result of adding an AC load resistance R_l to the (degenerated) SRPP. With a 10kΩ load the distortion is reduced, and experimentation showed that at 10V_rms the lowest distortion was achieved with an 8.1kΩ load (coincidentally, this is the same as the output impedance, so the gain was therefore halved to 12.3). The optimum load predicted by equation (8.9) is a little lower at about 7.5kΩ. Fig. 8.9 shows distortion versus loading to illustrate the point. Apparently the degenerated circuit exhibits strong distortion cancellation with a certain load but, interestingly, the bypassed case does not. This is the opposite of what we found for the totem-pole amplifier from the previous section. As usual, the distortion spectrum was almost exclusively second harmonic until it became suppressed within the distortion null.

For interest, fig. 8.10 shows an SRPP using a 6SN7GT –a fashionable valve in a fashionable configuration. The degenerated gain was 9.5 and output impedance 4.7kΩ. The theoretical optimum load should be about 5.3kΩ, but the experimental optimum was 5.7kΩ, causing the gain to drop to 5.3. The bypassed gain was 14.6 and output impedance 2.3kΩ. The affection for the 6SN7GT remains justified, as the distortion for high-impedance loads is pleasingly low.

297

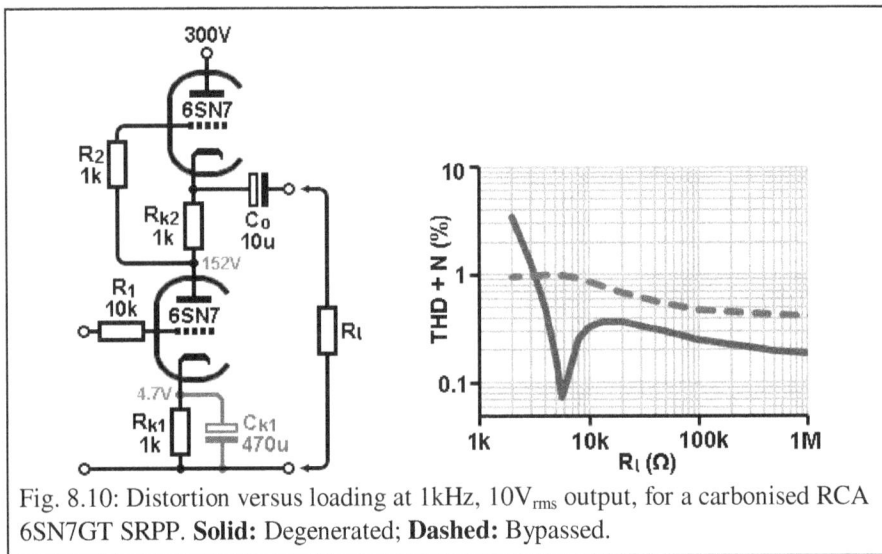

Fig. 8.10: Distortion versus loading at 1kHz, 10V$_{rms}$ output, for a carbonised RCA 6SN7GT SRPP. **Solid**: Degenerated; **Dashed**: Bypassed.

Since the SRPP shares so much in common with the White cathode follower it too is sometimes used to drive heavy loads such as headphones. To complete this section then, fig. 8.11 shows an ECC88 headphone driver biased to the same quiescent current as the White cathode follower example from section 7.14.3. In practice, a build-out LR network and protection Zeners should be added, but they are omitted here for brevity. Fig. 8.12 shows distortion versus output power for both the degenerated and bypassed case. The optimum load predicted by equation (8.9) is about 750Ω, while the experimental value was 1.2kΩ for the degenerated case. This circuit is therefore not strictly optimised for typical headphone impedances as R$_{k2}$ is too large. Nevertheless, it does manage to deliver enough output power for the job, but the

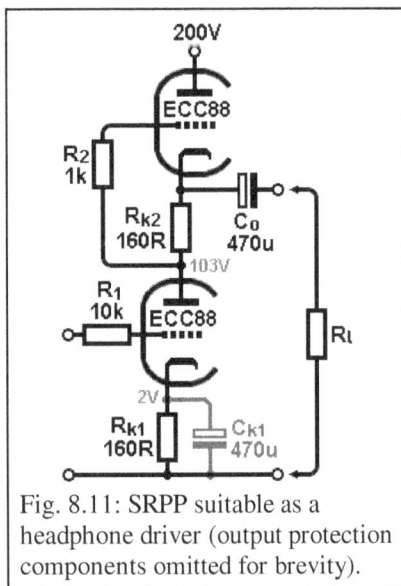

Fig. 8.11: SRPP suitable as a headphone driver (output protection components omitted for brevity).

distortion is about an order of magnitude worse than for the White cathode follower. Interestingly, the bypassed SRPP produces less distortion with heavy loads than the degenerated case, although its behaviour is more complex.

The degenerated output impedance was 1.5kΩ and the unloaded gain was 15.3, while for the bypassed case the figures were 810Ω and 23 respectively. Since both the distortion and output impedance are much too high for direct use with

298

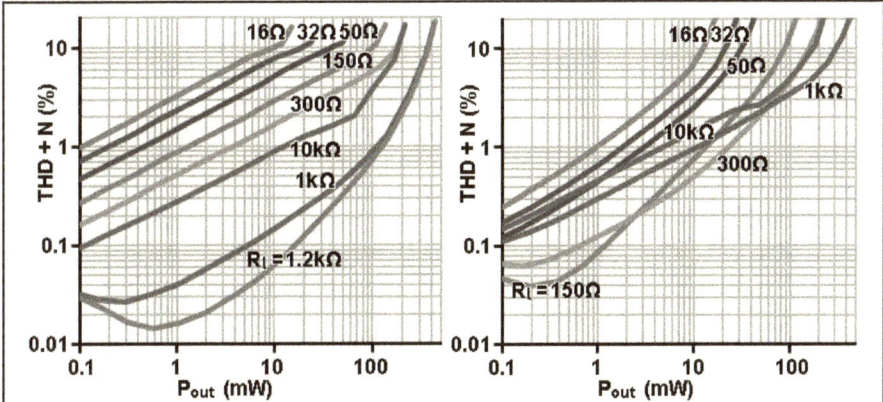

Fig. 8.12: Distortion versus output power at 1kHz for the SRPP circuit in fig. 8.11 using a Mullard ECC88. **Left:** Degenerated; **Right:** Bypassed.

headphones, global feedback would have to be wrapped around the SRPP and an earlier gain stage. However, since this combination would achieve more loop gain than a similar gain stage feeding a White cathode follower, we could conceivably achieve better overall performance from the SRPP headphone amp. The SRPP's greater resilience to blocking distortion is a further advantage, although its PSRR is, of course, worse.

8.1.3: The μ-Follower

With the SRPP we saw that the internal resistance of the current source created by the upper valve was:

$$r' = r_{a2} + R_{k2}(g_{m2}r_{a2} + 1)$$

This is not large enough to be called a *constant* current source because R_{k2} is used for biasing V_2 and so cannot be very large. Or can it? Perhaps we could use a much larger resistance than is needed for biasing, then simply tap off the required bias voltage from a suitable point along it, just like the cathode-biased cathode follower from section 7.5. We still need to couple the signal from the anode of V_1 to the grid of V_2, so a coupling capacitor and grid-leak must be added too. This produces the circuit in fig. 8.13 which has come to be known as a μ-follower. In practice R_{k2} will be split into a pair of resistors rather than being an actual pot.

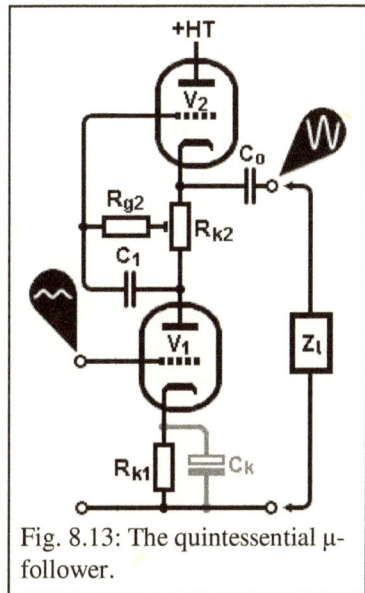

Fig. 8.13: The quintessential μ-follower.

The μ-follower is an SRPP taken to extremes by using a large resistance for the bootstrapped resistor R_{k2}. The design equations (8.5) to (8.7) for the SRPP therefore apply to the μ-follower too (assuming the grid leak R_{g2} is much larger than R_{k2}, which is normally the case). Again, any external load appears effectively in parallel with r_{a2}. By making R_{k2} a fairly large value, the active load can achieve a very high internal resistance and may approach a true constant current source as far as AC signals are concerned. The gain of the lower valve will therefore approach μ_1 while the upper valve functions as a quasi-cathode follower* and provides a very low output impedance, hence the name. Although this circuit evolved naturally from earlier totem-pole and SRPP circuits, credit for publishing the earliest truly recognisable μ-follower appears to belong to Cooper and Nixon for a patent applied for in 1948.[9] However, the name 'μ-follower' seems to be a recent development.

Although both valves necessarily pass the same quiescent anode current they don't necessarily need the same bias voltage. Nevertheless, giving them the same bias does rather ease the design process since it enforces the same quiescent voltage drop across each one. We are then free to apportion the available HT voltage as seems appropriate. The author decided to use the same 6SN7GT and 300V HT voltage as in the previous section. Allowing 125V across each valve (a perfectly reasonable figure) leaves 50V available to be dropped across the bootstrap resistor. If we choose to allow 5mA of anode current then R_{k2} must equal 50V/5mA = 10kΩ. It is then easy to find the bias point from the anode characteristics graph, as shown in fig. 8.14, and since we stipulated that both valves will be biased the same we only have to do the job once. In this case the bias voltage appears to be about 3.4V, so the bias resistors must each be 3.4V/0.005A = 680Ω.

The upper cathode will rest at roughly 125V+50V = 175V which exceeds the 100V $V_{hk(max)}$ rating of the valve. The heater supply must therefore elevated to about 85V,

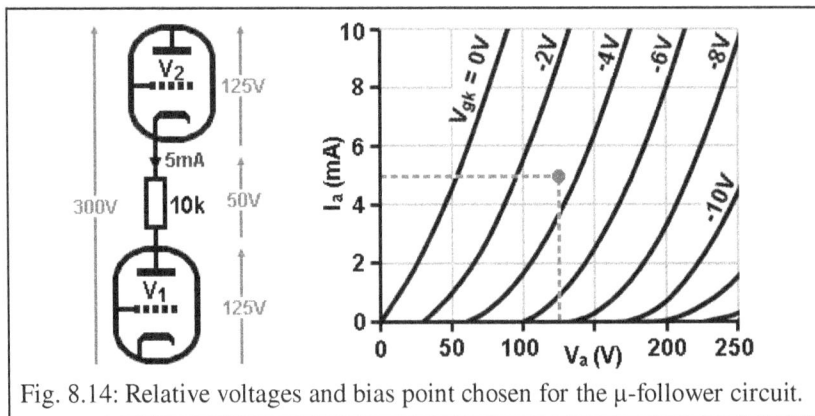

Fig. 8.14: Relative voltages and bias point chosen for the μ-follower circuit.

---

* The upper valve is only a *quasi*-cathode follower because it does not enjoy full 100% negative feedback like an ordinary cathode follower; its feedback fraction is at best only $R_{k2}/(R_{k2}+r_{a1})$ when R_{k1} is bypassed.

[9] Cooper, V. J. & Nixon J. E. A. (1951). Improvements in or relating to Thermionic Valve Amplifiers. British patent 657312. See also: Cooper, V. J. (1951). Shunt Regulated Amplifiers. *Wireless Engineer*, May, pp132-45.

putting it within ±90V of both cathodes. A generic value of 1MΩ was used for the upper grid leak. Since the upper valve behaves like a cathode follower this will appear bootstrapped to several times its real value, so there is no need to be overzealous with the coupling capacitance C_1. The 6SN7GT has an ideal μ of 20 so the lower valve should end up seeing an impedance that exceeds $20 \times 10k\Omega = 200k\Omega$ (at least when driving high-impedance loads). This is so much larger than r_{a1} (~10kΩ) that the gain should approach μ.

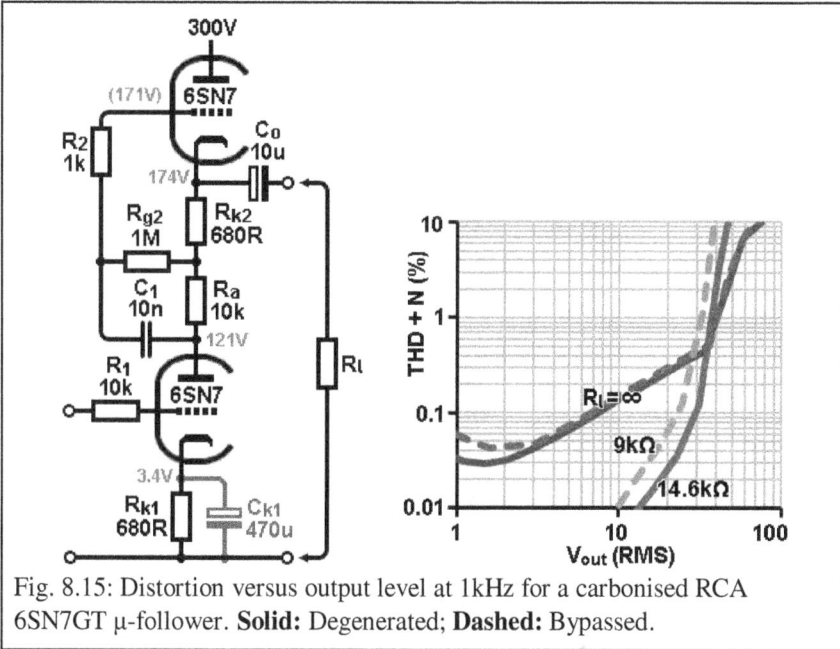

Fig. 8.15: Distortion versus output level at 1kHz for a carbonised RCA 6SN7GT μ-follower. **Solid:** Degenerated; **Dashed:** Bypassed.

Fig. 8.15 shows the final circuit with measured voltages, which are very close to the design values (the grid voltage of the upper valve is inferred as it cannot be measured with an ordinary multi-meter). Note that the total bootstrapped resistance is split into the upper bias resistor R_{k2} and the 10kΩ resistor R_a. The unloaded gain was found to be 17.2 when degenerated or 18.1 when C_{k1} was fitted –a very small difference that confirms the internal resistance of the current source is very large. The output impedance was 1.1kΩ and 700Ω, respectively. A corollary of allowing V_1 to achieve high gain is that the noise contribution of V_2 becomes negligible; the resulting EIN is practically the same as an ordinary gain stage followed by a cathode follower. For this the EIN was 1.2μV (20Hz–20kHz) and was exactly the same when the lower triode was arranged as a simple resistor-loaded gain stage. This is a distinct advantage over the SRPP and means the μ-follower can safely be used as the input stage for a low-noise preamplifier. But do not forget that its maximised gain also results in the maximum Miller capacitance; in this case about 130pF.

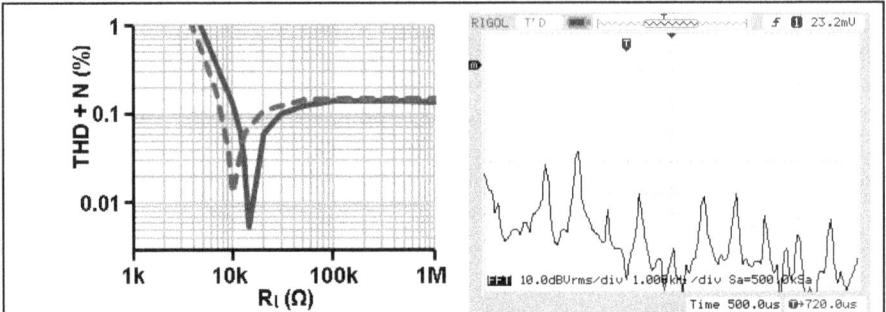

Fig. 8.16: **Left:** Distortion performance of the circuit in fig. 8.15 at 1kHz, $10V_{rms}$ output, when the resistance of R_l is varied. **Solid:** Degenerated; **Dashed:** Bypassed. **Right:** Harmonic spectrum within the distortion null for the degenerated case. At this point THD+N = 0.0053%.

Fig. 8.16 shows that the (unloaded) distortion is low and very similar to the transistor-CCS loaded example from section 6.2.7. Needless to say, it was constant with frequency. Furthermore, since the μ-follower is effectively a bootstrapped pair (section 7.9), albeit stacked on top of one another, it should come as no surprise to discover that it too shows a distortion null when suitably loaded. Fig. 8.16 shows the result of loading the circuit with the optimum R_l values of 14.6kΩ (degenerated) and 9kΩ (bypassed) as determined by experimentation, whereupon the distortion plummets to record low values. As usual, the null is achieved primarily by suppressing the second harmonic, as indicated by the rather exciting spectrum. Unfortunately, fig. 8.16 also demonstrates that the null is so narrow that there is little hope of maintaining it over time or with different valve samples. For example, a Russian 6H8C/6N8S in the same circuit produced similar results with values of 11kΩ (degenerated) and 7kΩ (bypassed).

The obvious disadvantage of the μ-follower is that the supply voltage needs to be quite high to accommodate the drop across two valves and the bootstrap resistance, and the strain on the heater-cathode insulation is similarly high (sometimes a separate elevated heater supply for the upper valve will be unavoidable). The appearance of a coupling capacitor is also sure to offend a few readers and admittedly makes the circuit appear rather busy. A modern alternative is to replace the upper triode with a MOSFET –such as the DN2535– to create a hybrid μ-follower (fig. 8.17) which mollifies all these criticisms in one fell swoop. Although it looks superficially like an SRPP, it is not, as the MOSFET has much higher g_m and internal resistance than a triode, so it behaves as a good constant-current source even though its source resistor (replacing R_{k2}) is quite small.

Fig. 8.17 shows an example circuit built around a 12AX7/ECC83 with the MOSFET's source resistor R_s adjusted to produce an anode current of 1.4mA. The unloaded gain was found to be 95.5 whether C_k was fitted or not, and the distortion, also shown in the figure, is extremely low, as we should expect from a CCS-loaded 12AX7. The figures do begin to degrade above 2kHz, as shown in fig. 8.18, as the

302

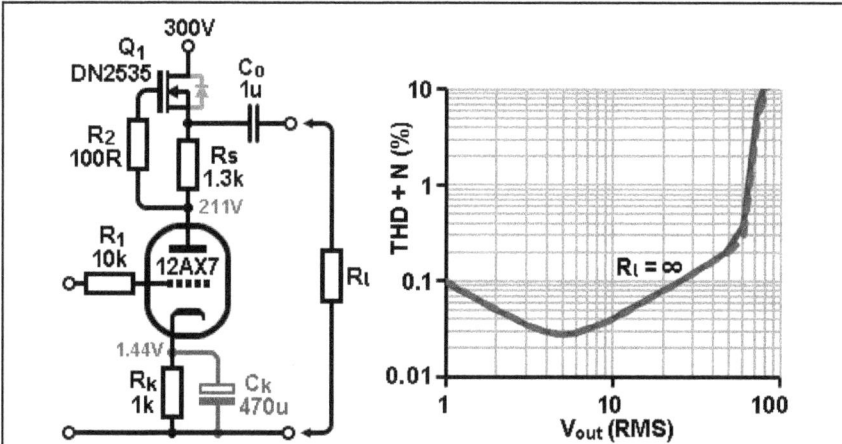

Fig. 8.17: Distortion versus output level at 1kHz for a Sovtek 12AX7LPS hybrid μ-follower. **Solid:** Degenerated; **Dashed:** Bypassed.

internal resistance of the current source begins to decline, even though the frequency response was still perfectly flat. The output impedance of the circuit was 5.6kΩ (degenerated) and 2.1kΩ (bypassed), and the Miller capacitance for this particular specimen was around 300pF; this high figure was due in part to the output coupling capacitor being rather close to the grid wiring.

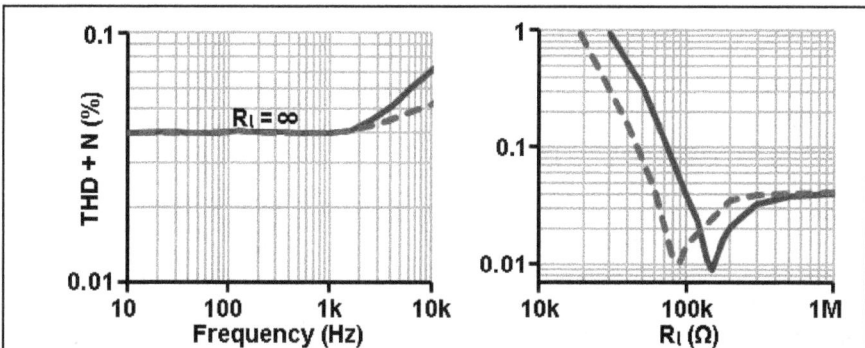

Fig. 8.18: **Left:** Distortion versus frequency for the circuit in fig. 8.17 at $10V_{rms}$ output. **Right:** Distortion versus loading at 1kHz, $10V_{rms}$ output. **Solid:** Degenerated; **Dashed:** Bypassed.

MOSFETs have larger stray capacitances than triodes, but the DN2535 has a surprisingly small gate-drain capacitance of <5pF. Its gate-source capacitance is an enormous 200pF but in this application it functions as a quasi-source-follower so this is bootstrapped to a much smaller value. The bandwidth of the circuit therefore remains wide even though the internal resistance of the valve may be quite large. With the grid-stopper shorted the bandwidth was somewhere around 200kHz when

303

bypassed, and 85kHz even when degenerated. The EIN (20Hz–20kHz) was 0.88µV when bypassed or 1.3µV when degenerated. There is little reason to leave such a circuit degenerated in practice, and LED bias would make for a supremely simple circuit that makes the most of every part. Such a circuit using an ECC88/6DJ8 operating within its noise minimum at 3mA would make a low-cost yet excellent input stage for a phono preamplifier, for example.

8.1.4: The Cascode

The cascode (fig. 8.19) is a totem-pole amplifier that uses a pair of triodes to simulate many of the properties of a pentode. The first cascode patent was applied for in 1928 and was intended as an improved radio-frequency amplifier (it even used a third triode as an active anode load),[10] but it did not catch on; eclipsed instead by the recently-introduced pentode valve. It was then re-invented a decade later for use as the error amplifier in valve-regulated power supplies and it was here that it was given the name 'cascode'[11] to imply **casc**aded-triodes having similar characteristics to a pen**ode.**[12] By the 1950s its advantages over the pentode were finally being exploited in television circuits, and several late-generation dual triodes were specifically designed for this application, including the ECC84/6CW7, 6BK7, 6BQ7, 6BZ7, and of course the popular ECC88/6DJ8. Transistor cascodes are now widely used in RF circuits so it has finally lived up to its inventor's expectation. But how does the cascode perform for audio?

Referring to fig. 8.19, V_1 is an ordinary common-cathode gain stage whose output directly feeds into the cathode of V_2. The grid of V_2 is fixed at some convenient DC voltage and is effectively grounded as far as AC is concerned, which makes V_2 a **common-grid gain stage**. The grid of V_2 serves a very similar function to the screen grid of a pentode, so for brevity it may as well be called the screen grid; it shields everything 'below it' from the electric field of the anode, thereby minimising the Miller effect in V_2. Now, remember that when looking into a cathode we usually see impedances divided by µ+1, so as V_1 looks into the cathode of V_2 it sees a relatively small load impedance of $(R_a+r_{a2})/(\mu_2+1)$. Consequently, the load line for V_1 will be near vertical, so it develops very little voltage gain. This means the Miller effect in the lower valve is also minimal, so the overall circuit can achieve a very wide bandwidth, which is why it is so useful for radio circuits. What little

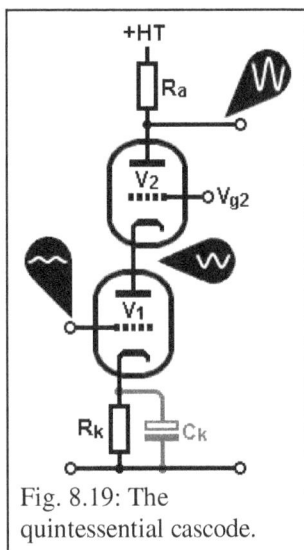

Fig. 8.19: The quintessential cascode.

---

[10] Francis, O. T. (1932). Radio. US Patent 1886386.
[11] Hunt, F. V. & Hickman, R. W. (1939). On Electronic Voltage Stabilizers, *Review of Scientific Instruments*, **10**, January, pp6-20.
[12] 'Cathode Ray' (1955). The Cascode. *Wireless World*, August, p397.

voltage gain *is* developed by V_1 is multiplied by the gain of V_2, so the overall gain is actually quite high, being equal to:

$$A_1 = -\mu_1 \frac{R_a(\mu_2 + 1)}{R_a + r_{a2} + r_{a1}(\mu_2 + 1)} \approx g_{m1}R_a \tag{8.10}$$

This is the same simplification as for a pentode and indicates that the characteristics of the upper valve are of little importance, so it doesn't need to be the same type. A cascode can therefore deliver the high gain and low Miller capacitance of a pentode but without the inherent drawbacks of a real pentode such as screen-grid current, partition noise, and strong microphonics. If R_k is unbypassed then it will degenerate the transconductance by a factor of $1/(1+g_{m1}R_k)$ and the gain will be reduced by approximately the same factor, but since low gain rather defeats the object of using a cascode in the first place, degeneration is rarely used.

When looking down into the output of the cascode the upper triode looks like a degenerated (common-*cathode*) gain stage, where the degenerating resistance is formed by the internal resistance of V_1. The total resistance seen looking in the anode is hence: $r_{a2} + r_{a1}(\mu_2 + 1)$. The output resistance is simply this in parallel with R_a, but since the latter is normally so much smaller it will dominate the total:

$$R_o = \frac{R_a[r_{a2} + r_{a1}(\mu_1 + 1)]}{R_a + r_{a2} + r_{a1}(\mu_1 + 1)} \approx R_a \tag{8.11}$$

This is again the same simplification as for a pentode. Although this means the output resistance of the cascode may be relatively high, it is at least very well defined and unlikely to change much with age. These are useful characteristics when driving equalisation networks, for example.

In most traditional cascode circuits the screen grid is fully bypassed to ground with a capacitor. By considering the potential divider formed by R_a and the internal resistance of the cascoded valves (and then turning the equation upside-down) it is easy to find the PSRR:

$$PSRR = \frac{R_a + r_{a2} + r_{a1}(\mu_2 + 1)}{r_{a2} + r_{a1}(\mu_2 + 1)} \approx 1 \tag{8.12}$$

In other words, virtually all of the noise on the power supply will appear unattenuated at the anode, so the cascode has a diabolical PSRR of ~0dB, just like a pentode, *if the screen grid is bypassed to ground*. However, unlike a pentode, the upper grid does not draw any current so there is no screen degeneration to worry about, and we will see later how leaving the grid *un*bypassed can be a benefit for PSRR.

Cascoded valves can be treated like a single compound valve, and the overall anode characteristics can be quickly estimated from those of the lower triode

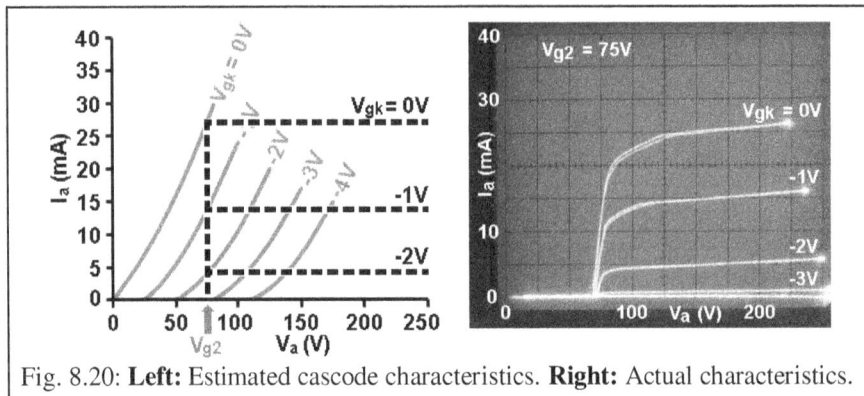

Fig. 8.20: **Left:** Estimated cascode characteristics. **Right:** Actual characteristics.

(even if the two valves are different). Draw a vertical line at V_{g2}, and where the grid curves intersect this line, draw horizontal lines out to the right. These are the new cascode grid curves. Fig. 8.20 shows an example using an ECC88 with a screen voltage of 75V together with the actual measured characteristics for comparison; the agreement is good enough for most purposes.[*] The characteristics look very much like those of a pentode except they are shifted to the right by an amount equal to V_{g2}, after all, the anode of the upper triode cannot swing below its own grid.

There are no strict rules about how to apportion the available voltage across each of the stacked elements, but things usually work out fairly nicely if the screen voltage is set to about one third of the HT voltage. Having already set V_{g2} to 75V, a supply voltage of $3 \times 75V = 225V$ will do in this case. A convenient value for R_a might be 10kΩ, thereby putting the load line somewhat below the knee of the characteristics.

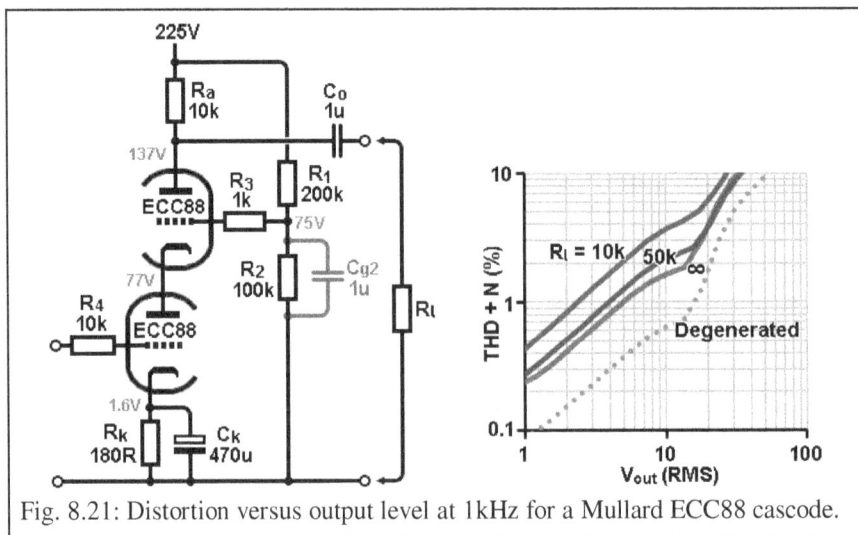

Fig. 8.21: Distortion versus output level at 1kHz for a Mullard ECC88 cascode.

---

[*] A more accurate method can be found in: Grant, W. (1957). Cascode Characteristics. *Wireless World*, January, pp33-36.

This allows for an AC load line to rotate and pass closer to the knee itself, which should afford good headroom. Choosing a quiescent anode current of 8mA implies a bias voltage of about −1.5V, so a 180Ω cathode bias resistor will do. The screen voltage will be set by a potential divider from the HT. Fig. 8.21 shows the complete circuit with measured voltages.

The unloaded gain was 87 while the Miller capacitance was a mere 80pF –much better than a pentode would achieve. However, since V_1 operates into a near-vertical load line its linearity is only as good as the linearity of g_m –which isn't very linear. As fig. 8.21 demonstrates, the distortion is very high (pure second harmonic) despite using a valve reputed for good linearity, and the maximum output before clipping is severely limited. Omitting C_k reduced the gain to 32.4 and the distortion by the same factor, as indicated by the dotted line in the figure. If anything can be said for the high distortion it is that at least it can't get much higher. For example, loading the stage with a ridiculously low 10kΩ only caused distortion to double. In short, the cascode is suitable only for handling very small signals such as those from a phono pickup or microphone.

In most traditional cascode circuits we find the screen fully bypassed to ground. However, this is an opportunity missed, as an unused grid is nearly always good for something. Treating V_2 as a degenerated gain stage, any signal applied to the screen grid will be amplified by a factor of:

$$A_2 = -\mu_2 \frac{R_a}{R_a + r_{a2} + r_{a1}(\mu_2 + 1)} \qquad (8.13)$$

This will be very small –rarely more than 3. Since we know that nearly all the power supply noise will appear at the anode it could conceivably be cancelled out by feeding forward to the screen-grid a fraction of the noise equal to $1/A_2$. Simply leaving out the screen grid bypassed capacitor will go some way to achieving this, as illustrated in fig. 8.22.

For example, for the cascode in fig. 8.21 the PSRR was measured at a measly 1.7dB with C_{g2} fitted. In other words, of the noise on the power supply, $10^{-1.7/20} = 0.82$ times as much appears at the anode, i.e. almost all of it. The gain of V_2 was found to be 2.54, so by omitting C_{g2} a fraction of power supply noise equal to $100k / (100k + 200k) = 0.33$ will be fed forward to the screen grid and promptly amplified and inverted by V_2 to produce a fraction $0.33 \times 2.54 = 0.85$ at the anode. This subtracts from the 0.82 already there, leaving −0.03. The negative sign indicates that we are actually feeding a little too much noise into V_2, leading to an inversion of the

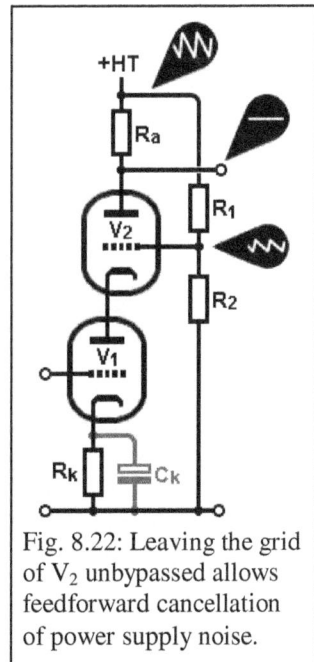

Fig. 8.22: Leaving the grid of V_2 unbypassed allows feedforward cancellation of power supply noise.

307

noise at the anode. Nevertheless, the PSRR is thus increased to 30dB (confirmed by measurement) –a pleasant reward just for leaving out a capacitor. A more accurate null could be achieved by feeding less noise to the screen grid, although this would mean reintroducing a capacitor (to block DC) and a critically adjusted resistor, which rather spoils the simplicity of the circuit, so the author was not inclined to chase the optimum.

Critics may point out that by leaving the screen grid unbypassed, Johnson and excess noise from the divider resistors R_1/R_2 will be amplified by V_2. However, its gain is so small that this does not add much to the total output noise. The same argument applies to noise generated inside V_2 itself; in other words, the overall noise performance of a cascode is usually determined only by the lower device (ignoring power-supply noise, of course). For the circuit in fig. 8.21 the EIN (20Hz–20kHz) was found to be 0.86µV with C_{g2} fitted, or 0.9µV without –exactly the sort of figure we would expect from a *single* ECC88 triode at this anode current.

A simple way to achieve an extra-low noise figure is to use a JFET as the lower device, so creating a hybrid cascode. It is advantageous to use a JFET with a very small cut-off voltage, quoted on the data sheet as $V_{gs(off)}$ (this is the FET equivalent of grid base), because this implies only a very small source resistor will be needed for biasing. This can then be left unbypassed to reduce distortion and degenerate the FET's otherwise excessive transconductance, but without introducing too much Johnson noise. A popular JFET choice is the 2SK170 (obsolete but still available as NOS) or LSK170, which have a quoted $V_{gs(off)}$ ranging from –0.2 to –2V; the author's sample measured –1.1V. The FET needs a few volts across itself to function, which can be accommodated by the ordinary bias voltage of a triode with a fairly large grid-base such as an ECC82/12AU7. The other characteristics of the upper device have little effect on circuit performance and can happily be ignored –there is no point using an expensive or exotic part here.

Fig. 8.23 shows the author's test circuit. The source resistor R_s was adjusted on test to obtain a current of 5mA, at which point the Sovereign 12AU7WA settled with a bias voltage of 7.8V, so the FET is safely shielded from high voltages. The gain was found to be 77.4 and the Miller capacitance was around 60pF. The output impedance was 9.7kΩ. The EIN was 0.3µV which is below the noise achievable with even the best all-valve circuit, and it was substantially free from microphonics too. It is no coincidence that circuits like this are popular as input stages for moving-magnet phono preamplifiers (chapter 10). Fig. 8.24 shows distortion versus level which is not much different from the ECC88 cascode previously, but with a typically triode-like harmonic spectrum. It is worth pointing out that a phono

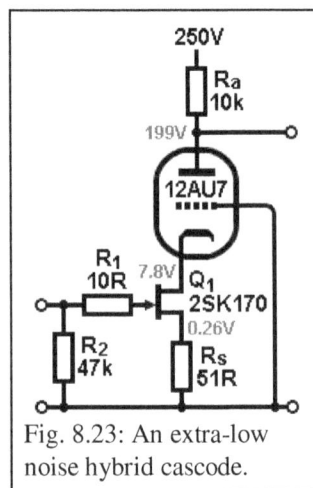

Fig. 8.23: An extra-low noise hybrid cascode.

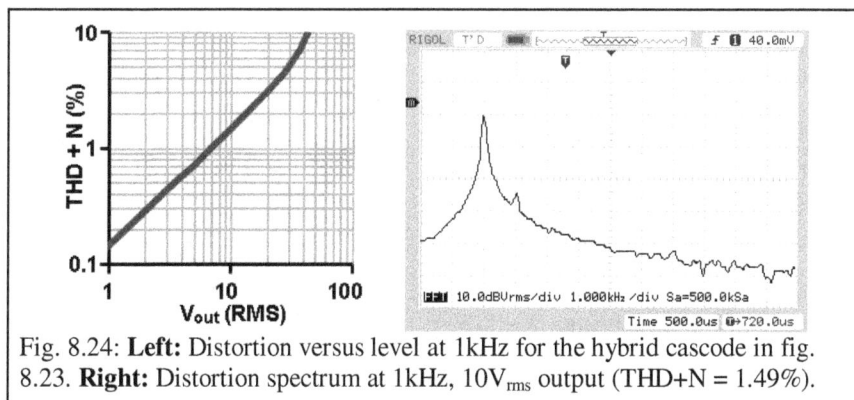

Fig. 8.24: **Left:** Distortion versus level at 1kHz for the hybrid cascode in fig. 8.23. **Right:** Distortion spectrum at 1kHz, $10V_{rms}$ output (THD+N = 1.49%).

pickup might produce a nominal audio level of only $5mV_{rms}$ which would be amplified to 386mV. By extrapolating beyond the left-hand side of the graph it would presumably experience about 0.06% THD most of the time.

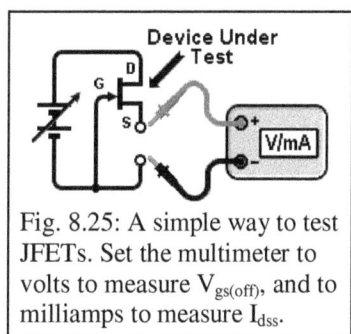

Fig. 8.25: A simple way to test JFETs. Set the multimeter to volts to measure $V_{gs(off)}$, and to milliamps to measure I_{dss}.

A practical disadvantage of this circuit is that FET characteristics vary enormously between samples, and to build a stereo amplifier we would need a matched pair.[*] Fig. 8.25 shows a simple method for measuring and matching JFETs using a digital multimeter. The power source can be a bench supply and should be set to about ten volts (a 9-volt battery is also ideal). Setting the meter to measure volts will virtually cut-off the FET and will indicate $V_{gs(off)}$ (a positive number will be displayed if the leads are connected as shown in the figure). Setting the meter to measure milliamps will turn the FET hard on and will indicate the saturation current, I_{dss} (10mA for the author's sample). Try to match these figures within ±5% between different samples. When the cascodes are finally built, any residual gain difference can be trimmed out by adjusting one of the source resistors.

8.2: The Cathode-Coupled Amplifier

The essential cathode-coupled amplifier is shown in fig. 8.26. It consists of two valves whose cathodes are tied together, hence the apt name bestowed on it by Schroeder[13] of RCA (interestingly, Schroeder's patented circuit was essentially identical to one patented a few years earlier by Crosby[14], also of RCA). From the

---

[*] The LSK389 is a dual matched version of the LSK170.
[13] Schroeder, A. C. (1949). Cathode-Coupled Wide-Band Amplifier, US Patent 2460907.
[14] Crosby, M. G. (1942). Limiting Amplifier, US Patent 2276565.

figure it is easy to see that V_1 functions as a simple cathode follower, so it suffers no Miller effect and does not invert the signal. V_2 is a common-grid gain stage (like the upper stage of a cascode) so it too enjoys minimal Miller effect and does not invert the signal. When a positive input is applied, V_1 conducts more current and so produces a positive signal at the cathode. This in turn causes V_2 to reduce its conduction, causing its anode voltage to rise, so the two valves operate in anti-phase (but not equal in magnitude). Thus overall the circuit is non-inverting, and the lack of Miller effect means it can achieve a very wide bandwidth. However, the internal cathode

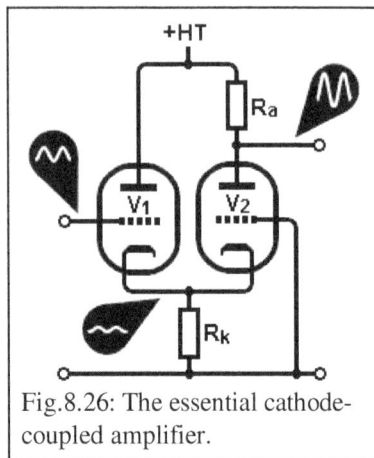

Fig.8.26: The essential cathode-coupled amplifier.

resistance of V_2 is relatively small and of similar order to the cathode resistance of V_1, so there is considerable attenuation as the signal is passed from one cathode to the other. The total gain of the circuit is therefore not very high. On the other hand, significant distortion cancellation can take place between V_1 and V_2 since they operate in anti-phase. The scope for variation in this circuit is very large; different valve types may be used, with bi-polar power supply rails, active loading and so forth –enough possibilities to fill another book. We will cover only a few first-principle examples here.

Putting the gain of the circuit into one big formula is not very enlightening. It is simpler to view it as the gain of the cathode follower (equation (7.1)) multiplied by the gain of the common-grid stage:

$$A = \frac{g_{m1}(r_{a1} \| R_k \| r_{k2})}{1 + g_{m1}(r_{a1} \| R_k \| r_{k2})} \cdot \frac{(\mu_2 + 1)R_a}{R_a + r_{a2}} \tag{8.14}$$

The cathode impedance r_{k2} is equal to $(R_a + r_{a1})/(\mu_2 + 1)$. Similarly, writing the output resistance all in one go results in an unwieldy formula. It is simpler to note that when looking into the anode of V_2 it appears to be a slightly degenerated gain stage, degenerated by the resistance formed by $r_{k1} \| R_k$, where r_{k1} is equal to $r_{a1}/(\mu_1 + 1)$. The overall output resistance is therefore:

$$R_o = \frac{R_a \left(r_{a2} + (r_{k1} \| R_k)(\mu_2 + 1) \right)}{R_a + r_{a2} + (r_{k1} \| R_k)(\mu_2 + 1)} \tag{8.15}$$

For the simplistic circuit in fig. 8.26, both valves share the same cathode resistor and therefore have the same bias voltage. However, V_1 has no anode resistor, so practically all the supply voltage is dropped across it, meaning it will end up running at a much higher anode current than V_2 (if the two valves are alike). For example, fig. 8.27 shows the ECC82/12AU7 anode characteristics with a typical 250V HT and 47kΩ load line. In order to keep the anode dissipation of V_1 within reasonable limits

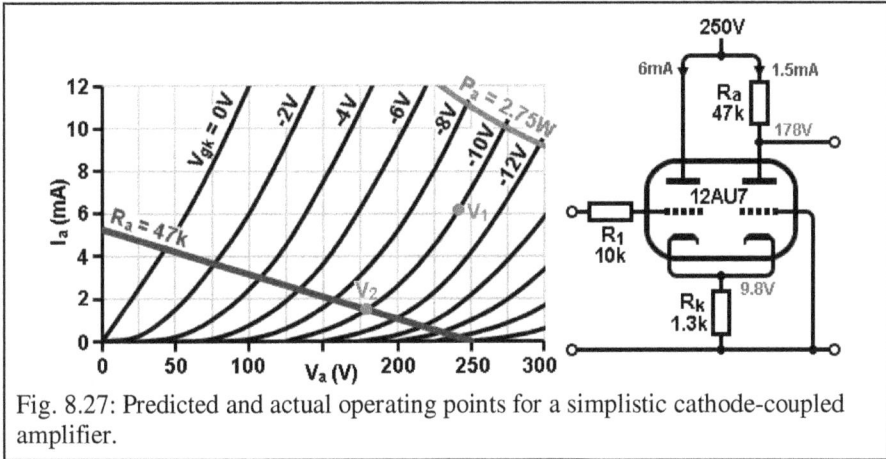

Fig. 8.27: Predicted and actual operating points for a simplistic cathode-coupled amplifier.

it will have to be biased to not less than about −10V, 6.1mA, leaving $250 - 10 = 240$V across it, as indicated by the dot in the figure. V_2 has no option but to share the same bias voltage, which is consequently on the cold side at about 1.6mA. The total current flowing in R_k is therefore 7.7mA meaning a $10V / 7.7mA = 1.3k\Omega$ resistor is required. Measured voltages for the actual test circuit –using a Sovereign 12AU7WA– are also shown in the figure.

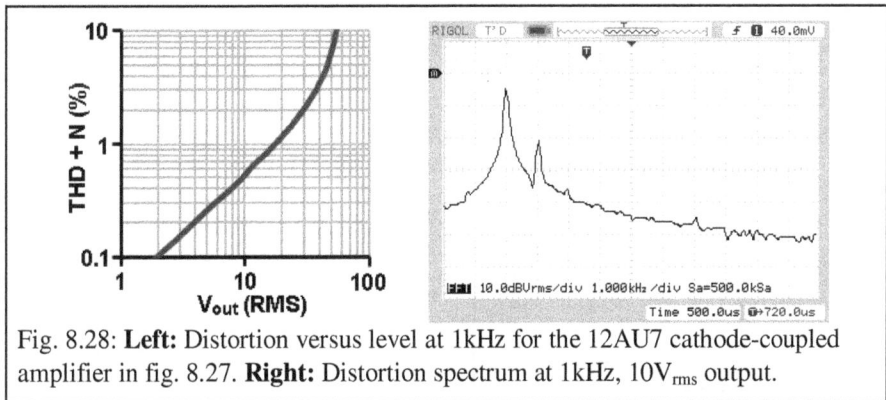

Fig. 8.28: **Left:** Distortion versus level at 1kHz for the 12AU7 cathode-coupled amplifier in fig. 8.27. **Right:** Distortion spectrum at 1kHz, 10V$_{rms}$ output.

The gain of the circuit was found to be 6.7. This low figure is the result of the cathode follower having to drive such a small load resistance, so it produces a gain to the cathode of only about 0.6. This is then amplified by the common-grid stage which has a gain of about 11, as confirmed by the load line. The input capacitance was about 22pF. Fig. 8.28 shows distortion versus level, which is rather better than would be expected for a cool-biased 12AU7 driving a 47kΩ load. The distortion spectrum indicates that the third harmonic is only about 15dB below the second, which is relatively strong for a triode, suggesting some distortion cancellation is taking place as expected.

311

The fact that V_1 hogs so much anode current at the expense of V_2 is unfortunate, so many designs aim for smaller, more equalised anode currents in each device. One solution is to use a valve with a larger grid base for V_2, while another is simply to use separate

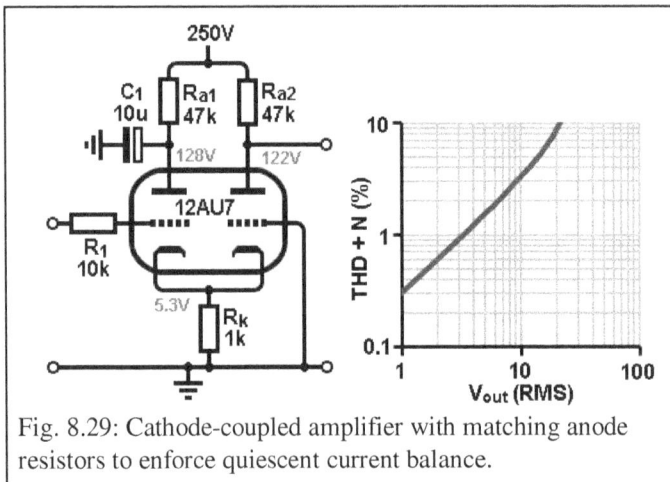

Fig. 8.29: Cathode-coupled amplifier with matching anode resistors to enforce quiescent current balance.

cathode resistors and to couple the cathodes together via a capacitor rather than directly. However, probably the most common choice is to retain matched valves and to add a matching anode resistor to V_1, thereby enforcing symmetry on the circuit. By itself this would reintroduce the Miller effect, so the anode of V_1 must also be decoupled to ground with a capacitor, which slightly improves the PSRR of the circuit too.

Since both valves now operate on identical *DC* load lines a more conventional bias voltage can be chosen, say –5V in this case. This should result in about 2.5mA in each valve, thereby demanding a 1kΩ shared bias resistor. Fig. 8.29 shows the test circuit –note that the anode of V_1 is bypassed to ground with C_1. The gain was unchanged at 6.7 but what we have saved in anode current we have lost in distortion, which is now very poor, presumably because the linearity of the cathode follower is much worse now that it is operating at lower current and driving an even heavier load.

Since reducing the current in V_1 to match V_2 has made things worse, perhaps we should try a contrary approach: bias V_2 hotter

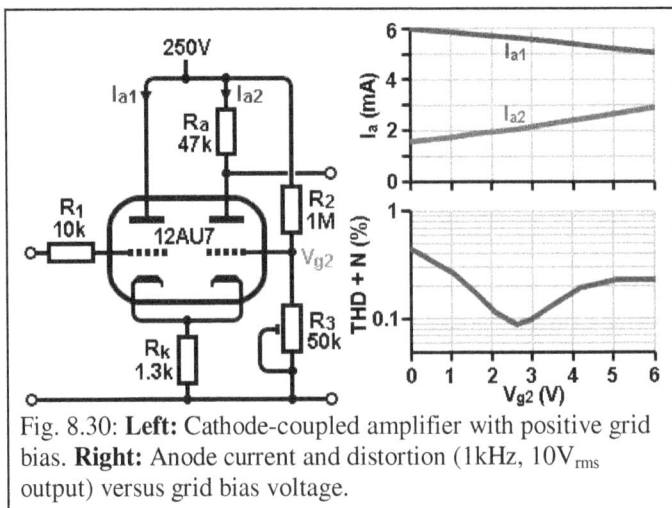

Fig. 8.30: **Left:** Cathode-coupled amplifier with positive grid bias. **Right:** Anode current and distortion (1kHz, 10V$_{rms}$ output) versus grid bias voltage.

312

instead. Going back to the circuit of fig. 8.27, current in V_2 could be increased by applying a positive voltage to its grid. As the grid voltage is made more positive and the current in V_2 increases, the voltage across R_k also increases, thereby reducing current in V_1 at the same time. Eventually the two could be made equal, although there is no particular advantage to this; as shown in fig. 8.30, distortion improves and eventually levels off before current balance is achieved. For this example a cancellation null occurs at 2.6V, but this will vary with different valve samples.

Yet another modification to the circuit might be to apply the positive voltage not from the HT but from the anode of V_2 itself, thereby creating a local feedback loop. However, let us go one better by additionally buffering the output with a conveniently DC-coupled cathode follower, as shown in fig. 8.31. This is a classic topology that actually combines two compound amplifiers –the second and third triodes form a compound pair known as a 'cathode repeater'.[15]

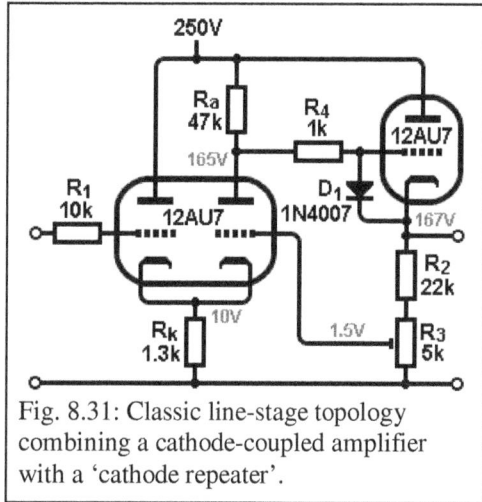

Fig. 8.31: Classic line-stage topology combining a cathode-coupled amplifier with a 'cathode repeater'.

In this case the feedback loop affects DC conditions as well as AC, so there is considerable interaction between the three stages as the trimpot R_3 (i.e. the amount of feedback) is varied. Increasing the feedback voltage not only affects distortion cancellation in the first two triodes but also reduces the anode voltage of the second stage, which in turn affects the biasing of the third stage.

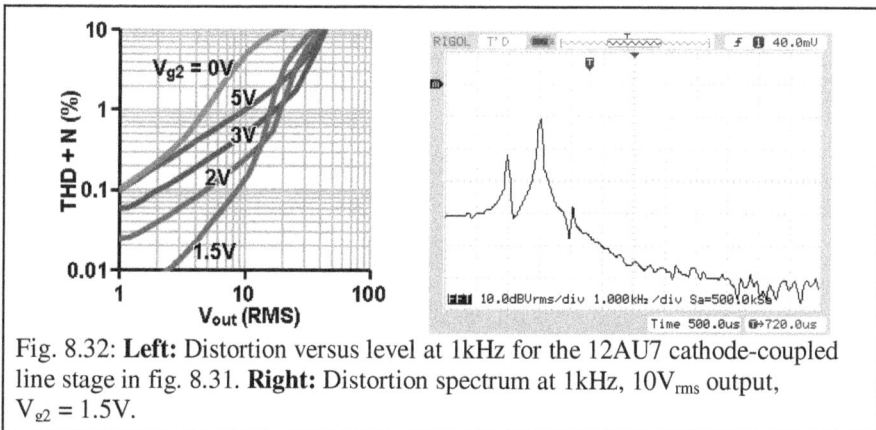

Fig. 8.32: **Left:** Distortion versus level at 1kHz for the 12AU7 cathode-coupled line stage in fig. 8.31. **Right:** Distortion spectrum at 1kHz, $10V_{rms}$ output, $V_{g2} = 1.5V$.

[15] Cooper, V. J. (1950). A New Amplifier: The Cathode Repeater, *Marconi Review*, **13** (97), p72.

Fig. 8.32 shows how distortion varied as the grid voltage of the second triode was increased. A null occurred at 1.5V resulting in dominant third harmonic, as shown in the figure. Note that this is not the same as the 2.6V optimum for the circuit in fig. 8.30 because we are now combining various distortion cancellation mechanisms with negative feedback too (a pair of German RFT ECC82s produced a similar null with a grid voltage of 2.1V, so anything in this range would be worth aiming for in practice). At this point the overall gain was 6.0 and the output impedance was 320Ω.

Such a low gain, wide-bandwidth circuit could form the basis of an excellent line stage, since a $100k\Omega$ volume pot could be placed at the input where it would have no trouble with the low input capacitance, and modern signal sources rarely need much amplification. In practice R_1 would be reduced to perhaps $1k\Omega$, the two bottles would probably be supplied from separate power supply smoothing filters (chapter 11), and output protection components would be required (section 7.13).

Chapter 9: Controls

This chapter covers some fundamental aspects of volume, balance, and tone controls. Admittedly, 'tone control' is fairly a vague term which suggests a vast range of possibilities, so here we will cover only a couple of variants suitable for modern, domestic hi-fi reproduction, rather than the more elaborate sort used for music mastering, or the crude sort found in old radios.

9.1: Source Selection

CD players and computer sound cards can usually deliver a maximum level of $2V_{rms}$. In practice, when playing the most heavily-compressed modern music styles, it is not unusual to measure an average level of around $400mV_{rms}$. This is enough to drive a typical power amplifier to full output without needing any additional gain. For MP3 players and cell-phones an absolute maximum level of $1.2V_{rms}$ is a fair assumption, with actual music (at the full-volume setting) hovering around $200mV_{rms}$. Phono preamps are highly variable and might produce nominal levels anywhere from 150mV to $500mV_{rms}$.

An input selector switch is simple in concept but not necessarily in practice, at least if we expect the very best performance. The high-impedance nature of valve circuits rules out electronic switching options such as FETs and CMOS analog switches. These invariably require low-impedance or virtual-earth techniques to achieve acceptable distortion, crosstalk, and offness performance (though they can be used in lo-fi or musical-instrument applications). We are therefore forced to use mechanical switching solutions.

Apart from the obvious need for clean, reliable contacts (implying gold plating to prevent oxidation), we still have to contend with stray capacitance between contacts, as illustrated in fig. 9.1. Each dashed capacitor may amount to half a picofarad or so in practice. With a simple input selector arrangement this is enough to allow noticeable crosstalk (increasing with frequency) from any unselected channels, unless the source impedance of the selected channel, R_s, is small enough to shunt the interfering signal away. There is no guarantee that this will be the case, particularly in the high-impedance world of valve technology, so more exclusive switching arrangements are worth exploring.

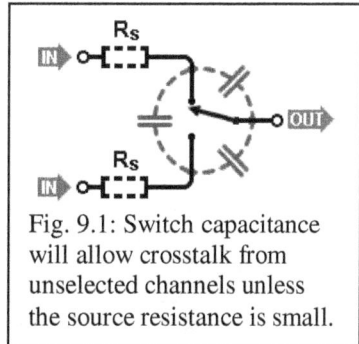

Fig. 9.1: Switch capacitance will allow crosstalk from unselected channels unless the source resistance is small.

Jones[1] has suggested using a rotary switch with alternate contacts grounded to reduce the stray capacitance between the remaining used contacts. Self[2] has reported

---

[1] Jones, M. (2012). *Valve Amplifiers* (4th ed.), Elsevier, Oxford. p549.
[2] Self, D. (2010). *Small-Signal Audio Design*. Elsevier, London. p394.

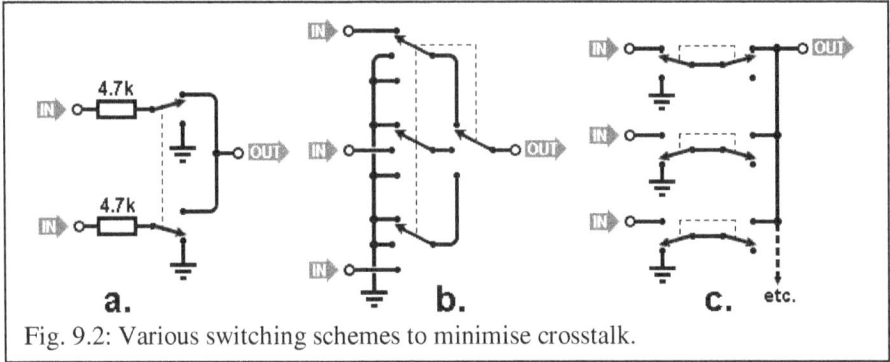

Fig. 9.2: Various switching schemes to minimise crosstalk.

a 5dB improvement in crosstalk with this approach, but it of course has the aesthetic disadvantage of redundant mute positions on the switch. An alternative solution is to arrange the selector to ground any unused inputs automatically. However, to avoid shorting the signal source directly to ground, a series resistor must be added, as shown in fig. 9.2a. This resistor is an unfortunate addition since it adds noise and reduces the bandwidth of the channel in use, though this may still be bearable compared with the other limitations in a valve system.

Fig. 9.2b shows a different scheme using a commonplace 4P3T (four pole, three throw) rotary switch, which allows selection between three channels and does away with the resistors. Any crosstalk now has to navigate the capacitance of two switches in series, the mid-point of which is grounded when not in use. Provided the source impedance is low (not more than a few kilohms, say) this scheme can achieve >100dB separation between inputs.

If more inputs are needed then it may be difficult to find a rotary switch with enough poles and throws, so we might resort to relays. Fig. 9.2c shows how this could be arranged using one DPDT (double pole, double throw) relay per input, and it accomplishes the same functionality as previous. This has the added advantage that the relays can be placed exactly where they are needed on the PCB while the selector switch or buttons can be positioned anywhere. The quality of the selector switch is no longer a concern since it only controls the relay coils. The relays *do* have to be of good quality, however, so they must be the hermetically-sealed kind; anything less will oxidise and become intermittent on small signals. The EA2 series of miniature relays made by NEC/TOKIN is exceptionally good.

9.2: Volume Control

Most hi-fi systems will need a volume control, which is unfortunate since it does nothing to improve fidelity. Attenuating a signal with a resistive divider always worsens the signal-to-noise ratio and introduces bandwidth limitations, so it is worth keeping in mind the following fundamental aspects of attenuator design.

Assuming the volume control is implemented with an ordinary resistive divider or potentiometer, the output resistance will reach its highest value at the −6dB setting.

At this point, half the total resistance lies on either side of the wiper. This occurs at the 12 o'clock position for linear pots and about 3 o'clock for logarithmic ones. Since the resistances are in parallel as far as the Thévenin equivalent is concerned, as shown in fig. 9.3, the output resistance at this setting will be one quarter of the total resistance. At all other settings the output resistance will be less. At the −6dB position, therefore, the Johnson noise added by the attenuator is greatest. For example, a 100kΩ pot has a maximum output resistance of 25kΩ which will add 2.8μV of audio-band Johnson noise to the attenuated signal, which is probably more than the following valve stage will add.

The attenuator also forms an RC filter with any load capacitance, shown dashed in fig. 9.3. At the −6dB setting the cut-off frequency will be lowest, and naturally we do not want this to encroach too much into the audio band. If we specify a

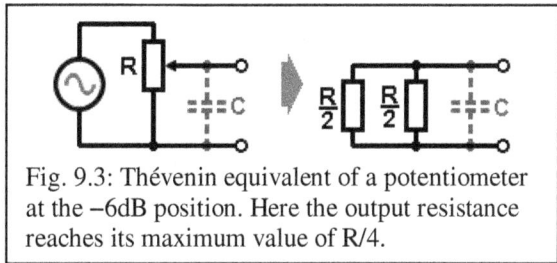

Fig. 9.3: Thévenin equivalent of a potentiometer at the −6dB position. Here the output resistance reaches its maximum value of R/4.

maximum loss of −0.1dB at 20kHz, say, then using equation (1.49) it is easy to find the lowest allowable cut-off frequency of the filter. A loss of 0.1dB means the gain of the filter is 0.989 at 20kHz, implying a cut-off frequency of:

$$f_c = \frac{Bf}{\sqrt{1-B^2}} = \frac{0.989 \times 20k}{\sqrt{1-0.989^2}} = 133\text{kHz}$$

Using a 100kΩ pot, with its maximum output resistance of 25kΩ, the load capacitance must therefore be no greater than:

$$C = \frac{1}{2\pi fR} = \frac{1}{2\pi \times 133000 \times 25000} = 48\text{pF}$$

The Miller capacitance of most gain stages will be greater than this, so the attenuator would probably need to be buffered with a cathode follower or similar. However, if we are willing to relax the limit to −0.5dB at 20kHz then this can tolerate a bandwidth of 57kHz and load capacitance of 111pF. This is much more manageable and could be further justified by the fact that volume controls are very rarely left at the −6dB position anyway.

9.2.1: Potentiometers

The most obvious way to implement a volume control is with a potentiometer. Typically this will be a dual-gang device so both stereo channels can be controlled with one knob. Since the human ear responds roughly logarithmically to sound intensity, the volume pot needs to have a logarithmic taper; as it is turned down the voltage is attenuated very rapidly at first, then more slowly as it approaches the quieter end of things. A volume control for hi-fi use will need to cover a range of at least 50 to 60dB, usually with a full mute at the final setting.

The trouble with most pots is that they have poor tolerance and poor matching between gangs, and the effects of this are all the greater when the tapers are not linear. At a given control knob position it is common to find one channel is attenuated somewhat more than the other. This is likely to be worse at low-volume settings and will pan the stereo image to one side or the other. Similarly, since the tracks will have slightly different total resistances, the source resistance of each pot will not be the same, leading to differences in bandwidth. But let's be realistic; a pot of reasonable quality will not be so bad as to pan the stereo image *wildly* to one side, or to make the bandwidth of one channel so narrow that it sounds like a subwoofer. A few degrees of pan is not the end of the world and can easily be corrected with a balance control (section 9.3), and as long as the pot is not loaded by too much capacitance the bandwidth will always be wide enough, whatever the matching. There is no shame in using an ordinary (but good quality) pot, and beginners should not feel immediately bullied into using an expensive stepped attenuator.

In vintage valve designs the volume control was often $1M\Omega$ but these days it is common to aim for $100k\Omega$, which is a compromise between heavier loading of the previous stage but much improved noise and bandwidth. So-called 'passive preamps' are nothing more than a volume pot in a box, which means the pot is not buffered from cable capacitance. The pot resistance in a passive preamp is therefore rarely more than $20k\Omega$ which is a very tough load for most vale preamps to drive with low distortion. Solid-state circuits are more capable in this regard and should be quite happy driving $10k\Omega$ or even $5k\Omega$ pots. In other words, passive preamps are not a good match for valve sources.

Although a volume pot can serve as the grid-leak path for the following valve, carbon tracks wear out and grid current grows worse over time, eventually leading to intermittency and scratching sounds when operating. It is therefore a good idea to add a dedicated grid-leak resistor. The extra loading will alter the pot taper somewhat, and will also cause the input resistance to vary with pot setting, falling to a minimum at full-volume when both resistances are directly in parallel. Having said that, the grid-leak resistor can be larger than usual since it is only a scratch-suppression measure. For example, a $4.7M\Omega$ grid-leak resistor will not cause a noticeable change in the taper of a $100k\Omega$ pot and only causes the input resistance to fall to a minimum of $97.9k\Omega$.

In some cases the law-bending effect of a loading resistor is deliberately exploited to create a logarithmic taper from a linear pot, as shown in fig. 9.4. Unfortunately, this only gives a useful logarithmic response down to about $-20dB$ (indicated by the dotted line in the graph), below which the volume rapidly drops off. This technique is therefore of little use as a hi-fi volume control, but it can be used for other purposes such as balance control. It is also commonly used in studio equipment where matching between channels is more important than taper (not only are linear pots more consistent than logarithmic ones, they are cheaper too, and studio equipment needs dozens of controls). With such equipment it is understood that the pot will nearly always be used in the favourable zone. In practice the slugging resistor

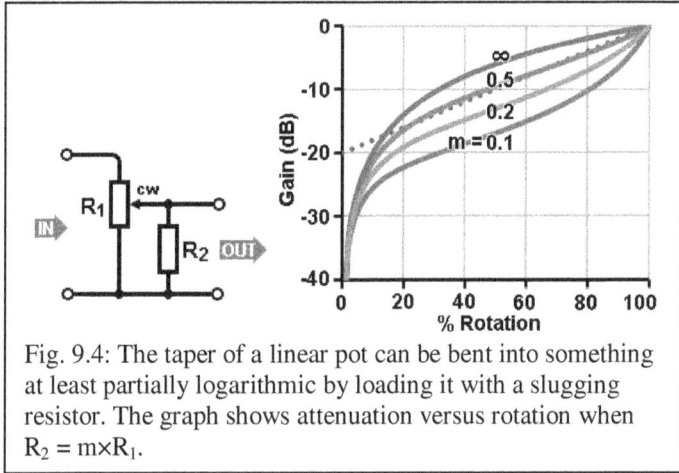

Fig. 9.4: The taper of a linear pot can be bent into something at least partially logarithmic by loading it with a slugging resistor. The graph shows attenuation versus rotation when $R_2 = m \times R_1$.

R_2 will often be a trimpot to allow factory compensation of pot tolerance.

9.2.2: Stepped Attenuators

When an ordinary potentiometer is deemed unsatisfactory for a volume control, the traditional, high-quality alternative is a stepped attenuator. This is simply a good-quality rotary switch with multiple throws, combined with a potential divider made from fixed resistors. Fig. 9.5 shows the basic principle using a 6-throw switch. A practical control will of course need many more positions for finer adjustment, and two gangs for stereo control. Commercial examples commonly have twenty-four throws, although Elma make rotary switches with up to forty-seven throws.

Since the resistors can be close-tolerance we can achieve very accurate matching between channels and have any taper we like. For example, a 24-throw switch might be arranged to give 55dB attenuation in twenty-two 2.5dB steps, with the bottom position grounded for full mute. However, a switch with fewer throws would have to use larger increments, and even 2.5dB steps might be considered too coarse. Therefore, it is not unusual to arrange for the step size to be ≤2dB at high settings where it will spend most of its time, with coarser steps at low volume settings.[3]

Working out the resistor values is straightforward, if a little tedious. It normally begins by creating a diagram like fig. 9.5 and labelling the desired

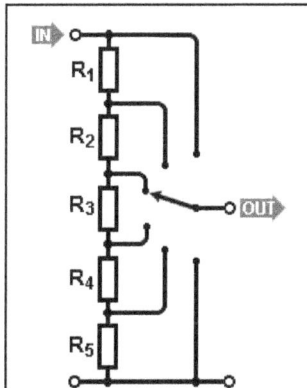

Fig. 9.5: The basic principle of a stepped attenuator.

---

[3] Leach, W. M. (1979). Build a Stepped Volume Control, *Audio*, February, pp44, 46.

319

attenuation at each step. If we call total resistance of the attenuator R_{total} (e.g. 100kΩ), and the resistance across which the output is taken at each step R, then R can be found from:

$$R = R_{total} 10^{\frac{dB}{20}}$$

(9.1)

Where 'dB' is the attenuation at each step, in decibels. We would then work down the ladder, subtracting each value of R from the one we had previously until all the (ideal) values are known. A spreadsheet can speed up this process as well as making it easy to scale the answers by any desired factor in order to push more of them towards standard values. There are also calculators on the internet which will do most of the work. If the attenuator is loaded by a grid-leak resistor then this will throw off the step sizes somewhat, unless it is taken into account at the calculation stage. But as already mentioned, the grid-leak can be several megohms which will not cause noticeable discontinuities in the step size of a 100kΩ attenuator.

Good quality switches with two gangs and plenty of throws are expensive. A low budget alternative is to use a pair of commonplace 2P6T rotary switches. One switch can provide coarse control and the other fine, effectively duplicating the functionality of a

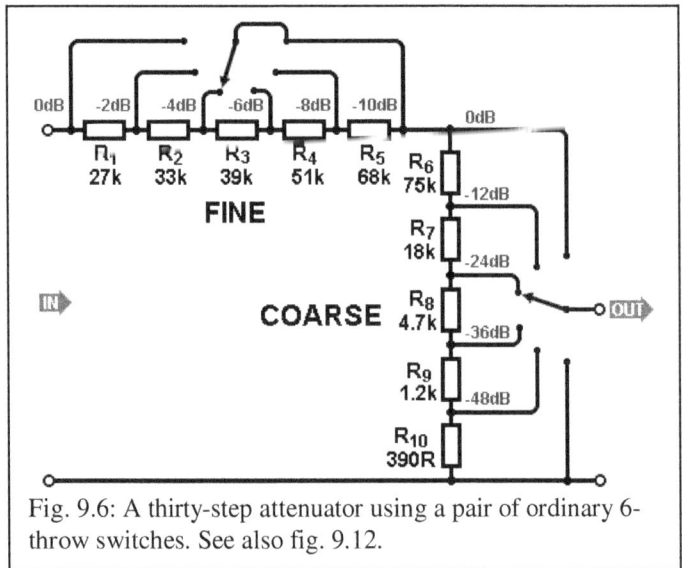

Fig. 9.6: A thirty-step attenuator using a pair of ordinary 6-throw switches. See also fig. 9.12.

$6 \times 6 = 36$ position switch. There are various ways this could be arranged, and fig. 9.6 shows one example. This design provides up to 58dB attenuation in 2dB steps, and despite using ordinary E24 resistors the maximum step error is only 0.3dB (and occurs in only three positions). Disadvantages of this design are that the input resistance is not constant, varying from 99kΩ up to 315kΩ, although this could equally be regarded as a 100kΩ attenuator that 'improves' at lower settings. Of greater concern is the output resistance which reaches a maximum of 69kΩ, meaning load capacitance must be kept below 20pF to guarantee >100kHz bandwidth, which certainly means buffering it with a cathode follower. Also, because it was felt desirable to have a total mute at the bottom position, five possible switch combinations are unused, but the author considers these acceptable shortcomings, considering the excellent value for money offered by the design.

320

9.2.3: Relay Attenuators

A further alternative to an expensive rotary switch is to use relays. These can be controlled by a cheap rotary encoder interfacing with a microcontroller, which could also handle infrared remote control and who knows what else. But rather than use twenty-four relays to duplicate a twenty-four position switch, the most economical option is a binary ladder attenuator in which a few relays are controlled in a binary sequence, e.g. 000, 001, 010, 011 etc. This allows a large number of steps using only a few relays.

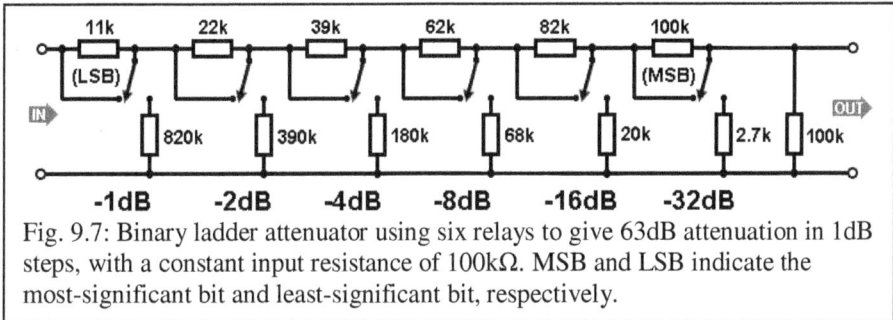

Fig. 9.7: Binary ladder attenuator using six relays to give 63dB attenuation in 1dB steps, with a constant input resistance of 100kΩ. MSB and LSB indicate the most-significant bit and least-significant bit, respectively.

Fig. 9.7 shows an example giving –63dB attenuation in $2^6-1 = 63$ steps, i.e. 1dB each, with a constant input resistance of 100kΩ (notice that each section of the attenuator has the same 100kΩ input resistance). A further relay would be needed for a total mute. A small disadvantage of the design is that the output resistance reaches 50kΩ at the worst setting, unlike an ordinary 100kΩ potential divider which would reach only 25kΩ. A less obvious but more awkward problem is that the variable timing of the relays can lead to unpleasant volume glitches when switching, particularly when stepping from 011111 to 100000, say. This makes a microcontroller essential to manage a clever timing schedule, but is rather beyond the scope of this book.

9.3: Balance Control

A balance control is used to adjust the relative volumes of the left and right stereo channels. This has the effect of panning the stereo image one way or the other and is normally used to correct for differences in loudspeaker sensitivity or non-ideal listening position. This is not quite the same as the panoramic (panpot) control found on recording equipment, however, as a balance control cannot send information from either channel to the opposite speaker.

A balance control does not need to cover a wide range; as little as 10dB amplitude difference is enough to give the impression of a pan fully to one side. In other words, if the right channel is made 5dB louder and the left channel 5dB quieter, the sound will appear to come almost completely from the right speaker –there is no need to silence one channel completely. Whether or not a balance control should maintain the same volume at all settings –by increasing one channel by the same amount that

321

the other is attenuated– is largely a matter of preference. Since it is only occasionally used for fine corrections there seems to be no unequivocal reason why a constant-volume characteristic should be adhered to.

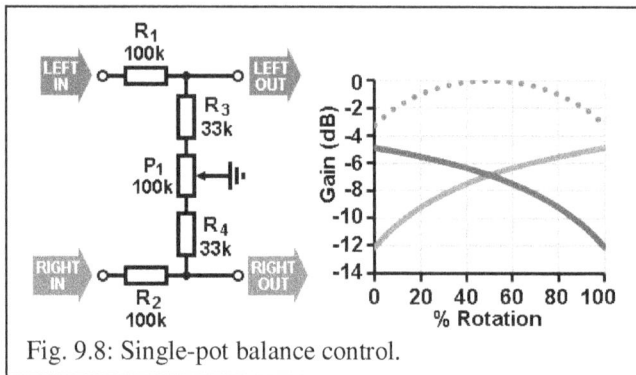

Fig. 9.8: Single-pot balance control.

Fig. 9.8 shows one implementation of a balance control which is appealing because it uses a single-gang linear pot. The gain response of the two channels is plotted, and the dotted line shows the apparent change in overall volume (i.e. the sum of the two channels, normalised to the centre positon) which falls by a tolerable ~3dB at the extreme settings. As with volume controls, it is wise not to rely solely on the pot for a grid-leak path, so additional grid-leak resistors should be added as required, though they can be several megohms. The main handicap of this circuit is that the output resistance varies from $25k\Omega$ to $57k\Omega$, so the load capacitance must be kept below 28pF for >100kHz bandwidth.

Fig. 9.9 shows an active balance control based on the same principle as previously. With the pot at the centre setting both valves provide equal gain. When turned to one end, one cathode bypass capacitor is connected to ground via the pot wiper, so the corresponding valve achieves maximum gain, while the opposite stage becomes more degenerated and its gain decreases (it is worth remembering that the output resistances of the two stages will vary too). To obtain a smooth

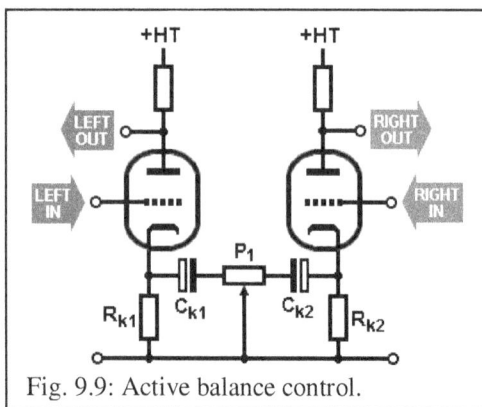

Fig. 9.9: Active balance control.

range of control with a linear pot, without too much drop in overall volume at the extremes, the balance pot P_1 should have a value of around one to two times larger than either cathode resistor. The capacitors can be made arbitrarily large so they provide bypassing down to sub-audible frequencies.

322

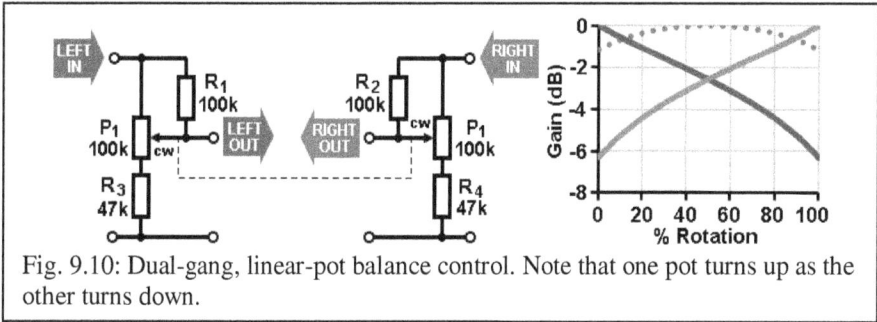

Fig. 9.10: Dual-gang, linear-pot balance control. Note that one pot turns up as the other turns down.

Another approach to a balance control is to use a slugged linear pot, as mentioned earlier. This can give a very good approximation to a logarithmic response over a limited range of attenuation, and a balance control only needs to cover a limited range. A dual-gang pot is used for stereo, wired so that one channel is turned up as the other is turned down, so a more constant volume characteristic is obtained. Fig. 9.10 shows an example of this. Pull-up slugging (R_1 and R_2) is used because it results in less attenuation at the centre setting than pull-down slugging. The input resistance varies with setting, falling as low as 97kΩ in this case, which is an unavoidable consequence of using a slugged pot.

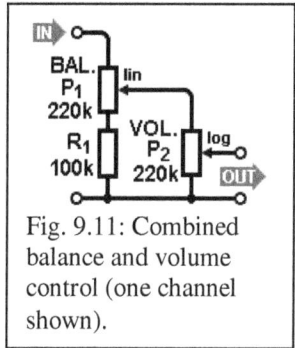

Fig. 9.11: Combined balance and volume control (one channel shown).

A popular variation on the previous theme is to combine volume and balance controls by letting the volume control do double-duty as the slugging resistance, as shown in fig. 9.11. Note that the pots have been increased to 220kΩ so the input resistance does not drop too low at the highest gain setting, which also means the Johnson noise and bandwidth worsen. The balance

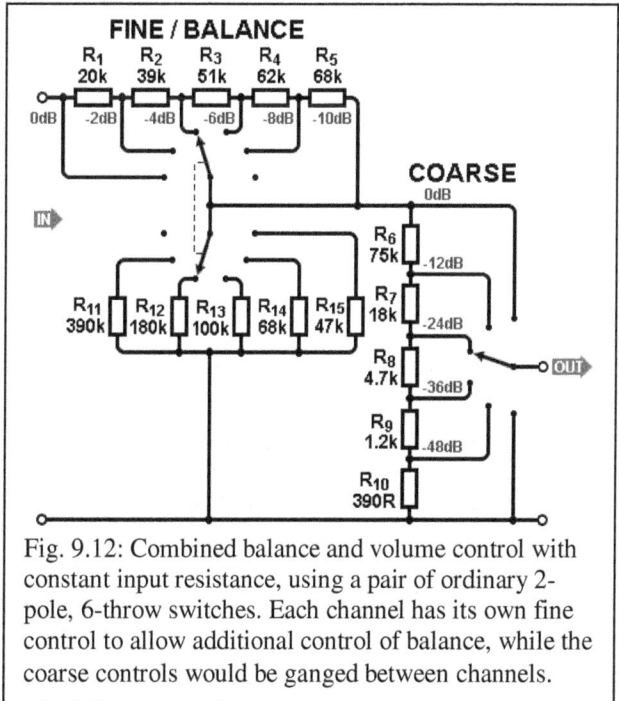

Fig. 9.12: Combined balance and volume control with constant input resistance, using a pair of ordinary 2-pole, 6-throw switches. Each channel has its own fine control to allow additional control of balance, while the coarse controls would be ganged between channels.

323

control presents a variable source resistance to the volume control which will alter its taper slightly (although pots are shown, they could of course be stepped attenuators). Equally, the balance may vary slightly with volume setting. However, these errors are very small and not noticeable in practice.

One final option is worth mentioning. The stepped attenuator from fig. 9.6 used one switch for coarse control and one for fine, which suggests the possibility of using separate fine controls for the left and right channels so they serve the secondary purpose of balance control. This does mean manipulating three knobs in a somewhat interactive way, but some users may be willing to accept this in exchange for the excellent economy of such a design. Using separate fine controls also leaves us with a spare gang on each, which could perhaps be used to control a string of LEDs to indicate channel imbalance. Alternatively, fig. 9.12 shown a modified version of fig. 9.12 where the spare gang is used to switch in resistors R_{11}–R_{15}; these are designed to give the whole circuit a constant input resistance of 101kΩ ±2%. Pleasingly, the maximum output resistance of the circuit is also reduced to 34kΩ.

9.4: Tone Controls

Tone controls have become unfashionable, it being argued that if the components of the system are of sufficient quality there should be no need for tone correction. This is either a clever way of persuading consumers they need to upgrade expensive equipment to fix minor tonal imbalances, or a monumentally naïve view of acoustics. Certainly it is easy enough to make an *amplifier* with a razor-flat frequency response, but it is quite impossible to do the same with loudspeakers or headphones. Room acoustics add a whole new dimension of frequency-response anomalies. Tone controls cannot hope to correct for everything but they can certainly make broad-brush improvements to the listening experience. Provided they give a flat response at the centre setting, or can be switched out, we can always access the unadulterated signal when required.

Early tone controls invariably provided an enormous amount of boost and cut, which might be necessary for sound recording but not modern hi-fi reproduction. For the latter, only subtle tonal shifts are needed to compensate for such things as:
- Room acoustics;
- Mismatching in loudspeaker frequency response;
- Frequency response anomalies due to finite loudspeaker damping;
- The falling treble response of well-used vinyl.

A review of early literature will show that for many years a veritable menagerie of tone control designs were used in audio circuits, many of them bafflingly overcomplicated. A hint of rationality came in 1949 when a combined bass and treble circuit was published –apparently independently and at almost the same time–

by E. James[4] in the UK and James Faran[5] in the US.[*] The circuit could in principle provide symmetrical cut and boost, and a flat response at the centre setting. But almost immediately the famous Peter Baxandall[6] pointed out that passive tone controls nearly always had to be followed by a gain stage to make up for the loss, and that these functions could be combined. In other words, by taking the James tone controls (albeit slightly refined) and placing them in the feedback loop of a gain stage, whatever gain is thrown away is traded for increased linearity and headroom. The Baxandall tone control has been the *de facto* basis for most hi-fi tone controls ever since.

9.4.1: Passive James/Baxandall Tone Controls

Although the Baxandall feedback approach is the most popular and unquestionably the most sensible for solid-state circuits, there is still a place for passive tone controls in valve equipment. Passive controls cannot suffer from the headroom and stability problems of feedback circuits, and also have the small advantage that they do not introduce a phase inversion. Moreover, it is easier to appreciate the design limitations of feedback-based controls after studying them first in passive isolation.

The original James tone control arrangement is shown in fig. 9.13, and the easiest way to optimise it for hi-fi use is to exploit its symmetry. By viewing the controls as a pair of potential dividers it is obvious that a perfectly flat response will be obtained if the time constants in the upper and lower arms are matched. This is particularly easy to achieve if we make the top and bottom arms identical, in which case the overall insertion loss will be −6dB at the centre of rotation (using linear pots). Up to 6dB of boost will then be available, which is quite enough for modern hi-fi use. In order to maintain

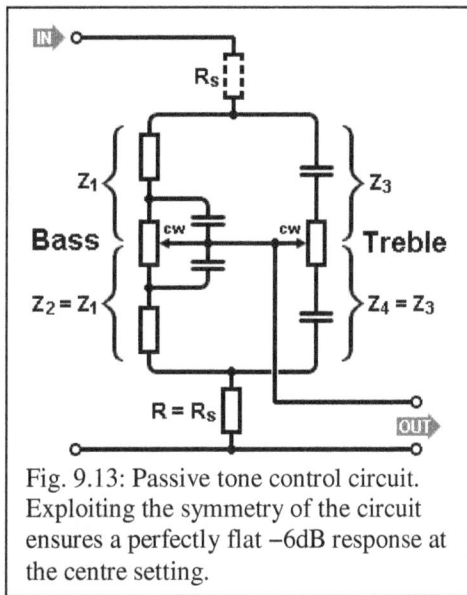

Fig. 9.13: Passive tone control circuit. Exploiting the symmetry of the circuit ensures a perfectly flat −6dB response at the centre setting.

---

[4] James, E. J. (1949). Simple Tone Control Circuit, *Wireless World*, February, pp48-50.

[5] Faran, J. J. (1949). Low Loss Tone Control, *Audio Engineering*, June, p31.

[*] Williamson of the famous Williamson amplifier almost hit upon the same circuit in the same year. Williamson, D. T. N. (1949). Design of Tone Controls and Auxiliary Gramophone Circuits, *Wireless World*, October, pp365-9.

[6] Baxandall, P. J. (1952). Negative-Feedback Tone Control, *Wireless World*, October, pp402-5.

the symmetry of the upper and lower halves of the circuit an additional resistor has been added to the original design, which must be equal to the source resistance R_s. Without it there will be a droop in the treble response at the supposedly-flat setting. Even with this addition, the source resistance should be kept small or the range of treble control (using convenient values) will be limited.

Having set the requirement for matched impedances in the upper and lower arms of the dividers we find that we will need to match several capacitors, either by using precision devices or by measuring them by hand. Trouble can be saved by using the alternative version of the circuit in fig. 9.14, which uses only two capacitors (they will still need to be matched between stereo channels, though). The change to the treble control means R_3 becomes necessary to minimise interaction between

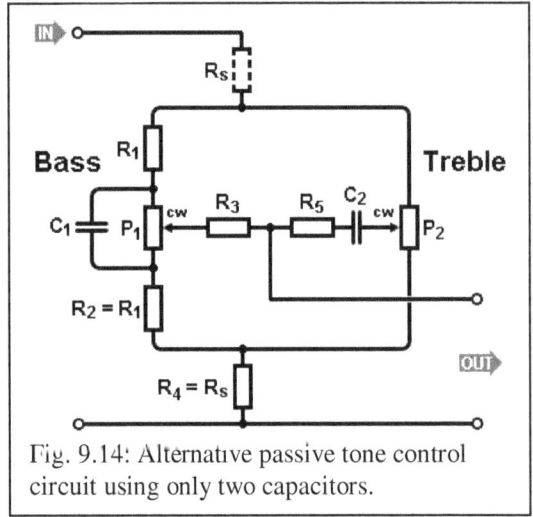

Fig. 9.14: Alternative passive tone control circuit using only two capacitors.

the two controls, and it will often be roughly equal to P_1. R_5 is not essential but is often included to produce equal boost and cut from both controls, mainly for aesthetic reasons (i.e. the graph looks nicer in magazines).

The end-stop resistors R_1 and R_2 (assumed equal) set the maximum amount of boost and cut. The main design limit here is that we cannot let R_1 and R_2 be zero ohms as this would place C_1 directly in parallel with the driving stage, which would produce excessive distortion due to the falling reactance at high frequencies. In other words, R_1+R_2 largely determine the worst-case input impedance, so they will probably need to be at least 20kΩ each. The maximum bass cut is simply $R_2/(R_1+R_2+P_1)$, but this range will shrink if the circuit is loaded by, say, a grid leak resistor (see shortly).

At maximum boost, the bass zero or 'pivot' frequency has a time constant of $(R_2\|P_1)C_1$. In other words, for a desired pivot frequency f:

$$C_1 = \frac{1}{2\pi f(R_2 \| P_1)} \qquad (9.2)$$

This will normally be set somewhere between 500Hz and 1kHz.

Assuming $R_1 = R_2$, at maximum boost the treble zero or pivot frequency has a time constant of $(R_1+2R_3+R_4+P_1/2)C_2$. In other words, for a desired pivot frequency f:

$$C_2 = \frac{1}{2\pi f (R_1 + 2R_3 + 2R_4 + P_1/2)} \qquad (9.3)$$

This will also normally be set somewhere between 500Hz and 1kHz, so there may be some mid-range overlap between the controls.

Provided the source impedance is negligibly small, the same maximum boost and cut is produced from both controls when:

$$R_4 = \frac{R_1(R_1 + 2R_3 + P_1/2)}{P_1} \qquad (9.4)$$

If the input impedance of the circuit is too low then it will load down the preceding stage. Therefore, in a valve circuit the pots will usually be between 470kΩ and 1MΩ so the other impedances will also work out suitably large. The equations given above are idealised since they assume no source or load impedance; these are best dealt with using a computer circuit simulator.

Fig. 9.15 shows a practical version of the circuit and deserves further comment. Although a grid-leak path is already provided via the bass pot, carbon tracks wear out. A dedicated grid-leak resistor, R_6, has therefore been added. However, by itself this would load the circuit and upset the flat response at the centre setting, so an identical resistor R_7 is added to the upper half of the circuit to maintain symmetry. In this case these components are 1MΩ and load the circuit fairly heavily, compressing the total range of control to about 8dB. This is still sufficient for the subtle adjustments needed in a hi-fi system but, if a wider range is needed, increase R_6 and R_7.

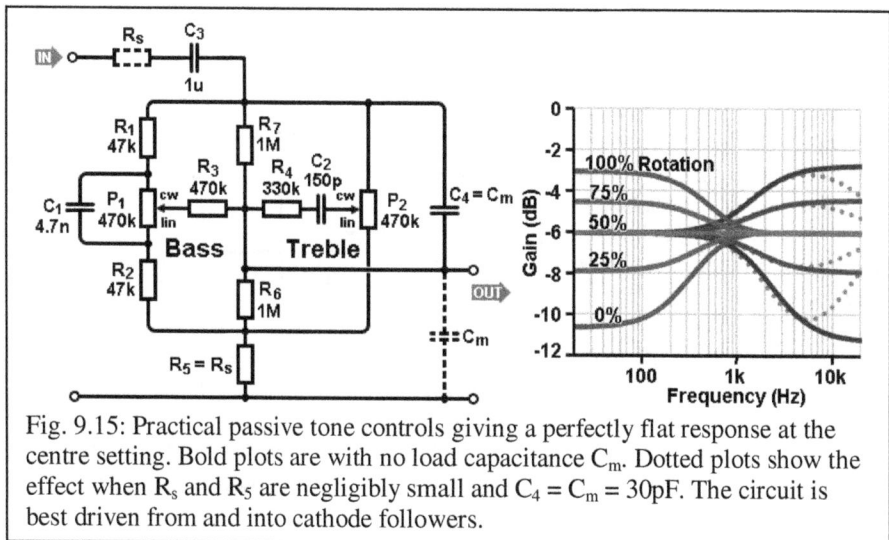

Fig. 9.15: Practical passive tone controls giving a perfectly flat response at the centre setting. Bold plots are with no load capacitance C_m. Dotted plots show the effect when R_s and R_5 are negligibly small and $C_4 = C_m = 30pF$. The circuit is best driven from and into cathode followers.

327

There are two further components that unavoidably upset the symmetry of the circuit and *will* degrade its flatness. The first is the input coupling capacitor C_3 which is required to keep DC off the pots and the subsequent valve grid. Fortunately, it is not difficult to make this arbitrarily large so that it does not impair the low frequency response. The other hidden component is the input (Miller) capacitance of the following valve, C_m. In theory, symmetry can be restored by adding an equal capacitance C_4 to the upper half of

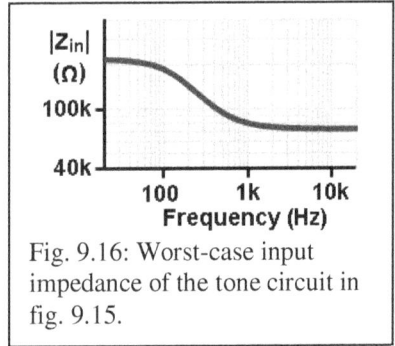

Fig. 9.16: Worst-case input impedance of the tone circuit in fig. 9.15.

the circuit, but this only works if R_s and R_5 are negligibly small. Even if this is the case, the added capacitances bypass the treble pot and will truncate the range of control at very high frequencies (although this is not necessarily a bad thing[7]), as indicated by the dotted lines on the frequency response graph. Therefore, to maintain flatness, the circuit needs to be driven either from a very low source resistance or be loaded with very little capacitance. This implies a cathode follower on one side or the other, or both. The pots could of course be replaced by stepped attenuators, and ordinary 6-throw switches would probably work well. The worst-case input impedance occurs when the bass control is at maximum and the treble control at minimum. This is plotted in fig. 9.16 and falls to $73k\Omega$ at high frequencies, mainly determined by R_1+R_2.

9.4.2: Passive Tilt Control

A delightfully simple method of making subtle adjustments to equalisation is a 'tilt' control, which has a see-saw frequency response. It boosts treble and cuts bass at one extreme, and does exactly the opposite at the other extreme, returning to flat at the centre setting. Tone controls based on this principle have been used in guitar amplifiers since at least the 1960s, but their application to hi-fi appears to have been delayed until 1970 when Ambler[8] eventually inserted one into a

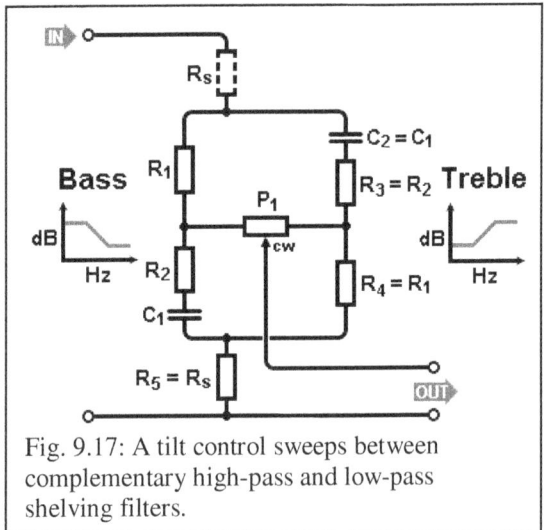

Fig. 9.17: A tilt control sweeps between complementary high-pass and low-pass shelving filters.

[7] Thomas, M. V. (1974). Baxandall Tone Control Revisited, *Wireless World*, September, pp341-3.
[8] Ambler, R. (1970). Tone Balance Control, Wireless World, March, pp124-6.

feedback loop, just as Baxandall did with the James tone controls twenty years earlier. The use of only one pot (dual gang for stereo) is an obvious economic advantage over Baxandall controls.

Fig. 9.17 shows the prototypical circuit which consists of high-pass and low-pass (shelving) filters with a pot connected between them to allow continuous adjustment from one extreme to the other. It is again desirable to exploit the symmetry of the circuit (diagonally this time) to ensure a truly flat response at the centre setting. Although flatness is maintained by making $R_5 = R_s$, the source resistance should be kept below $10k\Omega$ or it will compress the range of control and make the circuit more sensitive to load capacitance. An easy way to proceed with the design is to make P_1 around one to two times the value of R_1, then calculate R_2. At high frequencies the input impedance will be dominated by $(R_1+R_2)/2$, so we will probably want to make these resistors at least $220k\Omega$ to avoid excessive loading of the preceding stage.

Assuming $R_1 = R_4$, to achieve equal boost and cut at each extreme:

$$R_2 = R_3 = \frac{R_1^{\,2}}{2R_1 + P_1} \tag{9.5}$$

To a rough approximation the crossover frequency, f, will have a time constant of $C_1(R_1+R_2)$, or in other words:

$$C_1 = C_2 = \frac{1}{2\pi f(R_1 + R_2)} \tag{9.6}$$

The design can be further refined using computer simulation or adjustment on test.

Fig. 9.18 shows a practical circuit. Again, it is desirable to provide a dedicated grid-leak resistor rather than to rely only on the pot. This is accomplished by adding R_7

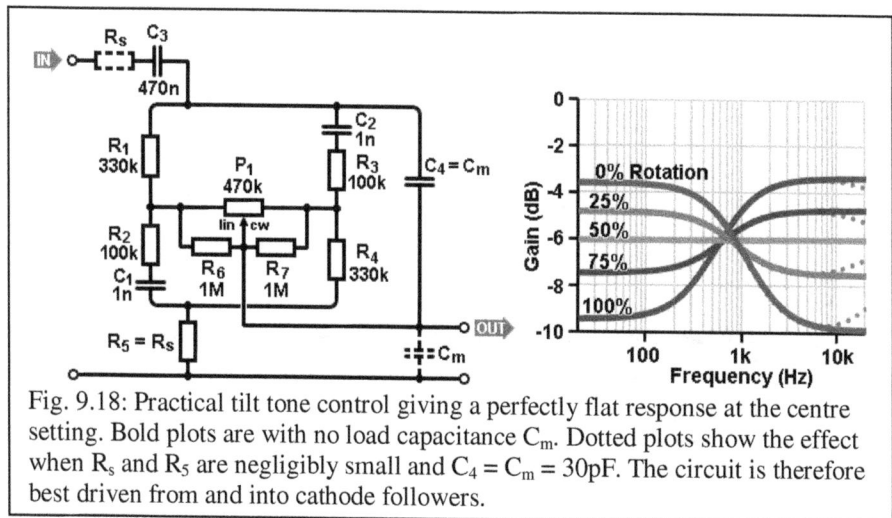

Fig. 9.18: Practical tilt tone control giving a perfectly flat response at the centre setting. Bold plots are with no load capacitance C_m. Dotted plots show the effect when R_s and R_5 are negligibly small and $C_4 = C_m = 30pF$. The circuit is therefore best driven from and into cathode followers.

and R_8 in the positions shown, to maintain symmetry and a flat centre response. This is an ideal arrangement because these resistors can be then taken into account when calculating R_2 and R_3, since they can be lumped in with P_1. As described previously, any load capacitance will drag down the high-frequency response, but symmetry can be maintained by adding an equal capacitance C_4, provided R_s and R_5 are negligibly small. The input impedance is worst when set to maximum bass boost. This is plotted in fig. 9.19 and falls to 190kΩ at high

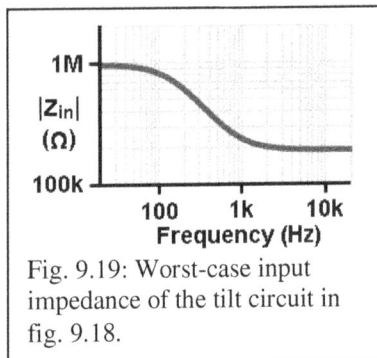

Fig. 9.19: Worst-case input impedance of the tilt circuit in fig. 9.18.

frequencies, which is not quite so taxing on the driving stage as the previous James/Baxandall controls.

9.4.3: Active James/Baxandall Tone Controls

The gain thrown away in passive tone controls usually has to be made up again with another stage of gain. Baxandall popularised the idea that by placing the tone controls in a feedback loop around that stage, we trade the gain for greater linearity and headroom. As a bonus, it is easier to achieve large amounts of boost/cut this way than with passive controls, and an interesting corollary is that it tends to force the equalisation curves to be more evenly spaced (in decibels), which is aesthetically pleasing if nothing else. Fig. 9.20 shows the transformation from the raw, passive tone control circuit into the feedback or 'active' tone control (ground symbols are included to give a sense of familiarity). The output terminal of the passive circuit is fed into the input of an inverting amplifier, and the amplifier's output feeds back into what used to be the passive circuit's common or ground terminal. Baxandall's 1952 design provided an unwieldy ±20dB of boost/cut and used an SP61 pentode to muster the loop gain. This is far more control than any modern audiophile needs, and by relaxing the range of boost it is possible to use a triode for the job.

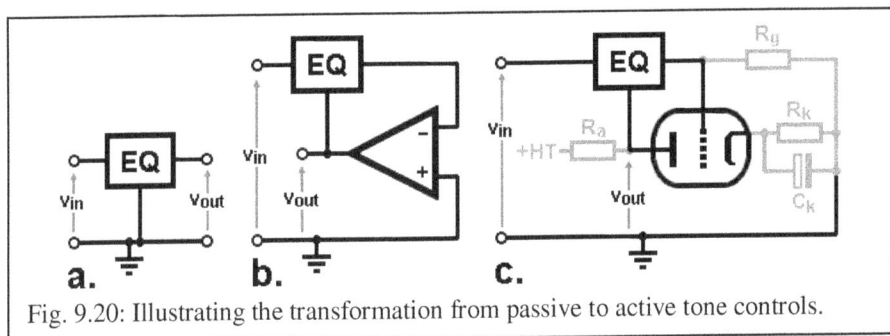

Fig. 9.20: Illustrating the transformation from passive to active tone controls.

Controls

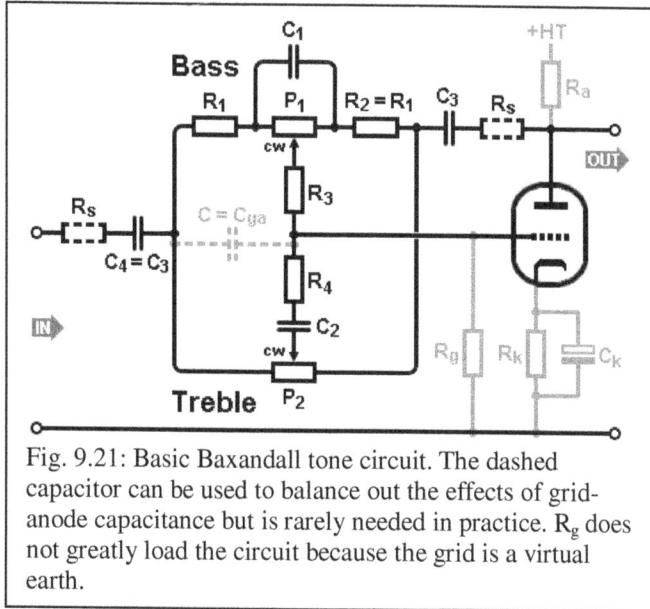

Fig. 9.21 shows such a circuit, drawn in the usual orientation. Because this is an inverting feedback amplifier the grid should behave as a virtual earth, so any impedance connected between grid and ground will have very little signal voltage across it and will appear bootstrapped to a high value. Consequently, the grid-leak resistor R_g does not

Fig. 9.21: Basic Baxandall tone circuit. The dashed capacitor can be used to balance out the effects of grid-anode capacitance but is rarely needed in practice. R_g does not greatly load the circuit because the grid is a virtual earth.

significantly load the circuit, so it should not require 'balancing out' to maintain a flat response at centre setting, unlike the passive version of the circuit. Exceptions may occur with very low-gain valves like the 6SN7 or ECC82 which may still require balancing (with a resistance equal to the bootstrapped value of R_g). For similar reasons, Miller capacitance is also not a problem. In theory, its effects can be balanced out by connecting a capacitor equal to C_{ga} between input and grid (shown dashed), but C_{ga} is such a small value that stray capacitance alone will go some way to doing this, so this component is never needed in practice. The source resistance R_s should be small (e.g. a cathode follower), but if it is not, an equal resistance should

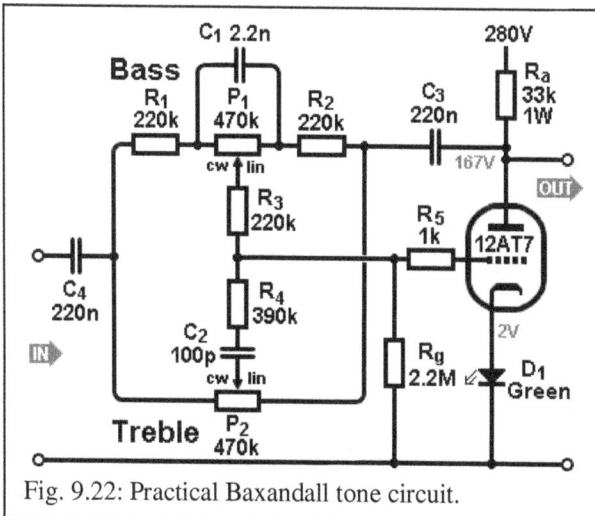

Fig. 9.22: Practical Baxandall tone circuit.

be added in the feedback path (regardless of the output resistance of the valve stage itself), as shown in the figure. Similarly, the input and feedback coupling capacitors should be equal in value. The worst-case input impedance occurs at maximum treble boost and, perhaps ironically, it falls as R_3 *increases*. It is therefore advisable not to make R_3 any larger than P_1

331

even though this means a certain amount of interaction between the controls.

Setting the range of bass control is easy by using the approximation that the gain of a shunt feedback amplifier is equal to $-R_f/R_{in}$ (though we shall discard the minus sign). At very low frequencies where C_1 and C_2 appear to be open circuit, the maximum gain is therefore $(R_2+P_1)/R_1$ and the minimum is $R_2/(P_1+R_1)$. To ensure a flat response at the centre setting it is standard practice to set $R_1 = R_2$ which also results in symmetrical boost and cut. At the centre setting the input and feedback impedances are equal so the gain will ideally be unity, though slightly less in practice, owing to the limited loop gain provided by the valve.

For the treble control, provided $R_1 = R_2$, the maximum boost is equal to $(R_1+2R_3+R_4+P_1/2)/R_4$. Maximum cut is the inverse of this since we have already enforced symmetry. Notice that this does not depend on the treble pot at all. The same boost and cut is produced from both controls when:

$$R_4 = \frac{R_1(4R_3 + 2R_1 + P_1)}{2P_1} \tag{9.7}$$

The capacitors can be selected using equations (9.2) and (9.3).

The ECC81/12AT7 is an ideal candidate for an active tone control because it can achieve relatively high open-loop gain, and can operate with enough anode current to overcome any possibility of slew-rate limiting, unlike the ECC83/12AX7 or 6SL7, for example. It may not have the best open-loop distortion figures, but this is a feedback circuit so the closed-loop performance should still be quite good. LED biasing is also ideal as it ensures constant open-loop gain all the way down to DC (there is no need to bypass the LED with a capacitor since this is not a low-noise application).

Fig. 9.22 shows a practical circuit tested by the author using a Sovereign 12AT7WA. Fig. 9.23 shows the frequency response contours as the controls are varied from minimum to maximum in quarter-rotation intervals (driven by the very low output impedance of the AP1). At the centre

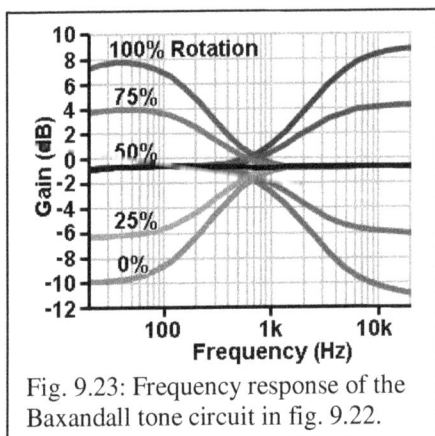

Fig. 9.23: Frequency response of the Baxandall tone circuit in fig. 9.22.

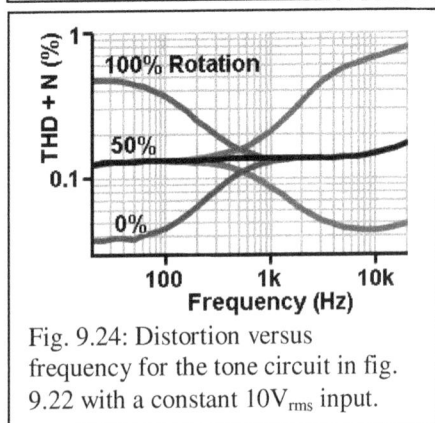

Fig. 9.24: Distortion versus frequency for the tone circuit in fig. 9.22 with a constant $10V_{rms}$ input.

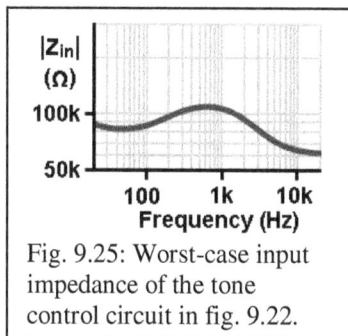

Fig. 9.25: Worst-case input impedance of the tone control circuit in fig. 9.22.

setting the gain was exceptionally flat at −0.7dB ±0.1dB between 18Hz and 40kHz. The coverage is also attractively symmetrical, and the cut/boost is more even (in decibels) with pot rotation than the passive control from section 9.4.1. Fig. 9.24 shows distortion versus frequency for a constant *input* level of $10V_{rms}$. At full boost there is less feedback and a larger output level, so distortion increases, and *vice versa*. But even at full boost the figures are quite low, considering the signal level, and in a typical hi-fi system the circuit would probably be handling much smaller input levels so the distortion would be proportionately lower during actual use. The worst-case input impedance occurs with both controls at full boost and is plotted in fig. 9.25 (simulated), falling to a minimum of 62kΩ.

9.4.4: Active Tilt Control

The passive tilt control from section 9.4.2 can also be placed in a feedback loop, and exactly the same design principles apply as discussed in the previous section. If the source resistance cannot be kept below of couple of kilohms then an equal resistance should be added in the feedback path. Provided R_g is relatively large then it is bootstrapped to the point where a balancing resistance between input and grid is not required to maintain flatness.

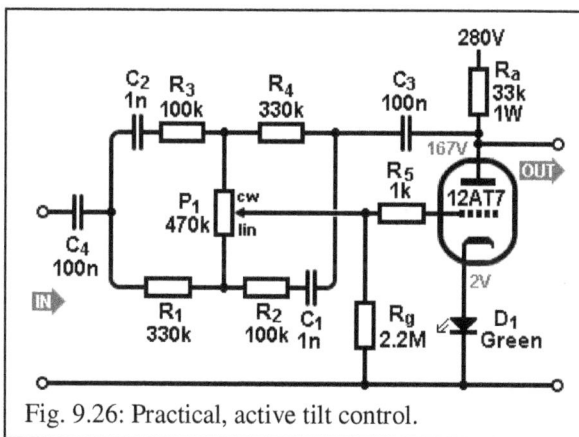

Fig. 9.26: Practical, active tilt control.

Maximum bass boost is equal to $(R_4+P_1)/R_1$, and equal boost and cut obtain using the same equation as for the passive circuit. The worst-case input impedance occurs at maximum treble boost and is a little more than half R_3, so it is desirable to make this larger than 100kΩ. Since this also determines R_2, and assuming we already have a potentiometer in mind, we can then find R_1.

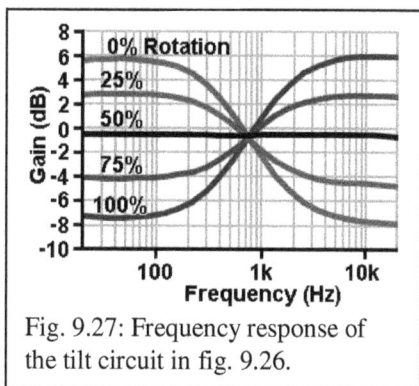

Fig. 9.27: Frequency response of the tilt circuit in fig. 9.26.

333

Fig. 9.26 shows a practical circuit with component values unabashedly borrowed from the passive version, and fig. 9.27 shows the frequency response contours. At the centre setting the gain was −0.5dB ±0.1dB from <10Hz to 21kHz, and the range of boost/cut is more than twice what it was for the passive circuit. Fig. 9.28 shows distortion versus frequency for a constant input level of $10V_{rms}$. At the centre setting the distortion was dominated by the second harmonic, with the third being about 22dB lower. Boosting either way caused the odd harmonics to increase relative to the second. The worst-case input impedance occurs at full treble boost and falls as low as 62kΩ, as shown in fig. 9.29 (simulated).

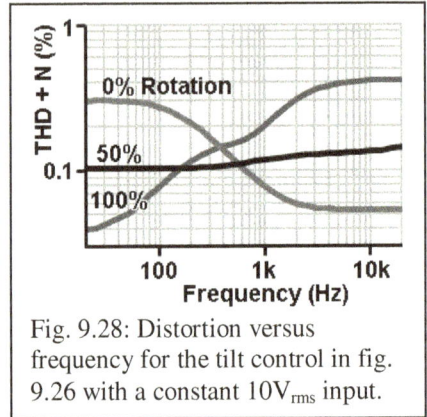

Fig. 9.28: Distortion versus frequency for the tilt control in fig. 9.26 with a constant $10V_{rms}$ input.

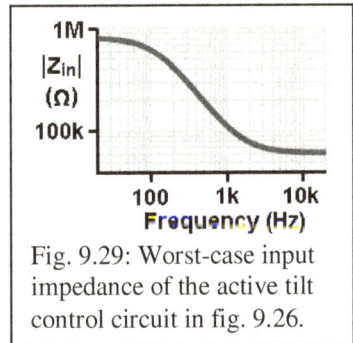

Fig. 9.29: Worst-case input impedance of the active tilt control circuit in fig. 9.26.

Chapter 10: The Phono Stage

A phono stage is a preamplifier that accepts the tiny signal from a phonograph pickup cartridge and boosts it to a line level suitable for driving the rest of the audio chain. Unlike most hi-fi devices, however, it does not have a flat frequency response, but applies very specific equalisation to compensate for the deliberately uneven frequency response of the vinyl disc and pickup. The requirements of the phono stage are arguably the most stringent in the average hi-fi reproducing chain; it has to handle the smallest signals with the lowest possible noise, provide the highest gain, *and* apply this strict equalisation. Modern vinyl achieves a dynamic range of about 50dB and an SNR of at least 60dB. This level of performance is no match for compact discs, but it is astoundingly good considering the number of reproducing steps involved in creating a vinyl record (look it up), and does not seem to be a barrier to listening enjoyment.

10.1: RIAA Equalisation

Most disc cutting lathes have a 'constant-velocity' characteristic, that is, the maximum velocity with which the cutting stylus can oscillate back and forth is the same whatever the frequency. This is equivalent to saying the recorded signal has the same maximum slew-rate limit at all frequencies, so the cutting stylus cannot move as far laterally at high frequencies as it can at low frequencies. If nothing was done about this it would mean bass frequencies would be cut with much greater amplitude than high frequencies. This would not necessarily matter because magnetic pickup cartridges are also constant velocity devices –but in the opposite direction– so their output increases with frequency and complements the record cut. However, lathing low frequencies with such high amplitude would waste space on the disc as well as force the highs to suffer a much worse signal to noise ratio. Therefore, to compensate for these effects, the low frequencies are attenuated and the high frequencices are boosted before being cut into the disc. This is referred to as pre-emphasis. An exactly complementary de-emphasis must be applied on playback, thereby bringing everything back to a flat frequency response.

Early recording companies each had their own preferred pre-emphasis curves. Although most were fairly similar, they were sufficiently unalike that the more up-market phono preamplifiers had to provide a range of switchable de-emphasis curves. By the 1950s the situation was getting out of hand, so several standards bodies around the world agreed to adopt the same curve. This was close enough to several of the pre-existing ones that it would work for those too, if you weren't too picky, so it had some backwards compatibility. Over time this has come to be known as the RIAA equalisation curve after the Recording Industry Association of America –the standards body with the catchiest name. Fortunately, this standard was adopted shortly before the stereo LP arrived in 1958.

The RIAA equalisation curve is shown in fig. 10.1 and the idealised Bode version is shown faint. The curve is defined by three cut-off frequencies (two poles and one zero), which are more usually quoted as time constants:

- $\tau_1 = 3180\mu s$ (50.05Hz)
- $\tau_2 = 318\mu s$ (500.5Hz)
- $\tau_3 = 75\mu s$ (2122.1Hz)

Remember that $\tau = \dfrac{1}{2\pi f}$.

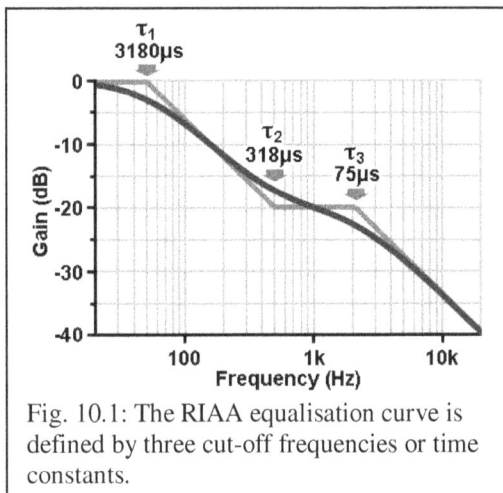

Fig. 10.1: The RIAA equalisation curve is defined by three cut-off frequencies or time constants.

These time constants plug into the transfer function that describes the curve:

$$H(s) = \frac{(1+s\tau_2)}{(1+s\tau_1)(1+s\tau_3)} = \frac{s\tau_2+1}{s^2\tau_1\tau_3 + s(\tau_1+\tau_3)+1} \tag{10.1}$$

It is then a matter of designing a filter that matches this transfer function. This is not as tedious as it sounds, as most of the hard work has been done for us by earlier practitioners. As described in section 10.3, there are a number of textbook circuits with simple design equations that can be used without having to grapple directly with the equation above.

10.1.1: The IEC Amendment

In 1972 the IEC published its own version of the RIAA curve, shown in fig. 10.2. It is identical to fig. 10.1 except for a low-frequency roll-off defined by a pole at 7950µs (20.02Hz). This is intended to reduce subsonic signals, although the exact motivation for this change is unclear. Self[1] suggests it may have been a reaction to the brief fashion for noise reduction systems which did not work well with subsonic content. Whatever the reason, the IEC amendment has been widely criticised by audiophiles for being ineffectual, as it provides barely 6dB

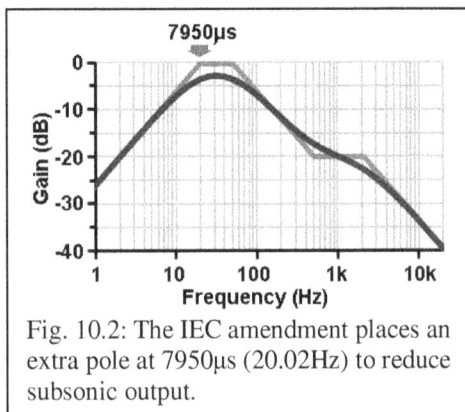

Fig. 10.2: The IEC amendment places an extra pole at 7950µs (20.02Hz) to reduce subsonic output.

---

[1] Self, D. (2014). Optimising RIAA Realisation, *Linear Audio*, 7, pp43-62.

of rejection at typical rumble frequencies, yet introduces an uncomfortable 3dB loss at 20Hz plus equally uncomfortable phase delay within the audio band. Many designers therefore choose to make it a switchable option, if it is included at all. More effective subsonic attenuation can be achieved with an active (usually Butterworth) filter.

10.1.2: The Neumann Fallacy

Logically, the pre-emphasis applied at the disc cutting stage cannot continue boosting treble forever; it must level off at some point. In the 1990s an erroneous rumour propagated that the most popular Neumann lathes placed this 'unofficial' pole at 50kHz, and that this should therefore be corrected by adding a 50kHz zero to the RIAA de-emphasis curve. This is a peculiarly naïve recommendation, however, since it ignores the many other bandwidth limitations in the tape machine, pre-emphasis equaliser, cutter amplifier and cutting head itself. If we are supposed to compensate for one, should we not compensate for all? It turns out the answer is emphatically no, because it is all taken into account in the cutting process.

For example, the Neumann SAB74B equaliser (with which most records will have been cut) levels off its RIAA pre-emphasis curve at about 100kHz, but added to this is the falling open-loop gain of the (LF356) opamp used in the circuit. At the input of the unit there is also a 49.9kHz second-order Butterworth filter and a further pole at 482kHz to reject RF. All these factors combine to

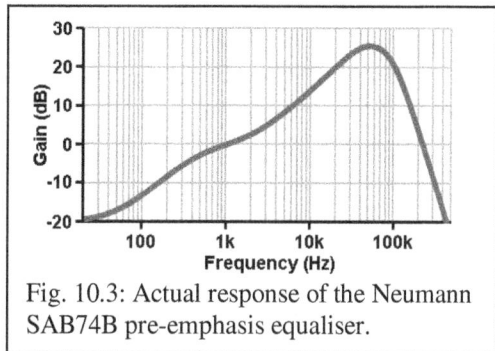

Fig. 10.3: Actual response of the Neumann SAB74B pre-emphasis equaliser.

produce the response shown in fig. 10.3. Clearly, above 50kHz the response does not merely level off but drops at a rate approaching third-order, so adding a single 50kHz zero to the de-emphasis curve is not going to make much of a correction. In fact, it makes the replay response about 0.6dB *worse* at 20kHz because it does not properly cancel the complex poles of the Butterworth filter, although phase delay is negligibly affected and remains no worse than 4µs at 20kHz relative to 20Hz. The response of the pickup cartridge alone will utterly swamp these trifling numbers.

That aside, it must be appreciated that the cutting engineer applies *ad hoc* manual equalisation to the master, whilst monitoring the recording on a standard RIAA-equalised playback system (gentle 20Hz/20kHz boost/cut controls are included in the SAB74B). Adding another zero to the RIAA equalisation curve is therefore second-guessing something which the cutting engineer has already compensated for aurally (or perhaps we need to compensate for his ears too?). Since different cutting lathes use different filters operated by different engineers, trying to justify the addition of a zero to the de-emphasis curve, solely on the basis of one incorrectly-analysed

337

Neumann machine, is plain silly. If it is felt that there is still something lacking from the final presentation then it is because the replay system needs some tone controls, not because secret knowledge has been withheld by the cutting industry.

10.2: Phono Stage Design

When designing a phono stage there is a basic choice between using passive or active (i.e. feedback) equalisation. Employing a bit of both is also possible, but is seldom a useful approach. For reasons that will emerge, passive equalisation is by far the most sensible choice for valve circuits. There is then a decision to be made about whether to implement the equalisation with a single all-in-one-go filter, or to split the job up into two separate filters with gain in between.

We also need to know how much gain will be needed, as this will influence the choice of valves and topology, among other things. The first stage needs to provide significantly more gain than is lost in RIAA equalisation, otherwise mid-and high-frequency signals reaching the second stage may end up smaller than when they left the cartridge. This would mean second stage noise would begin to dominate, which complicates valve choice. In practice, with all-in-one-go equalisation, the first stage gain should be at least ×30 (~30dB), in which case only frequencies above about 6kHz will suffer any significant noise degradation in the second stage, and most musical content is below this frequency. With split equalisation the first stage gain can be relaxed, which permits a wider choice of valves. The interface between the pickup cartridge and the first valve is explored later.

Assuming a nominal input reference level of $5mV_{rms}$ at 1kHz (see section 10.4.1), the *net* gain of the phono stage needs to be 30dB to bring the output up to $158mV_{rms}$, which is a reasonable line level. Increasing the gain to 40dB would accommodate less sensitive pickups. But since we need to overcome the 20dB (1kHz) lost in RIAA equalisation, the valves actually need to achieve a combined gain of 50 to 60dB (×316 to ×1000), which is easily attainable with two gain stages. The second stage will often be arranged for less gain than the first, which concentrates the noise performance in the first stage where it belongs. Lower gain also implies more headroom in the second stage, exactly where it is needed. Moving-coil pickups produce about ten times less output than moving magnet types, so they need an extra stage of amplification to bring them up to a suitable level before being fed into the rest of the phono stage. This is dealt with in section 10.5.

Unless the circuit is being built into an integrated power amp, the output must be capable of driving a reasonable length of cable and possibly a rather low load impedance. Too many modern valve phono designs seem to forget this. At the very least we should design for a 500pF ‖ 100kΩ load, and also ensure that the output impedance is low enough to reject external hum fields. Even vintage designs had the decency to provide a cathode follower output stage.[2]

---

[2] Beggs, G. E. (1952). Precision Pre-amplifier, *Electronics*. 25(7), July, pp121.

10.2.1: Active Equalisation

With passive equalisation the input stage needs enough headroom to accept the full un-equalised signal from the pickup. We might therefore shrewdly observe that if we apply equalisation by wrapping feedback around the input stage, the apparent headroom increases in proportion to the feedback, not to mention the reduction of distortion that feedback also brings to the table. The trouble with this approach is that if the feedback RIAA filter is carefully designed using standard equations, so that it is the exact inverse of the desired RIAA curve, we will find the response of the final amplifier is *not* correct, because feedback moves the poles around (zeros are unaffected by feedback, however).

One way around this problem is to design the feedback network to be slightly 'incorrect', so the lack of loop-gain actually bends the final response back to the desired shape. This is so tedious to calculate that it is often easier to apply a cut-and-try approach using computer circuit simulation. The alternative option is to stick with the convenient RIAA filter equations and instead minimise the final error by ensuring the amplifier has more than 20dB excess gain at all frequencies before feedback is added (i.e. >20dB loop gain). This is easy with solid-state circuits. Also, since silicon tends to run at low voltages and therefore has less inherent headroom than valve circuits, and poorer open-loop linearity too, it is logical to use a feedback approach for RIAA equalisation in a solid-state design. Indeed, it is the only tenable position these days; one good opamp is all that is needed to make an excellent if unimaginative phono stage.

Valve circuits are the opposite. With RIAA equalisation, if we want the closed-loop gain to be a bare-minimum 30dB at 1kHz then it must be 50dB at 20Hz, meaning we would need >70dB open-loop gain at 20Hz to avoid serious RIAA error when using the standard design equations. Achieving this much gain means using multiple gain stages, which means interstage coupling, which leads to a poor phase margin and difficulty stabilising the circuit. Therefore, valve designs invariably settle for less open-loop gain and instead have to fiddle with the RIAA filter to get close to the desired response. But even if this is achieved, the meagre loop gain makes the RIAA

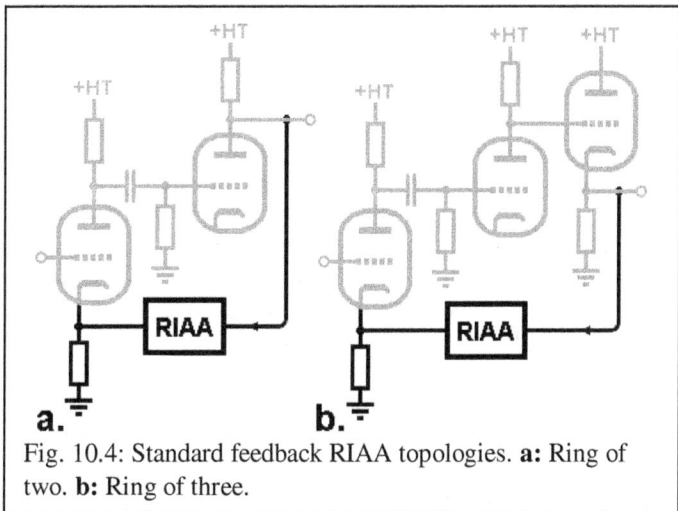

Fig. 10.4: Standard feedback RIAA topologies. **a:** Ring of two. **b:** Ring of three.

accuracy prone to drift as the valves age.

The standard topology for active RIAA equalisation is always a non-inverting or series-feedback amplifier (because the alternative inverting arrangement has poorer SNR). A series-feedback amplifier can be built from two cascaded gain stages with feedback returned to the cathode of the input valve as in fig. 10.4a –an arrangement known as a **ring of two**. The next level of sophistication is to add an output cathode follower to minimise the open-loop output resistance, which eases the design of the feedback network and makes the circuit more immune to aging. This is shown in fig. 10.4b and is classically called a **ring of three**. In either case the input valve cathode is substantially unbypassed, resulting in more noise than if the same valve were used in a passively-equalised design with full cathode bypassing, and valves are not particularly quiet to start with. In short, feedback equalisation suits solid-state designs, but passive equalisation is better for valve designs.

10.2.2: RIAA Accuracy

To accomplish RIAA equalisation we need a circuit that matches the transfer function described by equation (10.1) (the classic reference is a paper by Lipshitz[3] which presents design data for several possible arrangements). The equalisation should follow the RIAA curve with good accuracy, and it is not difficult to get within ±0.5dB using standard methods. The modern consensus seems to be to aim for ±0.1dB accuracy, which can be more challenging in a valve circuit, and is probably unnecessary anyway. It is more important that the equalisation is well matched between the left and right stereo channels, which may require measuring and matching components by hand.

Resistors are far more accurate than capacitors. It is easy to buy resistors with 1% or even 0.1% tolerance, and the E24 range gives us plenty of values to choose from. Capacitors, on the other hand, become expensive in 1% tolerance or better, and are often limited to values in the E12 or even the E6 range. Precision devices are also rarely available above 10nF. Most of the design effort will therefore go into choosing the capacitors. Those worried about capacitor distortion should use polystyrene or polypropylene, which most precision capacitors are anyway. In practice, valve distortion will always swamp capacitor distortion, so even polyester capacitors are essentially blameless provided they have a high voltage rating (at least 200V, say), as demonstrated in chapter 2.

To obtain awkward values we will often need to use series/parallel combinations of components, and this has the added advantage that it reduces the variance of total value. To see why this is so, consider two 1nF 1% capacitor in parallel. With perfect devices the total capacitance would be 2nF, but if both devices happen to be 1% high or low, the total could range from 1.98nF to 2.02nF. Therefore, if we demand 100% confidence, the total 'composite component' still has 1% tolerance. However, the chances of getting *two* 1nF capacitors that are both as bad as they can possibly be,

---

[3] Lipshitz, S. P. (1978). On RIAA Equalization Networks, *JAES*, 27(6), pp458-81.

and both in the same direction, are less than the chances of getting a single 2nF 1% capacitor that is as bad as it can be. By the magic of statistics it turns out that when equal-valued components are used in series or parallel there is roughly a 95% chance of the total being within $T/\sqrt{N}$ of the nominal value,[*] where T is the tolerance of the individual devices and N is the number of devices used. In other words, our two 1% capacitors together amount to a $1\%/\sqrt{2} = 0.71\%$ composite capacitor, provided we are willing to relax our confidence level to 95%. Three such devices would make a 0.58%-tolerance component, and so on. This is a cheap way to build an accurate circuit and may avoid the need to hand-measure components. Using multiple capacitors in series has the added bonus of reducing the signal voltage dropped across each one, thereby reducing any supposed capacitor distortion.[†]

10.3: Realising Passive Equalisation

The most direct way to carry out RIAA equalisation is with a single all-in-one-go filter. As already mentioned, there are several networks that will do this, but the one in fig. 10.5 is by far the most practical choice. It is then a simple matter of choosing component values that satisfy the design equations in the figure. However, we must remember to take into account any source resistance and load capacitance. For example, fig. 10.6 illustrates how part of R_1 is actually made up from the source resistance R_s in parallel with a necessary grid-leak resistor R_g, and C_1 is partly made up from the Miller capacitance of the following stage. The trouble with the source resistance of a valve stage is that it depends on r_a, which varies between samples and increases with age. It is therefore common (though not essential) to place a cathode follower before the RIAA filter since its small source resistance contributes little to the accuracy of the equalisation curve.

Assuming we know the source resistance and Miller capacitance, we would begin by estimating a value for the

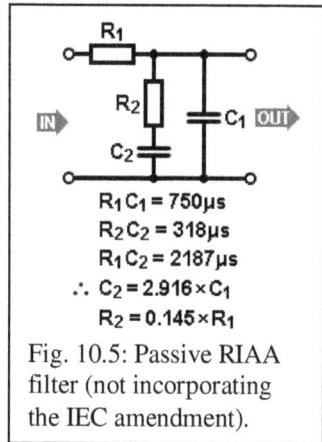

$$R_1 C_1 = 750\mu s$$
$$R_2 C_2 = 318\mu s$$
$$R_1 C_2 = 2187\mu s$$
$$\therefore C_2 = 2.916 \times C_1$$
$$R_2 = 0.145 \times R_1$$

Fig. 10.5: Passive RIAA filter (not incorporating the IEC amendment).

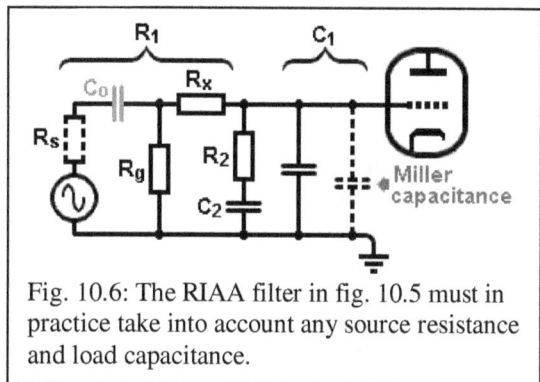

Fig. 10.6: The RIAA filter in fig. 10.5 must in practice take into account any source resistance and load capacitance.

---

[*] This assumes the variance in component value has a Gaussian distribution, which is not entirely correct but seems to be a reasonable approximation to reality.

[†] Contrary audiophiles will insist that several capacitors must be several times worse.

341

series resistor R_x, as this sets the minimum input impedance of the network at high frequencies and should not be too small or it will load down the driving stage and increase distortion. After making this first choice we could then find C_1, followed by C_2 which must be 2.916 times larger. Having found the approximate size of these capacitors we would see if they can be made more easily from some combination of standard values, then go back and recalculate the resistors, which certainly *can* be made from combinations of standard values. It may take several iterations to massage the design into something accurate and practical, but the process is at least very simple. An example appears in section 10.6.3.

A coupling capacitor C_o will usually be necessary too, and it must be large enough not to interfere with the equalisation, which is no trouble if the grid-leak is the usual 1MΩ. Alternatively, the cut-off frequency can be set to 20.02Hz to implement the IEC amendment, if $C_o(R_s+R_g) = 7950$μs. In some designs C_o and the grid leak are placed *after* the RIAA filter to minimise its effect on low-frequency accuracy, but this also introduces attenuation, i.e. it worsens the SNR, which is probably more troubling. Also, precision capacitors rarely come with high voltage ratings, so by putting C_o (which is a less critical device) before the filter it can have a high-voltage rating to block DC while the other capacitors only have to handle start-up and signal voltages, so can perhaps be only 63V-rated devices.

The all-in-one-go filter in fig. 10.5 can alternatively be split into two parts separated by a stage of amplification, as shown in fig. 10.7. This makes the choice of filter components easier since the time constants don't interact with one another. The most logical arrangement is to place the 75μs filter first, so it easily incorporates the Miller capacitance of the intermediate gain stage (this also attenuates dust and scratch clicks before they hit this valve). There is usually no need for another stage of gain, so the 3180μs/318μs shelving filter will normally feed an output cathode follower which has very little input capacitance to worry about. This creates a classic topology, illustrated in fig. 10.8. A possible

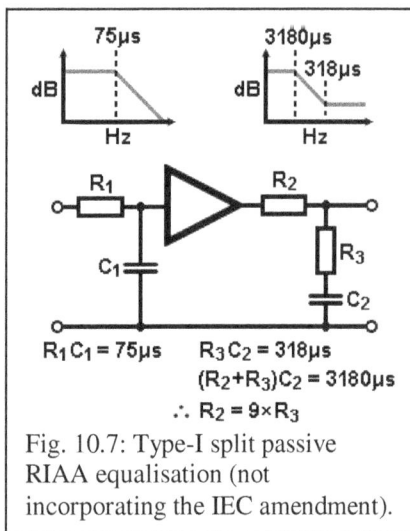

$$R_1C_1 = 75\text{μs} \quad R_3C_2 = 318\text{μs}$$
$$(R_2+R_3)C_2 = 3180\text{μs}$$
$$\therefore R_2 = 9 \times R_3$$

Fig. 10.7: Type-I split passive RIAA equalisation (not incorporating the IEC amendment).

disadvantage of this arrangement is that the second gain stage must have plenty of headroom since only frequencies above 2.122kHz are attenuated before hitting its grid, and most musical information is below this. We will call this 'type-I' split equalisation to distinguish it from what follows.

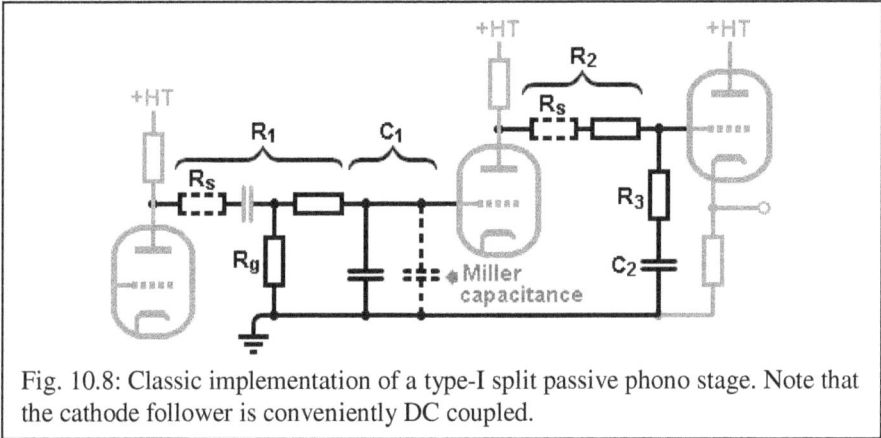

Fig. 10.8: Classic implementation of a type-I split passive phono stage. Note that the cathode follower is conveniently DC coupled.

An alternative split equalisation arrangement is shown in fig. 10.9, which we will call type-II. This has some advantages over the previous type-I scheme. Frequencies below 2.122kHz – where most information lies– are now attenuated by up to 12dB before reaching the second stage, which makes more effective use of its headroom. This also means the first (input) gain stage can provide less gain without seriously spoiling the EIN. R_2 serves as a ready-made grid leak, and the IEC amendment may be implemented with a coupling capacitor such that $R_2C_o = 1875\mu s$ (*not* $7950\mu s$). The final low-pass filter attenuates everything above 50.05Hz and so provides a significant reduction of noise generated in all previous stages.

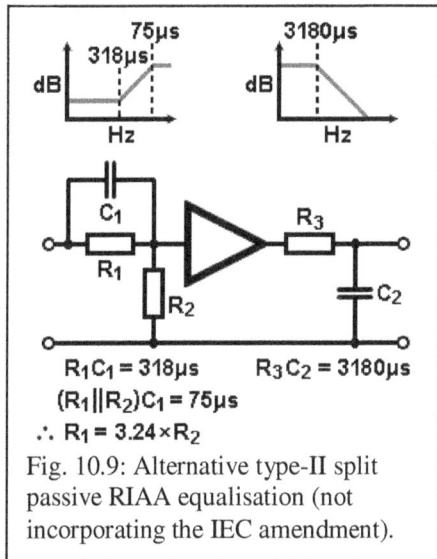

$R_1C_1 = 318\mu s$ $R_3C_2 = 3180\mu s$
$(R_1 \| R_2)C_1 = 75\mu s$
$\therefore R_1 = 3.24 \times R_2$

Fig. 10.9: Alternative type-II split passive RIAA equalisation (not incorporating the IEC amendment).

The disadvantage of this scheme is that source resistance and Miller capacitance do not neatly merge with the 318μs/75μs shelving filter and so will introduce an unwanted extra pole (we can't swap the positions of the filters because this would degrade the SNR by attenuating everything too much, too soon). The unwanted pole will occur at:

$$\tau = \frac{4.24 \times R_2 R_s C_m}{R_1 + R_2 + R_s}$$ microseconds, so the easiest way to deal with it is to drive the

shelving filter from a cathode follower, as illustrated in fig. 10.10. The source resistance will then be negligible, pushing the unwanted pole up towards 1MHz. We then only need to take Miller capacitance into account, which appears in parallel

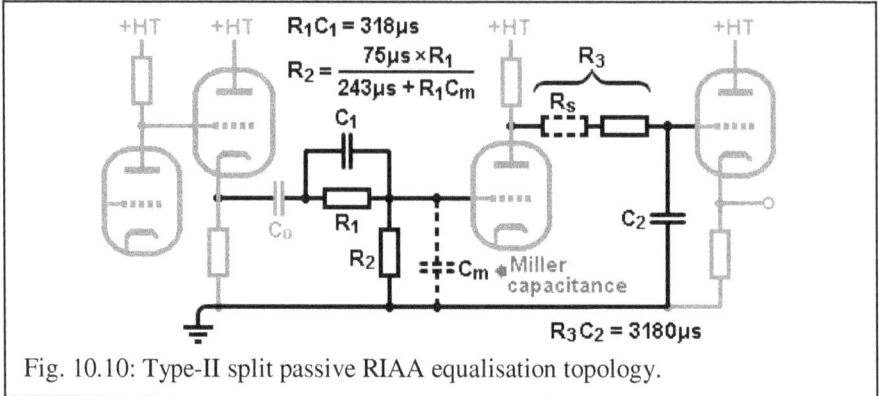

$$R_1C_1 = 318\mu s$$

$$R_2 = \frac{75\mu s \times R_1}{243\mu s + R_1C_m}$$

$$R_3C_2 = 3180\mu s$$

Fig. 10.10: Type-II split passive RIAA equalisation topology.

with C_1 as far as the Thévenin equivalent is concerned, and the design equations become as shown in the figure.

10.3.1: An Inverse RIAA Filter

A convenient way to test a phono stage using an ordinary audio oscillator is to insert an inverse-RIAA filter between the two, so the overall output from the phono stage can be checked for flatness. However, the RIAA curve continues falling forever, but the inverse response cannot continue rising forever – it must level off at some point. Other authors –misguided by the Neumann fallacy– have suggested setting this pole at 50kHz, but this leads to almost 1dB error at 20kHz. The circuit in fig. 10.11 improves on this by placing the awkward pole above 100kHz which ensures superior results on square-wave tests.

With a 4.08V input, the MM output provides 5mV at 1kHz, and the MC output 455μV, which are representative of real pickups. The output resistances are small enough that ordinary cartridge loading networks do not affect the circuit's accuracy. Modern signal generators usually have a source impedance of 50Ω or 75Ω which

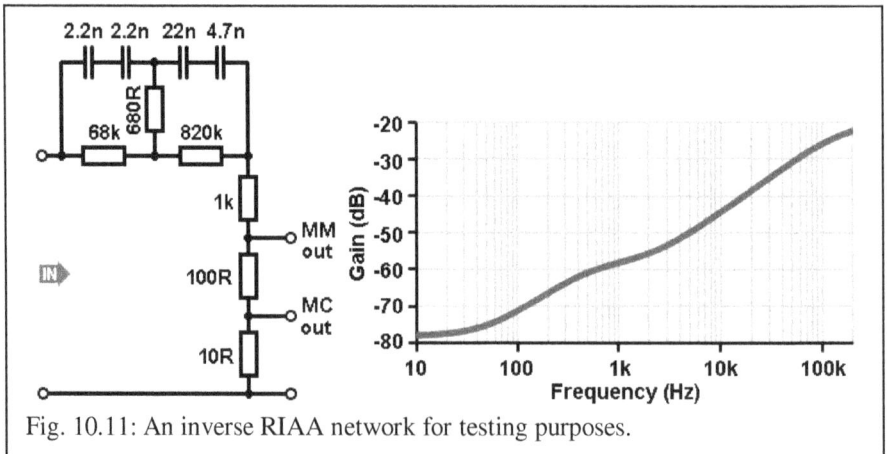

Fig. 10.11: An inverse RIAA network for testing purposes.

344

also has negligible effect on the error. However, vintage test equipment may struggle to drive the network as its input impedance falls to 9.4kΩ at 20kHz, or 2.2kΩ at 100kHz.

The 680Ω resistor may look out of place in an RIAA network but in this case it improves error without having to resort to rare capacitor values. As shown in fig. 10.12, with ideal components the error is well within ±0.05dB across the audio band. Using 1% components the worst-case error is ±0.1dB around the median, and even 5% capacitors result in little more than ±0.3dB error. Such unfortunate component combinations are unlikely to arise in practice.

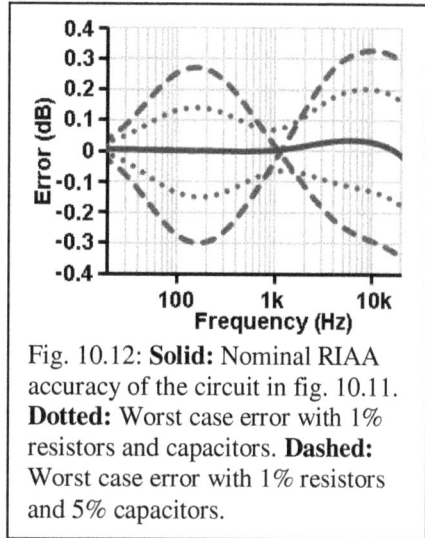

Fig. 10.12: **Solid:** Nominal RIAA accuracy of the circuit in fig. 10.11. **Dotted:** Worst case error with 1% resistors and capacitors. **Dashed:** Worst case error with 1% resistors and 5% capacitors.

10.4: The Input Stage

The proper design of the input stage of a phono preamplifier is crucial, as a poor design decision will quickly spoil the signal-to-noise ratio. There is also a balancing act to maintain between maximising gain while achieving appropriate cartridge loading and headroom. Proper grounding and a neat, tight layout are essential to avoid hum, and anything we can do to maximise PSRR is helpful. Wires leading from the input sockets to the first valve stage should be twisted or shielded to reject hum.

10.4.1: Moving-Magnet Pickup Cartridges

The nominal output from a pickup cartridge is usually specified as an RMS voltage when tracking a 1kHz signal at a groove velocity of 5cm/s. Under these conditions, hi-fi moving-magnet pickups usually generate between 2mV and 5mV. Pickups designed for disc jockeys are typically higher sensitivity, occasionally exceeding 10mV.

The maximum recorded velocity on a record rarely exceeds 30cm/s,[4] so a nominally 5mV (5cm/s) pickup should rarely deliver more than 5mV × 30/5 = 30mV$_{rms}$ or 42mV$_{pk}$ to the phono stage. This is consistent with Jones[5] who has reported observing rare maximum peaks at 16dB above the 5cm/s level, which is 31.5cm/s. However, he also reports clicks due to dust and scratches reaching 22dB above the 5cm/s level, which would amount to 89mV$_{pk}$ using the same pickup. Since a click

---

[4] Amos, S. W. (ed.) (1977). *Radio, TV and Audio Technical Reference Book.* Newnes, London.
[5] Jones, M. (2012). *Valve Amplifiers* (4th ed.), Elsevier, Oxford. p623.

could cause clipping, the circuit must be designed for polite overload recovery, i.e. good immunity to blocking distortion. Channel separation for a pickup is usually an embarrassing 20dB to 30dB so this at least is not something we need to worry about in the rest of the circuit.

A moving-magnet (MM) pickup can be modelled fairly realistically as an inductance of around 400mH to 800mH in series with a resistance of

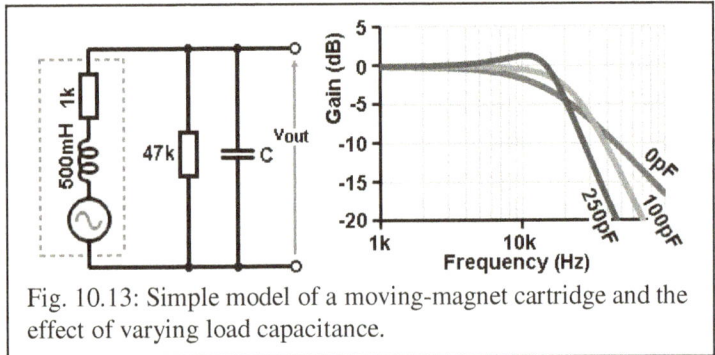

Fig. 10.13: Simple model of a moving-magnet cartridge and the effect of varying load capacitance.

around 500Ω to 1.5kΩ. The recommended load resistance is more-or-less standardised at 47kΩ, so by itself this would create an LR filter with a −3dB frequency anywhere from 11kHz to 19kHz, which is too low for hi-fi use. To overcome this premature roll-off the cartridge must additionally be loaded with some capacitance, creating an RLC circuit. The damping factor can then be tweaked to get a wider bandwidth at the expense of a more rapid roll-off above the audio band, and a possible resonant peak. One might expect the most desirable response to be maximally flat or Butterworth, which requires a capacitance equal to $L/(2R^2)$. However, manufacturers consistently recommend using more than this, typically 250pF, resulting in a couple of decibels of peaking, which everyone is apparently happy with. Fig. 10.13 illustrates this with a typical cartridge model. At least 50 to 100pF will be provided by the interconnecting cable, so the Miller capacitance of the first amplifier stage should be kept below 150pF where possible. This is not difficult to achieve but it does rule out certain high-gain valves with excessive C_{ga} such as the otherwise promising EC91 (C_{ga} = 2.5pF).

10.4.2: Valve Choice

As a rough guide, the first gain stage should be able to handle at least 200mV$_{pp}$ input from the cartridge. This allows for >6dB excess headroom with even high sensitivity cartridges, and accommodates most likely click levels (anything higher is certainly a click so it doesn't matter if it is clipped off). This is no problem for any valve, so there is no need to worry about grid base. Similarly, at these signal levels, linearity will take care of itself. The main criterion for the input valve is instead noise, which must be low. Modern vinyl achieves an SNR of at least 60dB, and an MM cartridge will generate 0.4µV to 0.7µV of Johnson noise by itself, so we

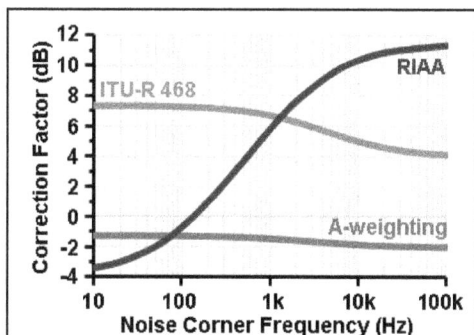

Fig. 10.14: Correction factors for converting unweighted 20Hz–20kHz noise into weighted noise in the presence of pink noise, assuming RMS metering. Reproduced from fig. 5.9.

don't want to add much more to this. A final, unweighted SNR of 70dB relative to 5mV can be considered a good effort for a pure-valve phono preamp.[*]

Recall that RIAA equalisation greatly worsens the noise figure if there is much pink-noise content, as shown by fig. 10.14. The input valve therefore needs to be one with both low shot noise *and* low flicker noise. Unfortunately, these are conflicting requirements, because shot noise is reduced by maximising g_m (and therefore I_a), whereas flicker noise is reduced by

minimising I_a. This implies using a high-g_m device at only a moderate anode current; 2 to 3mA is usually the best compromise. The need for very low noise and microphonics immediately rules out most pentodes unless they are triode-connected. Normally we would also consider using several devices in parallel to reduce the EIN, but this is not always possible with triodes because it multiplies the already-high Miller capacitance.

Fig. 10.15 shows the equivalent input noise voltage spectral density of some triodes. The ECC81/12AT7 and ECC82/12AU7 are clearly the noisiest, mustering an EIN rarely better than 1μV in the audio band, which is about as high as we should dare to go. The 6J52P/6Ж52П (triode connected) appears the best, but the hidden variables are that such high-g_m pentodes suffer acutely from microphonics and parameter spread, so will need to be hand-selected, and they invariably have unworkably high Miller capacitance too. The ECC88/6DJ8 is a popular choice as it

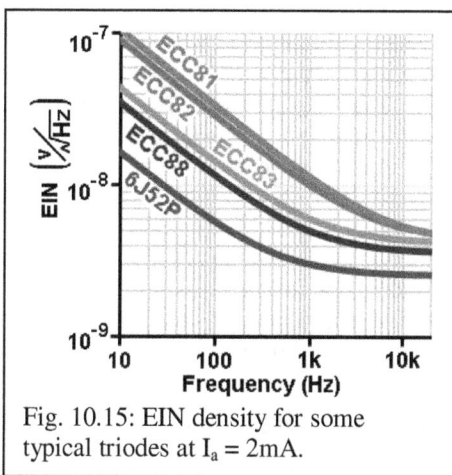

Fig. 10.15: EIN density for some typical triodes at $I_a = 2$mA.

can achieve an EIN of about 0.6μV on a good day, with Miller capacitance around 100pF, though it too suffers noticeably from microphonics. The ECC83/12AX7 is a

[*] Be aware that many commercial manufacturers use weighted or invalid measurement techniques to boost their advertised figures.

347

good choice for low-cost designs –despite being a low-g_m device– as it has tolerable noise, just-bearable Miller capacitance (typically 200pF), high gain, but low microphonics. Its widespread use in guitar amps makes it cheap and plentiful so it is not too painful to try out a few samples for the best performance. The 6H2Π/6N2P is almost the same as the ECC83 except for a remarkably low C_{ga} of 0.7pF, so it can achieve <150pF Miller capacitance in most cases. A further option for low Miller capacitance is to use a pentode in low-capacitance triode mode (section 4.8.1). In fact, the Miller capacitance may then be so small that two devices can be used in parallel for lower noise. Nevertheless, it is very difficult to get the EIN of the input stage below 0.5µV whatever valve is chosen, so if you expect better, be prepared to discard a crate of them in the attempt.

The topology of the input stage is dictated mainly by noise considerations too. It needs to have enough gain that it will dominate the noise behaviour of the whole preamp, otherwise we will have to consider noise from other sources too, which is an unnecessary complication. In practice this means the input stage will need to provide a gain of at least 30dB for all-in-one equalisation; at least 26dB with type-I split equalisation; and at least 20dB with type-II. The cathode bias resistor or LED (if used) must be fully bypassed or its intrinsic noise will be amplified. A CCS anode load is a popular choice but makes the output resistance equal to r_a and therefore more subject to device variation, which is not good if the first stage is expected to drive the RIAA network directly. A CCS will also degrade the EIN when using a high-r_a triode like the ECC83 (compared with a resistive load). The half-µ and SRPP stages are out of the question as their EIN is always about 3dB worse than a single triode, but the µ-follower is a contender. It provides a very low output impedance and near optimum EIN, but at the cost of high Miller capacitance, high supply voltage and heater-cathode stress. A cascode can provide low input capacitance and, if the lower device is a JFET (section 8.1.4), very low noise too, but its distortion and PSRR are poor. Taking all this into account we are forced to conclude that it is difficult to improve upon the plain old resistor-loaded, triode gain stage; it is inherently low noise, simple, and well behaved. A clean HT and DC heater supply are, of course, essential in every case. A grid stopper should always be included but it must be small enough not to add any further noise to the system. A value of 100Ω should be safe in any valve design since it is less than the internal resistance of the pickup and generates less than 0.2µV Johnson noise, well below the EIN of even the luckiest valve choice.

10.5: Moving-Coil Pickup Cartridges

Moving-coil (MC) pickups are traditionally regarded as the highest-quality sort of pickup, although these days moving-magnet cartridges have become so good that the quality division is no longer clear. Nevertheless, they are likely to remain popular. The output voltage from an MC pickup is usually an order of magnitude lower than for an MM type, at around 100µV to 500µV. This is too low to plug directly into a phono stage designed for a moving-magnet cartridge, so an extra stage of very low-noise amplification is needed. Traditionally this is provided by a special step-up transformer, but active solutions are also possible. A switch is normally provided to select either MM or MC input, but since this can result in an almighty pop if it is thrown while the amp is on, it will usually be positioned somewhere discreet such as on the back of the unit.

A useful survey of MC output voltages is provided by Self and is reproduced in fig. 10.16; the high-output Ortofon Turbo types have

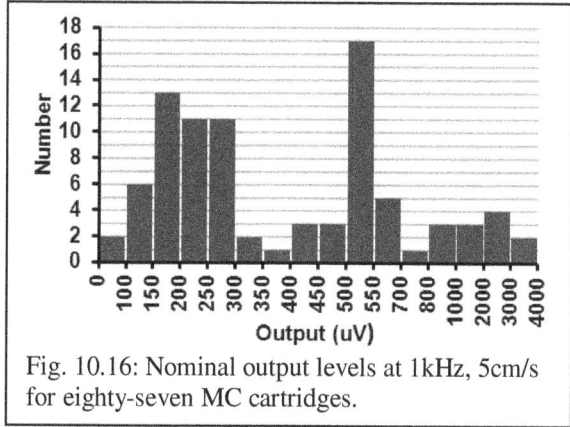

Fig. 10.16: Nominal output levels at 1kHz, 5cm/s for eighty-seven MC cartridges.

also been added to the chart. Interestingly, the series is bimodal, so the vast majority of cartridges can be accommodated by a preamplifier designed to handle 250μV and 500μV input levels (e.g. with switchable gain), although minimalist designs will settle for one or the other. Assuming the MC head amplifier is to be bolted onto the front of a conventional phono stage designed for a nominal input of about 5mV, the head amplifier will therefore need a gain of about ×10 (20dB), possibly switchable to ×20 (26dB). The very high-output MC pickups require no additional gain and can be used directly with an MM phono stage.

Because moving-coil pickups consist of only a few turns of wire they have negligible inductance and very small internal resistance, usually ranging from 1Ω to 10Ω, but as high as 160Ω for Denon's high-output types. There is therefore no need for any special load capacitance (they can tolerate several nonofarads); a load resistance is all that is required. The manufacturer will quote a recommended minimum value, usually around five times the cartridge resistance, but 100Ω to 500Ω is suitable for most types.

10.5.1: Moving-Coil Input Transformers

The traditional way to boost very small signals with minimal additional noise is with a specially-designed transformer. To minimise Johnson noise, MC input transformers are designed to have only a few ohms of primary winding resistance, and extremely good magnetic shielding. At these signal levels, distortion is no great problem, and bandwidth can also be very wide thanks to the low impedance nature of the circuit. The most well-known manufacturers of MC transformers are Sowter in the UK, Cinemag and Jensen in the US, and Lundahl in Sweden. For reasons stated previously, the step-up turns ratios are usually around 10 to 20, and some devices provide multiple windings which can be connected in series or parallel to allow different combinations and switchable gain. All wires leading to and from the transformer must be twisted to reject hum.

The transformer must have a suitable load resistance placed across its secondary (R_3 in fig. 10.17), which is reflected across to the primary and therefore provides the

349

correct cartridge loading. As an example, the Ortofon Quintet cartridge has a nominal output of 500µV, an internal resistance of 7Ω, and a recommended load resistance of >20Ω. The output can be boosted to 5mV using a transformer with a step-up ratio of 10, which is an impedance ratio of $10^2 = 100$. In other words, the secondary load will appear one hundred times smaller from the point of view of the cartridge. To load the cartridge with >20Ω the secondary load resistance must therefore be >2000Ω. A higher value improves the SNR, but inevitably reduces the bandwidth, owing to interwinding capacitance and finite primary inductance. The transformer manufacturer should provide some indication of what is achievable. For example, the Sowter 9570 has a step-up ratio of 10 and claims to be able to maintain up to 1kΩ primary impedance. It could therefore be loaded with the 47kΩ that is already needed for MM cartridges, which will be reflected across to the MC cartridge as 470Ω. Alternatively, we might choose to buy a little more bandwidth by using something like 10kΩ, so the MC will see 100Ω.

In addition to the main resistive load, an RC Zobel network is usually needed too. The transformer manufacturer should provide information about this, but if not, it is fairly easy to find a suitable network with a little experimentation. Drive the transformer from a 1kHz square-wave generator while adjusting the Zobel network until the most visually

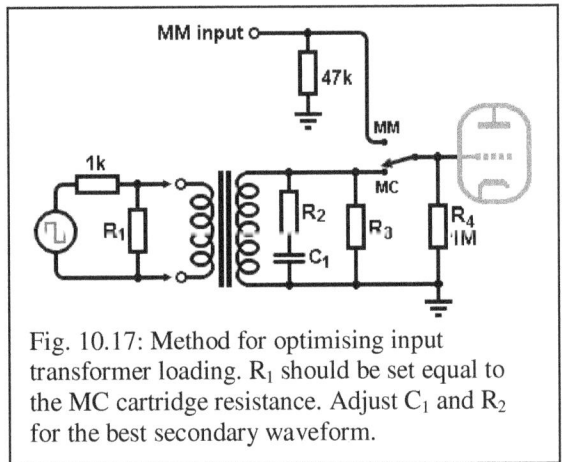

Fig. 10.17: Method for optimising input transformer loading. R_1 should be set equal to the MC cartridge resistance. Adjust C_1 and R_2 for the best secondary waveform.

satisfying secondary waveform is obtained, e.g. no overshoot, or a small overshoot but no undershoot. An input attenuator will be necessary to reduce the output of the generator to suitable MC test levels (a few millivolts peak) as well as simulating the source resistance of the intended cartridge, as illustrated in fig. 10.17. Assuming the waveform is monitored prior to RIAA equalisation there is no need for an inverse RIAA filter when performing this test. This method could also be used with the transformer out of circuit, provided the phono stage Miller capacitance and any other loading is duplicated with dummy components.

10.5.2: Active Moving-Coil Head Amplifiers

MC transformers are unavoidably expensive and can easily represent the most expensive part of the phono stage. For example, the previously mentioned Sowter 9570 costs £68 at the time of writing. And you need two for stereo. Understandably, many builders will be interested in cheaper, active solutions.* For

---

* Every now and then someone suggests using a common-grid triode stage for an MC input, since the input impedance of a common-grid stage is very low and

350

our purposes a single low-noise opamp
is all that is needed to make a
respectable head amplifier, and fig.
10.18 shows this in essence (this is not
yet a complete circuit). R_1 provides a
suitable load for the pickup while R_2
and R_3 set the gain of the stage to:

$$A = 1 + \frac{R_2}{R_3} = 1 + \frac{82}{10} = 9.2 \ (19.3\text{dB})$$

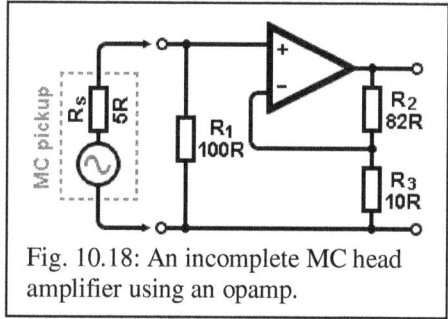

Fig. 10.18: An incomplete MC head
amplifier using an opamp.

These resistors must be small in value or they will contribute excessive noise.
However, as will be shown, they do not need to be impossibly small because the
opamp will ultimately dominate the noise performance. Setting R_3 to 10Ω is a
convenient standard value.

Remember the familiar formula:

$$V_{johnson} = \sqrt{4kTRB}$$

Where:
k = Boltzmann constant, 1.38×10^{-23} J/K
T = absolute temperature in kelvin
R = resistance in ohms
B = bandwidth in hertz

The pickup is represented as having a 5Ω source resistance. Using the above formula
this is found to generate 41nV Johnson noise in the audio band. Compared to a
typical reference signal level of 500µV this yields a maximum theoretical SNR of
81.7dB, assuming nothing else adds noise. Loading the pickup with 100Ω degrades
this by less than 0.1dB. Now let's see how much the rest of the circuit spoils this
figure.

The source resistance seen at the non-inverting input of the opamp is R_s in parallel
with R_1, or 5∥100 = 4.76Ω, and generates 39.6nV of Johnson noise. The source
resistance seen at the inverting input is R_3 in parallel with R_2, or 10∥82 = 8.9Ω, and
generates 54.1nV of noise. Now for the sucker punch. The best low-voltage-noise
opamps have an EIN voltage density of about 1nV/√Hz. When integrated across the
audio band this amounts to an equivalent input noise voltage of 1nV×√19980 =
141.4nV, which is a good deal more than either of the previous figures mentioned.
There is also noise current flowing into the inputs, but this usually has a density
around 2pA/√Hz which when flowing through the source resistance is entirely
negligible. The total equivalent input noise voltage of the whole circuit in fig. 10.18

apparently an ideal match for an MC pickup. Unfortunately, the EIN of a common-
grid stage is identical to an ordinary common-cathode stage, and the quietest valve is
still too noisy for MC use, even if you can find one that isn't too microphonic.

is therefore going to be close to:

$$v_N = \sqrt{39.6^2 + 54.1^2 + 141.4^2} = 156.5nV$$

When amplified by the head amplifier's gain of 9.2, the output noise sent to the phono stage will be 1.44µV. Since the phono stage will probably have an EIN of >0.5µV, this makes a total of >1.52µV. And since the 500µV signal voltage has also been amplified to 4.38mV, the SNR is now <69.2dB. RIAA equalisation may improve or degrade this figure depending on the noise corner frequency (fig. 10.14).

Losing 12.5dB of SNR between the pickup and the phono stage might sound like a lot, but it is actually not too bad. The final SNR is quite respectable for what

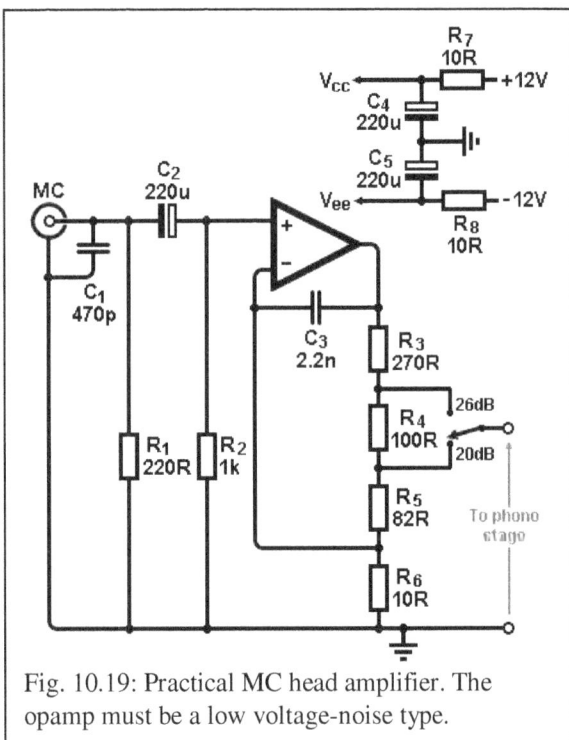

Fig. 10.19: Practical MC head amplifier. The opamp must be a low voltage-noise type.

is purportedly a *valve* phono preamplifier, and at least we only have one device to blame for it: the opamp. Even if the opamp were completely noiseless and the only additional source was the phono stage, the final SNR would probably not exceed 74dB, so in real terms we have only lost 6dB or so. Squeezing out better performance requires a jump to more complex hybrid circuits using BJTs, and this is not the right book to discuss them. For interested readers, an excellent example is presented by Self.[6]

Having satisfied ourselves that a single (but well chosen) opamp can perform the task, the circuit in fig. 10.18 can be turned into something more practical, as shown in fig. 10.19. C_1 serves only to bypass RF to ground and should be soldered directly to the input socket. C_2 is essential to keep any DC bias current out of the cartridge, and in combination with R_2 has a cut-off frequency of 0.7Hz.[*] R_1 provides a suitable load for the pickup as well as providing a complete charging path for C_2.

[6] Self, D. (2010). *Small-Signal Audio Design*. Elsevier, London.
[*] Those who think this will have some detrimental effect on microvolt-level audio signals should contact the tooth fairy for further advice.

A shortcoming of fig. 10.18 is that the total resistance of the feedback loop was only 92Ω which is a very heavy load for an opamp to drive, even at small output levels. The easy way around this is to add some extra resistance in series with the output of the opamp so it sees a higher total. This is what has been done in fig. 10.19, bringing the load up to a more bearable 462Ω. Of course, this means the opamp is now operating at a much higher gain of $1+452/10 = 46.2$, but this does not matter because it is never going to be handling any large signals. For example, a 500µV cartridge picking up a 22dB dust click will produce $8.9mV_{pk}$ which would be amplified to $411mV_{pk}$ at the output of the opamp, nowhere near any sort of headroom limit. Of course, we don't actually want this much gain, so the output signal is tapped off a lower point on the feedback resistance, and this also presents a convenient way to switch between two (or indeed more) gain settings. In practice this will probably be a small DIL switch or jumper, mounted on the PCB to prevent idle toying. C_3 rolls off the gain above 160kHz to improve stability. Suitable opamps with ~1nV/√Hz voltage noise, available in a through-hole package, include the AD797, LT1028 and LT1115. There are many more available in surface-mount packages.

The power supply rails are bipolar and should be regulated, although the exact voltages are not at all critical since the opamp is never going to produce more than $1V_{pp}$. The head amplifier is not likely to need more than 10mA per opamp and a convenient way to integrate the circuit into the phono stage is to use the bipolar supply to power the valve heaters too, since they need a DC supply anyway.

10.6: A Moving-Magnet Phono Design Example

There can be no universal cook-book formula for designing an amplifier, phono or otherwise, given the infinite variety of design constraints, performance specifications, budget and so forth. Every engineer has his own preferences, prejudices and design flair, but the following example may give newcomers some insight into the thought processes that take place.

The inspiration for this project came after the author acquired a box of assorted television valves which happened to include some PC97s. An internet search revealed that the PC97 is an unusual 7-pin triode with high µ, high-g_m and, crucially, a grid-anode capacitance of just 0.5pF. This remarkable figure is achieved by employing a specially shaped 'bathtub' anode, plus a pair of metal plates (which must be connected to cathode) to screen the grid side rods from the anode. Its linearity is poor (borderline variable-g_m), but its other properties suggest that it might be worth trying as the input stage for a phono preamplifier. Incidentally, it was introduced by Mullard in 1962 as a cheaper alternative to the (by then) standard cascode front end used in television tuners, and it does not appear to have an American equivalent.

10.6.1: The Input Stage

The PC97 datasheet does not include anode characteristic curves so the author had to measure them, as shown in fig. 10.20. Rather than dive right into the whole preamp design the author was interested first in testing a PC97 gain stage alone. It was decided to bias the valve using a cheap red LED (1.8V), and perusal of the curves suggested that a 47kΩ anode load with a 250V HT would put the quiescent current at about 2.5mA and the gain around 40. At this point the r_a appears to be about 13kΩ so the output resistance should be 13k∥47k = 10.2kΩ.

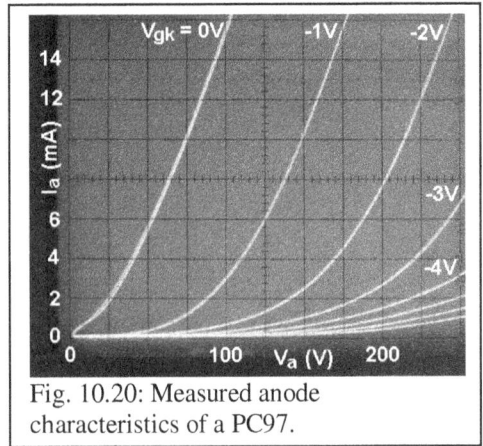

Fig. 10.20: Measured anode characteristics of a PC97.

Fig. 10.21 shows the circuit in which six specimens were tested. The voltage gain varied from 37 to 40 and the output resistance from 10.5kΩ to 14kΩ, which are very close to the predicted figures. Distortion was

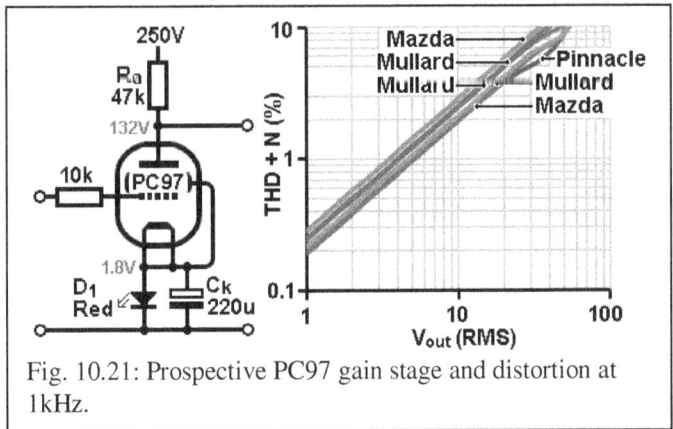

Fig. 10.21: Prospective PC97 gain stage and distortion at 1kHz.

fairly poor, as expected. However, fig. 10.21 demonstrates that at least the consistency is good, with several traces lying on top of one another. As an MM input stage it will have to operate around 5mV×40 = 200mV, so in actual use the distortion should be about 0.05% which the author grudgingly accepts as bearable. The average EIN was 0.95µV in the audio band, which is higher than might be expected for such a high-g_m valve, but good enough for the application (leaving the LED unbypassed increased this figure by almost 4dB). Only one sample (Pinnacle) showed any trouble with microphonics, and this one also had noticeable popcorn noise, producing a much worse EIN of about 2µV. Every barrel has a rotten apple. However, the real reason for using this valve is its Miller capacitance, which was found to be stunningly low at 56pF for every sample. Allowing 4pF for C_{gk} this implies a total C_{ga} of about (56–4)/(40+1) = 1.3pF, so the external stray capacitance must only be about 0.8pF. This is plausible since the anode and grid pins are on

354

opposite sides of the valve socket. At this point the author was satisfied that the PC97 lives up to expectations, so the rest of the design can progress.

10.6.2: Fleshing-Out The Design

Since the input stage will provide a gain of 32dB (×40) and RIAA equalisation will introduce a loss of 20dB at 1kHz, a further 18dB (×8) to 28dB (×25) will be needed to bring the total up to 30 to 40dB. Some obvious choices that fall within this range are the ECC82/12AU7, 6SN7, and ECC88/6DJ8. A degenerated ECC81/12AT7 might also work, but in this case we will use the ECC82. This decision is not at all influenced by the fact that the author has an awful lot of them in his spares box.

The next question is how to implement the equalisation. The PC97 provides enough gain to use the all-in-one go approach, which makes the most of the available headroom in the second stage. The preamp will need to be buffered from interconnects, and one possibility is to make the second stage a μ-follower which would provide the necessary low output impedance. However, this would also make the gain of the circuit highly dependent on valve characteristics, which is not ideal from the point of view of channel matching. We will therefore use an ordinary gain stage with a balance trimpot to allow minor adjustment of gain (along the same lines as the balance control described in section 9.3). This stage will then be DC-coupled to a cathode follower output stage. Many valve types would work here, but we will use another ECC82 (it's a big spares box).

A rough sketch of the circuit now looks like fig. 10.22, and it is worth taking a moment to consider the power supply (of which more in chapter 11). The input stage is already presumed to be supplied with about 250V which will come via an RC smoothing filter. To get excellent smoothing and decoupling from the output stage – without resorting to an inordinately large capacitor– a relatively large 10kΩ dropping resistor will be used. Since the input stage will consume 2.5mA this will drop 25V meaning the second stage will hence run off 275V. By making the second and third stages into a constant-current draw amplifier (section 7.10) we

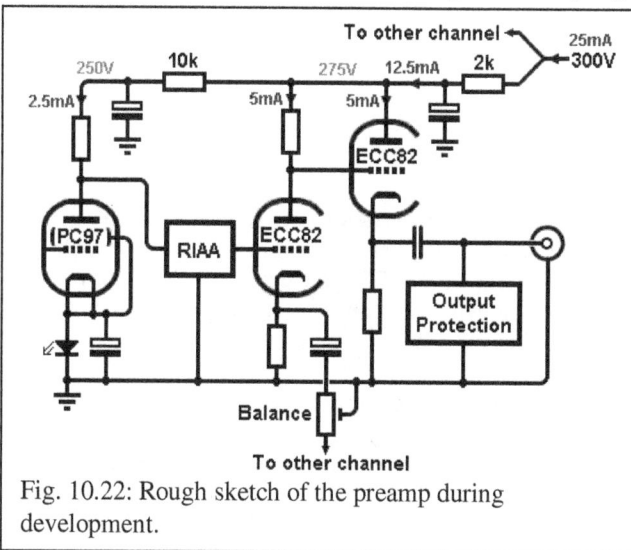

Fig. 10.22: Rough sketch of the preamp during development.

side-step any pressing need to decouple these valves from one another, so they can both be supplied from the same power supply node. Admittedly the current balance of these two stages will be upset somewhat by the unknown load impedance of whatever amplifier it happens to be plugged into, but we will presume this won't be less than 100kΩ which is not excessively heavy. Letting the cathode follower run at 5mA –and presumably something similar for the preceding stage since they will have equal load resistors– brings the estimated total demand for one channel up to 12.5mA. A 2kΩ dropping resistor could therefore be fed from a 300V common supply, which is a nice round number.

10.6.3: The Second Stage

Returning to the ECC82 second stage, the 27kΩ load line in fig. 10.23 suggests a fully-bypassed gain of 13 with fair linearity (which is not a problem since the PC97 is sure to dominate the distortion anyway). A bias of –5V will result in $I_a \approx 4.8$mA implying a 1042Ω cathode bias resistor, so naturally we will use 1kΩ. At this point

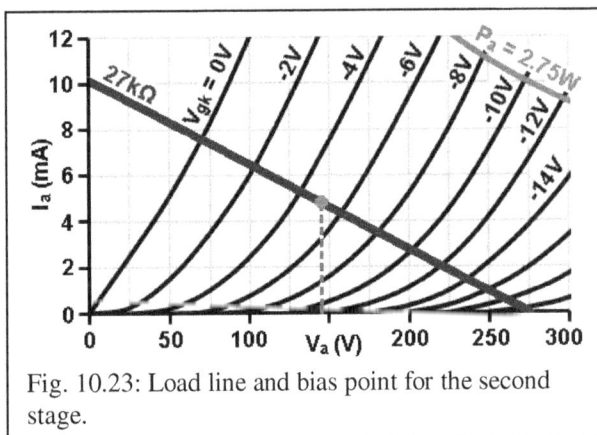

Fig. 10.23: Load line and bias point for the second stage.

$r_a \approx 10$kΩ and $\mu \approx 17.5$, so by using a 500Ω balance pot the gain can be reduced from its maximum value down to:

$$A_{(degenerated)} = \mu \frac{R_a}{R_a + r_a + R_k(\mu+1)}$$

$$A_{(degenerated)} = 17.5 \times \frac{27k}{27k + 10k + 0.5k \times (17.5+1)} = 10.2$$

This is a 2dB range and should be sufficient to compensate for any channel imbalance. The bypass capacitor must be large enough not to introduce unwanted bass attenuation. Applying the formula $1/(2\pi R_k C_k)$ indicates that 220μF will push the zero down to 0.7Hz which is not likely to offend anyone. Adding the trimpot resistance in series with the capacitor only lowers this further.

We are now in a position to design the RIAA network, reproduced in fig. 10.24. The series resistor R_x needs to be large enough that the PC97 is not grossly loaded at high frequencies. A tentative value of 200kΩ is a good start since it is several times larger than the anode resistor and generates about 8μV of Johnson noise which is much less than the ~40μV appearing at the anode of the input stage, so this is nothing to worry about.

The formulae in the figure indicate that $C_1 = 750ns/200k\Omega = 3.75nF$, and therefore C_2 is $2.916 \times 3.75 = 10.94nF$. Remember, these are just rough figures to get us started on choosing practical capacitors. We have come close to a standard value of 10nF, so let's run with it (an obvious alternative would be two 22nF capacitors in series). Working backwards we need C_1 to be $10nF / 2.916 = 3.43nF$, but this will partly be made up of the input (Miller) capacitance of the second stage.

The datasheet quotes $C_{ga} = 1.6pF$ and $C_{gk} = 1.8pF$. Adding 1.5pF to these figures to allow for strays, the total input capacitance is likely to vary from:

$$C_{in} = C_{gk} + C_{ga}(A+1) = 3.3 + 3.1 \times (13+1) = 46.7pF$$

down to: $3.3 + 3.1 \times (10.2+1) = 38pF$, depending on the setting of the balance pot. This will have negligible impact on the RIAA response since it is only an 8.7pF variation in the total. Taking the average as 42pF we need a fixed capacitance of $3.43 - 0.042 = 3.388nF$, which can be closely met using 3.3nF and 82pF in parallel.

R_2 must be $318\mu s/10nF = 31.8k\Omega$ and can be made exactly from 30kΩ in series with 1.8kΩ. We're now back where we started with R_1 which must be $31.8k / 0.145 = 219.31k\Omega$, some of which comes from the output resistance of the first stage, R_s, in parallel with grid leak R_g. As we saw earlier, this varies with valve sample but is around 12kΩ on average. We can safely ignore R_g by making it much larger than R_s, e.g. 1MΩ. The series resistance R_x must hence be $219.31k - 12k = 207.31k\Omega$ and can be made from 200kΩ in series with 7.5kΩ. All these components must be 1% or hand selected devices. If we wish to implement the IEC amendment too then the coupling capacitor C_o must be $7950\mu s/(R_s+R_g)$ or 7.86nF. However, we will

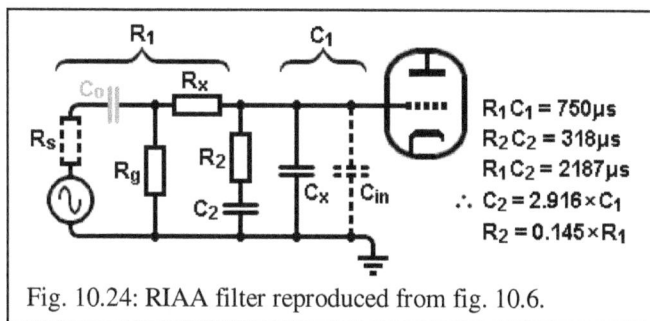

Fig. 10.24: RIAA filter reproduced from fig. 10.6.

use a larger value of 100nF, which is a compromise between extending the bandwidth and reducing phase delay while retaining a modicum of rejection of disc warp.

10.6.4: The Output Stage

From the load line we know that the anode voltage of the second stage will be approximately 145V which, using the same 27kΩ load line for the cathode follower, leads to a bias of about −4V and $I_a \approx 5.5mA$. The g_m is then about 2mA/V so the output resistance should be less than $1 / 0.002 = 500\Omega$, although added to this will be a 100Ω build-out resistor to ensure stability when driving cable capacitance. An obligatory grid-stopper and grid-cathode protection diode will also be needed.

An output coupling capacitor of 1μF would create a healthy 1.6Hz cut-off frequency when driving a 100kΩ AC load and might sound like enough. However, its reactance at 50Hz is 3183Ω which is uncomfortably large and could make the cable susceptible to picking up mains hum. Instead, a 10μF capacitor will be used, followed by a second capacitor to eliminate DC offset caused by leakage current. We can perform a neat trick by placing output clamping Zeners between these two coupling capacitors, so the second one can be a low-voltage part which saves space and money. Fig. 10.25 shows the completed preamp circuit. Audio ground must be bonded to chassis earth right at the input connectors to minimise susceptibility to external hum. Although the design is intended to have very low input capacitance, more can always be added to tune the pickup response to taste, as represented by the faint capacitor C_{in}.

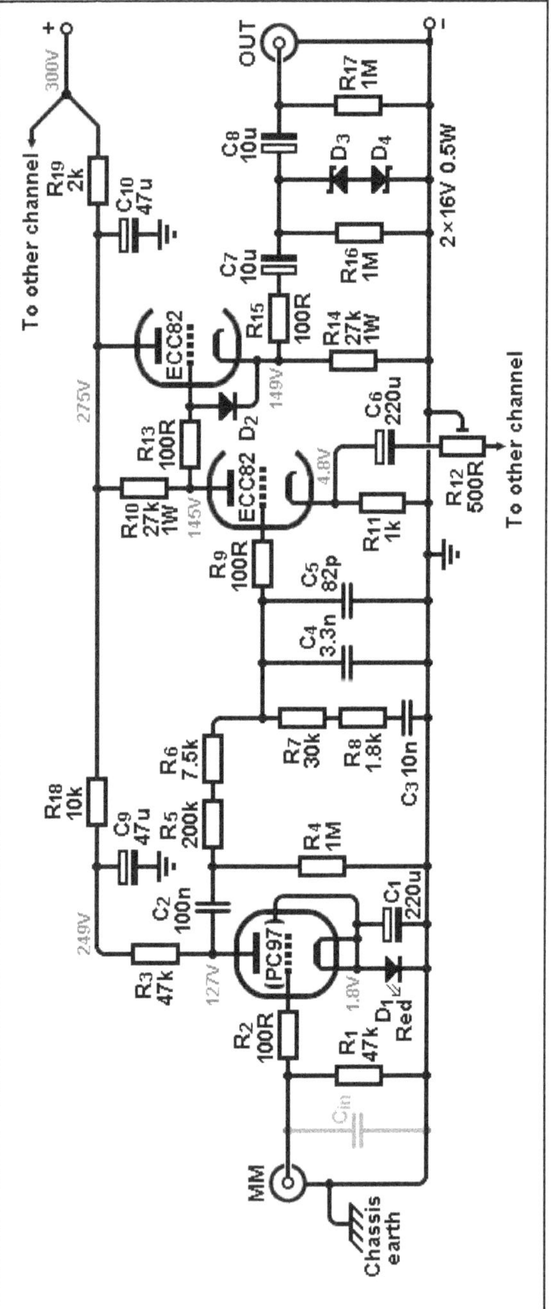

Fig. 10.25: Final PC97 phono preamplifier with measured voltages.

358

10.6.5: The Power Supply

The description of the power supply will be kept fairly brief, as the design of power supplies in general is dealt with in detail in chapter 11. Readers may therefore want to skip ahead at this point, or simply be patient.

In fig. 10.25 the two smoothing capacitors C_9-C_{10} must be large enough to ensure good decoupling between the gain stages and to prevent unwanted bass boost. A 47μF capacitor has a reactance of 169Ω at 20Hz which is negligible compared to the anode resistors, and 47μF happens to be one of the most common values in high-voltage electrolytics, so this will do fine. But will they provide enough smoothing of the ripple voltage from the rectifier? Estimating quite crudely that the EIN of the whole preamp will be about 1μV, this will be amplified and appear at the output as about 40μV. We will therefore declare that the power supply ripple at the top of C_9 should be no greater than this figure or hum may begin to dominate the noise floor. In reality we can tolerate more than this because we have ignored the circuit's PSRR, but we now at least have a simple and conservative number to work towards. The power supply will use a full-wave rectifier so the dominant ripple frequency will be twice the mains frequency or 100Hz.

The gain of the smoothing filter formed by R_{19}-C_{10} is roughly X_c/R or $34Ω / 2000Ω = 0.017$ at 100Hz. We can therefore allow the incoming ripple from the 300V supply to be up to $40μV/0.017 =$ 2.4mV. Now let's work in the other direction and state that we will use a 47μF reservoir capacitor too (both because 47μF capacitors are common and because it leaves the possibility for using a valve rectifier which can't tolerate a large capacitive load). There is a standard formula, (equation (11.3)), for estimating the maximum ripple voltage

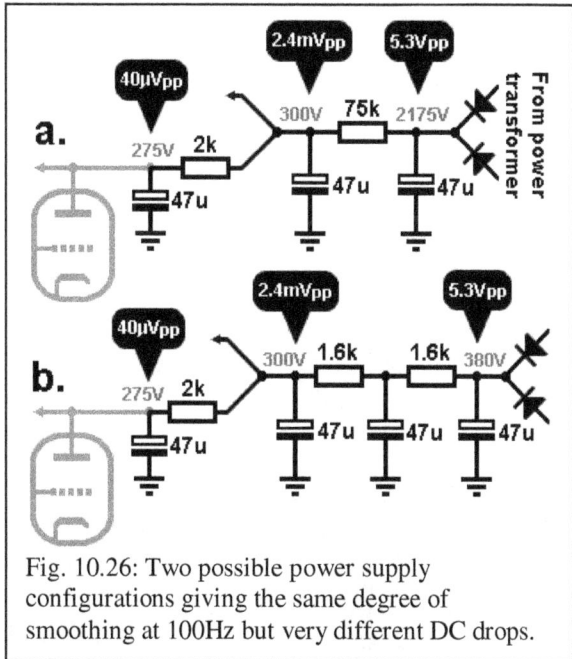

Fig. 10.26: Two possible power supply configurations giving the same degree of smoothing at 100Hz but very different DC drops.

from a full-wave rectifier. We know the total load current for the preamp will be 25mA, and the mains frequency is 50Hz, so the initial ripple voltage will be up to:

$$V_{pp} = \frac{I_{dc}}{2fC} = \frac{0.025}{2\times50\times47\times10^{-6}} = 5.3V_{pp}$$

359

In practice it will be less than this so we are adding another yet another layer of conservatism to the calculations. To bring this down to $2.4mV_{pp}$ using another RC filter would require a dropping resistance that is $5.3V / 0.0024V = 2208$ times larger than the reactance of the smoothing capacitor. This could be achieved with $75k\Omega$ and $47\mu F$, for example, as illustrated in fig. 10.26a. Unfortunately, with 25mA flowing in it such a resistor would drop 1875 volts and dissipate 47 watts! Clearly this isn't a viable option.

If instead we used two cascaded RC filters then the dropping resistors would only need to be $\sqrt{2208} = 47$ times larger than the capacitive reactances, so a pair of $1.6k\Omega/47\mu F$ RC filters would work, each dropping 40V and dissipating 1W, as illustrated in fig. 10.26b. Alternatively, a 2.5H choke has $1.6k\Omega$ reactance at 100Hz and could be used in place of one or both of the resistors, subject to the resonant caveats of LC smoothing as explained in section 11.6.4. These are viable solutions, but even so, several cascaded filters is aesthetically unappealing if nothing else. Let us instead abandon this approach and use a voltage regulator, which has the added benefit of eliminating variation due to mains voltage fluctuations.

A little experimentation showed that a simple Zener follower (section 11.7.5) was not up to the task, owing to the finite PSRR of the MOSFET source follower (although two cascaded would no doubt do the job). By contrast, a Maida regulator yielded output noise and ripple of $<1\mu V_{rms}$ which is far better than required. The regulator needs a little more than 10V across itself before it will drop out of

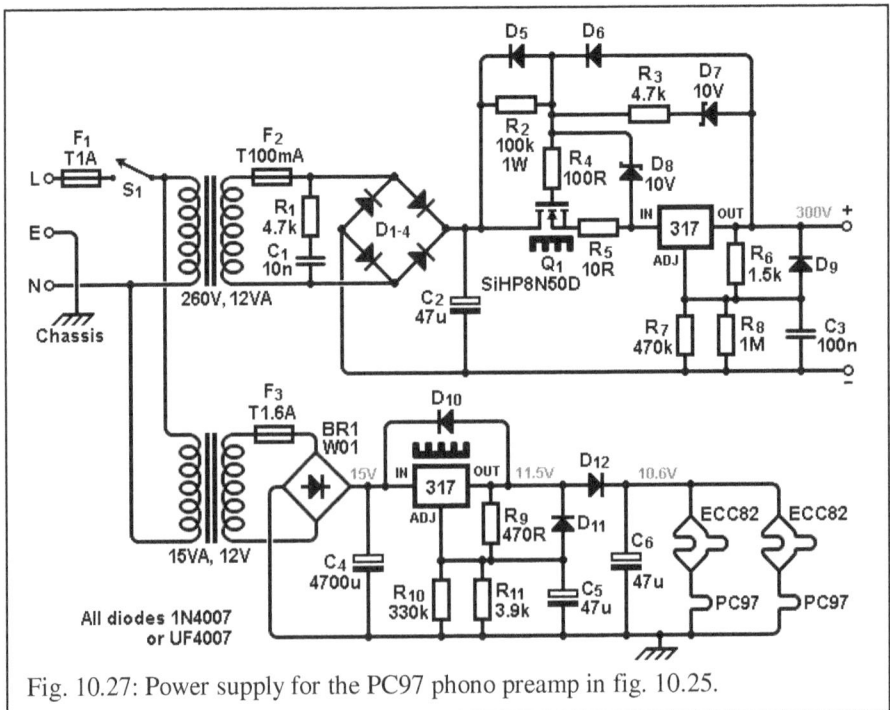

Fig. 10.27: Power supply for the PC97 phono preamp in fig. 10.25.

regulation, say 15V. Therefore, a raw nominal DC supply voltage of 340V is in order since this will accommodate almost ±10% variation in mains voltage, i.e. as low as 315V or as high as 370V. At its maximum this would leave 70V across the regulator so the MOSFET would have to dissipate almost $70V \times 0.025A = 1.75W$, worst case. This is small enough to be handled with a clip-on heatsink.

It is easy enough to obtain $340V_{dc}$ with a $340V / 1.3 \approx 260V_{ac}$ transformer and solid-state rectifier, or perhaps $270V_{ac}$ with a valve rectifier such as an EZ80. Any excess voltage can always be burnt off with a dropping resistor prior to the reservoir capacitor. The author happened to have a Danbury VT1551 transformer which has a $240V_{ac}$ 125mA secondary –far more powerful than needed. This means that with the lighter 25mA load, the voltage will relax to around 10 to 15% higher, easily meeting the $260V_{ac}$ requirement. It also has a $6.3V_{ac}$ heater winding but this was not used. Fig. 10.27 shows the whole power supply design. The transformer was mounted on a rubber gasket at the far end of the long (19 inch) chassis, to minimise vibration and stray coupling to the audio circuit.

The ECC82 has a 6.3V 300mA heater (or 12.6V 150mA), while the PC97 has a 4.5V 300mA heater. The author therefore considered running all the heaters in series, probably from a $20V_{ac}$ transformer and a LM317 constant-current regulator. However, he happened to have a $12V_{ac}$ 15VA transformer already at hand. After rectification this will provide about $12V \times 1.3 = 15.6V_{dc}$, minus diode drop. This leaves enough voltage for an ordinary LM317 voltage regulator to supply the heaters with 10.8V 600mA in a series/parallel arrangement, as shown in fig. 10.27. Although this implies a total power consumption of $15V \times 0.6A = 9W$, the transformer must be rated for almost twice this figure, owing to poor power factor (section 11.1.4), which is why a 15VA transformer happened to be ideal for the job. With 4700μF of reservoir capacitance the ripple voltage was found to be $\sim 1V_{pp}$ before the regulator, which has to dissipate 2.1W and can be managed with a good sized clip-on heatsink, or by mounting it to the chassis with insulating hardware. Heater elevation was deemed unnecessary as the ECC82 has a $V_{hk(max)}$ of 180V.

10.6.6: Measured Performance

After completion the output impedance of the preamp was found to be better than predicted at 510Ω. Fig. 10.28 shows the measured frequency response together with the overall response when preceded by the inverse RIAA filter from section 10.3.1 (gain normalised to 1kHz). The equalisation accuracy above 100Hz is excellent, while the 1.5dB droop down to 20Hz is due to the coupling capacitor C_2. This might trouble a few commentators, but the overall phase shift at 20Hz was only 10°. The two traces show how the balance trimpot allows the gain to be varied over a 1.5dB range –a little less than expected, but still enough.

With the input shorted the total output noise in the audio band was $95μV_{rms}$. Dividing by the 1kHz gain of 51 gives an effective EIN of 1.9μV. Relative to 5mV this is a passable SNR of 68dB (omitting C_1 degraded this by a further 3dB). Now,

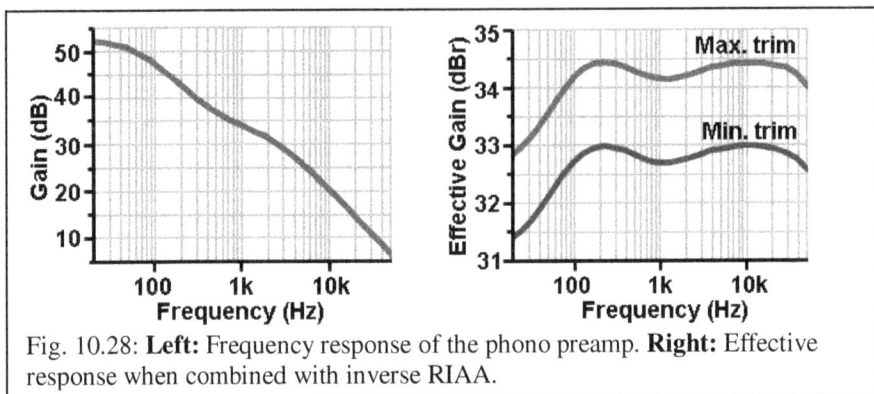

Fig. 10.28: **Left:** Frequency response of the phono preamp. **Right:** Effective response when combined with inverse RIAA.

the EIN of the input valve itself was measured earlier as ~1µV, so this figure appears to have been degraded by 5.6dB as a result of RIAA equalisation. Referring to fig. 10.14 this suggests a noise corner frequency of about 1kHz which is entirely plausible. The SNR could potentially be improved by using two PC97s in parallel since their Miller capacitance is so low.

We have to be very careful when interpreting distortion measurements for a circuit with a non-flat frequency response. In this case, for a given output level the distortion was roughly constant up to 2kHz, rising thereafter. A small part of this will be due to the declining load impedance on both the first stage and cathode follower (which was driving the 100kΩ ‖ 500pF input of the AP1). Mainly, however, it is because with increasing frequency the input stage has to handle an ever increasing signal amplitude to achieve the given output level. Fig. 10.29 shows distortion versus level at three spot frequencies, and extrapolation beyond the noise floor suggests that at the nominal output level of 250mV the distortion is about 0.025% at low frequencies. This is better than predicted from the PC97 alone, presumably due to distortion cancellation, although it was visually pure second harmonic at all levels. Also shown in fig. 10.29 is distortion with a constant 20mV *input* level, which paints a different picture. Distortion now appears to fall with frequency as harmonics are suppressed by the RIAA filtering.

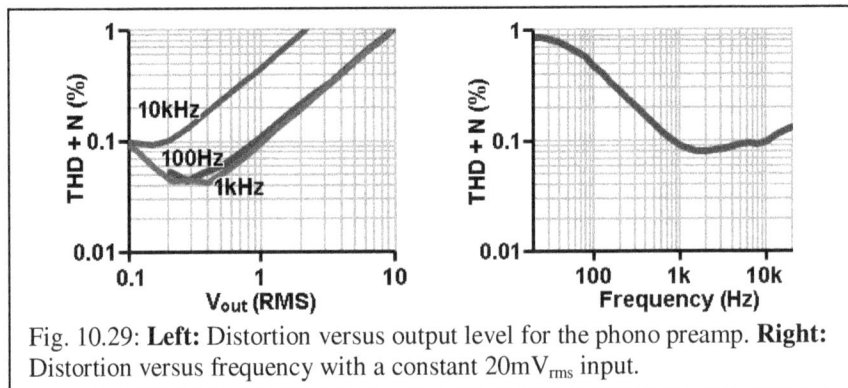

Fig. 10.29: **Left:** Distortion versus output level for the phono preamp. **Right:** Distortion versus frequency with a constant 20mV$_{rms}$ input.

Since the input stage is LED biased and directly coupled to the cartridge, it should be entirely immune to blocking distortion (not that the cartridge is likely to produce voltages high enough to cause this). The series resistance R_{5-6} in the RIAA network also goes a long way to reducing blocking in the second stage. As noted in section 10.4.1, the cartridge may produce click voltages as high as ~90mV_{pk}, which is not high enough to cause actual clipping at any point in

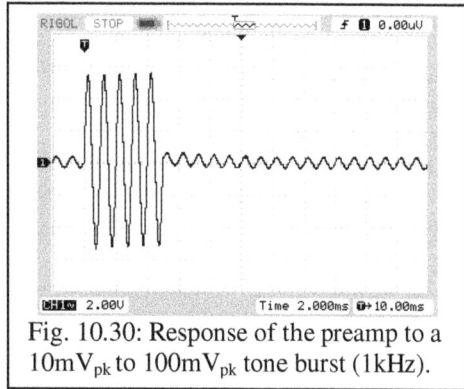

Fig. 10.30: Response of the preamp to a 10mV_{pk} to 100mV_{pk} tone burst (1kHz).

this circuit. To confirm this the author fed a 1kHz burst waveform into the inverse RIAA filter and thence into the preamp. The levels we adjusted so that the nominal level at the grid of the input valve was 10mV_{pk}, interrupted by a 100mV_{pk} burst to simulate a severe click or scratch. The oscillogram in fig. 10.30 confirms that no actual clipping is produced, and there is no significant recovery sag. The tiny hint of recovery sag just after the burst is not an overload effect but is due to the small error in the RIAA equalisation which slightly alters the waveform envelope. Altogether the amplifier is accurate, quiet, and thoroughly pleasant to listen to!

363

Chapter 11: Power Supplies

The power supply is sometimes perceived as rather boring when compared to the amplifier circuit proper, yet the power supply is the heart and lungs of any amplifier. Indeed, the 'amplifier' is really just a clever means for modulating the amount of power being drained from the supply and redirecting it somewhere else, such as into a loudspeaker. It is the power supply that places the ultimate limit on what we can get out of a design, and it therefore needs to be considering while (and indeed before) tackling the rest of the circuit.

Power supplies can be divided into two main types: linear power supplies and switch-mode power supplies (SMPSs). Linear supplies are the traditional kind using expensive and heavy transformers with big capacitors and inefficient brute-force regulation techniques. By contrast, SMPSs are cheap, lightweight, and ubiquitous to modern appliances. Both types can provide galvanic isolation from the mains by using transformers, and this is absolutely necessary for audio equipment since the user regularly comes into physical contact with the audio circuit (e.g. when touching audio connectors). The difference is that SMPSs can use much smaller transformers because they operate at much higher frequency.

Linear supplies are conceptually simple and inherently robust. They can easily run for decades, provided they are not abused, and repair is usually straightforward. SMPSs are more complicated, and while they are not necessarily less reliable, the motivation to make them small and cheap inevitable leads to borderline heat sinking and sweating capacitors, so they have become associated with short lifetimes. Repair is often a specialist job. In principle, SMPSs are more efficient than linear supplies, and this is particularly true at the low-voltages and high-currents needed by most modern appliances. But at higher voltages and modest currents the difference is sometimes exaggerated; SMPSs *can* be made more efficient, but it requires considerable optimisation. Often they are criticised for being noisy, though the state of the art has progressed to the point where they can now compete with linear supplies in this regard. However, they do require more close attention to shielding if a product is to meet RFI and EMI regulations. There is also the worry that if multiple SMPSs are used in the same system then their unavoidably different switching frequencies will lead to audible intermodulation products, even if the supplies themselves are relatively quiet. If multiple or unusual voltages are required then this tends to force the designer to use a multi-output (bespoke) supply running at a single frequency. The design of SMPSs is an engineering specialism by itself (about which this author does not pretend to be an expert), which is not an inviting prospect for hobbyists. And SMPSs are presumably the sworn enemy of those audio peculiarists who eschew silicon of all kinds. This chapter will therefore focus mainly on linear supplies.

The power transformer is usually the most expensive and awkward item to obtain, as there is an understandably limited range of off-the-shelf devices tailored to valve circuits (particularly outside the US/Canada). For the hobbyist on a budget it is often more prudent to begin by considering what power supply is actually feasible, and

364

what might be done with it, rather than to create a grand amplifier design and then try to find a suitable power supply to bolt-on as an afterthought. In fact, calling it 'the power supply' is really just a conceptual contrivance. In truth there is no convenient dividing line where the supply ends and the amplifier begins; they form a single device, so at every stage of amplifier design we should be considering and optimising how it will merge with its supply.

11.1: Essentials of Rectification

Except for a few special applications such as valve heaters, most circuits require a DC supply. It is the job of the rectifier to take the AC produced by a transformer and convert it into DC. The simplest example is a **half-wave rectifier** as shown in fig. 11.1. When the transformer voltage is large enough to overcome the forward voltage of the diode (about 0.7V for a silicon power diode) current is able to flow as indicated by the arrow. But when the voltage is negative, the diode blocks current. The voltage appearing across the load is simply the positive halves of the input sine wave (hence the name of the circuit), so while

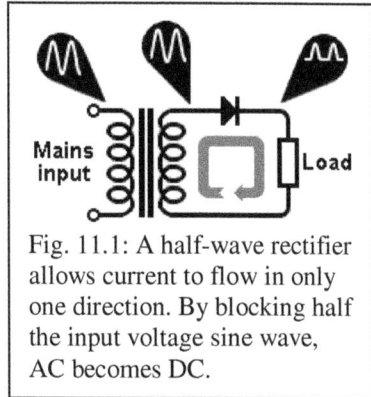

Fig. 11.1: A half-wave rectifier allows current to flow in only one direction. By blocking half the input voltage sine wave, AC becomes DC.

the current *is* unidirectional, it is not 'smooth'. Another problem with this rectifier is that the average current flowing in the transformer is also unidirectional, which can quickly lead to saturation and overheating of the transformer core. Half-wave rectifiers are therefore found only in primitive, low-current supplies.

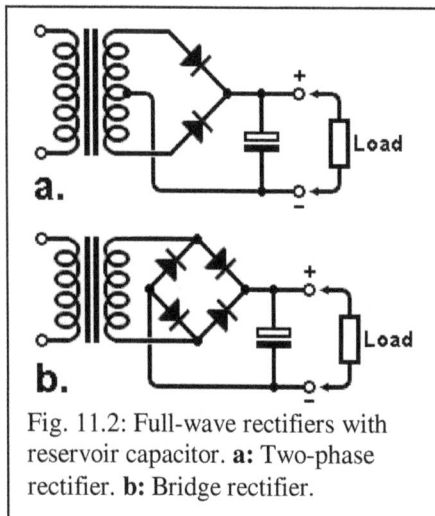

Fig. 11.2: Full-wave rectifiers with reservoir capacitor. **a**: Two-phase rectifier. **b**: Bridge rectifier.

We can improve efficiency and avoid transformer saturation by using both halves of the AC waveform, in which case we have a **full-wave rectifier**. There are two ways to do this, shown in fig. 11.2. The **two-phase rectifier** in fig. 11.2a is really two half-wave rectifiers, each using one half of the transformer secondary winding and alternately directing current into the same load. Since current now flows in *opposite* directions in each half of the transformer winding, the average core flux cancels out to zero, so there is no longer a risk of core saturation. With the **bridge rectifier** in fig. 11.2b, diodes conduct in alternate pairs to direct current into the load. This sort of rectifier makes more efficient use of the transformer winding than the two-phase type, but it incurs two diode drops.

While we are at it, we can also improve the smoothness of the DC by adding a capacitor across the output of the rectifier, also shown in fig. 11.2. When the rectifier conducts it has no choice but to dump current into the capacitor, charging it up to the peak AC voltage (minus the diode drop). When the rectifier switches off, the capacitor will begin to discharge via the load but, if we use enough capacitance, it will not have time to discharge very much before the rectifier tops it up again; hence the voltage across it will remain relatively constant. This capacitor therefore acts like a reservoir of energy from which the load can draw, while the rectifier periodically refills it, which is why it is called the **reservoir**

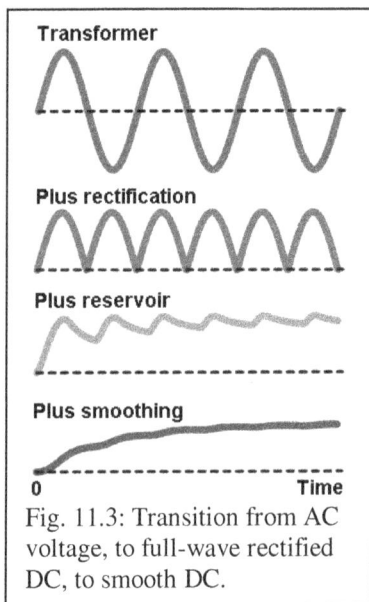

Fig. 11.3: Transition from AC voltage, to full-wave rectified DC, to smooth DC.

capacitor. In practice, however, it may take an inordinately large reservoir capacitance to get acceptably clean DC, so further stages of filtering or **smoothing** may be necessary too (section 11.6.1). This process is summarised in fig. 11.3.

The power transformer, rectifier, and reservoir capacitor, form a holy trinity. They should be thought of as one unit performing the basic task of converting AC into roughly smoothed DC. Wherever possible they should be laid out in a physically compact way, to minimise loop area and therefore noise emission. Once we have this fundamental circuit block taken care of then we are then free to attach refinements such as additional smoothing and/or regulation, as discussed later.

11.1.1: Average DC Output Voltage

One very important feature of the power supply is, of course, the average DC voltage that it delivers. As designers we are inevitably going to ask: what transformer voltage do we need in order to obtain a certain DC voltage, or alternatively, given a transformer, what DC voltage will it give us? The first-principle answer is to say the reservoir capacitor will be charged up to the peak AC voltage of $\sqrt{2} \times V_{rms}$. For example, a $12V_{rms}$ transformer ought to give us $\sqrt{2} \times 12 = 17V_{dc}$, or perhaps closer to $16V_{dc}$ after subtracting diode drops. Some text books leave it at that. Unfortunately, this simple approach is really quite inadequate for dealing with real world transformers, because the secondary voltage also depends on how much current we demand. A *fully loaded* $12V_{rms}$ transformer is actually more likely to give around $15V_{dc}$ minus diode drop. But then, how often do we load a transformer to its very limit?

366

The transformer secondary voltage varies under load owing to the simple fact that the transformer has winding resistance, which leads to voltage drop. Unfortunately, it is not a simple potential divider problem since the rectifier/reservoir form a highly nonlinear circuit. What's more, the sudden pulses of charging current occur near the peaks of the voltage sine wave, causing sudden voltage drop whenever the rectifier conducts, i.e. the secondary waveform becomes a clipped sine wave (see fig. 11.51 for example). These two effects combine to ensure that we never get the ideal $V_{dc} = \sqrt{2} \times V_{rms}$ quoted in the ordinary text books, except under no-load conditions; but transformer manufacturers almost never say what the no-load voltages are!

But let us continue undeterred. What information do we really *need* to predict power supply performance with accuracy? Quite simply:
- Primary winding resistance;
- Secondary winding resistance;
- No-load primary voltage, i.e. mains voltage;
- No-load secondary voltage.

If we already have the transformer in hand then we can easily measure the winding resistances, then connect it to the mains supply (taking great care of course) and measure the off-load voltages. To simplify things further, we can then refer the primary resistance to the secondary to find the total source resistance of the transformer from the point of view of the rectifier, which we will call R_s. This is equal to the secondary resistance plus the resistance reflected across from the primary by the square of the voltage ratio (turn ratio):

$$R_s = R_{sec} + R_{pri} \times \left(\frac{V_{sec}}{V_{pri}} \right)^2 \qquad (11.1)$$

Here the voltages are the *no-load* voltages, remember.

If we don't have the transformer handy then the most we can hope for is that the manufacturer will provide the regulation percentage –usually only vaguely. With this we can roughly estimate what the no-load voltage and secondary-referred resistance are. As covered in section 2.4.2, the percentage regulation is given by:

$$\%_{regulation} = 100 \times \frac{V_{no\ load} - V_{full\ load}}{V_{no\ load}}$$

Using this to find the transformer's secondary-referred resistance gives:

$$R_s = \frac{\%_{regulation} \times V_{full\ load}}{(100 - \%_{regulation}) \times I_{full\ load}}$$

And the off-load secondary voltage:

$$V_{\text{no load}} = \frac{100 \times V_{\text{full load}}}{100 - \%_{\text{regulation}}}$$

To predict the DC output voltage we also need to know how much current will be demanded by the load. With computer simulation we could go as far as to model the whole amplifier, but for more immediate gratification it is enough to represent the load as a simple resistance. In other words, simply estimate the DC voltage ($\sqrt{2} \times V_{\text{rms}}$ is close enough to start with) then divide by the expected load current to find the effective load resistance, which we will call R_l.

We now have enough information to create a very simple model of the power supply, as shown in fig. 11.4 (note that for a two phase rectifier, R_s is the source resistance of one *half* of the secondary windings). The graph in fig. 11.5 can now be used to predict the output voltage; it plots the variation in DC output voltage relative to the no-load transformer voltage as a function of R_s/R_l. Two curves are plotted, corresponding to different values of $f \times C \times R_l$, where f is the mains frequency (the doubling of the ripple frequency is already

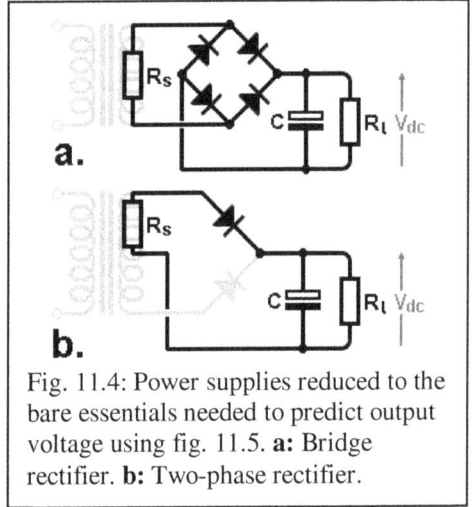

Fig. 11.4: Power supplies reduced to the bare essentials needed to predict output voltage using fig. 11.5. **a:** Bridge rectifier. **b:** Two-phase rectifier.

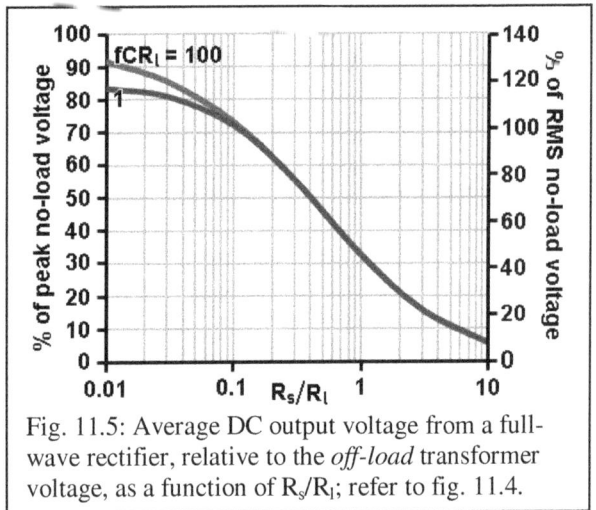

Fig. 11.5: Average DC output voltage from a full-wave rectifier, relative to the *off-load* transformer voltage, as a function of R_s/R_l; refer to fig. 11.4.

accounted for in the graph). Nearly all practical circuits will fall within the range shown. Notice that increasing the reservoir capacitance a hundredfold gives only a minor increase in the average output voltage, and only at small values of R_s/R_l. Using a very large reservoir capacitance mainly just improves the ripple voltage (next section). With a little care it is easy to use this graph to get within a few percent of reality.

But if all this seems like too much work then some rough approximations may suffice:

- When using silicon diodes and a >100VA transformer (which is likely to have good regulation), the *off-load* DC voltage can be expected to be around 1.4 times the rated (i.e. full load) RMS secondary voltage, and the *full-load* DC voltage is likely to be about 1.3 times the rated secondary voltage.

- When using a <10VA transformer (which is sure to have poor regulation) the off-load DC voltage will be around 1.8 times the rated secondary voltage, and the full-load DC voltage roughly 1.0 times the rated secondary voltage.

Intermediate sized transformers should fall somewhere between these extremes. Finally, one should remember that the DC voltage will also vary in sympathy with the mains voltage, and conservative designers will allow for ±10% variation.

11.1.2: Ripple Voltage and Reservoir Capacitance

Unless we have an infinite reservoir capacitance there will always be some residual ripple voltage riding on top of the average DC, as illustrated in fig. 11.6. When approaching a new design we need to know how much reservoir capacitance is needed to achieve the desired peak-to-peak ripple. Ripple is usually expressed as a percentage of the maximum (peak) DC voltage, and a typical figure might be 5% (though this is highly dependent on individual circuit requirements, of course). For example, if we were aiming for a 400V_{dc} supply with 5%

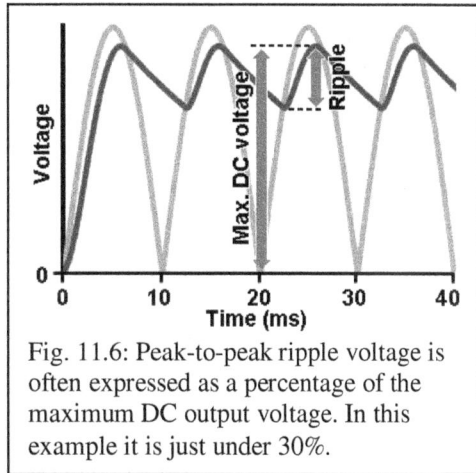

Fig. 11.6: Peak-to-peak ripple voltage is often expressed as a percentage of the maximum DC output voltage. In this example it is just under 30%.

ripple then this implies a ripple voltage of 400×0.05 = 20V_{pp}.

If we assume the load current is constant then the charge drained off the capacitor is Q = It, where I is the average load current and t is the time it flows for, which is the time between refills when the diodes are not conducting. The rectifier must supply an equal amount of charge during the period when the voltage is ramping back up from valley to peak, which is the ripple voltage, and we know that Q = CV. Equating the two we have CV = It, so the peak-to-peak ripple voltage is therefore:

$$V_{pp} = \frac{I_{dc}t}{C}$$

(11.2)

If we further assume that the capacitor spends *all* the time discharging and can somehow be recharged instantaneously, then t is equal to the time period of the raw rectified waveform. For full-wave rectification this is equal to twice the mains frequency: $t = \frac{1}{2f}$. Substituting this leaves us with:

$$V_{pp} = \frac{I_{dc}}{2fC} \quad \text{or} \quad C = \frac{I_{dc}}{2fV_{pp}} \tag{11.3}$$

In practice the ripple voltage will be less than this, both because the capacitor does not discharge linearly or for as long as we supposed, and because the source impedance of the transformer provides an additional filtering effect. Equation (11.3) is therefore a very conservative estimate, which is a good thing considering the poor tolerance of capacitors.

11.1.3: Ripple Current

However much charge is drained off the reservoir capacitor by the load, the rectifier has to re-supply an equal amount to keep the average DC voltage constant. Unfortunately, while the load can drain current off the reservoir at a steady rate, the rectifier only has a very short time in which to refill it. The rectifier therefore has no option but to conduct in brief, heavy pulses, called **ripple current**. This is so vital to the understanding of linear power supplies that it is worth saying twice: The current which flows around the transformer and rectifier, into the reservoir capacitor, flows in large, ugly, periodic pulses, while the current that flows out of the reservoir into the load is fairly even. This is shown figuratively in fig. 11.7. Fig. 11.8 shows the shape of the

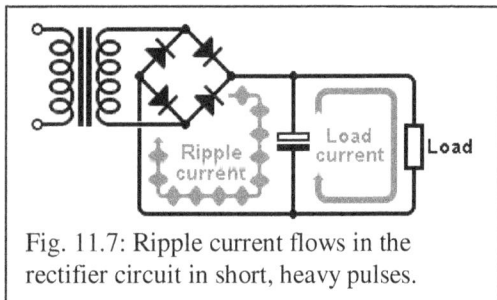

Fig. 11.7: Ripple current flows in the rectifier circuit in short, heavy pulses.

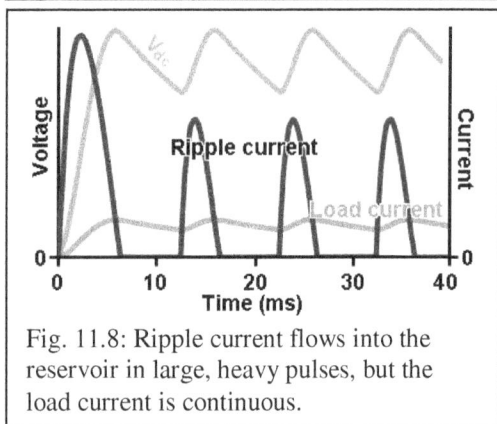

Fig. 11.8: Ripple current flows into the reservoir in large, heavy pulses, but the load current is continuous.

current waveforms in more detail, and the DC voltage (on a different scale) is shown faint for reference –notice that the ripple current pulses occur only when the reservoir is charging up, but the load current flows all the time. The very first pulse of ripple current after switch-on is also much larger than the steady-state pulses, owing to capacitor inrush (section 11.5.1).

Recalling a maxim from chapter 1 that the *average* current in a capacitor is always zero, the average incoming ripple current must hence be equal to the average outgoing load current. Unfortunately, the same cannot be said of the *RMS* values, because these depend on the *shape* of the waveforms. The RMS ripple current is always greater than the average, which is important because it means heating in the transformer wire resistance and in the capacitor's equivalent series resistance (ESR) will be higher than we might otherwise expect. Fuses

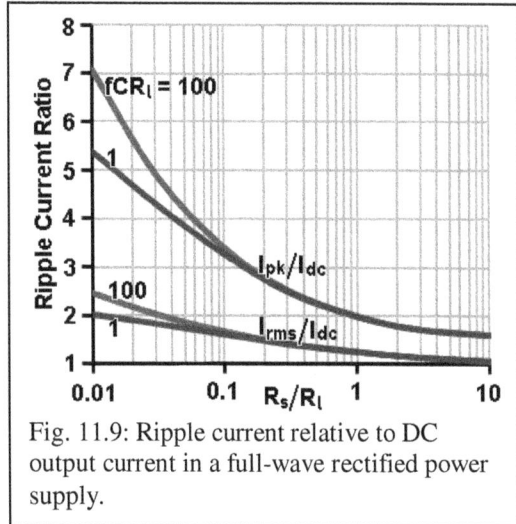

Fig. 11.9: Ripple current relative to DC output current in a full-wave rectified power supply.

work by resistive heating, so they have to tolerate the RMS ripple current too.

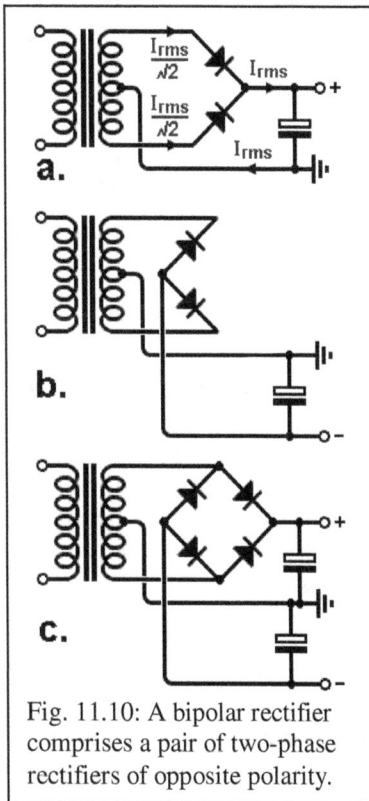

Fig. 11.10: A bipolar rectifier comprises a pair of two-phase rectifiers of opposite polarity.

Ripple current is a powerful source of hum and buzz. Any nearby circuit loops will pick up its ugly magnetic field, and any quiet circuits inadvertently connected to the 'dirty' wiring before the reservoir capacitor could end up with ripple current leaking into them. This is part of the reason for keeping the holy trinity of transformer-rectifier-reservoir, in tight physical formation; it minimises loop area and discourages you from making wayward connections to the rest of the circuit.

Trying to calculate the exact value of ripple current from first principles is tedious in the extreme, owing to the non-linear nature of the rectifier circuit. But it is easy to simulate circuits on computer, and it can also be found graphically. Fig. 11.9 shows the ratio of peak and RMS ripple current to the average load current, as a function of R_s/R_l, similar to fig. 11.5 earlier. The reservoir capacitor should have a ripple current rating at least equal to the total RMS ripple current (this is a safe estimate since the ripple current in the capacitor is actually slightly less than in the transformer/rectifier, as a portion flows into

371

the load instead). For an ordinary power supply the RMS ripple current can be expected to be around 1.5 to 2.5 times the DC load current, while the peak value could be up to seven times the DC current. This large peak, current multiplied by the transformer source resistance, causes sudden voltage drop on the peaks of the secondary waveform, which is why we never achieve the ideal $\sqrt{2} \times V_{rms}$ DC output voltage.

It is also worth pointing out that for a two-phase rectifier the RMS ripple current in each half of the secondary is not simply half the total indicated by fig. 11.9, but is closer to $1/\sqrt{2}$ times the total (but the *peak* value is the same wherever we choose to measure it), as shown in fig. 11.10a. A bipolar rectifier –although often built using a bridge-rectifier package– is really a pair of two-phase rectifiers, as illustrated in fig. 11.10, so the same principle applies. A centre-tapped power transformer therefore needs thicker wire –making it larger and more expensive– than a single-winding transformer using a bridge rectifier.

11.1.4: Power Factor and VA Rating

The fact that the transformer has to supply a larger RMS ripple current than is actually used by the load brings us to the concept of power factor. Of the total power delivered by the transformer, only some of it is dissipated in the load as heat or light, or sound, or whatever. The rest exists uselessly in the reservoir capacitor as *reactive* power. The *total* power handled by the transformer is called the **apparent power** and is equal to the *no-load* RMS secondary voltage multiplied by the RMS secondary current (i.e. ripple current), and has the units of volt-amps (VA). The *average* power delivered by the transformer is what does actual work in the load and is called the **real power**, in watts. The ratio of the two is called the power factor, PF:

$$PF = \frac{\text{Real power (watts)}}{\text{Apparent power (VA)}} \qquad (11.4)$$

If the transformer supplies nothing but a resistive load, such as valve heaters, then the power factor is equal to 1, i.e. all the power is burned off as heat in the load. But as soon as we introduce the reservoir capacitor into the proceedings we get ripple current which, being oddly shaped, comprises many harmonics of the mains frequency. Only the fundamental frequency delivers real power, the rest is simply an extra burden on the transformer and ultimately on the electricity supplier whose wires must be built to withstand the full RMS current (for this reason, appliances demanding more than about 500W may need to be fitted with power-factor correction circuitry in order to comply with national regulations).

For ordinary, linear power supplies the power factor can be determined from fig. 11.9. For example, if the RMS current in the transformer is determined to be 1.5 times the DC load current, then the power factor is $1/1.5 = 0.67$. In other words, we would need to use a transformer with a VA rating that is at least 1.5 times greater than the average load power. With low-voltage, high-current supplies like DC heaters, the ripple current usually ends up around twice the DC load current, so the power factor is much worse at about 0.5. The transformer would then need a VA rating at least twice as great as the load power; indeed, this is usually a good estimate for any conservative design.

11.1.5: Diode Ratings

When choosing rectifier diodes we must select devices which can withstand the reverse voltage and handle the forward current. The maximum reverse voltage that a rectifier diode can safely block will be quoted on the datasheet as its *reverse-repetitive maximum* (V_{rrm}), or the *peak inverse voltage* (PIV) on older datasheets. The datasheet will also quote the maximum allowable *average* forward current, $I_{F(av)}$. The manufacturer takes into account typical usage when quoting this figure, so we don't have to worry about the RMS or peak ripple current, even though we know they will be greater than the average. A maximum *non*-repetitive surge current (e.g. inrush) rating will be quoted, but for modern diodes it will be so far in excess of anything we have to deal with in a conventional power supply that this is not something we have to worry about either. Now, for a full-wave rectifier (not a voltage multiplier –section 11.2) the average current in each diode will be half the DC load current, since each diode only has to work half the time. Therefore, a good design rule is to use diodes with an $I_{F(av)}$ rating at least equal to the expected DC load current, since this automatically ensures a ×2 safety margin. Of course, there is no reason why we can't use even higher-rated diodes for extra robustness.

When silicon rectifiers are subjected to excessive reverse voltage they have a tendency to fail short, with disastrous results for the rest of the circuit. It is therefore worth being conservative with diode voltage ratings. The popular 1N4007 and UF4007 are rated for a reverse voltage of 1000V and an average forward current of 1A (up to 70°C), which is enough to accommodate many HT requirements. DC heater supplies are likely to need diodes with much higher forward current ratings. Discrete high-current diodes tend to be physically so large that it is often more convenient to use a bridge-rectifier package even if you don't intend to use all four of the diodes contained.

For a bridge rectifier the reservoir will charge up to the peak AC voltage, but the transformer secondary can only swing negative by a volt or so before the next pair of diodes begins to conduct. Each rectifier diode therefore needs to be rated to withstand only the peak AC voltage. For example, a transformer delivering $250V_{ac}$ would require diodes with a V_{rrm} rating of more than $\sqrt{2} \times 250 = 354V$, plus another 10% to allow for mains voltage variation, plus perhaps another 10% to allow for transformer regulation, bringing the total figure to 428V. The 1kV-rated 1N4007 or UF4007 would therefore be more than adequate.

For a two-phase rectifier the reservoir will charge up to the peak AC voltage, and the transformer secondary voltage will also swing down to the negative peak, so each rectifier diode has to withstand the full *peak-to-peak* voltage across the whole secondary. For example, a transformer delivering 350-0-350V would require diodes with a V_{rrm} rating of more than $\sqrt{2} \times 700 = 990V$, or more like 1198V after allowing

373

for mains variation and transformer regulation. A 1N4007 would therefore be inadequate. Higher-voltage diodes are available (e.g. the RGP02 series) but a common dodge is to use two or three diodes in series to increase the voltage blocking capability. A 10nF to 100nF ceramic capacitor should be connected in parallel with each diode to encourage equal voltage sharing (high-value resistors would be better, but it is a lot easier to find kilovolt-rated ceramic capacitors than kilovolt-rated resistors). Note that when using a *pair* of two-phase rectifiers –to create a bipolar supply, for example– the diodes must still be rated for the peak-to-peak voltage even though the circuit looks superficially like a bridge rectifier.

11.1.6: Switching Noise and Fast Rectifiers

The audio press often cites diode switching noise as a malevolent spectre of power supplies, though often with only a nebulous explanation of what switching noise actually is. In fact, calling it 'diode switching noise' is a bit misleading as it makes it sound like the noise is coming from the diodes themselves, when in fact the problem has at least as much to do with the power transformer.

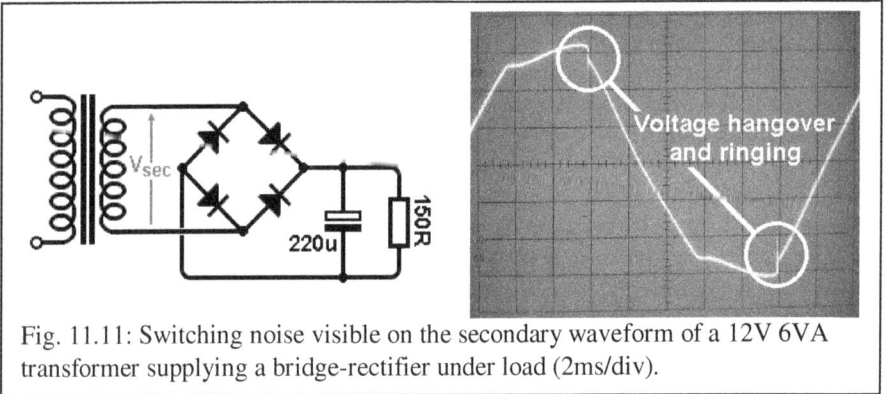

Fig. 11.11: Switching noise visible on the secondary waveform of a 12V 6VA transformer supplying a bridge-rectifier under load (2ms/div).

'Rectifier switching noise' is perhaps a better term as it encompasses the whole rectification process.

When a rectifier diode switches on and conducts, it allows current to flow from the transformer secondary and into the reservoir capacitor. This also causes the transformer's leakage inductance –typically around 10mH to 100mH– to become energised or 'charged up' if you like. Eventually the diode comes to switch off again, but the leakage inductance would rather keep the

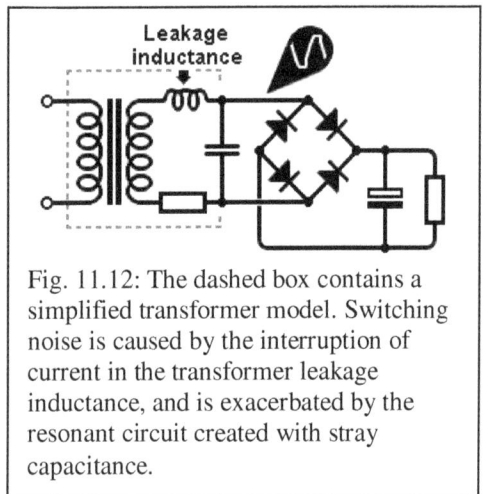

Fig. 11.12: The dashed box contains a simplified transformer model. Switching noise is caused by the interruption of current in the transformer leakage inductance, and is exacerbated by the resonant circuit created with stray capacitance.

374

current flowing just as it was, thank you very much, so it will generate a flyback voltage that keeps the diode turned on. In other words, the leakage inductance momentarily becomes the source of power as it releases its stored energy into the reservoir as an extra little dollop of ripple current. In this way the peak of the AC voltage waveform, instead of descending in a smooth way, is held artificially high by the flyback voltage until the stored energy is finally spent. This leads to a 'hang over' or 'cliff face' effect on the AC waveform. This can be seen in the oscillogram of fig. 11.11 which shows the secondary voltage of a small 12V transformer when driving a rectifier circuit –it is not much of a sine wave! The sharp, steep edge contains high frequency Fourier components which can couple into the audio circuit via stray capacitance, leading to irritating switching spikes appearing in the signal path. It will also couple across to other windings on the same transformer,

particularly from the HT to the heater supply, forging yet another path into the audio circuit.

But it gets worse. There is unavoidable stray capacitance across the transformer secondary, which includes the non-linear junction capacitance of the diodes themselves. We therefore have an RLC circuit formed by the transformer resistance, leakage inductance, and stray capacitance, as illustrated in fig. 11.12. The voltage hang-over is effectively a step input applied to this network, which will therefore resonate or ring. The effect is further worsened by the fact that a silicon diode does not switch off in an instantaneous way but will continue to conduct for a brief moment even after the voltage across it has become negative, which is called its **reverse recovery time**. The ringing will be high in frequency (visible only as spikes in fig. 11.11), so we have in effect a small RF transmitter, which can't be good news for nearby audio circuits. Fig. 11.13a shows a close-up of the ringing from fig. 11.11, revealing it to be around 125kHz.

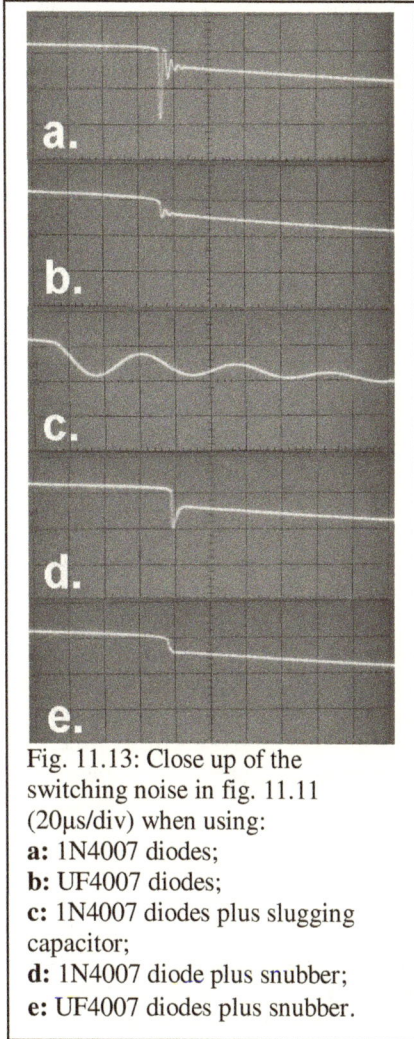

Fig. 11.13: Close up of the switching noise in fig. 11.11 (20μs/div) when using:
a: 1N4007 diodes;
b: UF4007 diodes;
c: 1N4007 diodes plus slugging capacitor;
d: 1N4007 diode plus snubber;
e: UF4007 diodes plus snubber.

There are three approaches to dealing with the switching problem: faster diodes, slugging, and snubbing. Simply substituting the ordinary 1N4007s with their fast-recovery cousins, UF4007s,

375

produced the oscillogram in fig. 11.13b, which is already much better. Incidentally, fast diodes tend to be more expensive, so a cheap dodge is to use ordinary diodes for the main rectifier, plus a single, fast diode immediately in series, as shown in fig. 11.14. The fast diode imposes its short recovery time on the whole supply. Schottky diodes also qualify as fast recovery types but are mostly limited to low voltage ratings.

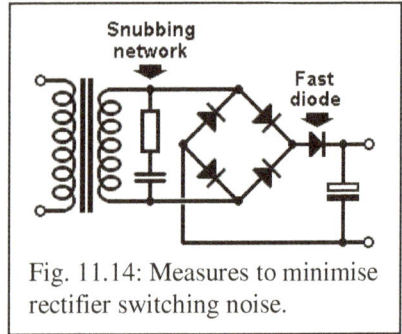

Fig. 11.14: Measures to minimise rectifier switching noise.

Slugging involves adding more capacitance to lower the resonant frequency and slow down the rate of change of voltage, so there is less energy likely to couple into other circuits. Often this takes the form of a capacitor connected in parallel with each rectifier diode. Confusingly, these are often referred to as snubbing capacitors when in fact they do no such thing. Snubbing implies burning energy off as heat, whereas slugging capacitors only spread the energy out, transferring it to lower frequencies. Also, only one capacitor is actually needed, in parallel with the transformer secondary. For example, four 100nF capacitors around a bridge rectifier are, as far as ringing is concerned, equivalent to one 100nF capacitor in parallel with the transformer. Such a capacitor added to the test circuit in fig. 11.11 produced the oscillogram in fig. 11.13c. The ringing frequency has been lowered to around 20kHz (suggesting the transformer has about 25mH leakage inductance).

Snubbing is a more sophisticated technique which involves adding an RC network across the transformer secondary, as indicated in fig. 11.14 (for a two phase rectifier, snubbing networks can be added to each half of the transformer secondary). The capacitor should be large enough that it swamps the effect of stray capacitance – about 10nF should be ample– leaving only the resistor to choose. The aim is to burn off the oscillatory energy as quickly as possible in the resistor, and there will be an optimum value that provides the most damping. Too little resistance and we will have slugging rather than snubbing; too much and the snubber does nothing at all. In some cases it is possible to monitor the circuit on an oscilloscope and so tune the snubbing resistance to the optimum value, but if not, a value between 1kΩ and 4.7kΩ ½W is usually a good compromise. Fig. 11.13d shows the effect of adding an optimised 10nF+1.5kΩ snubbing network with 1N4007s. The picture is quite good, but a small negative spike remains due to the diodes' lazy reverse-recovery time. Fig. 11.13e shows the result when using UF4007s with the snubber; the switching is now about as clean as we could hope for.

Simply adding resistance in series with the transformer can also be regarded as a sort of heavy-handed snubbing since it damps the resonant network and adds compliance. Valve rectifiers have a great deal of internal resistance and are often lauded as being 'quiet' for this reason. But a silicon rectifier with the same resistance deliberately added has much the same effect, as well as being a lot cheaper and more reliable. Either way, the extra series resistance wastes power, reduces the DC output voltage and degrades the 'regulation' of the whole power supply, so it is an inefficient route to take.

11.2: Voltage Multipliers

A particularly useful circuit for valve enthusiasts is the full-wave voltage doubler, sometimes called a Delon circuit in old books. As shown in fig. 11.15, it is usually drawn either as a bridge rectifier with two diodes replaced with capacitors, or as two half-wave rectifiers of opposite polarity with the load connected between them. The circuit provides twice the DC output voltage of an ordinary full-wave rectifier but, since we can't get something for nothing, the ripple current will also be

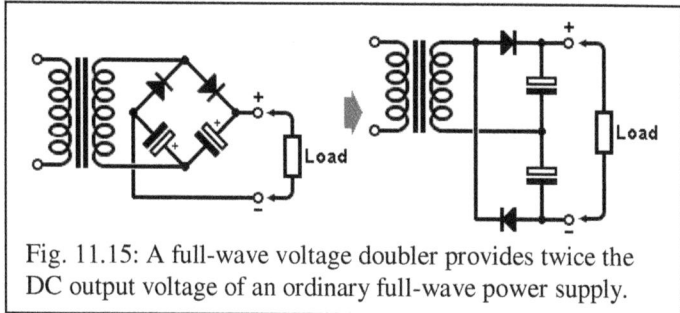

Fig. 11.15: A full-wave voltage doubler provides twice the DC output voltage of an ordinary full-wave power supply.

twice as great. In other words, the RMS ripple current in the transformer will be as much as five times the DC load current, and the average current in each diode will be equal to the DC load current. The diodes should therefore have an $I_{F(av)}$ rating of at least $2I_{dc}$, and also a V_{rrm} rating in excess of the peak-to-peak transformer voltage. In instances where high-current diodes are required it is sometimes easier to obtain them in a bridge-rectifier package and connect it as in fig. 11.16. Do not rely on this to double the effective current rating, however, as the parallel pairs of diodes will not share current equally.

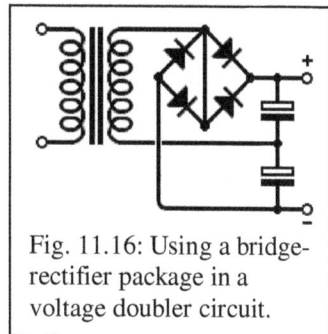

Fig. 11.16: Using a bridge-rectifier package in a voltage doubler circuit.

The two reservoir capacitors in the voltage doubler (which will normally be identical) are connected in series so can be viewed as one reservoir of half the capacitance. Thus one way to approach the design of a voltage-doubler is to begin by designing an ordinary bridge-rectifier circuit, then simply replace the transformer with one of half the voltage and twice the current rating, and replace the reservoir with two capacitors having twice the capacitance originally assumed (they can also have half the voltage rating). The final performance of the voltage doubler will then be substantially equal to what was originally designed.

As an example, fig. 11.17 shows an HT supply built using a readily available 50VA, 0–55, 0–55V transformer. The secondaries are connected in series to obtain $110V_{ac}$ which is then voltage doubled and will realistically produce about $290V_{dc}$ under load. Using 220μF capacitors, as shown, the ripple voltage will be about $7V_{pp}$.

377

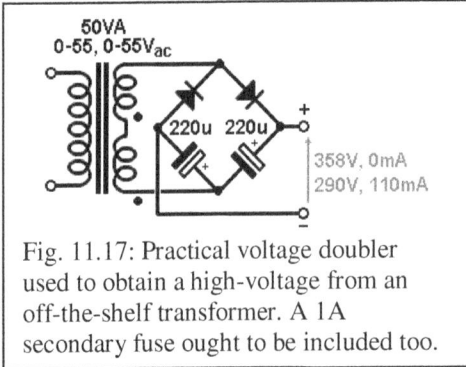

Fig. 11.17: Practical voltage doubler used to obtain a high-voltage from an off-the-shelf transformer. A 1A secondary fuse ought to be included too.

Notice that the maximum load power is 32W but a 50VA transformer is required, i.e. the power factor is 0.64, which is no different from an ordinary full-wave power supply. The diodes must be rated for $>380V_{rrm}$ and $>220mA$. The 1N4004 is rated for $400V_{rrm}$ and 1A, but the ubiquitous 1N4007 could of course be used too. Not only is this technique convenient because it uses an off-the-shelf transformer, but the capacitors need only be 200V rated, and we have the freedom to use a separate heater transformer of any desired voltage. All of this may be easier to obtain and perhaps cheaper than a traditional design using a dedicated valve transformer.

Another common application of the voltage doubler uses the junction of the two capacitors as the ground reference, thereby creating a bipolar supply (the ripple on each rail will then be at mains frequency even though it is a full-wave circuit as far as the transformer is concerned). This approach is often useful for equipment supplied by an external AC voltage adapter (US: wall wart). Fig. 11.18 shows a typical example using a 12V, 12VA transformer, and note that

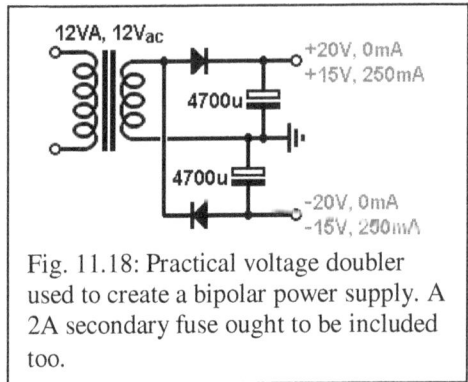

Fig. 11.18: Practical voltage doubler used to create a bipolar power supply. A 2A secondary fuse ought to be included too.

there is a fair amount of sag between no-load and full-load conditions, owing to the poor regulation of such a small transformer. The two supply rails should be equally loaded to avoid a net DC current in the transformer (core saturation). The diodes must be rated for $>40V_{rrm}$ and $>500mA$, so 1N4001s could be used ($50V_{rrm}$, 1A).

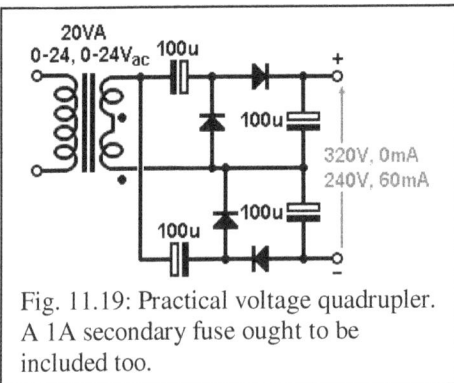

Fig. 11.19: Practical voltage quadrupler. A 1A secondary fuse ought to be included too.

Transformer voltages can be multiplied by more than two times with more extensive rectifier circuits, but at the expense of ever declining load regulation and power factor. Fig. 11.18 shows an example of a voltage quadrupler using a cheap 20VA, 0–24, 0–24V_{ac} transformer from which 240V_{dc} is obtained at 60mA maximum current. The regulation is poor, so this technique is best reserved for constant-current applications (into which category valve preamps usually fall). Using 100µF capacitors the ripple will be about $7V_{pp}$ (pay close attention to their orientation!). The diodes must be rated for more than the peak-to-peak transformer voltage and >120mA, so 1N4003s (or better) would suffice. The capacitors should be rated for more than half the output voltage, implying 200V devices. At full load they handle about half the total RMS ripple current, or about 220mA in this case.

11.3: Valve Rectifiers

The traditional method of rectification is of course the *valve* rectifier. Valve rectifiers normally contain one diode, or two with a shared cathode, conventionally intended for half-wave or two-phase rectification (fig. 11.20a), respectively. However, you can achieve bridge rectification at virtually no extra cost simply by adding a couple of silicon diodes, as in fig. 11.20b. These do not need to be fast rectifiers since the valve imposes its superior switching characteristics on the whole circuit. These days there is no *technical* justification for using a valve rectifier; ordinary silicon diodes are far more robust, efficient, and inexpensive. Power resistors can be deliberately added if we want to simulate the voltage

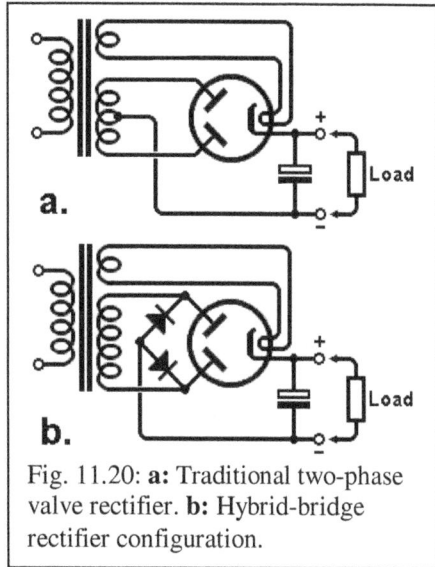

Fig. 11.20: **a**: Traditional two-phase valve rectifier. **b**: Hybrid-bridge rectifier configuration.

losses and reduced switching noise of a valve, and auxiliary circuits can provide a power-on delay if necessary, without adding great complexity. On the other hand, valves will always have superior aesthetic (and therefore marketing) appeal.

The high cost of the most popular rectifier valves has driven more enthusiasts to use 'junk box' alternatives, such as television damper/efficiency diodes like the PY500A. These were originally used as snubbing devices in televisions, but their high voltage and current ratings make them suitable for power supply duty too. In the author's opinion, however, valve rectifiers are suitable for low-current, light service only, where their inefficiency is of little consequence to the rest of the design. When used very conservatively their service life can be extremely long, so there needn't be any reason to worry about dwindling supplies of old-stock devices. Conversely, for high- or wildy-varying current applications like power amplifiers, valve rectifiers are too lossy, too expensive, and too unreliable to be justifiable. There is something quite perverse about paying the same money for a GZ34 as for an EL34 when the rectifier performs such a mundane job compared to an amplifier valve, and so badly compared to a bit of silicon.

Max. Rating	GZ34 5AR4	5Y3GT	EZ81 6CA4	EZ80	GZ32 5V4G	5U4GB
V_{heater}	5V	5V	6.3V	6.3V	5V	5V
I_{heater}	1.9A	2A	1A	0.6A	2A	3A
$V_{hk(max)}$	-	-	500V	500V	-	-
$r_{a\,(beam)}$	50Ω	350Ω	120Ω	300Ω	115Ω	150Ω
$V_{rrm(rms)}$	450V	450V	450V	350V	375V	450V
$I_{dc(max)}$	250mA	100mA	100mA	90mA	125mA	275mA
$I_{ripple(peak)}$	750mA	440mA	500mA	270mA	525mA	1A
$I_{surge(200ms)}$	3.7A	2.5A	1.8A	?	3.5A	4.6A
C_{max}	60μF	20μF	50μF	50μF	40μF	40μF

Table 11.1: Maximum ratings of some rectifier valves in a full-wave capacitor-input circuit. See also fig. 11.23.

11.3.1: Electrical Ratings

Capacitor-input rectification is a brutal task for a valve, owing to the high peak ripple current. There are severe limitations on what valve rectifiers can handle, and table 11.1 provides figures for some popular rectifier types. The datasheets may be quite extensive, including multiple graphs showing what operating conditions are permissible, and there is not space here to go into the various ways different manufacturers presented the information. Neither is there really any need to, because the true limitations are quite straightforward. Just like modern diodes, they have ratings for:

- Maximum repetitive reverse voltage (usually quoted as an RMS rather than peak value);
- Maximum average current (i.e. DC load current);
- Maximum surge current (e.g. inrush).

These things are self explanatory. But there is another, crucial rating: the maximum peak repetitive current, i.e. peak ripple current. This is taken for granted with silicon rectifiers, but valve cathodes can only handle so much current before they are at risk of saturation, degradation, and eventually flashover. This figure is not always stated explicitly on the datasheet but is bound up within the *minimum* **limiting resistance** and *maximum* reservoir capacitance ratings. These two things work together to keep both the peak ripple current and inrush current at safe levels.

The minimum limiting resistance is the total resistance which must be in series with each anode, for a given reservoir capacitance. Most of this resistance will come from the source resistance of the transformer itself, which was provided earlier by equation (11.1):

$$R_s = R_{sec} + R_{pri} \times \left(\frac{V_{sec}}{V_{pri}}\right)^2$$

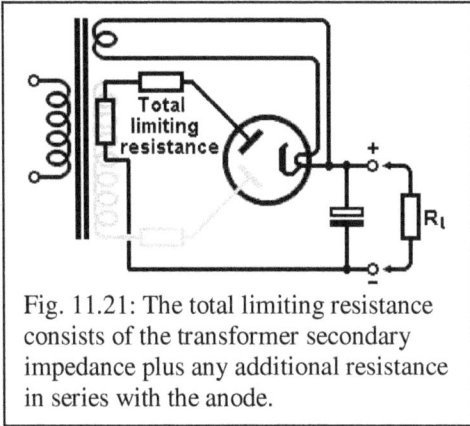

Fig. 11.21: The total limiting resistance consists of the transformer secondary impedance plus any additional resistance in series with the anode.

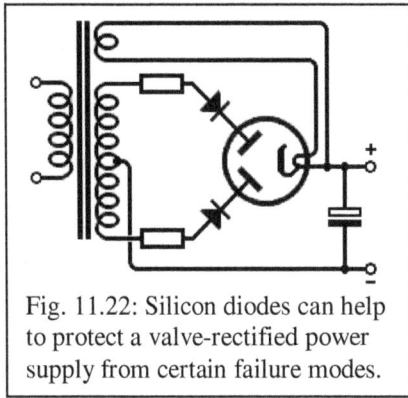

Fig. 11.22: Silicon diodes can help to protect a valve-rectified power supply from certain failure modes.

However, if the transformer does not provide enough source resistance by itself then more must be added to make up the deficit. Fig. 11.21 illustrates how these two things together form the total limiting resistance (per anode). Be aware that any added resistors must withstand the RMS ripple current, which for a two-phase valve rectifier is normally about 1.1 times the load current. Alternatively, a single limiting resistor can be added between the cathode and reservoir capacitor, or in the centre tap of the transformer, but it will have to handle about 1.5 times the load current. Furthermore, a common protective measure against valve rectifier failure is to add one or more silicon rectifier diodes in series with each anode, as in fig. 11.22. These diodes provide a fail-safe action if the valve happens to fail short, so protecting the reservoir capacitor and the rest of the amplifier from AC voltage.

Among designers more familiar with silicon, one of the most common abuses of valve rectifiers is to use too much reservoir capacitance in an attempt to make a nice, stiff power supply with low ripple. This inevitably leads to a short and painful life for the valve, ended by bright blue flashes inside the bulb as the cathode material evaporates and arcing occurs. However, it should be appreciated that it is not the reservoir capacitance *per se* that kills the valve, but the ripple/inrush current. The maximum value quoted on the datasheet is only a derived limitation based on the manufacturer's assumption that the device will be used in an ordinary, traditional power supply design. In theory, we can use any reservoir capacitance we like, provided we also take steps to keep the inrush and peak ripple currents within the recommended limits. As a rough rule of thumb, however much the reservoir capacitance is increased above the datasheet recommendation, the total series limiting resistance should be increased by at least the same factor. This of course incurs further voltage and power loss which is why manufacturers assume no one would want to do it. It is also why the author considers valve rectifiers to be suitable only for mild, unvarying current applications where extra smoothing or regulation can easily be added, removing the need for a big reservoir capacitor in the first place.

11.3.2: Voltage Drop

The voltage lost across a valve rectifier is equal to the peak ripple current multiplied by the internal (beam) resistance of the diode, which is of course relatively high. What's more, the voltage drop will vary in sympathy with changes in load current, so the HT voltage will rise when the current demands are low, and sag when the demands are high. This can be a problem in vintage amplifiers where the power supply capacitors are only rated to withstand the HT

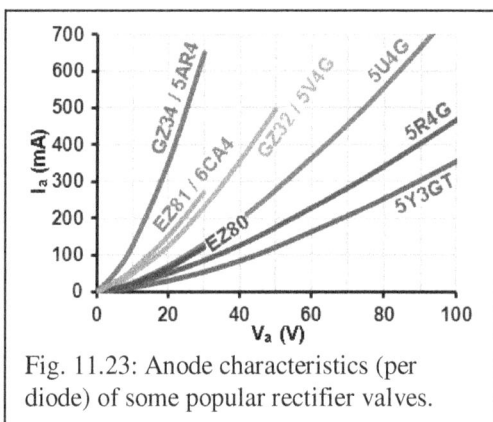

Fig. 11.23: Anode characteristics (per diode) of some popular rectifier valves.

voltage under *full load* conditions. If the amplifying valves are removed, so there is little or no load current, the supply voltage will climb to its maximum peak value since there will no longer be significant drop across the rectifier, possibly leading to bursting capacitors. A similar problem occurs if a valve rectifier is replaced with silicon, in which case it may be necessary to add some resistance to simulate the previous rectifier's internal anode resistance. Even then, a silicon rectifier has no warm up time, so the maximum voltage will be present for many seconds before the amplifier valves start conducting enough current to pull the voltages down to normal working levels.

Fig. 11.23 shows the anode characteristics of some popular rectifier valves. The internal beam resistance (per diode) can be determined from the curves in the usual way (section 3.2.1); table 11.1 summarises some typical values. This resistance adds to the source resistance of the transformer, from which circuit simulation or fig. 11.5 will yield the expected DC output voltage. Typically, R_s/R_l will end up in the range of 0.1 to 0.2 and the DC output voltage will be numerically close to the RMS transformer voltage. In other words, a 300-0-300V transformer with a valve rectifier is likely to produce close to $300V_{dc}$ at full load. With negligible loading, however, the same circuit will settle closer to $450V_{dc}$. If we further allow for 10% mains variation then we will need capacitors rated for at least 500V. This may require the use of multiple capacitors connected in series, with equalising resistors, as described in section 2.2.14.

11.3.3: Heater Supply

As a result of the need for a robust cathode, valve rectifiers tend to have quite hungry heaters, which is a serious disadvantage compared to silicon diodes. Some small rectifiers such as the EZ81/6CA4 have completely separate heater and cathode, just like most amplifying valves. They also have a deliberately high $V_{hk(max)}$ rating so they can be operated from the same heater supply as other valves in the amplifier, without needing heater elevation. Most rectifiers, however, are either directly heated or else have the heater connected to the cathode internally (in which case they are often referred to as 'directly heated' even when this is not strictly true). This forces the designer to use a separate heater supply just for the rectifier, which will float on top of the HT (e.g. fig. 11.22), thereby eliminating any possibility of abusing the heater-cathode insulation. Many valve power transformers have an extra 5V winding for just this purpose.

11.3.4: Hot Switching and Inrush Current

'Hot switching' refers to the practice of allowing the heater/cathode to warm up before throwing a switch to allow anode current to flow. This is the principle of most guitar-amplifier standby switches, for example. Depending on how such a switch is implemented, this may mean the rectifier valve will be fully warmed up before the reservoir capacitor is charged, so the valve will have to supply the full inrush current when the switch is finally thrown. This will cause momentary saturation which is very damaging to the cathode, and rectifier failure (arcing) in guitar amps is uncomfortably common for this reason. Hot switching of rectifier valves was expressly discouraged by valve manufacturers; the valve should always be allowed to charge the reservoir naturally from cold, as the heater warms up. Opening a standby switch can also induce a ghastly flyback voltage across the transformer winding, large enough to cause arcing in the valve. A precaution against this is again to add ordinary silicon diodes in series with each anode of the valve rectifier, as in fig. 11.22, to reduce the reverse voltage across it.

11.4: Choke-Input Rectifiers

All of the previous discussion has been about capacitor-input rectifiers, which are by far the most common type. However, an alternative arrangement is the choke-input rectifier, which has slightly different characteristics. Whereas a reservoir capacitor tries to maintain a constant voltage across itself by taking heavy pulses of ripple current, a 'reservoir choke' tries to maintain a constant current through itself by varying the voltage across itself. Choke-input rectifiers were more popular in the days of valve rectifiers when it was expensive to regulate the HT by electronic means. These days the reverse is true, and it has become more difficult to obtain suitably rated chokes for this purpose. Nevertheless, a few enthusiasts still embark on this path.

Fig. 11.24 shows the basic choke-input rectifier configuration. Since rectified half-cycles are unidirectional they contain a DC component, but also a whole series of harmonics which form an AC component. It is these harmonics which the choke should 'block' while allowing the DC component to pass as freely as possible. From chapter 1 we know that the DC component of a

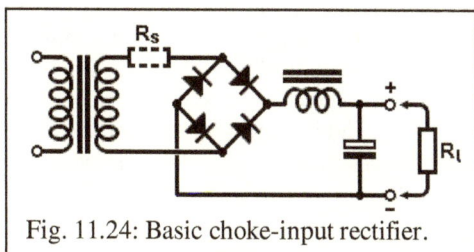

Fig. 11.24: Basic choke-input rectifier.

signal is equal to its mean average, and for a rectified sine wave this is $2 \times V_{pk}/\pi$. Thus if the choke blocks all the AC component then the output voltage would be pure DC equal to 64% of the peak AC voltage, or 90% of the RMS voltage (whereas a perfect capacitor-input rectifier produces a DC output voltage equal to the peak AC voltage). With infinite inductance this voltage would remain constant, regardless of how much load current was drawn, but in reality we must expect some practical limits to the choke-input power supply. The main one is that for a given inductance there is a critical value of load current which must be exceeded before the choke will properly regulate the output voltage. If the load current is less than this value, the DC voltage will climb towards the peak voltage, just as it does with a capacitor-input rectifier. It might be supposed that a bleeder resistor could be used to keep the current above the critical value at all times, but at switch-on the load voltage will oscillate at the resonant frequency of the inductor-capacitor combination, so it may still reach or even exceed the peak AC voltage. Therefore, with a choke-input filter, the power supply capacitors (and anything else) must be able to withstand the peak voltage *and more*, unless an automatic soft-start or over-voltage protection circuit is added. This is best modelled on a computer.

11.4.1: Design Calculations

For a given resistance in series with the choke, the minimum value of Inductance required for regulation is:[1]

$$L_{min} = \frac{R_s + R_l}{6\pi f} \tag{11.5}$$

Where:
R_s = the effective resistance in series with the choke, including transformer impedance, valve rectifier anode resistance (if used), and the DC resistance of the choke itself;
R_l = load resistance;
f = mains frequency.

L_{min} is the effective inductance of the choke *with the direct load current also flowing through it*. This is likely to be much less than the inductance we would measure without any direct current, so it is important to buy a choke designed for this sort of duty, which quotes the inductance with DC current (they are sometimes called 'reactors'). But that is not all. The current in the choke is the sum of the DC load current plus an AC current component which gets stored as a magnetic field in the choke's core. If the choke is not sufficiently rated to handle this current it will saturate, quickly leading to failure, sometimes even by catching fire. The current rating of the choke must be greater than the *peak* current if we are to avoid saturation, and it is this requirement above all that makes suitable chokes hard to find –an ordinary smoothing choke will not do.

---

[1] Schade, O. H. (1943). Analysis of Rectifier Operation. *Proceedings of the I.R.E*, July, pp341-61.

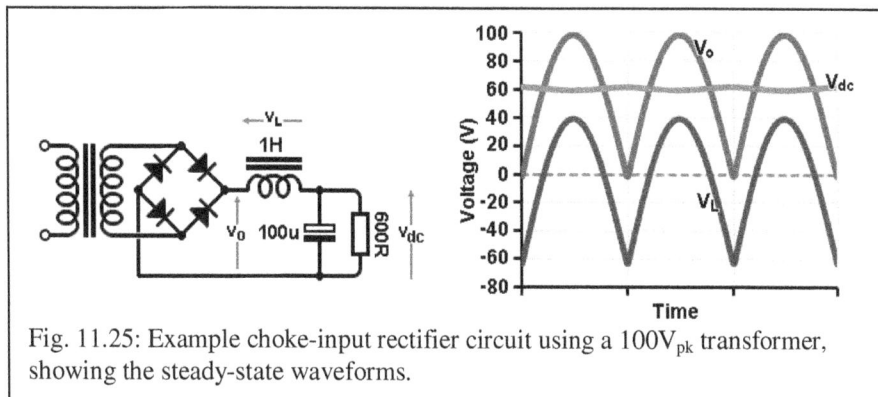

Fig. 11.25: Example choke-input rectifier circuit using a $100V_{pk}$ transformer, showing the steady-state waveforms.

To find the peak current, let us assume there is zero ripple voltage on the capacitor, so the whole AC component of the rectified sine wave is dropped across the choke. Fig. 11.25 shows an example circuit using a $100V_{pk}$ transformer (chosen because the voltages nicely translate into percentages) along with the resulting steady-state waveforms. From Fourier's theorem the frequency content of V_o can be written:

$$V_o = V_{pk}\left(\frac{2}{\pi}+\frac{4}{3\pi}\cos(2\omega t)-\frac{4}{15\pi}\cos(3\omega t)+\frac{4}{35\pi}\cos(4\omega t)...\right) \tag{11.6}$$

From the above we can see that the second harmonic of mains frequency dominates the rest, and since the reactance of the choke rises with frequency, all the higher harmonics can safely be ignored. The peak current in the choke is equal to the peak AC current component plus the load current, which works out as:

$$I_{L(pk)} = \frac{0.6V_{rms}}{2\pi fL}+\frac{0.9V_{rms}}{R_1} \tag{11.7}$$

Where f is *twice* mains frequency. This formula is a slight overestimation because it does not take into account R_s or the effect of higher harmonics, but it is better to err on the side of caution where the choke current rating is concerned.

The average current in the choke is relatively constant so we can easily predict the DC output voltage by treating R_s and R_1 as a potential divider (which we could not do with a capacitor input rectifier). Provided the load current is greater than the critical value, and ignoring diode drop, the DC output voltage will be:

$$V_{dc} = 0.9V_{rms}\cdot\frac{R_1}{R_1+R_s} \tag{11.8}$$

Fig. 11.26 shows a comparison of the load regulation of choke-input and capacitor-input rectifiers. The capacitor-input filter was set so that $fCR_1 =100$ when $R_s/R_1 = 0.1$. Two choke-input filters are represented, one set so the critical load resistance is $100R_s$, and another at $10R_s$. Notice that with the choke circuit the same degree of regulation is always achieved; increasing the inductance simply means the regulation is maintained down to smaller load currents. The amount of smoothing capacitance used after the choke does not improve the regulation either, it only reduces the ripple voltage.

Thanks to the constant-current action of the choke, the RMS current in the transformer is also the same as the DC load current, so the required transformer VA rating is simply $V_{rms} \times I_{dc}$. In other words, a choke-input rectifier provides us with a power factor of approximately unity (they are still used in kilowatt industrial applications for this reason).

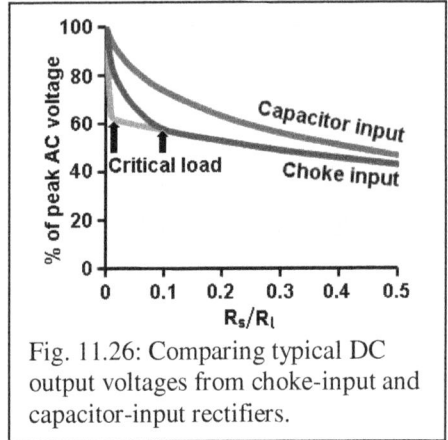

Fig. 11.26: Comparing typical DC output voltages from choke-input and capacitor-input rectifiers.

One further point worth mentioning is that when the rectifier diodes switch off, the choke will generate a considerable flyback voltage, in the same manner as the transformer leakage inductance described in section 11.1.6. It is therefore advisable to connect a 220nF to 470nF capacitor immediately prior to the choke –as shown in

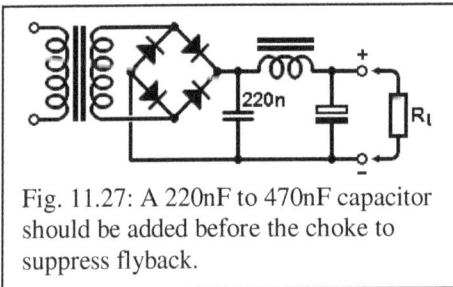

Fig. 11.27: A 220nF to 470nF capacitor should be added before the choke to suppress flyback.

fig. 11.27– to suppress transients. It is also worth pointing out that because the voltage across the choke occurs in rectified pulses, the electric field around it is very noisy, so it should not be placed too close to any sensitive circuitry unless it is electrostatically shielded. Any loose laminations will vibrate and buzz, so everything must be firmly clamped.

To sum up, the choke-input rectifier gives good voltage regulation and, because the current drawn from the transformer is roughly constant, the power factor is also very good. The disadvantages are the low output voltage and high cost of the choke, and the possibly wild oscillations and flyback voltages than can occur at start up, switch off, or if the load current drops below the critical value.

11.5: Power Supply Refinements

The following is a collection of power supply refinements that deserve a mention. Some, like fusing, are obligatory, while others are optional extras. In this book many of these features are omitted from circuit diagrams to save space and keep things clear, but this does not mean they cannot be added as the designer's whim commands.

11.5.1: Inrush Current Limiting

When power is first applied to a transformer there will be a brief but possibly enormous current surge that lasts for a few mains cycles. There are two reasons for this. Before power is applied, the transformer is not magnetised, i.e. it has zero core flux. If we are unlucky enough to switch on when the mains voltage is at the zero crossing then the core flux will increase to the point of saturation –and beyond. Momentarily, the only thing limiting the primary current is the winding resistance, so the primary current will be abnormally large for the first few cycles until the core flux settles to its steady state. Conversely, if the circuit is switched on at the peak of the mains cycle then there will be negligible magnetising inrush, but the near-instantaneous change in voltage across the reservoir capacitor will cause it to take several large gulps of charging current instead. One way or another, then, there will be an inrush current when the power supply is first switched on.

Transformers smaller than about 10VA use such fine primary wire that it will flex under the magnetic field of the inrush current. After a few hundred switching cycles this will lead to metal fatigue and eventually a break, so small transformers benefit enormously from some form of inrush limiting. Larger transformer can easily withstand the current, but we may still need inrush limiting measures to prevent otherwise reasonably-sized fuses from blowing every time we switch on.

Big power amplifiers commonly use a resistor to limit the current, which is automatically shorted out by a hefty relay or triac after a second or two (e.g. fig. 11.65). A less elaborate option is a negative temperature coefficient (NTC) thermistor in series with the mains input, as shown in fig. 11.28d. NTC thermistors have relatively high resistance when cold, but quickly fall to a fraction of an ohm as the device heats up with normal current flow. The thermistor should be chosen so that its cold resistance is a significant fraction of the transformer primary resistance, say 30 to 50%. It will get hot during normal running so be careful with its physical mounting. It is also worth noting that it takes time for the device to cool down after use, so a thermistor cannot protect against rapidly turning the supply off and on again.

11.5.2: Fuses

A fuse in series with the live wire where it enters the amp is absolutely mandatory.[*] This may be in a separate panel-mounted fuse holder or part of an IEC-inlet assembly. This main fuse must be of the ceramic type, also called a **high breaking capacity** or HBC fuse. These are filled with sand to dissipate heat and discourage vapourised metal from leaving a conductive path inside the cartridge, so they can safely interrupt very large fault currents. The fuse should be rated for around 1.5 to 2 times the normal operating current and must never be larger than the mains voltage divided by the transformer primary winding resistance, otherwise it can never blow. If this is insufficient to withstand inrush then some form of inrush limiting must be used, rather than a bigger fuse. Needless to say, the fuse must also have a voltage rating suitable for use on the mains, but this is no problem since most fuses are rated for mains use anyway.

In Europe there are two principle sorts of fuse to consider: **time delay** (T) and **fast acting** (F). The similar equivalents in America are **slow-blow**[†] and **quick-blow**. Fuses used anywhere in a power supply will nearly always be of the time-delay/slow-blow type since these can withstand a certain amount of inrush current. Nevertheless, inrush will still cause the fusewire to flex, and after enough cycles it will inevitably fail due to fatigue. This expectation is just something we have to live with. Only after replacing the fuse and finding that it immediately fails again can we be sure there is a genuine fault. It is also worth mentioning that European (IEC) fuses are specified by the maximum current they can handle *without* blowing, whereas American (UL/CSA) fuses are specified by the minimum current which *will* cause them to blow. In other words, a European 1A fuse will not blow when the current is one amp, but a US 1A fuse *will* (though it may take a fairly long time to do so). But in most applications the fuse has to be somewhat oversized in order to withstand inrush current and prevent false tripping, so this difference between ratings is of little practical consequence.

In addition to the main or primary fuse, most builders will want to add fuses to the transformer secondaries too, and in some countries this is apparently mandatory. Secondary fuses are mainly there to protect the precious transformer rather than downstream components, and can be of the glass type (non HBC) since fault currents are greatly limited by the transformer itself. They should be rated for around 1.5 to 3 times the rated transformer current, unless the transformer is capable of a lot more current than you're actually using, in which case you may want to use a smaller fuse rating more appropriate to the load current. However, one place where such fuses might *not* be used is on a purely AC heater supply, since there is very little to go wrong and the likelihood of a fault creating a permanent short is slim. Also, the inrush to the heaters is so severe that it may be difficult to find a fuse that will

---

[*] UK mains plugs contain fuses but they are there to protect against gross surges which can lead to fire, not to protect the equipment itself.
[†] Note that the alternative spelling 'slo-blo®' is not a technical term but a registered trademark of Littlefuse.

withstand it (automotive fuses are one possibility). However, if the heater supply has a rectifier for DC heaters or ancillary circuits then it should be fused, since rectifier diodes nearly always fail short.

When a fuse melts, an arc will be sustained until the gap widens enough to quench it, or until the voltage drops to the point where the arc cannot be sustained. With alternating current this happens very quickly since the voltage is continually passing through zero, but with DC the arc will be sustained for much longer. In other words, it is easier to interrupt AC than DC, so fuses rated for ordinary AC mains may have surprisingly paltry DC voltage ratings. If such a fuse is used in a DC supply beyond its voltage rating then its clearing time may not meet the manufacturers guarantee.[*] On the other hand, the safety implications for secondary fusing are much more benign compared with primary fusing, so an undefined clearing time may not matter. Nevertheless, it is always preferable to place a fuse in an AC rather than DC part of the circuit when possible, e.g. before the rectifier, as in fig. 11.28f.

It is also important to consider what might happen to the rest of the circuit when a fuse blows. For example, a blown fuse in the bias supply of a power output stage could lead to redplating and the destruction of the output valves. Similar problems could occur in a DC-coupled amplifier if a heater-supply fuse blows. This may require the addition of fast-acting/quick-blow fuses to other parts of the circuit to protect against such calamities.

11.5.3: Switches and Switching

A power switch in series with the incoming mains is usually fitted, and switches rated for AC mains are very common. If the mains cable is permanently wired to the amplifier (captive) then, to comply with BS EN UL 60065 both the live *and* neutral *must* be switched simultaneously using a double-pole switch (the earth is never switched), and the 'on' position must be indicated with a label or by illumination. However, if a detachable mains lead is used –e.g. an IEC lead or 'kettle lead'– then a switch is not legally required at all, and you are free to use a single- or double-pole device if you wish. Usually the switch will be placed after the fuse rather than before, although the author is not aware of any legal requirement for this.

Occasionally there is a need to provide a separate switch for the high-voltage DC supply. The most common reason for this is to implement a 'standby' feature which allows the heaters to warm up before the HT is applied, especially when using gas rectifier tubes. Conversely, ordinary high-vacuum rectifiers must *not* be hot switched (section 11.3.4). Receiving valves do not require pre-heating either, and specious fears about cathode sputtering may do more harm than good if the user leaves the amp in standby for too long, which promotes the growth of interface resistance. Furthermore, applying the HT to a pre-heated amplifier is more likely to cause an

---

[*] The Littlefuse® 477 series is one of the few types of 20mm-cartridge fuse with high DC voltage ratings: $500V_{ac}/400V_{dc}$.

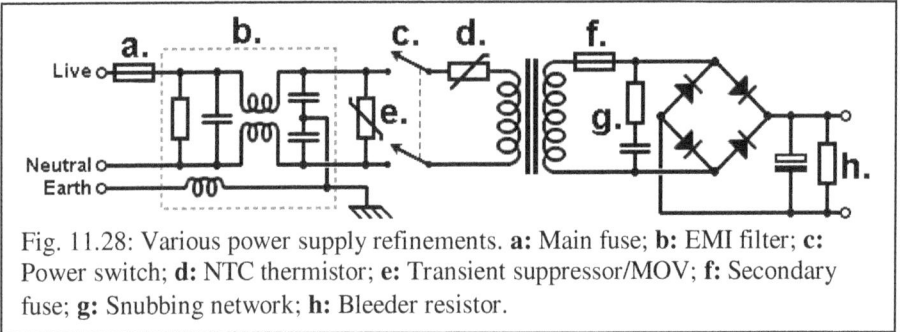

Fig. 11.28: Various power supply refinements. **a:** Main fuse; **b:** EMI filter; **c:** Power switch; **d:** NTC thermistor; **e:** Transient suppressor/MOV; **f:** Secondary fuse; **g:** Snubbing network; **h:** Bleeder resistor.

unpleasant thump in the speaker unless an auto-muting function is implemented, so you should think very carefully about whether an HT switch is genuinely necessary.

There is yet more to consider. Just as with fuses, switching a DC supply is much more stressful on a mechanical switch than switching an AC supply, owing to prolonged arcing. This leads to corrosion of the switch contacts and in extreme cases may even weld it shut. The switch will be subject to less arcing if it is placed in the AC part of the circuit, e.g. prior to the rectifier. However, in many valve circuits the HT voltage will considerably exceed the voltage ratings of common (and attractive looking) switches. Many designers simply use heavy-duty but otherwise ordinary mains switches and hope for the best. This seldom causes problems, but is a dubious practice if the safety inspector calls. If an HT switching scheme really is needed then an alternative is to let an optocoupler or solid-state relay (SSR) do the switching instead. Switching an SSR is a simple matter of turning on its internal LED. With this in mind, automatic start-up, shut-down, and fail-safe procedures could be implemented too, which is a more professional approach than expecting the user to follow a prescribed switching ritual.

11.5.4: Transient Suppression

The mains supply is occasionally polluted with transient voltage spikes caused by industrial motors, lightning strikes and so on. A protective measure against these spikes is a transient suppressor or metal oxide varistor (MOV), connected between live and neutral after the primary fuse, as in fig. 11.28e. 'Varistor' is short for variable resistor, although the device behaves more like a pair of back-to-back Zener diodes. Under normal circumstances the MOV has a very high resistance (several megohms), but if the voltage across it reaches a certain threshold –called the breakdown or clamping voltage– its resistance suddenly drops, thereby quashing the voltage spike and possibly blowing the fuse. Unlike a Zener diode, however, a MOV can withstand extremely large (but brief) current pulses.

Selecting a MOV simply requires choosing one with a continuous RMS working voltage that is 1.2 to 1.5 times the nominal mains voltage. However, sustained or repeated overloads can –and frequently do– cause MOVs to burn out, so they must be suitably positioned so as not to ignite anything else and create a fire hazard. It is perhaps worth noting that UL-rated equipment will place a thermal fuse in series with each MOV, mounted side by side, to detect heat and so interrupt the supply to the MOV. Thermally protected MOVs are also available, such as the Littlefuse® TMOV® series.

11.5.5: EMI Filtering

A common addition to any modern appliance is an electromagnetic interference (EMI) filter, where the mains enters the unit. Such filters are often built into IEC inlets and have become so ubiquitous that the author has never had to buy one, only scrounge them from discarded equipment. They usually consist of a common-mode choke with capacitors on each side to create a balanced CLC filter, and an inductor is sometimes included in the earth wire too, as shown in fig. 11.28b. A bleeder resistor is sometimes included to ensure the capacitors can discharge after the power cable is removed.*

Beginners sometimes assume that an EMI filter is supposed to turn an ugly mains supply into a sparkling clean sine wave, but we should be so lucky. They only do anything useful in the range of hundreds of kilohertz or megahertz, and even then they are largely intended to attenuate noise generated *inside* the appliance (particularly from SMPSs) from leaking *out* onto the mains (the mains cable can act as quite a good transmitting antenna, jamming local radio reception). Valve circuits generally use linear power supplies and seldom contain any high-frequency clocked logic, so they have minimal problems with EMI, either reception or transmission, but it can do no harm to include a filter. Be sure to use one that can handle the maximum expected current, which should be no problem in a valve preamp as even the most modest filters are rated for a couple of amps.

11.5.6: Bleeder Resistor

When a high-voltage power supply (>50V) is switched off it is desirable that the reservoir/smoothing capacitors have some means of discharging, otherwise they pose a shock hazard to unsuspecting users during repair or when replacing valves. Sometimes a discharge path is already provided by a heater-elevation divider or by equalising resistors across series-connected smoothing capacitors and so forth, in which case nothing more is required. But if not, it is a simple matter to add a bleeder resistor in parallel with one of the capacitors, and a value of 330kΩ to 1MΩ is typical, determined mainly by how much power we are willing to waste during normal running. Adding an LED in series with the resistor (visible from inside the unit) to give some indication of discharge is a nice touch for the repairman.

---

* According to BS EN UL 60065 the voltage must discharge to <60V within two seconds after switch off.

11.6: Smoothing, Bypassing and Decoupling

Despite a lot of discussion, so far in this chapter all we have managed to do is turn AC voltage into dirty DC voltage. The reservoir capacitor alone is rarely enough to provide the really clean DC needed by a preamplifier, so some further smoothing is necessary. This can be achieved with LC or RC filters, variously referred to as smoothing, bypass, or decoupling filters. These alternative names derive from the fact that there are really three interrelated jobs to be performed:

1. Smooth/filter out residual ripple voltage;
2. Bypass/provide a local energy supply for sudden current demands;
3. Decouple/isolate each amplifier stage from the rest.

All three of these tasks loosely revolve around making the power supply quiet and low-impedance or 'stiff', that is, not easily perturbed by constantly varying current demands. In other words, we want the power supply to be as close to an ideal voltage source as possible. Admittedly, the highest expression of design is to make an amplifier that doesn't need a good power supply in the first place, i.e. improve the circuit's PSRR to the point where it happily ignores the shortcomings of the supply. Unfortunately, this is rarely a trivial exercise, so we had better know about smoothing, bypassing and decoupling all the same.

11.6.1: Smoothing

Smoothing-out residual ripple voltage from the rectifier is the most obvious and easiest job to perform. In fact, if it was the *only* job then it might be performed with one resistor or inductor, plus one massive smoothing capacitor. It would not matter very much where we placed this filter –it could be housed in a separate box in the attic with long wires leading to the amplifier, if we so desired. Unfortunately, the more subtle requirements of decoupling and bypassing make such an approach impractical and uneconomical, so a network of more modest filters is normally used instead. We will start with the RC type since it is simpler and better behaved than the LC type.

Each RC smoothing stage is a low-pass filter whose attenuation increases with frequency. Of course, the only frequency we really want to pass is 0Hz or DC, but since infinite capacitors are only available for government work, we must compromise. For a given capacitance, a larger resistance lowers the cut-off frequency and therefore improves the smoothing.

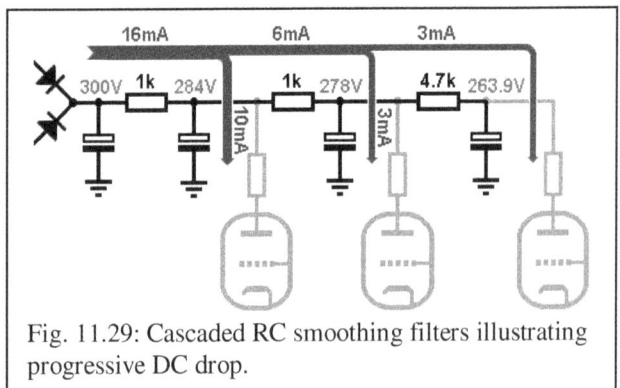

Fig. 11.29: Cascaded RC smoothing filters illustrating progressive DC drop.

However, there is also some DC drop across the resistor due to the load current flowing in it (fig. 11.29), so the choice of this **dropping resistor** is usually limited by how much voltage we can afford to throw away. Incidentally, the resistor must be capable of withstanding the full supply voltage and charging current at start up, which usually means using ½W devices or better, even though the steady-state dissipation may be minimal.

Provided the cut-off frequency is well below the frequency of interest, the gain of an RC filter can be approximated as X_c/R (or, if you prefer, the attenuation factor is R/X_c). Thus if we wanted to reduce 1V of ripple down to 1mV then the reactance of the capacitor at the ripple frequency would need to be one thousand times smaller than the resistance. For example, if the fundamental ripple frequency is 100Hz and the dropping resistor happens to be 1kΩ, this would require a smoothing capacitor of:

$$\frac{1}{2\pi \times 100\text{Hz} \times 1\Omega} = 1592 \times 10^{-6}\,\text{F} = 1592\mu\text{F}$$

This would be no problem on a low-voltage supply but is rather less trivial if we also need a 450V working voltage! This is just another reason why the 'all in one go' approach is not practical; we must consider distributed filtering. A cascade of filters is normally used so the ripple is progressively reduced. Less sensitive stages are fed from the earliest sections of the filter chain, while the most sensitive stages receive the quieter supply, but are also subject to the greatest voltage drop, as illustrated by fig. 11.29. Happily, this follows naturally from the fact that sensitive stages handle the smallest signals so do not require such high supply voltages, whereas later stages are less sensitive but handle the largest signals and therefore benefit from the higher voltages.

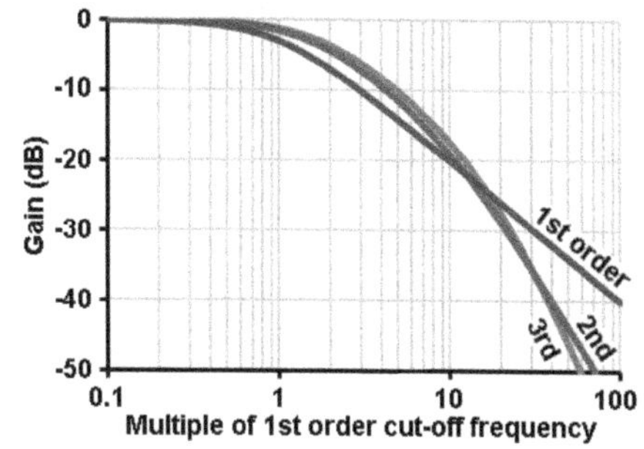

Fig. 11.30: Taking a total R and C and dividing it into equal RC sections only provides superior attenuation above a critical frequency.

Now, it might be thought that taking a *total* prospective dropping resistance and smoothing capacitance and dividing it up into several equal RC sections would give superior ripple attenuation since it increases the filter order (without increasing the DC drop). This assumption was explored in an interesting article by Scroggie[2] and, as fig. 11.30 shows, it is only true above a critical frequency. Also, something which Scroggie did not point out is that the output

---

[2] 'Cathode Ray' (Scroggie, M. G.) (1949). Smoothing Circuits: (1) Resistance-Capacitance. *Wireless World*, October, pp389-93.

impedance of such a network will be worse than that of the single RC case, so while smoothing might improve, bypassing efficacy is reduced (see shortly). Fortunately, capacitance is so cheap these days that there is no economic advantage in dividing up a fixed RC combination the way Scroggie did in 1949; we can use as much capacitance as we like.

11.6.2: Bypassing

A smoothing capacitor can also serve as a power-supply bypass capacitor, meaning it serves as a nearby or local supply of energy for a given amplifier stage. Sudden current demands (i.e. audio frequency signals) will drain their current from the bypass capacitor rather than from the raw supply whose source impedance is presumably larger, as illustrated in fig. 11.31. In other words, each bypass capacitor acts rather like a rechargeable battery, dedicated to each amplifier stage, while the raw power supply is the battery charger and supplies the average current. To be effective, the capacitive reactance must be small enough that it can be considered a short circuit to ground at the lowest frequency of interest; that is, it must be small compared to the *intended* impedance in series with the valve (e.g. the anode resistor). Too little capacitance and the reactance will rise too high at low frequencies, so the total impedance seen by the valve will increase, causing either a bass boost or cut, depending on circuit

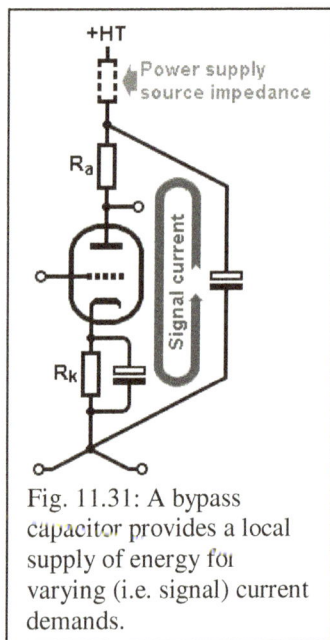

Fig. 11.31: A bypass capacitor provides a local supply of energy for varying (i.e. signal) current demands.

details. Putting it another way, the 'battery' would not have enough to juice to supply the prolonged current drain of bass frequencies.

To drive this point home, suppose we had a hair-raising 1kV raw power supply but we only needed 300V/1mA for an amplifying stage. This implies we could use a 700V / 1mA = 700kΩ dropping resistance. This is so large that we might think we can get away with a 1µF smoothing capacitor; after all, this would create a cut-off frequency of just 0.2Hz and provide about 52dB ripple rejection at 100Hz. But looking at it from the bypassing perspective, a 1µF capacitor has a reactance of 8Ω at 20kHz but 7958Ω at 20Hz, so the amplifier stage will see an ever increasing impedance at low frequencies. For a typical valve stage this could well be enough to cause an uneven frequency response. For this reason, valve stages will nearly always need at least 10µF of bypass capacitance, irrespective of smoothing and decoupling requirements (47µF 450V capacitors seem to be particularly widely stocked).

Ideally, every valve would have its own bypass capacitor (e.g. fig. 11.32b), mounted as close to the stage as possible to minimise the impedance in series with the capacitor. However, this is often impractical with bulky components, so a typical compromise is to provide one decoupling capacitor for every two gain stages. However, the two valves should be contiguous in the signal path. It would be a bad idea to allow, say, the first and last valves in a multi-valve chain to share the same bypass capacitor, as this would result in the worst possible decoupling.

11.6.3: Decoupling

If several amplifiers share the same power supply connection then there will be some degree of unwanted feedback (and indeed feedforward) between them. This is because the power supply does not have zero source impedance, so audio currents being drained from it will develop a corresponding voltage drop across it, which is shared by all the amplifying stages, as illustrated in fig. 11.32a. Another way to describe it is to observe that in fig. 11.32a the load resistors form potential dividers with the power supply impedance, so the output signal voltages from each stage will appear (attenuated) on the shared power supply rail, readily mixing and feeding into the anodes of all the valves. Since smoothing capacitors have rising reactance at low frequencies this tends to be a low frequency problem. A mild symptom might be increased LF distortion, while more severe positive feedback can cause sustained LF oscillation called **motorboating**, or sub-1Hz oscillation called **breathing**.

This problem is reduced by feeding each stage from its own RC or LC filter (with provisos –see next section) so that each stage is decoupled from the rest, and the lower the cut-off frequency the better. Any signals trying to creep along the power supply rail from other stages will then be attenuated by the local decoupling, thus reducing the loop gain around the amplifier via the power supply. Yes, this is similar in principle to smoothing, except we're now talking about program-induced ripple rather than predictable rectifier ripple. One massive capacitor in the attic could be thoroughly effective at smoothing rectifier ripple, but still be useless for decoupling since the wire leading from it to all the various stages would represent a shared (common) impedance across which a program-induced voltage would appear.

Ideally we would provide one filter for every amplifier stage, but the usual compromise of one decoupling

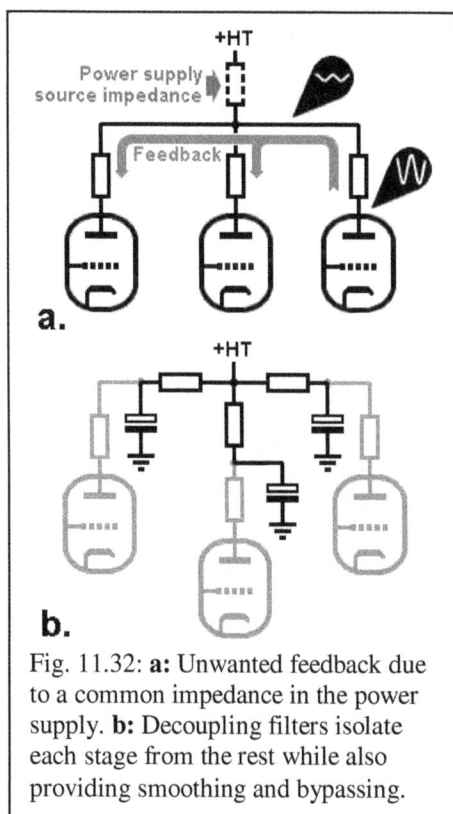

Fig. 11.32: **a:** Unwanted feedback due to a common impedance in the power supply. **b:** Decoupling filters isolate each stage from the rest while also providing smoothing and bypassing.

395

capacitor for every two stages usually suffices for decoupling too. If stereo channels share the same power supply then similar valves from opposite channels can share a filter, as in fig. 11.33, since positive feedback is not an issue with this arrangement (the problem is instead crosstalk, but as it should only exist at sub-audio frequencies it is irrelevant). However, wide-bandwidth or digital circuits *must* also have ceramic or plastic bypass capacitors mounted very close to *each and every* amplifying stage or IC, in addition to general electrolytic bypassing, as even a little inductance can form a common impedance that

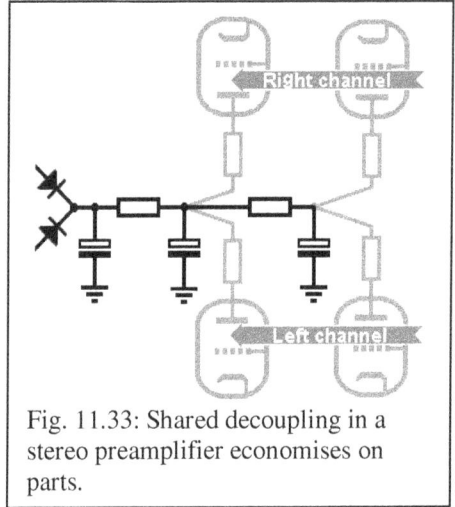

Fig. 11.33: Shared decoupling in a stereo preamplifier economises on parts.

leads to HF (rather than LF) instability. This is much less likely in audio circuits, especially the valve kind.

Whether to use a system of cascaded filters (e.g. fig. 11.33), or individual decoupling filters all connected to a common star point (e.g. fig. 11.32b), is at the designer's discretion. For similar component values a cascaded arrangement provides better smoothing whereas the star system gives better decoupling and greater freedom to choose DC drop. Of course, a combination of the two (a branching network) may also be used, and the whole system can be regarded as quite modular and flexible.

11.6.4: LC Versus RC Filters

The previous sections have shown RC filters, but in principle any or all of them could be replaced by LC filters. An LC filter is second order and can therefore provide a higher degree of ripple rejection than an RC filter. A more significant advantage is that a choke has low resistance, so it is possible to achieve a high degree of smoothing with less DC voltage drop. Output power efficiency was a particular concern in vintage power amplifiers, so it was common to use LC filtering for the power output stage but cheaper RC filtering elsewhere. This is much less of an issue now that capacitance and active regulators are comparatively cheap. The

disadvantages of LC filtering are that chokes are a lot more bulky and expensive than resistors, and they must also have an air gap and be suitably rated to withstand the full supply voltage between conductor and core, making them a specialist item. A further and more disconcerting disadvantage is that the choke will resonate with the capacitor at a

frequency of: $f_o = \dfrac{1}{2\pi\sqrt{LC}}$

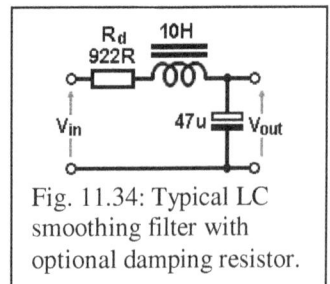

Fig. 11.34: Typical LC smoothing filter with optional damping resistor.

To illustrate this problem, fig. 11.34 shows an LC smoothing filter using typical component values. R_d is a damping (or dropping) resistor, which is usually absent from classical designs but may be added for reasons which will become clear. Fig. 11.35 shows the gain of the filter, and when $R_d = 0\Omega$, and the rather conspicuous feature is the resonant peak at 7.3Hz. At this frequency the filter magnifies rather than attenuates noise on the power supply. Of course, the rectifier ripple frequency will be much higher than this, so we might think there is no problem. However, as shown in the figure, the output impedance of the LC filter (i.e. when viewed from the load end) shoots up around the resonant frequency, so in this region we

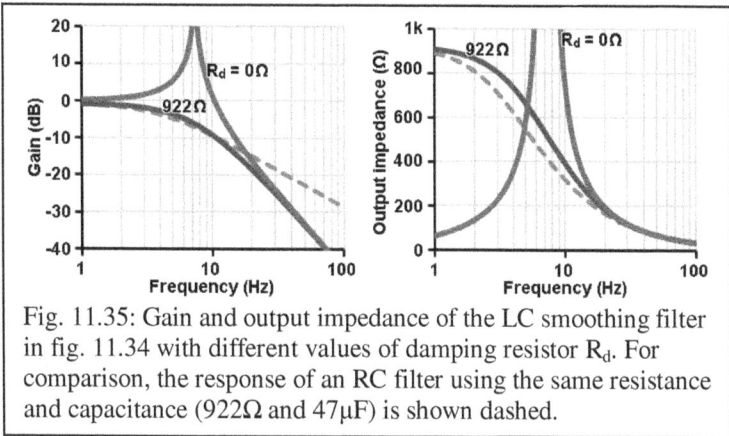

Fig. 11.35: Gain and output impedance of the LC smoothing filter in fig. 11.34 with different values of damping resistor R_d. For comparison, the response of an RC filter using the same resistance and capacitance (922Ω and 47µF) is shown dashed.

get little or no effective bypassing or decoupling, which may induce the amplifier circuit to motorboat. If the resonance happened to occur within the audible spectrum –which is quite possible with a small choke– then the situation would be even worse.

Resonance can be damped by adding resistance in series with the choke, some of which is already provided by the choke's own winding resistance. Ringing is completely suppressed when the circuit is critically damped or more, which requires a resistance equal to $2\times\sqrt{(L/C)}$ or more. In this example a value of 922Ω would provide critical damping, as shown in the figure (in practice we would probably use something between 470Ω and 1kΩ). Of course, adding resistance means introducing additional DC drop, which is unfortunate since minimal drop was one of the advantages of using a choke in the first place.

It is clear from the figure that R_d determines the output impedance at low frequencies. In other words, bypassing/decoupling close to the resonant frequency is improved by adding R_d, but at lower frequencies it is degraded. To add insult to injury it is, overall, slightly worse than a simple RC filter using the same resistance and capacitance (shown dashed)! When using an LC filter there is, therefore, a balance to be struck between acceptable DC drop and effective decoupling/bypassing. In general it is preferable to supply only one valve from an LC filter (or two valves but from opposite stereo channels), and to avoid cascades of undamped LC filters as this leads to multiple resonances.

11.6.5: The Gyrator or Simulated Inductor

An alternative to the traditional smoothing choke is a gyrator circuit, which simulates an inductor by 'inverting' the reactance of a capacitor. It deliberately exploits the fact that the output impedance of a cathode follower becomes inductive owing to grid-cathode capacitance, although these days a transistor does the job more efficiently. A practical circuit is shown in fig. 11.36, and the output impedance is equivalent to an inductance:

$$L_{eq} = C_1 \times R_1 \times R_3 \qquad (11.9)$$

In series with a damping resistance:

$$R_o = R_3 \left(\frac{R_1 + R_2}{R_2} \right) \qquad (11.10)$$

In this case amounting to 103H in series with 94Ω –much more impressive than the average smoothing choke.

The voltage across R_2 is equal to the gate-source voltage (typically around 4V) plus the drop across R_3. In this case $R_1 = R_2$, so the total DC drop across the gyrator is simply $2(V_{gs} + iR_3)$ and the total dissipation in Q_1 is $i(2V_{gs} + iR_3)$. If the load current is 100mA, say, then this would result in about 17.4V total drop and 1.27W dissipation in Q_1, the remaining 0.47W being dissipated by R_3. A TO-220 transistor package can dissipate little more than 1W by itself, so a clip-on heatsink would be needed (section 11.7.6).

There must be enough drain-gate voltage – i.e. the voltage across R_1– to accommodate the peak incoming ripple voltage, otherwise the MOSFET will switch off during the valleys of the ripple waveform. In other words, the peak-to-peak ripple voltage applied to fig. 11.36 must be less than $2(V_{gs} + iR_3)$. R_4 is the usual gate stopper and must be mounted very close to the MOSFET to discourage oscillation. D_1 protects the gate-source junction and C_1 from overvoltage, which also allows C_1 to be a small, low-voltage device. Most high-voltage MOSFETs will work in this application; the IRF820 and STP4NK50Z are 500V-rated options. The STP- types are especially handy

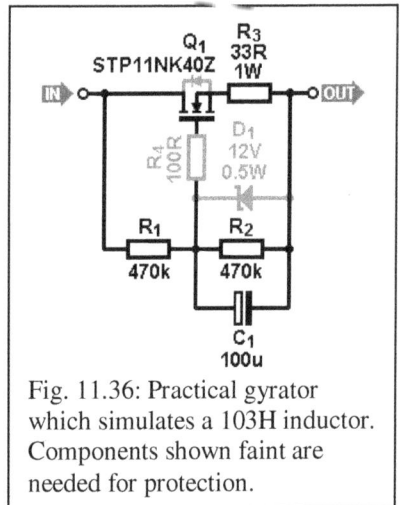

Fig. 11.36: Practical gyrator which simulates a 103H inductor. Components shown faint are needed for protection.

as they have built-in gate-source protection and are available in the all-plastic TO-220FP package, so can be bolted to the chassis without the need for insulation hardware.

11.6.6: The Capacitor Multiplier

A very simple and perhaps underappreciated circuit is the capacitor multiplier. Fig. 11.37 shows a practical example, and it can be seen that it is really just a source follower whose input is smoothed DC. The follower does its best to reproduce this smooth voltage at the output, minus V_{GS} (about 4V). The MOSFET gate has infinite input resistance so the only load on C_1 is R_2, thus the smoothing circuit can achieve a high degree of ripple rejection without dropping much DC voltage. This circuit is not a regulator or stabiliser; the output voltage is still free to rise and fall with changes in input voltage. However, this

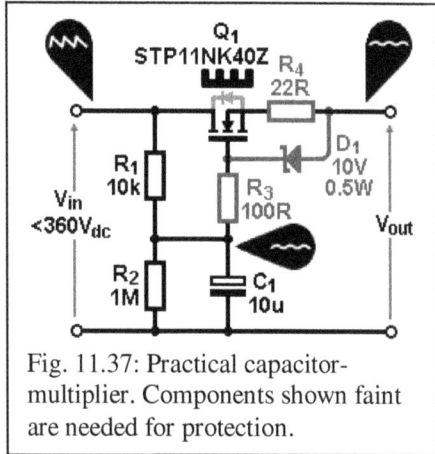

Fig. 11.37: Practical capacitor-multiplier. Components shown faint are needed for protection.

can be an advantage as it means dissipation in the transistor does not change significantly with changes in mains voltage. It is therefore ideal when the output voltage is not critical, and can also be used as a sort of pre-regulator to unburden a proper voltage regulator from the job of ripple rejection. Several capacitor followers can be cascaded together in exactly the same way as RC smoothing filters.

The purpose of R_2 is simply to form a potential divider with R_1, so the gate voltage is pulled below the input voltage, otherwise the transistor would switch off during the valleys of the incoming ripple voltage. As shown, the circuit will tolerate about 3% input ripple. During normal use the voltage across the transistor is equal to the voltage across R_1 plus V_{GS}. For load currents greater than about 80mA a TO-220 package will need heatsinking.

D_1 protects the gate-source junction from overvoltage and reverse voltage and, in combination with R_4, provides an essential soft current limit. Since the voltage across R_4 cannot exceed V_Z-V_{GS}, or about $10-4=6V$ in this case, if the load current tries to exceed $6V / 22\Omega = 0.27A$ the Zener will turn on and pull the MOSFET gate voltage down, preventing any further increase in current. This is necessary to protect the MOSFET from inrush currents at start-up, and from careless oscilloscope probing! The drain-source junction is inherently protected from reverse voltage by the MOSFET's internal body diode.

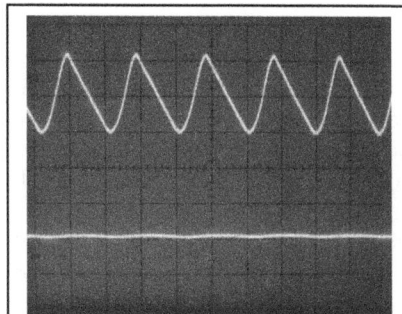

Fig. 11.38: Performance of the capacitor multiplier in fig 11.37 at 300V 80mA output, 5V/div vertical, 5ms/div horizontal.

Fig. 11.38 shows the performance of the circuit when supplying $300V_{dc}$ at 80mA from a $310V_{dc}$ raw supply. The input ripple is $\sim11V_{pp}$ while the output ripple is only $\sim180mV_{pp}$, and has much less harmonic content. This is exactly the amount of ripple rejection we would expect from a $10k\Omega/10\mu F$ smoothing filter, but to achieve the same DC drop at 80mA from an RC filter alone would have required 113Ω and $880\mu F$ (i.e. the same time constant as R_1C_1), hence the name 'capacitor multiplier'.

Practically any MOSFET with sufficient V_{DS} rating will work in this application. BJTs are also sometimes used, but BJTs cannot simultaneously withstand as much collector current and collector-emitter voltage as similar MOSFETs, owing to their 'secondary breakdown' phenomenon, making them more likely to fail at start up. By contrast, the STP11NK40ZA can withstand 100mA with a full 400V across it, until it gets too hot that is.

11.7: Voltage Regulation

When smoothing alone is not enough, we can resort to active voltage regulation. Voltage regulators are amplifiers in their own right, but their 'input signal' is a fixed DC reference voltage. Fifty years ago if you wanted a voltage regulator you had to design it yourself from scratch. These days a lot of the hard work is done for us by semiconductor manufacturers, at least for low-voltage applications.

11.7.1: Linear Regulator ICs

For regulating low voltages there is a whole world of integrated circuits to enjoy. Among the most common are '3-terminal' linear regulators, which will accept a dirty DC input voltage and produce a well regulated and fixed output voltage, though they can be used for adjustable voltage applications too. Nearly all have built-in short-circuit and thermal shut-down protection, making them virtually fool proof.[*]

Any foray into voltage regulator ICs is sure to begin with the old and reliable 78xx (positive voltage) and 79xx (negative voltage) series, produced by many manufacturers.[†] The part number indicates the output voltage, e.g. the 7805 is a +5V regulator, while the 7912 is a −12V regulator. Most versions will handle 1A and accept an input voltage up to 35V, but some will handle 1.5A and different (occasionally lower) input voltages. The series quotes about 60 to 70dB of ripple rejection, so 1V of input ripple will emerge as less than 1mV at the output, which is good enough for most audio applications and certainly for heater supplies. However, ripple rejection is a separate issue from a regulator's ability to maintain a constant output voltage in the face of long-term changes in input voltage, which is called its

---

[*] But as Douglas Adams noted, never underestimate the ingenuity of fools.
[†] Various letters are added to the part number to indicate different ratings or origin, such as LM7805, MC7805, KA7805 etc. Usually the differences are subtle and of little significance to the average user, but always check the datasheet.

line regulation. Similarly, its ability to maintain a constant output voltage in the face of changes in load current is called its **load regulation**. A modern regulator IC can typically hold its output constant within a few millivolts over the full range of expected input voltages and output currents, but this performance will degrade at higher (e.g. audio) frequencies. Application information on this and more is abundant on the internet, so we will cover only the essentials here.

Fig. 11.39 shows a textbook circuit with supporting components. Input capacitor C_1 is a local bypass capacitor, usually 100nF or more, which should be mounted close to the IC. Output capacitor C_2 is normally included to improve transient response and stability, and will be a few tens or hundreds of microfarads. An ordinary electrolytic or tantalum should be used here as the ESR is actually beneficial from the point of view of stability. D_1 is a rectifier

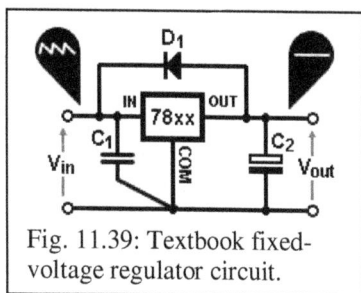

Fig. 11.39: Textbook fixed-voltage regulator circuit.

diode which protects the regulator from reverse voltage if, at switch-off, the input voltage happens to fall faster than the voltage stored on C_2.

Like any amplifier, a regulator requires at least some voltage across itself between input and output in order to function properly, and this is referred to as the **differential voltage**. If the input voltage drops too low then the device will not be able to maintain the correct output voltage, and is said to have 'dropped out' of regulation. The minimum necessary differential voltage to maintain proper functioning is therefore called the **dropout voltage**, and for the 78xx series it is about 2V. In other words, the input voltage must always be at least 2V higher than the output voltage, and ≥3V would be preferred for reliability.

The differential voltage is an unwanted necessity because it leads to power dissipation in the regulator, i.e. the voltage across the device multiplied by the current through it. Thus if a regulator has a dropout voltage of 2V and we want it to supply 1A, then at the very least it is going to have to dissipate 2W and will need a small heatsink (section 11.7.6). But the chances are that it will be supplied from a simple rectifier, so the average input voltage will actually need to be somewhat higher so the ripple valleys don't swing too low and cause momentary drop out. And we will probably want to supply *even more* voltage in order to accommodate mains variation. So what started off as 2W dissipation is now several times greater, and the necessary heatsink has grown in size and cost. This problem can (sometimes) be minimised by using low drop-out or switching regulators.

401

11.7.2: Adjustable Regulators

A 3-terminal regulator strives to maintain a constant voltage between its output pin and common pin, so if the common pin is raised or lowered by some externally-applied voltage, then the output voltage will track the change. For example, a favourite trick when powering 12.6V heaters is to use a 7812 to get the first twelve volts, then add a diode in series with the common pin, as shown in fig. 11.40. About 5mA of idle current flows out of the common pin and into the diode, which is

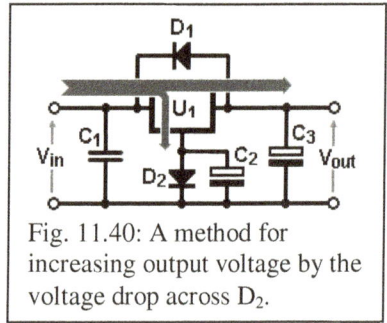

Fig. 11.40: A method for increasing output voltage by the voltage drop across D_2.

enough to ensure the diode drop is about 0.6V, so the output will be 'jacked up' to 12.6V (although as explained in section 2.6.12, it may be better simply to leave it at 12V). The common pin may also need to be bypassed with a few tens of microfarads to prevent oscillation with capacitive loads. Similarly, a 7805 can be jacked up using two diodes to bring the output close to 6.3V.

So-called adjustable regulators are really just 1.25V fixed regulators. What makes them 'adjustable' is the fact that the idle current flowing out of the common or 'adjust' pin is only a few microamps and more tightly controlled than in ordinary fixed regulators like the 78xx series. The output voltage can therefore be set accurately anywhere between 1.25V and some maximum limit, using resistors. Probably the most popular adjustable regulators are the 317 and its negative-voltage brother the 337. These devices are rated for 40V maximum input, 1.5A

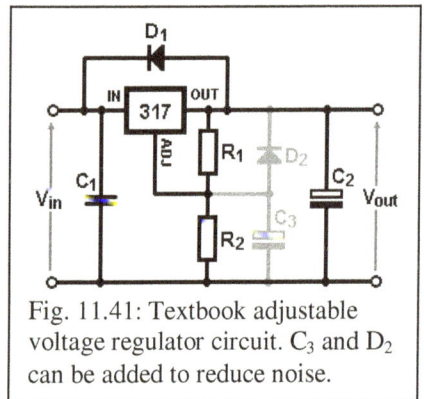

Fig. 11.41: Textbook adjustable voltage regulator circuit. C_3 and D_2 can be added to reduce noise.

current and even better ripple rejection than the 78xx series but, somewhat annoyingly, they have a different pinout from the 78xx series. Close cousins of these devices are the 350 (positive) and 333 (negative), both rated for 3A current, and the 338 (positive) rated for 5A.

Fig. 11.41 shows the quintessential adjustable regulator circuit. The IC will maintain 1.25V between its output and adjust pins, that is, across R_1. A constant current of $1.25/R_1$ therefore flows down the potential divider, and the default 1.25V output will be jacked up by the resulting voltage across R_2. Assuming the adjust-pin current is negligible, the voltage across R_2 will be $R_2 \times 1.25/R_1$, so the total output voltage will be:

$$V_{out} = 1.25\left(1 + \frac{R_2}{R_1}\right)$$
(11.11)

402

R_2 can of course be replaced by a potentiometer or trimpot, thereby forming a variable-voltage regulator. For minimum output noise R_2 should be bypassed by a few tens of microfarads, and a diode should then be added to protect the output-adjust junction from reverse voltage at switch off. These optional components are shown faint in fig. 11.41. If the load current is too small then the output voltage will tend to rise above the desired value, and for the 317 the minimum recommended load current is 3.5mA. Therefore, R_1 and R_2 are often chosen so they permanently drain this minimum current, but this is not necessary if a suitable external load is always present.

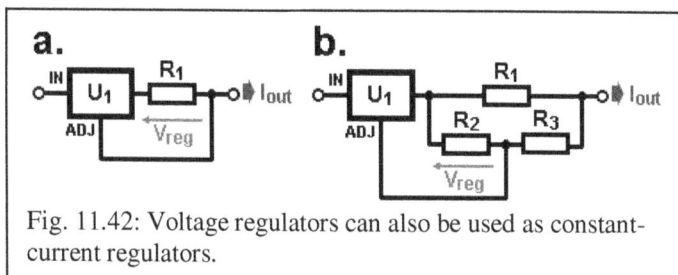

Fig. 11.42: Voltage regulators can also be used as constant-current regulators.

A voltage regulator can also be configured as a constant-current source, which is useful for series heater supplies among other things. In fig. 11.42a the output current is simply V_{reg}/R_1, and the resistor will have to dissipate I^2R watts. However, at high currents this arrangement can lead to an awkwardly small and precise value for R_1, not easily made using power resistors from the E6 range. A work-around is the arrangement in fig. 11.42b. A convenient power resistor somewhat larger than V_{reg}/I_o can now be chosen for R_1 (e.g. 10Ω), while R_2 and R_3 are 'programming resistors' that set the output current. R_2 can also be chosen to be a convenient value such as 1kΩ, leaving R_3 as:

$$R_3 = \frac{I_{out}R_1R_2}{V_{reg}} - R_1 - R_2 \qquad (11.12)$$

R_3 could be replaced with a pot of course, allowing the output current to be adjusted exactly.

11.7.3: Low-Dropout Regulators

Power dissipation in a regulator is the annoyance that turns what should be a cheap and simple circuit into a practical burden, because it demands heatsinking (section 11.7.6). As their name suggests, low-dropout regulators (LDOs) will function with much less voltage across themselves than ordinary regulators, so the input voltage can be lower and dissipation thereby reduced. Apart from this they are used in exactly the same way as ordinary regulator ICs, though internally they are rather different. To prevent oscillation, LDOs absolutely need an output bypass capacitor, and close attention should be paid to the datasheet recommendations.

The most common adjustable LDOs are the 1084 (5A), 1085 (3A), and 1086 (1.5A). They will withstand up to 29V input and have a dropout voltage of 1.5V. The Linear

Technology versions (LT1083/1084) are also available in comfortingly large TO-3P packages.

11.7.4: Switching Regulators

The ultimate way to minimise power dissipation is to use a switching regulator. These incorporate into one chip most of the parts needed to make a switch-mode power supply. The price we pay for this convenience, as with any SMPS, is a more involved design process and possible switching noise. The former can be avoided by buying a ready-made circuit, which are now cheaply available (often intended for LED lighting). The latter may be overcome with plenty of output filtering and possibly shielding of the whole circuit, but be prepared for plenty of manual tweaking.

There are hundreds of device types available, but a practical example for our purposes (e.g. heater supplies) is the LM2596 buck (step-down) regulator which, unlike most switching regulators, is available in a non-surface mount package (5-pin TO-220). The adjustable version (LM2596-ADJ) can provide 1.25 to 37V at up to 3A and will accept up to 45V input voltage.

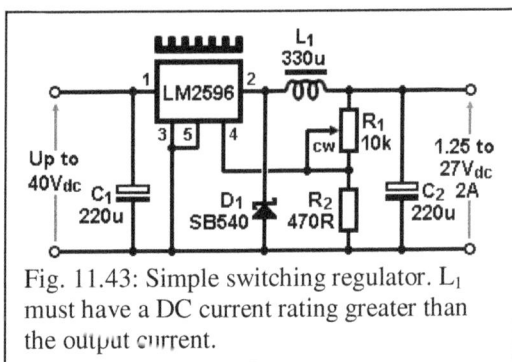

Fig. 11.43: Simple switching regulator. L_1 must have a DC current rating greater than the output current.

Fig. 11.43 shows a practical circuit using this device; all we have to provide is a Schottky diode, inductor, and a couple of resistors and capacitors. It is essential to consult the datasheet for specific application information as the manufacturer goes to great trouble to explain the proper choice of external components and layout. In particular, the inductor must have a DC current rating greater than the load current.

The switching frequency of fig. 11.43 varies from about 70kHz under light loading to about 30kHz at full load. This is high enough that further LC filtering can be applied using ferrite or even air-core inductors. The input voltage must be at least 1V higher the output, and the smaller the differential voltage, the better the efficiency. The author's test circuit varied from 80% to over 90% efficient at all load currents. The regulator may therefore have to dissipate up to $P_{out} \times (1-0.8)$ watts, so a heatsink is advisable for outputs greater than about 5W. The near-constant efficiency is a wonderful feature not just because it means less wasted heat, but because it makes the input voltage non-critical. We can supply whatever input voltage is convenient without having to worry about rectifier losses or exact choice of transformer voltage.

Simple boost (step-up) regulators are also available –such as the LM2577– but are usually limited to about 60V output, so cannot easily be used to generate HT voltages.

404

11.7.5: High-Voltage Regulators

In the world of regulator ICs, 'high voltage' means anything over about 20V. For example, the ambitiously named 317HV will accept a mildly tingling 50V input, impressing no-one from the world of valves. Slightly more respectable is the TL738 rated for 125V, 700mA. One potential disadvantage is that it requires at least 15mA load to maintain good regulation, although it is eminently useful for series heater chains. Above this the choices all but disappear. The VB408 and HIP5600 would accept $400V_{dc}$ or even $280V_{ac}$, and deliver about 30mA, but they are sadly obsolete. In current production is the LR8 rated for 480V but only 10mA, which

rather limits its usefulness. Nevertheless, fig. 11.44 shows a typical circuit, and it looks like any adjustable regulator but for one subtle feature: since this is a high-voltage circuit, D_1 must be included to prevent the output capacitor from discharging into the pot if its resistance is suddenly turned down, as this will otherwise burn out the carbon track. For the same reason the pot must not be bypassed with a large capacitance. However, if R_2 is a fixed resistor then this is not an issue.

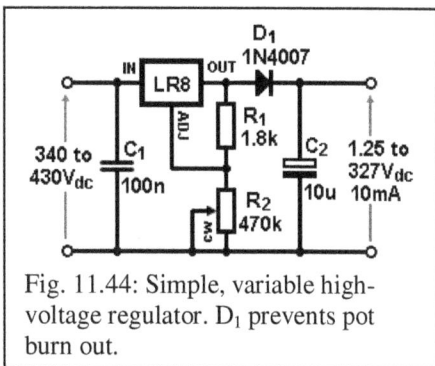

Fig. 11.44: Simple, variable high-voltage regulator. D_1 prevents pot burn out.

Given the lack of high voltage regulator ICs there comes a point where we have to be creative. A particularly simple and effective choice is a **Zener follower**, such as the one shown in fig. 11.45. Strictly speaking, this is a **voltage stabiliser** rather than a regulator, since it does not use global feedback, but the distinction is rarely observed these days. The Zener follower is closely related to the capacitor multiplier from section 11.6.6; it is a source follower whose input is a DC voltage that is stabilised by Zener diodes. The follower does its best to reproduce this voltage at the output, minus the gate-source voltage (about 4V). Unlike the capacitor multiplier, however, the output voltage is not free to follow the raw input voltage, but is fixed inasmuch as the Zener voltage remains fixed. Admittedly, Zener diodes (strictly avalanche diodes) have a positive tempco, so the voltage will increase typically by 2 to 3% for a 30°C rise in temperature. However, this is still better than mains voltage variation, and valve circuits almost never need anything more accurate.

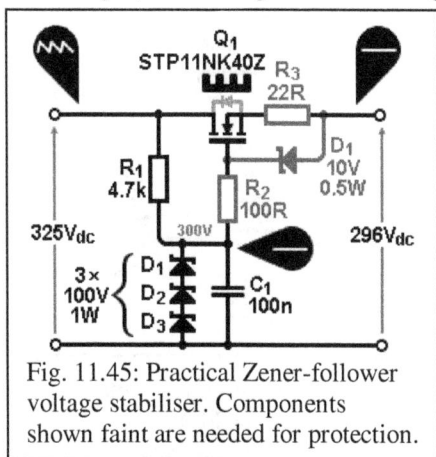

Fig. 11.45: Practical Zener-follower voltage stabiliser. Components shown faint are needed for protection.

R_1 needs to supply a few milliamps to the Zener string for good stabilisation.

A common refinement is to replace R_1 with a constant-current source, thereby eliminating variations in Zener current with changes in supply voltage, though this does rather spoil the simplicity of the circuit. Multiple transistors, each feeding a different part of the amp, could share the same Zener reference. However, if the supply wire leading to the regulator is more than a few centimetres long its inductance can lead to RF oscillation in the MOSFET. It is therefore advisable to add a local decoupling capacitor close to the MOSFET drain, say 100nF or more.

Fig. 11.46 shows an example of a variable-voltage Zener follower. This circuit was built as a very cheap and simple stand-alone bench power supply, using parts from the author's scrap box, but it contains a few subtleties of high-voltage power-supply design which may still be instructive. The small 10VA 240V isolating transformer was originally meant for a bathroom shaver unit. Since this can deliver little more than 25mA$_{dc}$ after rectification, very little current could be wasted in the stabiliser, which is therefore quite minimalist. R_1, R_2 and C_2 form a balanced filter. Not only does this reduce differential-mode hash from the mains, but it also filters the higher harmonics from the ripple voltage which would otherwise readily couple to the gate of the MOSFET via drain-gate capacitance. The raw DC is then stabilised to about 300V with a string of four 75V Zeners (plus an LED for power indication). This voltage is sampled by the pot which feeds the gate of the MOSFET. C_3 provides some further filtering, although a large capacitor cannot be used in this position because the charging current would otherwise burn out the carbon pot track. R_8 ensures the MOSFET gate is pulled down if the pot wiper fails to make good contact with the track.

The remaining parts are for short-circuit and reverse polarity protection, which is essential in any bench power supply. Q_2 is used for more accurate current limiting than in fig. 11.45. When the voltage across R_7 reaches about 600mV –i.e. when the output current is 27mA which is the maximum the transformer can handle

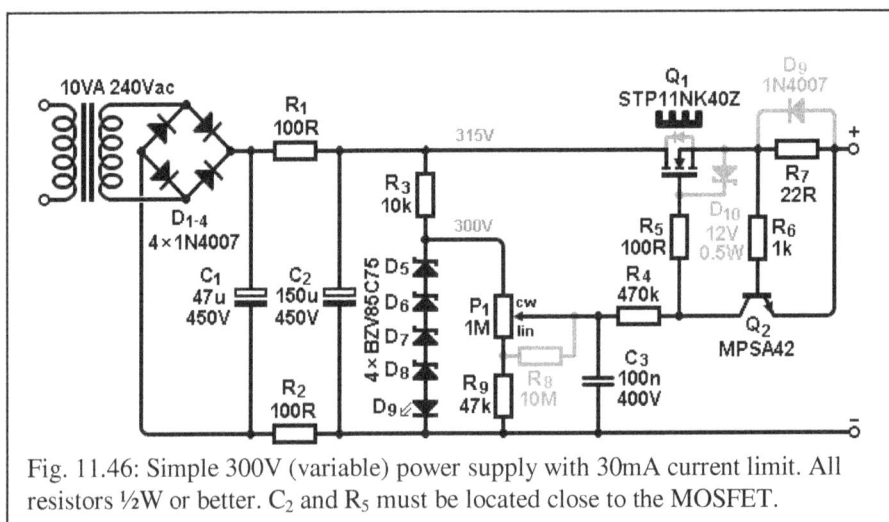

Fig. 11.46: Simple 300V (variable) power supply with 30mA current limit. All resistors ½W or better. C_2 and R_5 must be located close to the MOSFET.

continuously– Q_2 switches on and pulls current through R_4. This in turn pulls down the MOSFET gate voltage and so reduces the output voltage to prevent the output current from rising any higher. R_4 is fairly large and is necessary both to prevent C_3 from discharging excessive current into Q_2 during overload, and to prevent Q_2 from pulling too much current through P_1 (carbon pots really are quite fragile). R_6 is needed to protect Q_2's base from excessive current transients, especially when powering up into a short circuit. D_9 protects Q_2's base-emitter junction from reverse bias, such as when powering down with a capacitive load attached.

If the output is a dead short, the full supply voltage will appear across Q_1, so it will dissipate up to $310 \times 0.027A = 8.1W$ and therefore needs a heatsink (next section). In the author's version it was bolted to a piece of scrap copper bar, carefully selected because it 'looked big enough' (and indeed it was). This made the final unit completely immune to momentary and continuous output shorts, and it has proved to be a very handy piece of test equipment despite its cheapness. Most of the circuits in this book were tested using it. The ripple rejection turned out to be about 66dB at 100Hz, varying with setting somewhat.

The Zener follower is cheap, simple, and good enough for many valve applications, but it is possible to do better with only a little extra circuitry. Fig. 11.47 shows how an ordinary low-voltage regulator IC can be combined with a high-voltage transistor to create a versatile high-voltage regulator. This circuit has come to be known as a Maida regulator after an application note written in 1980,[3] although it is a simple adaptation of several older application notes.[4] U_1, R_3 and R_4 form

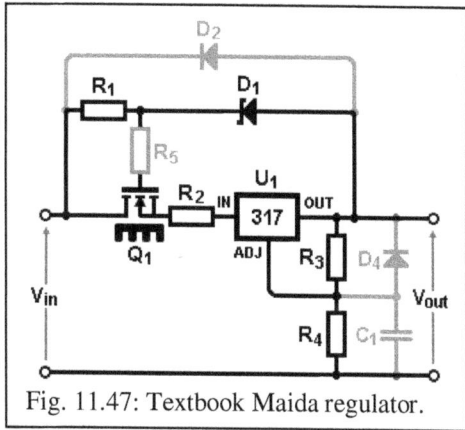

Fig. 11.47: Textbook Maida regulator.

a standard 317 voltage regulator, set up to produce the desired output voltage. To set a high voltage, R_4 will be relatively large, meaning the adjust pin current –which may be up to 100µA– will significantly affect the output voltage. R_4 may therefore require some adjustment on test. Meanwhile, Q_1 does the hard work of dissipating heat and withstanding the high input voltage. R_1 feeds a small current into D_1 which is a low voltage Zener, say 10V. Since V_{GS} of the MOSFET is usually about 4V, the voltage across U_1 therefore cannot exceed $10-4 = 6V$, so it is protected at all times, floating within a 6V 'window'. R_2 provides a crude current-limiting function by constricting this window as load current increases. Once $V_Z-V_{GS}-IR_2$ falls to about 2V, U_1 will drop out of

---

[3] Maida, M. (1980). High Voltage Adjustable Power Supplies. *National Semiconductor Linear Brief 47.*
[4] National Semiconductor (1980). *Voltage Regulator Handbook.*

407

regulation and prevent further increase in current. The parts shown faint are the usual protective components.

Regulators of this kind will normally withstand start-up conditions in a typical amplifier but may not survive an output short after the input voltage has risen to its full operating level, making them worryingly fragile when prototyping or experimenting. Since U_1 has built-in current and thermal limiting, neither of which are actually used here, this seems like an opportunity squandered.

With the previous shortcomings in mind the author developed the circuit variation in fig. 11.48 which is robust enough to withstand continuous shorts (with suitable heatsinking for the MOSFET), so it can even be used as a bench power supply. The secret to this circuit is R_2 and U_1, the latter being an LM317L. The 317L is the little brother of the 317 and comes in a TO-92 package. Consequently it has a much

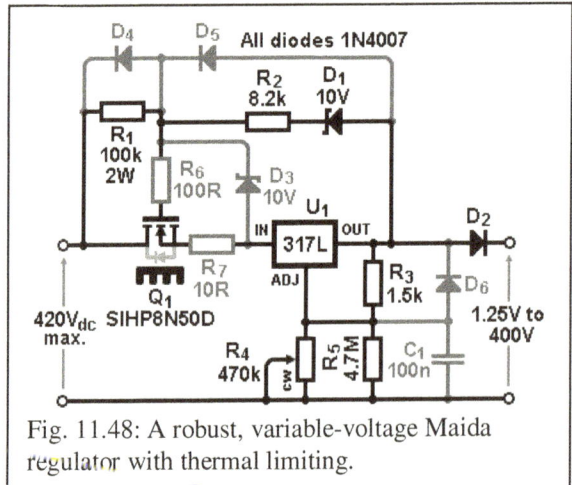

Fig. 11.48: A robust, variable-voltage Maida regulator with thermal limiting.

smaller built-in current limit of 200mA (enough juice for most valve projects) which is much more dependable than the simple current-limiting method used in fig. 11.47. R_7 is now included only to discourage oscillation rather than to provide current limiting.

R_3 allows 1.25/1.5k=0.83mA to flow down the voltage-setting divider, and another 100μA may flow from the adjust pin. This means dissipation in the pot is safely below 500mW. $R_4 \| R_5$ have a combined resistance of 427kΩ and set the maximum output voltage to 400V. However, the minimum load current specified in the 317L datasheet is 1.5mA, so the output voltage may be a little higher than expected, until the load starts drawing current.

The main reason for using the 317L, however, is that is has a built-in 2W thermal limit which, by judicious choice of R_2, is exploited to protect the MOSFET too. Ignoring R_7 which is small, the voltage across U_1 is equal to:

$$(V_{in} - V_{out} - V_Z)\frac{R_2}{R_1 + R_2} + V_Z - V_{GS}$$

Thus when the output voltage is set to the maximum value of 400V there is only about 6.8V across U1, and the full 200mA limit is available. But as the output

408

voltage is turned down, the drop across R_2 –and therefore across U_1– increases, causing U_1 to dissipate more heat (U_1 should *not* be attached to a heatsink). R_2 is selected so that if the output is a dead short, the voltage across U_1 reaches almost 40V (the maximum allowable) so it will limit the current to approximately 2W/40V = 50mA. The remaining voltage across the MOSFET will be about 420–40 = 380V so it will dissipate 380×50mA = 19W, which is manageable with a decent heatsink (<1.5°C/W; see next section). What's more, the thermal lag in the 317L means it will in fact deliver the full 200mA *briefly* even when the output voltage is set low. In other words, the circuit can deliver high voltage *and* high power under many practical conditions, but is smart enough not to overheat.

Various diodes have also been added for protection. Some of them are probably unnecessary but the author does not yet have the fortitude to find out which, having blown up several earlier circuit variations to reach this point. Diodes are cheap, and so far the circuit in fig. 11.48 has proved indestructible. It can of course be adapted for lower input voltages by increasing R_2 accordingly.

It is possible to devise much more sophisticated and 'high-end' discrete regulators with breathtakingly good regulation, transient response, and vanishingly low output impedance. However, chasing each decimal place of performance tends to lead to an exponential increase in complexity, and regulator design becomes yet another specialism. Such designs are likely to be unstable into certain loads, making them less universally applicable, and a firm understanding of loop compensation is required to optimise them. Since this book is aimed at readers of presumed mixed ability and experience, the regulators discussed here are deliberately simple and reliable. Simple circuits have simple shortcomings; the author much prefers a robust circuit that is good enough, to an excellent circuit that requires constant parental supervision.

11.7.6: Heatsinking

As valve enthusiasts we are often spoiled by the fact that valves don't need heatsinks; they are big enough to handle the anode dissipation limit quoted on the datasheet without assistance. Indeed, it is one of their few advantages for power amplification. It is understandable, therefore, when beginners buy a transistor boasting 100W dissipation and expect it to do just that, naively thinking that a piece of plastic the size of a thumbnail can do what a valve the size of a cola bottle can do. A few dead transistors later and they learn that even for moderate power dissipation, a semiconductor needs a helping hand from a slab of metal. The question is, how big does that slab need to be?

When a semiconductor turns electrical energy into heat, the heat energy has to flow from the silicon junction, through the case of the device, into the heatsink (if any), and finally into the surrounding air. The ease (or rather, difficulty) with which heat can flow through a material object is called its **thermal resistance** and is specified in degrees Celsius per watt (or sometimes kelvin per watt, which amounts to the same thing). In other words, it has how much the temperature of the material can be expected to rise for every watt of heat flowing through it. The situation is very much like an electrical circuit: heat energy (current) has to flow through a thermal resistance, thereby causing a temperature difference across it (voltage drop). Silver and copper have the lowest thermal resistance, aluminium is about twice as high, and steel is ten times worse again, but anything is better than silicon. Even fixing an M3×10 steel screw and nut to a TO-220 package will increase its power handling by about 100mW.

When a device is fixed to a heatsink we must add together the thermal resistances of each section of the 'thermal circuit' to find the total thermal resistance between the semiconductor junction and the air:

$$\theta_{ja} = \theta_{jc} + \theta_{cs} + \theta_{sa} \tag{11.13}$$

Where:
θ_{ja} = junction-to-air thermal resistance;
θ_{jc} = junction-to-case thermal resistance;
θ_{cs} = case-to-sink thermal resistance;
θ_{sa} = sink-to-air thermal resistance.

Figures will be provided on the device and heatsink datasheets. The total thermal resistance we *need* depends on how hot we are happy for the semiconductor to run. The maximum permissible junction temperature will be specified on the datasheet and is usually 125°C to 150°C. Of course, for reliable operation we should probably aim for something lower, say 100°C. Estimating the ambient temperature inside the chassis to be 40°C, the maximum allowable temperature rise of the junction would then be 100−40 = 60°C. If there is too much thermal resistance standing between the junction and the air (i.e. the heatsink is too puny) the junction will rise by more than this, because the heat can't get out of the device fast enough. Many voltage regulators have built-in thermal shut down, but a discrete transistor will simply burn to death. The heatsink must have a thermal resistance of:

$$\theta_{sa} = \frac{T_j - T_a}{P} - \theta_{jc} - \theta_{cs} \tag{11.14}$$

Where:
T_j = junction temperature;
T_a = ambient temperature;
P = power dissipated in the device.

For a bare metal-on-metal contact the thermal resistance of the case-to-sink interface, θ_{cs}, can be taken as about 1.5°C/W. Smearing a little thermal compound on the surfaces will reduce this to 0.5°C/W or even less, but be warned, it is *extremely* messy stuff and will ruin your shirt as soon as you open the tube. It is also slightly electrically conductive, so try not to smear it on anything else.

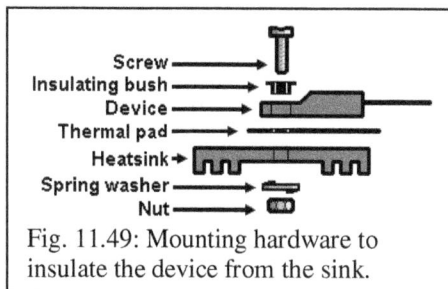

Fig. 11.49: Mounting hardware to insulate the device from the sink.

The metal tab of the device is normally connected internally to the junction, so if the heatsink is bolted to (or part of) the chassis, you cannot normally use a bare metal contact. Electrical isolation between device and heatsink is achieved using a mica or silicone thermal-pad (Sil-Pad®), plus a plastic bushing affectionately known as a 'top hat' to separate the screw from the tab, as illustrated in fig. 11.49. Unfortunately, despite what the advertising says, this hardware is a great spoiler of thermal resistance, amounting to 2 to 3°C/W. An all-plastic TO-220FP with thermal compound sometimes works out better but surprisingly few devices are sold this way.

It might be thought that the chassis would make an excellent and basically free heatsink, but while a sheet-metal chassis may have a lot of surface area, most of it is not useful because it is too far away from the device. By experiment the author has settled upon personal rules of thumb of 8°C/W for a typical mild steel chassis, and 2.5°C/W for a 1.5mm-thick aluminium one. Since insulation hardware is usually required too, the chassis is not necessarily very useful as a heatsink, and the hot spot could also be a user hazard.

Mounting devices on heatsinks while maintaining electrical isolation, and keeping the product easy to assemble and repair, is such a pain that professional designers will go to great lengths to avoid it. This may mean using several devices to share the dissipation so that each one requires only a small PCB-mounted or clip-on heatsink, or maybe nothing at all. Working out how much power a device can dissipate without any heatsink is easy, as the datasheet will quote the thermal resistance of the junction-to-air interface, θ_{ja}. For example, an LM7805 in a TO-220 package has θ_{ja}=65°C/W. Allowing a maximum junction temperature of 100°C in 40°C ambient, the maximum power that can be dissipated with no heatsink is:

$$P = \frac{T_j - T_a}{\theta_{ja}} = \frac{100 - 40}{65} = 0.92\text{W} \tag{11.15}$$

Most other TO-220 devices have similar θ_{ja}, so 'one watt for a TO-220' is an easy figure to remember when using such devices.

Although we should always do the maths, the following represent the author's own rules of thumb, assuming a TO-220 package and natural air convection.
- TO-220 package alone; up to 1W.
- Clip-on heatsink; up to 3W.
- 1mm-thick mild-steel chassis with insulation hardware; up to 6W.
- Finned aluminium heatsink the size of a matchbox; up to 7W.
- Finned aluminium sink the size of a deck of playing cards; up to 10W.

411

- 1.5mm-thick aluminium chassis with insulation hardware; up to 10W.
- Finned heat sink the size of a housebrick; up to 50W.

When power levels get serious we may have to add forced-air cooling to reduce the junction-to-air thermal resistance. In the author's own experiments using a 3-inch computer fan blowing air over various sinks from 150mm away, the effective thermal resistance of the sinks was roughly halved in each case (this does not mean twice the power can be dissipated, however, since the case-sink interface is unchanged). Even with the fan running at half-voltage a 1.5× improvement was usually possible, with concomitant reduction in vibration and electrical noise.

11.8: Heater Supplies

AC heater supplies are very simple, and the most common arrangement is to supply all the heaters in parallel, as in fig. 11.50. If more than one heater voltage is possible (e.g. 6.3V or 12.6V for an ECC83/12AX7), then the lower voltage is usually preferable to minimise hum. Although this means the highest current is needed, hum due to the magnetic field is usually a lesser influence than that due to the electric field. Proper layout of an AC heater supply is essential for the lowest-noise operation (section 5.9.6), and even with a DC heater supply it is worth treating the wiring as if it is AC.

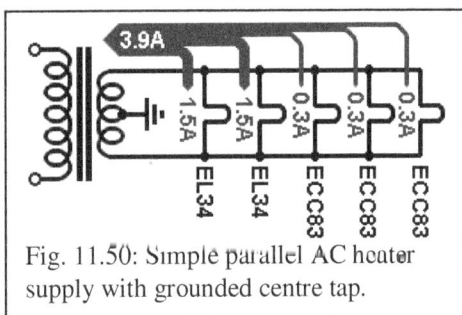

Fig. 11.50: Simple parallel AC heater supply with grounded centre tap.

11.8.1: Rectifier-Induced Hum

Since AC heaters operate at mains frequency which is quite low, hum due to electric-field coupling *ought* to be negligible if sensible lead dress has been observed (after all, even a whopping 10pF of stray capacitance has a reactance of over $300M\Omega$ at 50Hz). However, if the transformer supplies both the HT or some other DC power supply, *and* the heater supply, the clipped waveform produced by rectifier action will be reflected through the transformer to the heater winding too. The result is that the heater voltage will not be pure 50/60Hz but will contain all the high-frequency hash that rectifiers produce, plus any noise coming from the mains itself.

Fig. 11.51 shows an oscillogram of the heater-voltage waveform from a transformer which also provides the HT in a small valve amp, and it is obviously not a very good sine wave. The top of the wave is clipped and shows the classic 'hangover' explained earlier in section 11.1.6. The high dV/dt of this hangover, plus any rectifier switching noise that may exist too, will easily couple into the audio circuit via valve interelectrode capacitances. The lower trace shows the signal picked up on the grid of the input valve (which had a $1M\Omega$ grid leak) and transients coinciding with each hangover are clearly visible. It is these transients which often cause heater interference to sound buzzy, rather than the low hum we might otherwise expect. Heater balancing can suppress this differential-mode noise.

11.8.2: Electrical Heater Balancing

The heater supply must always have a DC path to ground, which may be a direct connection or an elevating circuit (next section). This is equally true for AC or DC supplies. Leaving the heater supply floating will result in almighty hum due to primary-to-secondary transformer leakage current, and is a common beginner's error. AC heater supplies should also be balanced to suppress the EM field. Not only does this reduce the magnitude of the voltage on each wire (e.g., each wire handles ±3.15V rather than one wire handling 6.3V and the other zero), but the opposing fields will tend to couple equal-but-opposite hum signals into the audio circuitry, which should cancel each other out.

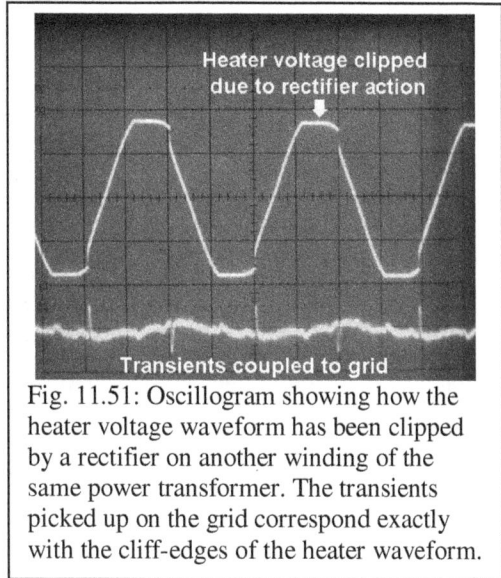

Fig. 11.51: Oscillogram showing how the heater voltage waveform has been clipped by a rectifier on another winding of the same power transformer. The transients picked up on the grid correspond exactly with the cliff-edges of the heater waveform.

If the transformer heater winding has a centre tap it can be grounded to create a balanced heater supply as in fig. 11.50, and this is by far the most common approach used in vintage equipment.

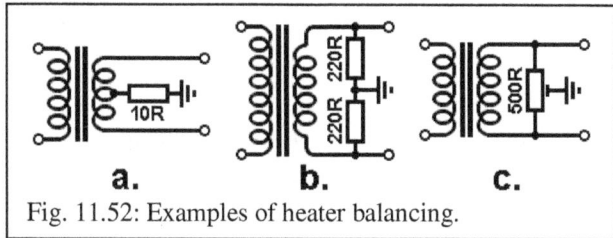

Fig. 11.52: Examples of heater balancing.

However, arcing between the anode and heater pins of valve sockets does occasionally happen (mainly with power output valves), and a direct connection to ground via the heater supply results in the worst possible fault condition. It is therefore preferable to connect the heater centre-tap to ground via a small (ideally fusible / flameproof) resistor, less than 100Ω say, as in fig. 11.52a. This is also a good habit for beginners to adopt, as they often seem to become fixated with grounding the centre-tap simply 'because it is there', even when some other ground reference is used. A burnt-out resistor is a much cheaper way to discover this confliction than a burnt-out transformer winding.

If there is no centre tap then the heater supply can be balanced using a pair of resistors instead, as shown in fig. 11.52b. They should be low-valued to encourage the shunting of transformer leakage current to earth, and values of 100Ω ½W to 220Ω ¼W are usual. There is no point in carefully matching these resistors since the coupling of hum into the audio circuit is itself not exactly balanced or predictable. With this in mind, an even better

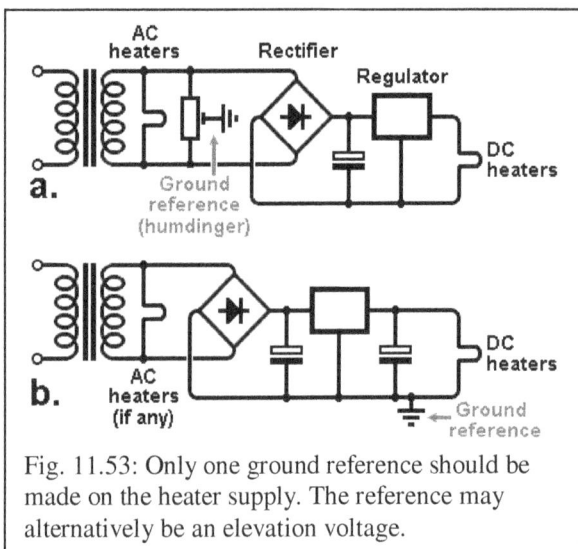

Fig. 11.53: Only one ground reference should be made on the heater supply. The reference may alternatively be an elevation voltage.

option is to use a **humdinger**. This is simply a trimpot connected across the heater legs, with the wiper connected to ground, as shown in fig. 11.52c. Again, a low value is preferable, and a 500Ω pot will dissipate less than 80mW at 6.3V. The pot can then be adjusted for minimum hum, which is not likely to occur at the exact centre setting.

Electrical balancing can also be used on DC heater supplies, although the benefits may not be so obvious. A common beginner's error is to try to add a ground reference on the AC side of the circuit and then another one on the DC side too. This will short out the rectifier! Only one ground reference should be used (usually the AC side results in the least hum); the other side of the circuit will still receive its reference through the rectifier.

11.8.3: Heater Elevation

Heater Elevation means referencing the heater supply to a DC voltage other than ground or zero volts. The heaters still operate at 6.3V or whatever, but this floats on top of the elevation voltage. As discussed in previous chapters, some valve stages such as cathode followers require the heater supply to be elevated to avoid exceeding the $V_{hk(max)}$ rating but, even when not explicitly required, elevation may reduce hum in AC-heated circuits.

All that is needed for elevation is a potential divider connected to the HT (any low-impedance DC source will also do).[*] The divider has the natural advantage that the elevation voltage will track any changes in the HT. No current flows 'into' the heater

---

[*] Jones has suggested a more complex elevation circuit which, despite its claims, provides neither a well smoothed voltage nor a low impedance path to ground, and is best avoided. Jones, M. (2012). *Valve Amplifiers (4th ed.)*. Newnes, London, p406.

414

supply from the divider; the heaters are simply jacked up by the elevation voltage. The DC voltage can be applied to a transformer centre tap, artificial centre tap, humdinger, or whatever reference connection the heater supply would normally have. Fig. 11.54 shows some examples. As noted earlier, in the case of DC heater supplies, only one reference should be used. The divider should have a fairly high resistance so as not to waste current, although R_2 should not be excessively large or $R_{hk(max)}$ may be grossly exceeded, so it is advisable not to make it greater than 100kΩ. C_1 provides decoupling/smoothing

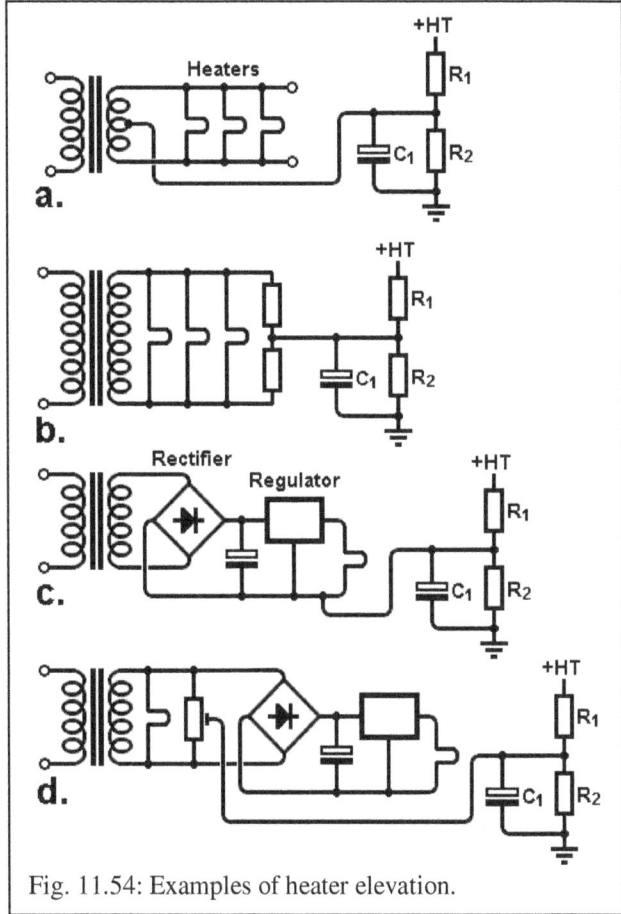

Fig. 11.54: Examples of heater elevation.

of the elevation voltage and can be arbitrarily large, say 10µF or more.

11.8.4: Common-Mode Noise

An unexpected source of buzz in an otherwise clean heater supply is common-mode noise. This is caused by mains noise leaking across from other windings on the transformer, via stray capacitance, and appearing *equally* on both heater wires. This noise will be seeking a path to mains earth, e.g. through heater-cathode

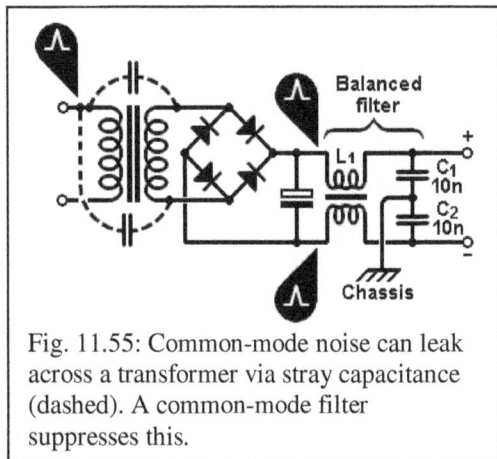

Fig. 11.55: Common-mode noise can leak across a transformer via stray capacitance (dashed). A common-mode filter suppresses this.

and heater-grid capacitances. But since the noise is common to both wires even a voltage regulator will be blind to it. This noise can be suppressed using a common-mode filter. Since heater current is usually quite large a common-mode choke would normally be used, rather than series resistors. This is followed by a pair of capacitors connected to chassis (i.e. mains earth, not audio ground) as shown in fig. 11.55. The choke L_1 is easily made by winding two wires a few times –and in the same direction– around a ferrite core. Such a filter could be used on an AC heater supply too but any benefit would probably be negligible compared to the differential-mode hum and buzz that come with an AC supply.

11.8.5: DC Heater Supplies

The ultimate way to eliminate hum is to use a DC supply, and this is usually mandatory for a modern phono or microphone preamp. However, a 6.3V transformer winding, when rectified and allowing for rectifier diode drop, will *not* produce enough DC voltage to supply most $6.3V_{dc}$ regulator circuits. It *might* work with Schottky diodes and a low dropout regulator, but leaves no room for mains brown outs. One solution is to use a voltage doubler which will generate close to $13V_{dc}$ at full load. Fig. 11.56 shows a practical example. At full load the 317 will dissipate a little over 4.7W so will need a small heatsink. Notice that by working at such low voltages and high currents the power factor is rather awful at just 0.53. In other words, the 800mA load amounts to 0.8/0.53 = 1.5A from the transformer, except we're using a voltage doubler, so we actually need a 3A transformer. C_1 and C_2 must have a ripple current rating of at least 2A. Diodes D_1 and D_2 could be replaced with a bridge rectifier package as in fig. 11.16 earlier.

The components shown faint provide an optional soft start. When power is first applied, Q_1 will short out R_2, but as C_6 charges up through R_3, the transistor will gradually switch off. Any general purpose transistor will do. D_5 provides base-emitter reverse voltage protection. With the values shown this will create a smooth ramp-up over about ten seconds. A problem occasionally encountered with circuits like this is that, when first switched on with the heaters cold (low resistance), the regulator will go into automatic shut down and refuse to start. For reasons that are

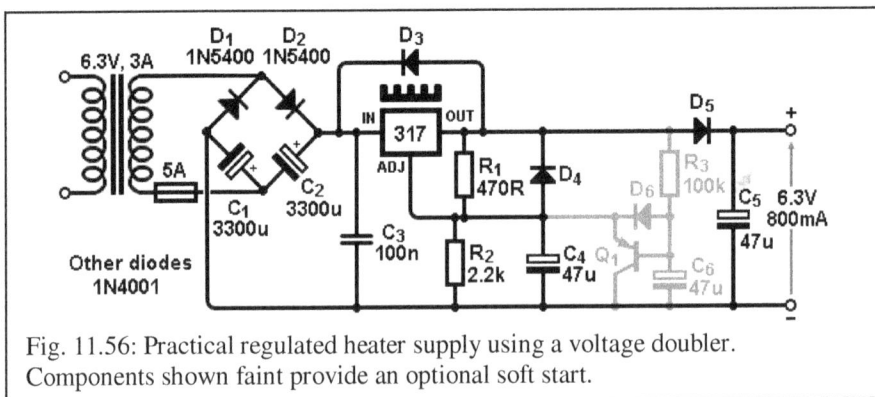

Fig. 11.56: Practical regulated heater supply using a voltage doubler.
Components shown faint provide an optional soft start.

416

not entirely clear, a diode in series with the output (D$_5$) sometimes cures this. For more hungry heater supplies is not sensible to try to derive regulated DC from a 6.3V winding. If the transformer happens to have a 5V winding that isn't needed then it can be connected in series with the 6.3V winding to make 11.3V, which will at least eliminate the need for a voltage doubler. Otherwise a separate heater transformer will be needed, which allows complete freedom of choice of voltage. As for the regulator we can opt for a 3A or even 5A device like the 1084. However, this will probably mean making the jump to thicker wire, a bigger heatsink, bigger rectifier (which may need heatsinking too) and a *lot* of reservoir capacitance. One big regulator may therefore be disproportionately more expensive than several smaller ones. In fact, several 317 regulators –each supplying only some of the heaters in the set– can add great flexibility:

- Several small or clip-on heatsinks could work out cheaper than one big heatsink, especially if they're home-made from scrap metal;
- We are not constrained to one global heater voltage;
- All the regulators could be supplied from the same transformer, or from separate transformers as required;
- Future conversion to valves with different heater requirements is likely to be easier;
- Whole regulator modules including their own rectifier, reservoir, and regulator, can be bought off the shelf quite cheaply.

Since a DC heater supply frees us from concerns about hum, it is more efficient to run at the highest voltage and lowest current possible, to minimise rectifier losses. It therefore makes sense to run heaters in series. Sometimes this will alleviate problems with heater elevation too, since the valves with the highest cathode voltages can be supplied from the higher-voltage end of the heater chain.

Different heaters might not have matching warm-up times, so to avoid power-hogging during start-up, they should be supplied from a constant-current regulator or current limiter. In theory, any

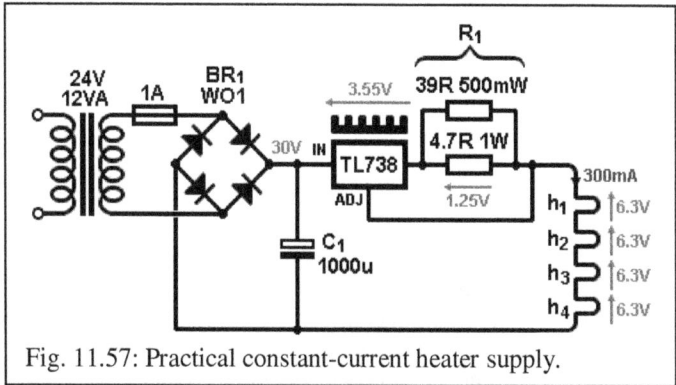

Fig. 11.57: Practical constant-current heater supply.

adjustable regulator could be pressed into service, but when power is first applied the heaters will be cold, so the full supply voltage will be imposed across the regulator. Since most regulator ICs are only rated for 35V or so, the 125V-rated TL738 stands out as prime candidate. Fig. 11.57 shows an example circuit supplying four 6.3V/300mA heaters. The current-programming resistance R$_1$ must be

417

1.25V/0.3A = 4.17Ω, which is created with 4.7Ω and 39Ω in parallel. As drawn, the regulator dissipates only 3.55V×0.3A = 1.07W and would be happy with a clip-on heatsink. Note that this circuit needs a 12VA transformer and supplies 7.6W of heaters, whereas the regulator in fig. 11.56 needed an 18.9VA transformer and a bigger heatsink just to supply 5W of heaters! The advantage of a series heater supply is clear. A free bonus of this circuit is that the constant-current action eliminates heater inrush, providing an inherent soft start.

In some applications the current-programming resistance in fig. 11.57 may be difficult to make using standard power resistors, and may also waste excessive power. Why waste power when it can be used to feed a heater? Fig. 11.58 replaces the power resistor with a heater itself – an arrangement described earlier in section 11.7.2 earlier. In this case a 6.3V heater (h₁) serves as 'master',

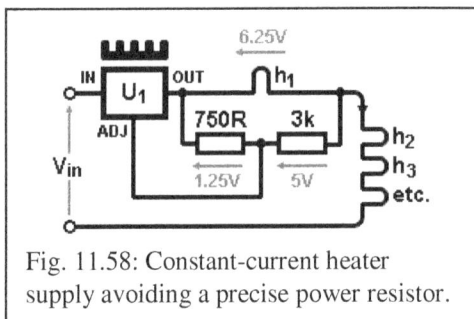

Fig. 11.58: Constant-current heater supply avoiding a precise power resistor.

and whatever current this requires will also feed the other heaters. However, this arrangement does not provide a soft start since the regulator will maintain the full voltage across h₁ from cold. Fortunately, indirectly-heated valves do not need a soft start anyway. Do not forget that all heater supplies must be referenced to ground somehow, as discussed in section 11.8.2.

11.8.6: Accommodating Different Heaters

Valves with unpopular heater voltages are often much cheaper than their 6.3V equivalents, while being otherwise pin compatible. It would be quite attractive, therefore, if an amplifier could accept different versions of the same valve without resorting to manual adjustment of the heater supply. Three methods are described here and all of them are somewhat wasteful of power. On the other hand, the freedom to use multiple valve types in the same circuit (so-called 'tube rolling') might outweigh this disadvantage.

The first method works for AC and DC and can be used when two valves are completely pin compatible. It involves using a carefully chosen heater supply voltage, plus an equally carefully chosen dropping resistor for each valve socket. With the right combination, two different heater types can receive the correct voltage (assuming one requires more voltage but less current than the other). To find the right combination, call the heater voltage of the first valve V₁, and its nominal heater resistance R₁. Likewise for the other valve type call its heater voltage V₂ and heater resistance R₂ (the nominal heater resistance is simply the rated heater voltage divided by the rated heater

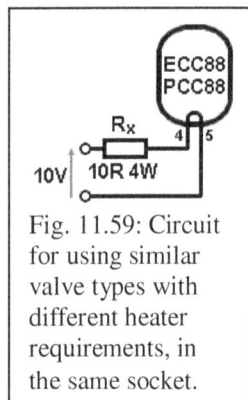

Fig. 11.59: Circuit for using similar valve types with different heater requirements, in the same socket.

418

current, given on the datasheet). The unknown dropping resistor R_x must then be:

$$R_x = \frac{R_1 R_2 (V_1 - V_2)}{R_1 V_2 - R_2 V_1} \tag{11.16}$$

And the heater supply voltage must be:

$$V_x = I_1 (R_1 + R_x) \tag{11.17}$$

Where I_1 is the heater current of the first valve type.

For example, the ECC88/6DJ8 needs 6.3V at 365mA so its heater resistance is $6.3/0.365 = 17.3\Omega$. The PCC88/7DJ8 needs 7V at 300mA so its heater resistance is $7/0.3 = 23.3\Omega$. Applying these figures to the previous formulae:

$$R_x = \frac{17.3 \times 23.3 \times (6.3 - 7)}{17.3 \times 7 - 23.3 \times 6.3} = 11\Omega$$

And:

$$V_x = 0.365 \times (17.3 + 11) = 10.3V$$

But the close standards of 10V and 10Ω work well enough. The power dissipated by R_x can be found from I^2R, where I is the largest heater current. In this case it would amount to $0.365^2 \times 10 = 1.3W$. Since it will have to burn continuously, using a 2W resistor would be a little unkind, so we probably ought to use a 4W resistor or better, as shown in fig. 11.59. Sometimes more than two valves can be accommodated this way. For example, the 6SN7 (6.3V/600mA), 8SN7 (8.4V/450mA), and 12SN7/12SX7 (12.6V/300mA) would each receive the correct voltage within 4% using a 22Ω (probably 15W) dropping resistance and 19V supply voltage.

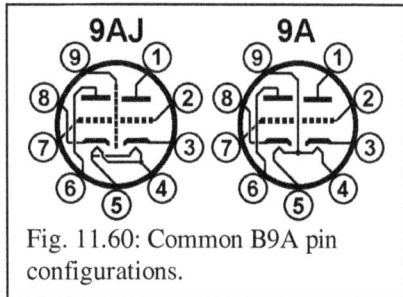

Fig. 11.60: Common B9A pin configurations.

The second method applies to 9-pin dual-triodes, most of which have one of two common pin configurations, shown in fig. 11.60. There are some types that are equivalent but for the different bases and heater powers, so they are not directly substitutable, which is a shame (they usually have different heater current requirements, so a constant-current heater supply is not an option). Example pairs include the ECC81/12AT7 (9A) and ECC85/6AQ8 (9AJ); and the ECC83/12AX7 (9A) and 6N2P/6H2Π (9AJ).

If you don't mind manual switching then this incompatibility can be overcome with the

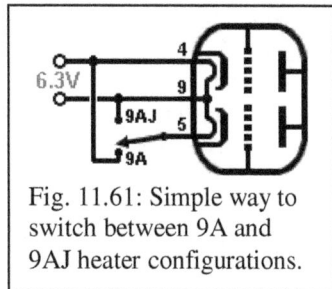

Fig. 11.61: Simple way to switch between 9A and 9AJ heater configurations.

419

arrangement in fig. 11.61 which works for both AC and DC (each valve will need its own switch). The system is fool proof since setting the switch in the wrong position will light only one heater, or neither, so in either case the mistake will soon be spotted with no harm done. Note that when a 9AJ-based valve is plugged in, the internal shield on pin-9 will be connected to the heater voltage, but since this is a low-impedance, low-voltage supply this should not affect shielding or hum.

An automatic (DC only) solution is shown in fig. 11.62. This uses pin-9 as a sort of 'identity detection pin'. When a 9A-based valve is plugged in, as shown, the base of Q_1 is connected to the heater common point and will act as an emitter follower. The regulator voltage will therefore be jacked up by the heater voltage of the 'lower' heater, plus V_{be}, plus the drop across D_1, which altogether works out

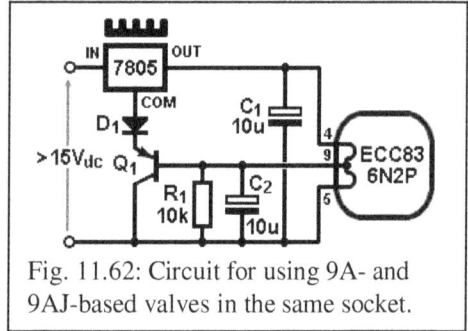

Fig. 11.62: Circuit for using 9A- and 9AJ-based valves in the same socket.

close to 12.6V. Conversely, if a 9AJ-based valve is plugged in, pin-9 will be connected to nothing but the internal shield. The base of Q_1 will therefore be pulled down by R_1 so the regulator voltage will only be raised by V_{be} plus the diode drop, producing 6.3V. C_2 is necessary to stabilise the regulator for capacitive loads (e.g. C_1), and it decouples the internal shield too. D_1 and Q_1 can be any general purpose devices as they only pass the few milliamps flowing out of the adjust pin. The regulator will dissipate almost 4W depending on the valve type plugged in, so it will need a small heatsink. The $\geq15V$ supply voltage could be derived from a slightly under-loaded $12V_{ac}$ transformer.

11.9: Back-to-Back Transformers

A popular way to make low-cost power supply for small valve projects is to use back-to-back transformers. In other words, use one transformer to step down to a low voltage ideal for heaters, then also use that low voltage to drive a second transformer 'backwards' to step back up again, producing a high voltage. However, this technique is very lossy, so do not expect to be able to feed a power amplifier this way. Back-to-back transformers tend to make the best use of parts when the first transformer is between 12VA and 50VA, and the voltage is stepped down as little as possible in between, e.g. avoid using 6V transformers if you can use a higher voltage.

The first transformer has to supply the heater power plus the power demanded by the second transformer, so the first one will often be a bigger device. On the other hand, sometimes it is easier to obtain two identical transformers than two different ones. Either way, the important thing is not to overload the first transformer. Small EI transformers (up to 50VA, say) will consume anywhere from 3 to 10VA of magnetising power. Small toroidal transformers demand less, up to perhaps 5VA. More powerful transformers are likely to require more magnetising power. This still applies when the transformer is being driven backwards, so the first transformer has to supply this magnetising power, thereby reducing its remaining 'useful' VA. As a rule of thumb, subtract 5VA from the first transformer's rating; whatever remains can be divided up between heater and HT power as required.

It is also important not to be too optimistic about the high voltage that can be obtained with this system, as transformers are deliberately overwound to compensate for their regulation figure. In other words, a small 12V transformer is really a 14V transformer (or thereabouts) with unavoidable resistance. Assuming it is optimised for 230V mains, driving it backwards with $12V_{ac}$ will therefore produce closer to $12/14 \times 230 = 197V_{ac}$ rather than the full $230V_{ac}$. We also have to put up with twice the normal source impedance, so the voltage will sag more under load. The author has used back-to-back EI transformers many times and it tends to yield about $200V_{dc}$ to $230V_{dc}$ under load. No doubt toroidal transformers would yield a little more, but nowhere near the $300V_{dc}$ figure that beginners might anticipate. Incidentally, do not be tempted to drive the second transformer with a higher-than-rated voltage as this will cause it to saturate.

Fig. 11.63 shows a practical supply using a couple of 12V transformers. Measured output voltages are shown, and the loading has been arranged to result in full use of the first (20VA) transformer. Note the monumentally poor regulation of the HT! Fortunately this would be of no consequence for a preamp with near constant current consumption. The low voltage supply might serve a 6V regulator, or a 12V low-dropout regulator if slightly under

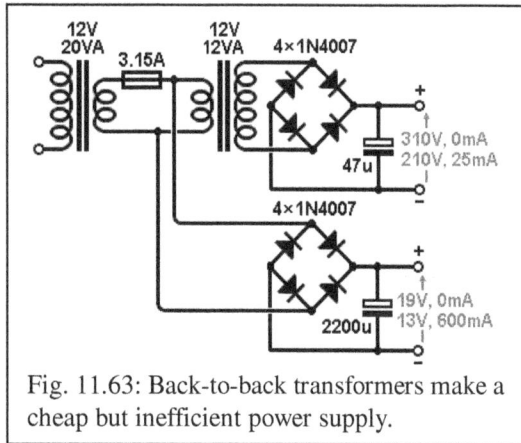

Fig. 11.63: Back-to-back transformers make a cheap but inefficient power supply.

loaded. Alternatively, up to 900mA of heaters could be supplied directly from the 12V AC without rectification.

11.10: Reducing Primary Voltage

It is sometimes desirable to reduce the primary voltage applied to a transformer. This might be the case when using transformers scrounged from vintage equipment originally designed for 110V or 220V mains. Most households receive more than this nowadays which can lead to transformer saturation or excessive heater voltages when using these old devices.

An efficient way to reduce the mains voltage is to use an autotransformer. This wastes little power and provides a constant voltage drop, independent of load current. It does *not* provide safety isolation from the

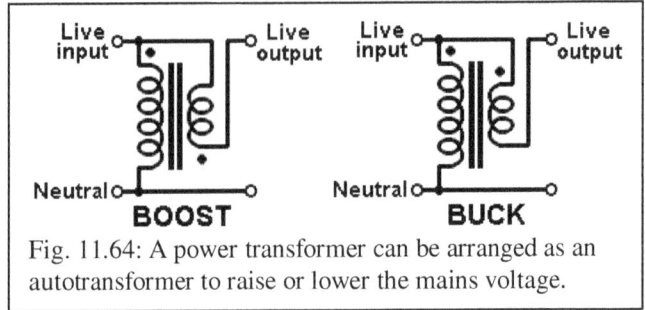

Fig. 11.64: A power transformer can be arranged as an autotransformer to raise or lower the mains voltage.

mains; it simply reduces its voltage. It is easy to make an autotransformer using an ordinary power transformer; simply connected the primary to the mains in the usual way, then connect the secondary in series with the live feed. Depending on the phasing chosen, this will either add or subtract the secondary voltage from the live voltage, as shown in fig. 11.64 (a quick check with a voltmeter will confirm which way around it has been connected). The transformer secondary must have sufficient current rating to handle the live current demanded by the following power supply. In other words, if an amplifier needs 1A and we want to drop the mains voltage by 20V, we will need a 20V 1A (20VA) transformer, or better.

11.11: Transformerless Supplies

In any audio equipment the signal ground will inevitably be connected to the power supply 'ground' terminal. Therefore, when the user comes into contact with the signal ground (e.g. by touching the input/output sockets), he comes into contact with the power supply. This is why, if the equipment is mains powered, it absolutely *must* use a power transformer. This may be a traditional mains transformer or an isolating SMPS (which is basically just a transformer running at high frequency); either way, the user is galvanically isolated from the dangerous mains supply. But we sometimes need auxiliary circuits that are entirely unconnected from the audio circuit (user), in which case it is sometimes possible to power them directly from the mains and so save on transformer VA. Applications include inrush-current

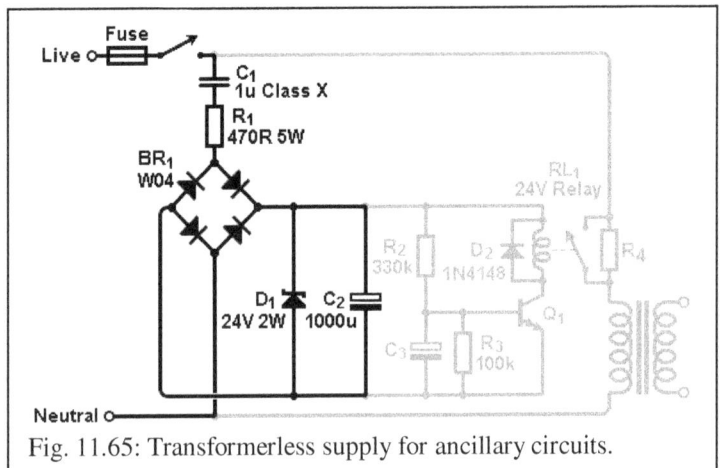

Fig. 11.65: Transformerless supply for ancillary circuits.

limiting relays, standby-control circuits (again using relays), illumination, and cooling fans (valve heaters *do* require a transformer, however, because a breakdown in heater-cathode insulation could connect the user to the heater supply).

Fig. 11.65 shows how low-voltage DC is obtained from the mains. The circuit is connected to the live wire via a 1µF capacitor which drops voltage without dissipating any heat, so it is sometimes called a **wattless dropper**. This *must* be a class-X safety capacitor as only these may legally be connected across the mains.[*] Because the capacitive reactance is much larger than the effective load resistance, the capacitor behaves like a constant-current source, always limiting the RMS current to less than:

$$\frac{V_{rms}}{2\pi fC} \tag{11.18}$$

Where V_{rms} is the mains voltage and f is the mains frequency,

R_1 is essential to limit the peak charging current at switch on. The bridge rectifier then diverts the AC current into a Zener diode to provide a stabilised DC voltage which can be anything up to about 30V. The Zener diode should be capable of handling the same amount of current as calculated above. The maximum DC current available for the load will be about twenty percent lower; trying to draw more will steal too much from the Zener, causing the DC voltage to droop. With a 1µF wattless dropper the circuit can provide up to about 55mA$_{dc}$ on 230V mains, or 30mA$_{dc}$ on 120V mains.

The components shown faint represent a typical circuit to be powered –in this case an inrush-limiting relay. At switch on, C_2 will charge up to the Zener voltage. C_3 then charges much more slowly through R_2 until the voltage across it is enough to turn on Q_1, energising the 24V relay and shorting out the inrush-limiting resistor R_4. As drawn, the time delay is approximately one second per hundred microfarads used for C_3, and a second or two should be more than enough to suppress inrush. C_1 discharges again through R_3 at switch-off. Q_1 can be any general purpose Darlington such as 2N6427, MPSA13, etc. Remember, this circuit is allowable only because there is *no way* for the user to come into contact with it. It should be considered part of the mains supply and must have no connection to the audio circuit.

11.12: External Power Supplies

Very sensitive equipment may require a separate power supply unit to achieve maximum separation from the (noisy) power transformer and rectifier. If this is the case then it is important to ensure that all the reservoir capacitance is contained within the power supply box. Fig. 11.66a shows a violation of this rule where a smoothing capacitor has been added to the amplifier, with nothing separating it from the reservoir capacitor. The second capacitor is therefore directly in parallel with –

---

[*] Technically, class-Y is also allowed, as they are even more robust than class-X.

and hence part of– the reservoir capacitance. A portion of the ripple current will flow down the cable and into the amplifier box, which can pollute the quiet amplifier ground, leading to hum and buzz. To avoid this a dropping resistor, smoothing filter or voltage regulator should be added to the output of the power supply unit to keep the ripple current contained inside it. An example is shown in fig. 11.66b where a common-mode smoothing

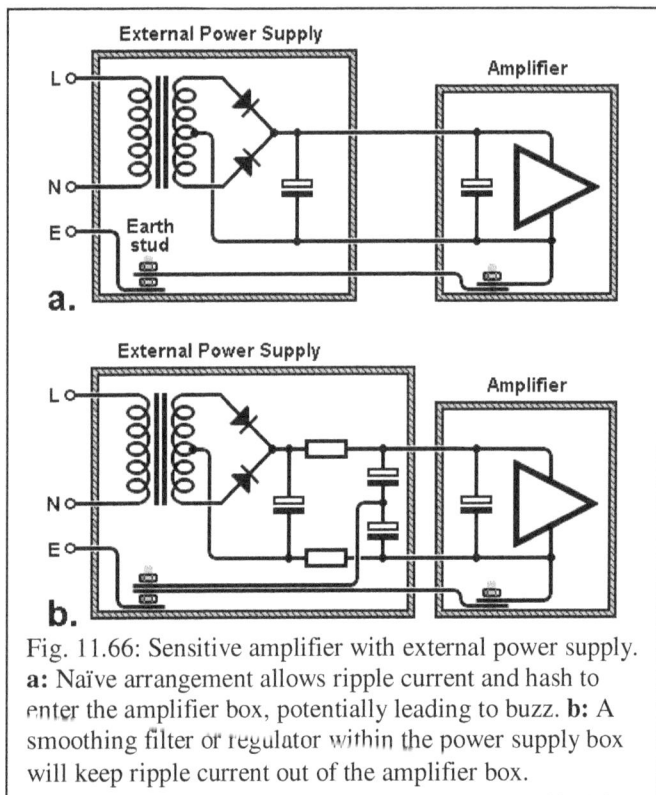

Fig. 11.66: Sensitive amplifier with external power supply. **a**: Naïve arrangement allows ripple current and hash to enter the amplifier box, potentially leading to buzz. **b**: A smoothing filter or regulator within the power supply box will keep ripple current out of the amplifier box.

filter has been added to serve the dual purpose of containing the ripple current and shunting common-mode hash to earth before it reaches the amplifier box. A common-mode choke where power enters the amplifier box is also favourable addition, especially when using an SMPS. The power supply chassis (if it is metal) must be connected to mains earth for safety, but in theory the amplifier chassis does not since mains does not enter it. Nevertheless, the amplifier chassis should in practice be earthed too for shielding reasons (the extra safety is a bonus).

To link the two boxes we will presumably need some connectors, but finding ones that are officially rated for high-voltage DC is very difficult, even though the currents involved in valve projects are usually small. However, since the voltages being handled here are isolated from the mains, the immediate danger to the user is much less than with mains voltage itself. Most builders will therefore resort to sensible-looking, heavy-duty connectors, even though they may not have 'official' high-voltage ratings (but to avoid dangerous confusion never use a standard mains connector –such as an IEC socket– for anything other than actual mains).
The author has used *unusual* miniature mains connectors such as those made by Bulgin. The 'Cliffcon' ZC (miniature multi-pole), and the S (loudspeaker) series, made by Cliff, are also likely candidates.

Index

Angular Frequency, 13, 47, 51

Anode, 98-9

  Dissipation, 105, 128, 173, 241, 310, 409

  Resistance, 117-21, 167, 174-5, 211, 224, 248, 252-3, 257, 293, 382, 384

  Load Resistor, 121-2, 127, 136-9, 182, 213, 215, 267-8, 310, 312, 359, 394

  Stopper, 152

  Voltage Rating, 105

Arc Protection, 163-4, 262, 265, 358

A-Weighting –See Noise

Band-Gap Reference, 227

Banxandall –See Control

Bessel, 44

Bias, 123

  Automatic-, 125

  Cathode-, 124-5, 140, 168, 243-4, 259-62

  Cold-, 133-4, 243

  Contact-, 124

  Diode-, 113, 184, 243-6, 260, 265, 284, 304, 332, 362

  Fixed-, 112, 124, 261-3, 275

  Grid leak-, 124

  Hot-, 135

  Noiseless-, 262

  Self-, 125, 256, 259

BJT (Bipolar Junction Transistor), 208-9, 224-9, 247, 352, 400

Blumlein, 247

Bode Plot, 45-7, 58, 60, 141

Boltzmann Constant, 189, 199-200, 212, 224, 351

Bootstrap, 150, 250-1, 260-2, 267-70, 273, 299-303, 331

  Pair, 267-70,

Breathing, 395

Build-Out Resistor, 152, 255-6, 260, 262, 264-5, 274-5, 286, 298, 357

Butterworth, 44, 52, 61, 94, 195, 337, 346

Capacitance, 18, 72

  Definition of, 18

  Interelectrode, 98, 149, 240-1, 249-51, 413

Capacitor,

  Air-, 76

  Aluminium Electrolytic-, 72-6, 80-6, 142, 158, 243, 263, 292, 359, 401

  C0G, 76-8

  Ceramic-, 72-3, 76-8, 83, 220, 374, 396

  Class-X, 423

  Class-Y, 423

  Compensation-, 164, 281

  Coupling-, 82-4, 123-5, 138-40, 158-62, 230, 259, 263-4, 269, 275-7, 281, 292, 299, 302-3, 328, 331, 342-3, 358

  Distortion, 77, 79, 84-6, 243, 340-1

  Multiplier, 399-400, 405,

  Polypropylene-, 78, 80, 230, 340

  Polystyrene-, 76, 78, 340

  Reservoir-, 359, 361, 365-72, 374, 377, 380-3, 387, 392, 417, 423-4

  Tantalum-, 72, 86, 401

  Screen-Grid Bypass-, 173, 178-81, 307-8

Silvered Mica-, 75-6
Smoothing-, 218, 248, 270, 359, 386, 391-5, 423
Snubbing-, 375-7, 390
Tolerance, 72, 81, 85-7, 142, 243, 340, 370
Cascode, 108, 207, 226-8, 237, 239, 304-10, 348, 353
Hybrid-, 308-9
Cathode, 99-104, 106
Bias Resistor, 112, 124-7, 141-3, 168, 243, 256, 259-61, 299, 301, 348, see also Bias
Bypass Capacitor, 84, 140-2, 146, 152, 158, 160, 163, 173, 176, 183, 243, 290-1, 322
Coupled Amplifier, 309-14
Follower, 56, 108-9, 163, 247-87, 290, 293-6, 298-301, 309-13, 317, 320, 327-9, 340-3, 355-7, 398, 414
Internal Resistance of, 142, 215-6, 252, 310
Load Line –see Load Line
Poisoning, 101-3
Repeater, 313
Saturation, 100-1, 106, 109, 115-6, 134, 380, 383
Sputtering, 101, 103, 389
Stripping, 101
Temperature, 100, 102, 200, 204-5, 212
CCS (Constant Current Source/Sink), 221-45, 269, 273, 275-279, 283, 289, 302, 348
Clipping, 16, 129-36, 143, 155, 159-61, 180-2, 222, 232-4, 241-2, 253, 267, 276-7, 282, 286, 291, 346, 363, 412-13
Common,
Cathode, 121, 1714, 305, 351
Grid, 304, 310-13, 350

Constant-Current-Draw Amplifier, 270-2, 355
Control, 315-34
Balance-, 318, 321-4, 355-7
Pan (Panoramic)-, 321
Tilt-, 328-30, 333-4
Tone-, 270, 315, 324-334, 338
Volume-, 31, 70, 314-24
Coupling, 158-65
AC-, 139, 158-60, 162, see also Capacitor
DC-, 109, 162-4, 251, 264-6, 268, 286, 313, 343, 355, 389
Direct-, 163-4, 264-6
Double-, 158, 358
Current, 2-5
Definition of, 2
Ripple-, 73-4, 82, 370-3, 375, 377, 379, 380-3, 423-4
Cut-Off (Anode Current), 102, 104, 115, 121, 130, 133-6, 155, 160-2, 169, 259, 286

Damping, 38-9, 83, 89, 152, 179, 185, 220, 376, 396-8
Critical-, 44, 52, 61
Factor, 38, 44, 346
Over-, 44-5, 53
Ratio, 38, 43-5, 52-3
Under-, 44, 52, 103
Decibel, 40, 55, 193, 320, 330
Degeneration, 53, 182, 222, 247, 254
Cathode-, 141-50, 157-8, 175-6, 178-9, 213, 215, 222, 243-5, 269, 272, 279, 290-2, 295-308, 310, 322, 355-6
Screen Grid-, 170-1, 173, 179-80, 184-5
Delon Circuit, 377
Dielectric Absorption, 73, 75-8

White Cathode Follower, 270, 281-7, 293-9

Zero –see Filter
Zobel Network, 94, 96, 350

www.ingramcontent.com/pod-product-compliance
Lightning Source LLC
Chambersburg PA
CBHW031405180326
41458CB00043B/6620/J

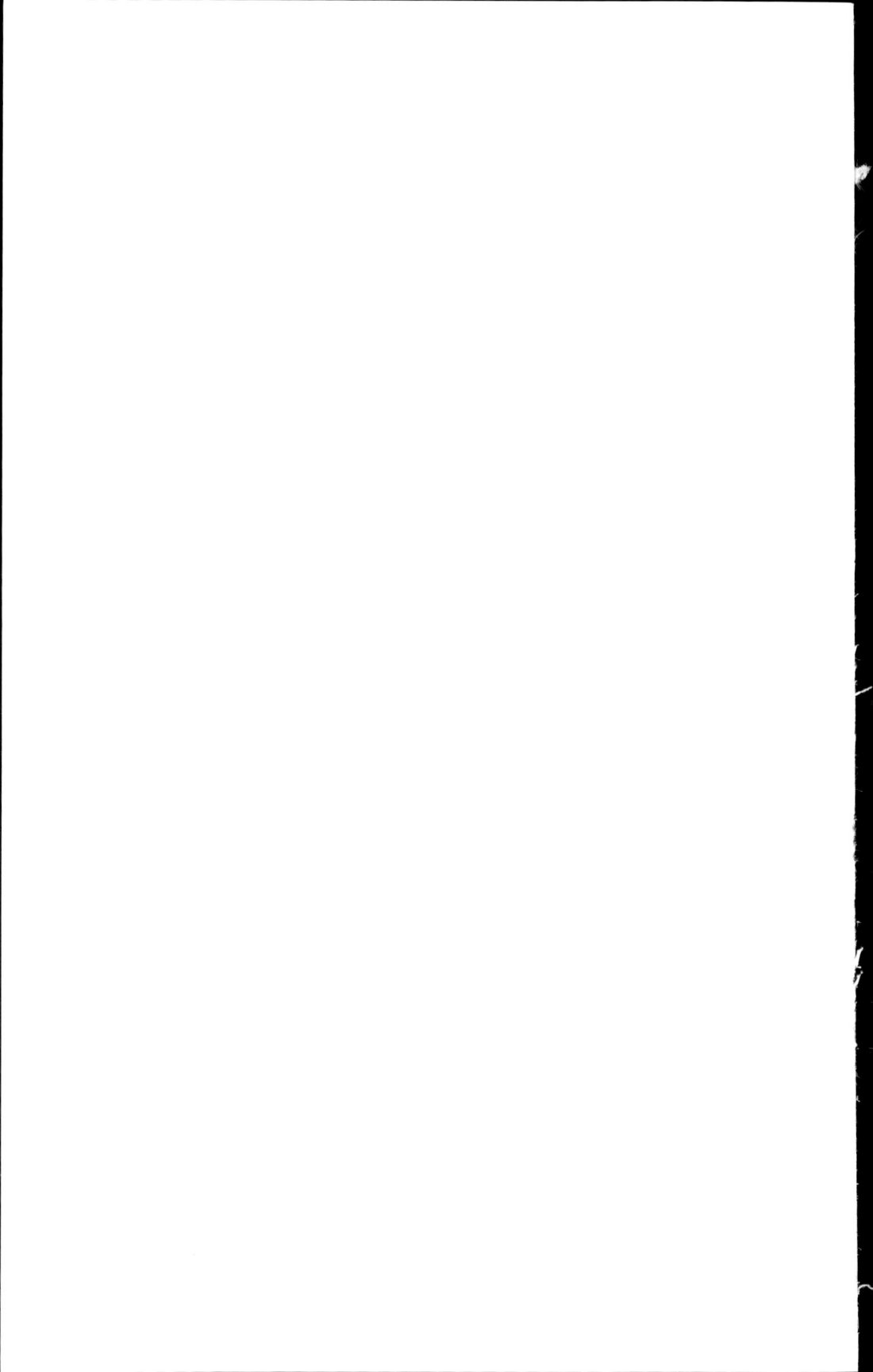